Clinical Microbiology

Clinical Microbiology

Second Edition

Keith Struthers

CRC Press
Taylor & Francis Group
6000 Broken Sound Parkway NW, Suite 300
Boca Raton, FL 33487-2742

Printed on acid-free paper

International Standard Book Number-13: 978-1-4987-8689-8 (Pack – Book and Ebook)
International Standard Book Number-13: 978-1-138-10190-6 (Hardback)

Visit the Taylor & Francis Web site at
http://www.taylorandfrancis.com

and the CRC Press Web site at
http://www.crcpress.com

Printed and bound in Great Britain by
TJ Books Limited, Padstow, Cornwall

Contents

Preface

Infectious diseases and medical microbiology embrace the broad discipline of clinical microbiology, a subject that highlights the importance of infection in every medical specialty. Diseases range from those associated with foreign travel, to the patient in the emergency department with a community acquired pneumonia, or the patient with antibiotic-associated diarrhoea on a surgical ward. While each example here is different in terms of the direct knowledge required to manage the patient safely, basic principles relating to epidemiology, transmission, clinical assessment, diagnosis and treatment are shared.

This is underpinned by an understanding of the pathogenic properties of organisms, their interaction with the human host and how they cause disease. This directs safe and prudent prescribing of anti-infective agents; correct infection control and public health practice completes the process.

I have written this book as a general introductory resource, to provide the background knowledge for a wider and deeper appreciation of this enthralling discipline. The medical student can use this knowledge as the basis for ward-based clinical training.

About the author

Dr Keith Struthers is a consultant microbiologist. He was Clinical Director of Medical Microbiology in its various forms in Coventry and the Coventry and Warwickshire Pathology network in the period 1998–2010. From 2010 to 2014 he worked at the Public Health England laboratory at Heart of England NHS Trust, Birmingham, being Director, Infection Prevention and Control in 2012–2014.

Disclaimer

This book is written as a general education text in clinical microbiology to provide a broad background of the subject at medical school. In clinical practice a wide range of organisms must be considered, and in all situations, the most recent local and national guidelines, policies and procedures on all aspects of diagnosis, treatment and infection control must be used in patient care and staff safety.

The information on antibiotics and antibiotic use in the text is for educational purposes. In the clinical setting, reference must be made to national documents such as the British National Formulary and local antibiotic guidelines, where dosages, length of treatment, contraindications, interactions, allergies and cautions are identified. The author names specific agents as examples in a class of antibiotics for use in particular clinical settings. As such the author does not have any commercial interest or conflict of interest in naming a particular antibiotic. This statement applies to vaccination, vaccines and diagnostic tests, as well as all relevant equipment used in clinical areas and the diagnostic laboratory.

Acknowledgements

I wish to thank Mr Michael Collins, Lead Biomedical Scientist, Department of Microbiology, Chesterfield Royal Hospital NHS Trust and Mrs Jane Shirley, Regional Director of Operations, West Midlands Public Health Laboratory, Birmingham, for enabling me to use images of laboratory equipment. I also thank Sarah Foulkes, Field Epidemiology Service, Public Health England, West Midlands, for the material presented in Figure 3.3.

Citation of the "Sepsis Six" (chapter 3) from the UK Sepsis Trust is gratefully acknowledged.

Material in this book includes information from: Struthers K, Westran R (2003) *Clinical Bacteriology*, Manson Publishing; Struthers K, Weinbren M, Taggart C, Wiberg K (2012) *Medical Microbiology Testing in Primary Care*, Manson Publishing.

Abbreviations

AAD	antibiotic-associated diarrhoea
ACDP	Advisory Committee for Dangerous Pathogens
AECB	acute exacerbation of chronic bronchitis
AFB	acid-fast bacilli
AIDS	acquired immunodeficiency syndrome
ALF	acute liver failure
ALT	alanine transaminase
AML	acute myeloid leukaemia
AMO	amoxicillin
ANC	Antenatal clinic
ANTT	aseptic non-touch technique
APC	antigen-presenting cell
aPTT	activated partial thromboplastin time
ARF	acute renal failure
ART	antiretroviral therapy
AST	aspartate transaminase
ATP	adenosine triphosphate
AUC	area under the curve
AZT	zidovudine
BAL	bronchoalveolar lavage
BBV	blood-borne virus
BCG	bacillus Calmette-Guérin
BHI	brain–heart infusion
BKV	BK virus
BMI	body mass index
BMS	biomedical scientist
BP	blood pressure
bpm	beats per minute
BSE	bovine spongiform encephalopathy
BV	bacterial vaginosis
cAb	core antibody (of HBV)
cAg	core antigen (of HBV)
cAMP	cyclic adenosine monophosphate
CABG	coronary artery bypass graft
CAP	community-acquired pneumonia
CAPD	chronic ambulatory peritoneal dialysis
CCHFV	Congo Crimean haemorrhagic fever virus
CDT	*Clostridium difficile* toxin
CDAD	*Clostridium difficile*-associated diarrhoea
CFTR	cystic fibrosis transmembrane regulator
CIP	ciprofloxacin
CJD	Creutzfeldt–Jakob disease
CLI	clindamycin
CLM	cutaneous larval migrans
CLW	contact lens wearer
CMC	cytoplasmic membrane complex
CMI	cell-mediated immunity
CMIA	chemiluminescent microparticle immunoassay
CMV	cytomegalovirus
CNS	central nervous system
CNS	coagulase-negative staphylococci/central nervous system
COA	co-amoxiclav (amoxicillin/clavulanic acid)
COPD	chronic obstructive pulmonary disease
CPE	carbapenemase-producing Enterobacteriaciae
CPK	creatinine phosphokinase
CRP	C-reactive protein
CSF	cerebrospinal fluid
CSU	catheter specimen of urine
CURB-65	confusion, blood urea, respiratory rate, blood pressure, age ≥65 y
CT	computed tomography
CTP	cytidine triphosphate
CTX	ceftriaxone
CVC	central venous catheter
CXR	chest X-ray
DAP	daptomycin
DIPC	Director, Infection Prevention and Control
DNA	deoxyribonucleic acid
DNase	deoxyribose nuclease
DOT	directly observed therapy
DTH	delayed-type hypersensitivity
eAb	e antibody (of HBV)
eAg	e antigen (of HBV)
EBV	Epstein–Barr virus
EBNA	Epstein–Barr (virus) nuclear antigen
ECS	endo-cervical swab
EDTA	ethylenediamine tetra-acetic acid

EEG	electroencephalogram
EEV	Eastern equine encephalitis virus
EIA	enzyme immunoassay
ENT	ear, nose and throat
EPP	exposure-prone procedure
EPS	expressed prostatic secretions
ERT	ertapenem
ERY	erythromycin
ESBL	extended-spectrum β-lactamase
ESR	erythrocyte sedimentation rate
ETT	endotracheal tube
EVD	external ventricular drain
Factor X	X haemin
FBC	full blood count
FFP	filtering face piece (mask)
FUO	fever of unknown origin
GABA	γ-amino-n-butyric acid
GAS	*Streptococcus pyogenes*
GBS	*Streptococcus agalactiae*
GCS	*Streptococcus dysgalactiae*
GCU	gonococcal urethritis
GDH	glutamate dehydrogenase
GEN	gentamicin
GFR	glomerular filtration rate
GI	gastrointestinal
GISA	glycopeptide-intermediate *Staphylococcus aureus*
GM	Gram (stain)
GRSA	glycopeptide-resistant *Staphylococcus aureus*
GTP	guanosine triphosphate
HAP	hospital-acquired pneumonia
HAV	hepatitis A virus
HBcAg/cAg	hepatitis B c antigen
HBeAb/eAb	hepatitis B e antibody
HBeAg/eAg	hepatitis B e antigen
HBsAg/sAg	hepatitis B s antigen
HBsAb/sAb	hepatitis B s antibody
HBIG	hepatitis B virus immunoglobulin
HBV	hepatitis B virus
HCC	hepatocellular carcinoma
HCl	hydrochloric acid
HCP	health care professional
HCV	hepatitis C virus
HDU	High-dependency unit
HDV	hepatitis D virus
HEPA	high-efficiency particulate arresting (filter)
HEV	hepatitis E virus
HHV6	human herpes virus 6
Hib	*Haemophilus influenzae* type b (vaccine)
HIV	human immunodeficiency virus
HPV	human papilloma virus
HR-CT	high resolution computed tomography
HRPII	histidine-rich protein II
HSV	herpes simplex virus
HTLV	human T cell lymphotropic virus
HUS	haemolytic uraemic syndrome
ICC	Infection control committee
ICD	infection control doctor
ICED	implantable cardiac electronic device
ICN	infection control nurse
ICT	Infection control team
ICU	Intensive care unit
IDU	injecting drug user
IE	infective endocarditis
γ-IFN	gamma interferon
Ig	immunoglobulin
IG	hyperimmune globulin
IGT	interferon γ test
IL	interleukin
INH	isoniazid
INR	International normalized ratio
IRES	internal ribosome entry site
IS	insertion sequence
IUGR	intrauterine growth retardation
IV	intravenous
JCV	JC (John Cunningham) virus
JEE	Japanese encephalitis virus
KS-AHV	Kaposi sarcoma-associated herpes virus
LEV	levofloxacin
LFT	liver function test
LIN	linezolid
LJ	Lowenstein–Jensen
LP	lumbar puncture
LPS	lipopolysaccharide
LRTI	lower respiratory tract infection
MALDI-TOF	matrix-assisted laser desorption/ionization time of flight
MBC	minimum bactericidal concentration
MBL	mannose-binding lectin/ metallo-ß-lactamase
MC&S	microscopy culture and sensitivity
MDR-TB	multi-drug-resistant tuberculosis
MER	meropenem
MERS	Middle East respiratory coronavirus
MET	metronidazole
MEWS	modified early warning score

MHC	major histocompatibility	PVE	prosthetic valve endocarditis
MIC	minimum inhibitory concentration	PVL	Panton Valentine leucocidin
MMR	measles, mumps, rubella	PZA	pyrazinamide
MRAB	multi-resistant *Acinetobacter baumannii*	RBC	red blood cell
MRI	magnetic resonance imaging	RF	replicative form
mRNA	messenger RNA	RFLP	restriction fragment length polymorphism
MRSA	methicillin-resistant *Staphylococcus aureus*	RVFV	Rift Valley fever virus
MSM	men-who-have-sex-with-men	RIF	rifampicin
MSSA	methicillin-sensitive *Staphylococcus aureus*	RNA	ribonucleic acid
MSU	midstream urine	rRNA	ribosomal ribonucleic acid
NA	nuclear antigen	RSV	respiratory syncitial virus
NAAT	nucleic acid amplification test	RT	reverse transcriptase
NAD	factor V (nicotinamide adenine dinucleotide)	RTA	road traffic accident
NAG	N-acetylglucosamine	RT-PCR	reverse transcriptase PCR
NAM	N-acetylmuramic acid	RUQ	right upper quadrant
NGU	non-gonococcal urethritis	sAb	surface antibody (of HBV)
NHL	non-Hodgkin's lymphoma	sAg	surface antigen (of HBV)
NN	neonatal period	SBP	spontaneous bacterial peritonitis
NSAID	non-steroidal anti-inflammatory drug	SIRS	systemic inflammatory response syndrome
NVE	natural valve endocarditis	SNP	single nucleotide polymorphism
OHD	Occupational health department	SOT	solid-organ transplant
OM	outer membrane	SSI	surgical site infection
OPAT	outpatient parenteral antibiotic therapy	STD	sexually-transmitted disease
PBP	penicillin-binding protein	STI	sexually-transmitted infection
PE	pulmonary embolus	SVR	sustained virological response
PBP	penicillin-binding protein	TB	tuberculosis
PCR	polymerase chain reaction	TCR	T-cell receptor
PCV	pneumococcal conjugated vaccine	Th	T helper (cell)
PDA	patent ductus arteriosus	TK	thymidine kinase
PEG	polyethylene glycol	TNF	tumour necrosis factor
PEN	benzylpenicillin	TOE	transoesophageal echocardiography
PEP	post-exposure prophylaxis	TP	terminal protein
PG	peptidoglycan	TPN	total parenteral nutrition
PHE	Public Health England	tRNA	transfer RNA
PICC	peripherally inserted central catheter	TSE	transmissible spongiform encephalopathy
PID	pelvic inflammatory disease	TSS	toxic shock syndrome
PM	plasma membrane	TST	tuberculin skin test
PN	perinatal period	TTE	transthoracic echocardiography
PO	by mouth	TV	*Trichomonas vaginalis* or trichomoniasis
PPD	purified protein derivative	U+E	urea and electrolyte
PPE	personal protective equipment	URT	upper respiratory tract
PPI	proton pump inhibitor	UTI	urinary tract infection
PPV	pneumococcal polysaccharide vaccine	UTR	untranslated region
PrP	polyribose ribitol phosphate	UV	ultraviolet
P/T	piperacillin/tazobactam	VAP	ventilator-associated pneumonia
PUO	pyrexia of unknown origin	VAN	vancomycin
PV	prosthetic valve	vCJD	variant Creutzfeldt–Jakob disease
PVC	peripheral venous cannula		

VDRL	Venereal disease research laboratory
VHF	viral haemorrhagic fever
VIP	visual infusion phlebitis (score)
VLM	visceral larval migrans
VP	ventriculoperitoneal
Vpg	virus protein genome
VRE	vancomycin-resistant enterococci
VU	voided urine
VZIG	varicella zoster virus immunoglobulin
VZV	varicella zoster virus
WBC	white blood cell
WCC	white cell count
WEE	Western equine encephalitis virus
WGS	whole genome sequencing
XDR-TB	extensively-drug resistant tuberculosis
ZN	Ziehl–Neelsen

Chapter 1

Introduction to Clinical Microbiology

INTRODUCTION

The structure and function of organisms determines their classification and provides an appreciation of the diversity of the diseases they cause in terms of epidemiology, transmission, diagnosis and treatment. A simplified and abbreviated classification of bacteria, viruses, fungi and parasites is the basis of this chapter.

BACTERIA

In clinical practice, bacteria can be broadly classified as gram-positive, gram-negative and anaerobes (**Figures 1.1a–c**); mycobacteria, mycoplasma, ureaplasma, spirochaetes and obligate intracellular organisms are included here in a miscellaneous group (**Figure 1.1d**).

The gram-staining character of bacteria is dependent on the structure of the cell wall. The stain enables determination of overall shape and size; bacteria are usually rod-like (bacillary), round (coccoid) or in the case of *Haemophilus influenzae*, cocco-bacillary (**Figure 1.2**).

The Gram stain involves spreading a loop-full of specimen on a glass slide, which is then heat-fixed and subjected to various stains (**Figure 1.3**). Gram-positive bacteria retain the crystal violet/iodine complex and stain blue-black. With gram-negative bacteria, the crystal violet/iodine complex is eluted out when the outer lipid layer of the cell wall dissolves in the acetone; they then

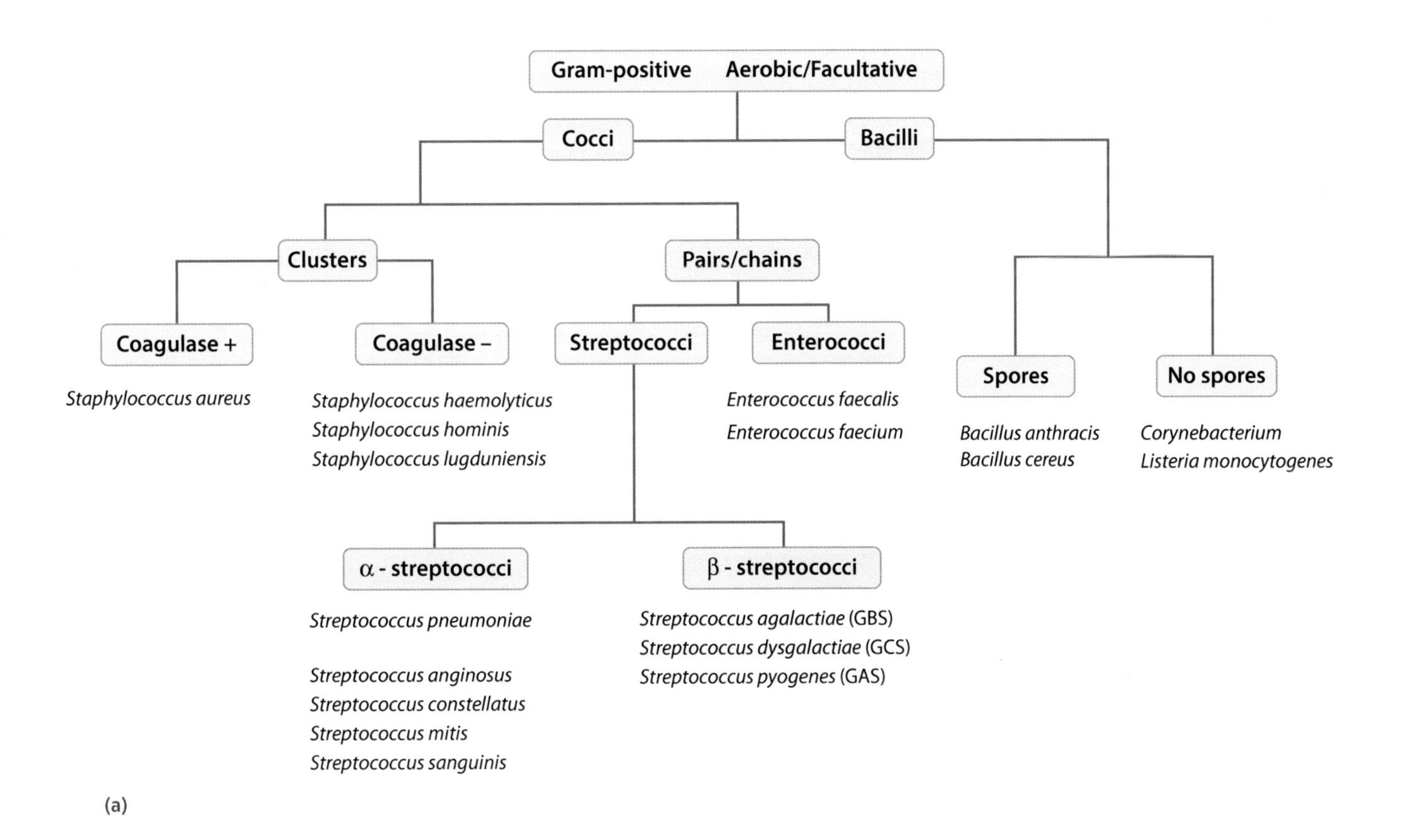

(a)

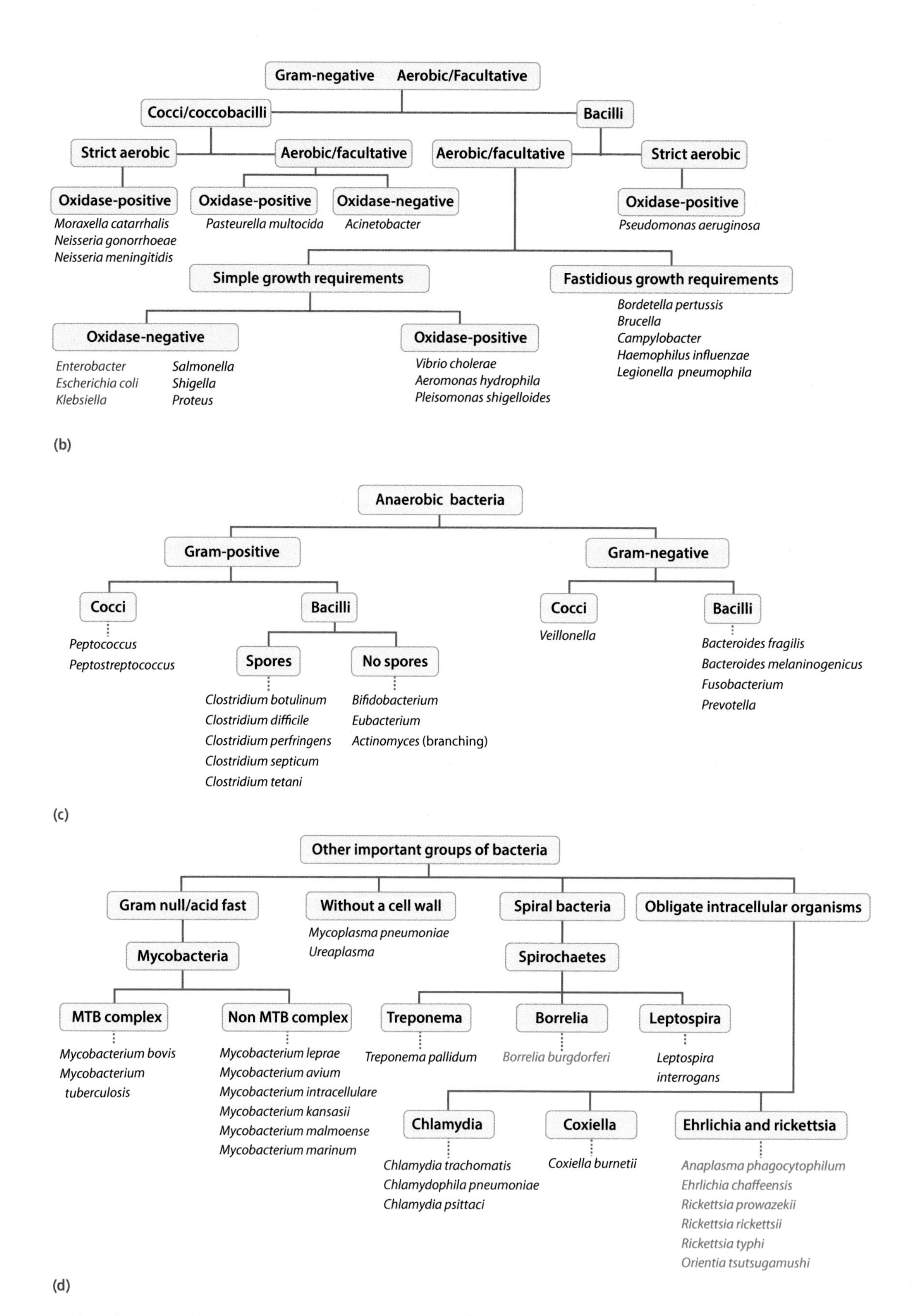

Figure 1.1 Classification of bacteria: (a) gram-positive aerobic/facultative; (b) gram-negative aerobic/facultative; (c) anaerobic; (d) a group of miscellaneous bacteria. (MTB: *Mycobacterium tuberculosis*. Organisms highlighted in red are "lactose fermenters". Organisms highlighted in green are arthropod-borne.)

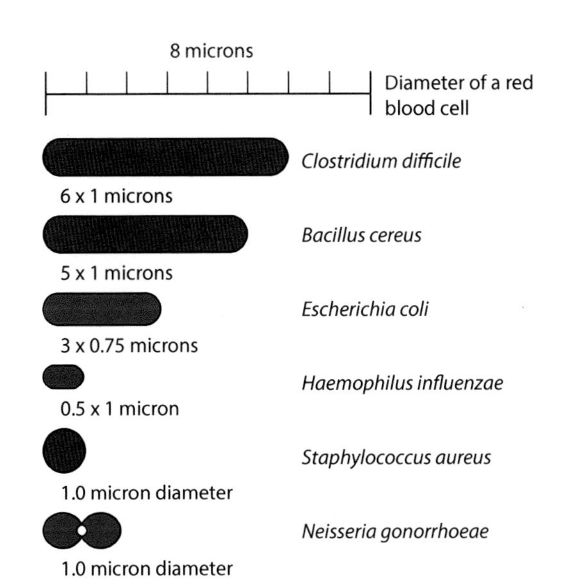

Figure 1.2 The size of selected bacteria in relation to the diameter of a red blood cell. Both gram-positive bacteria (black/dark blue) and gram-negative bacteria (red) can be bacillary, coccoid, or cocco-bacillary in shape.

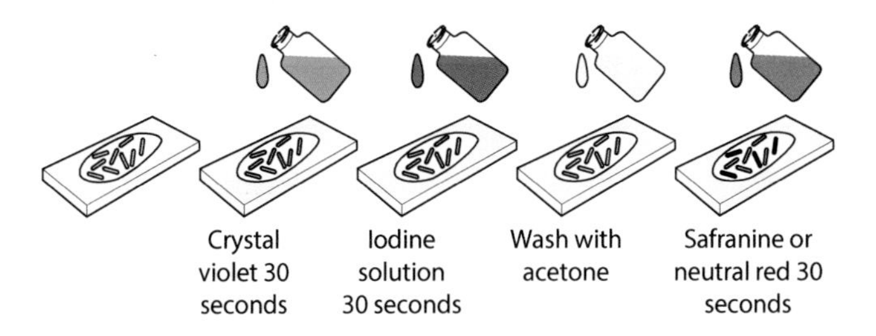

Figure 1.3 The Gram stain procedure.

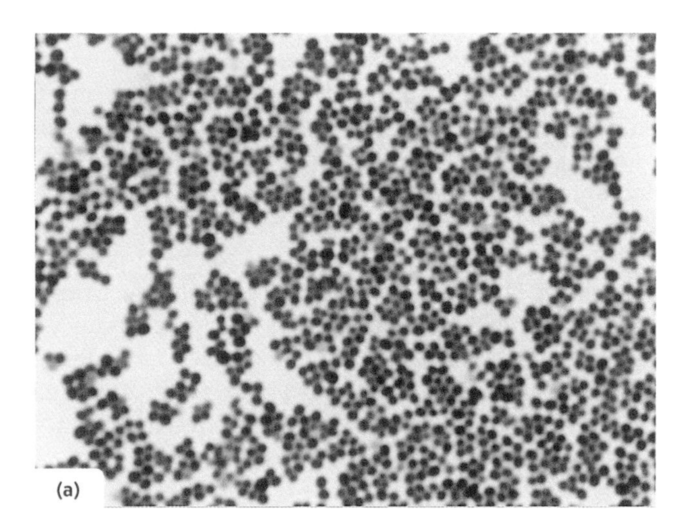

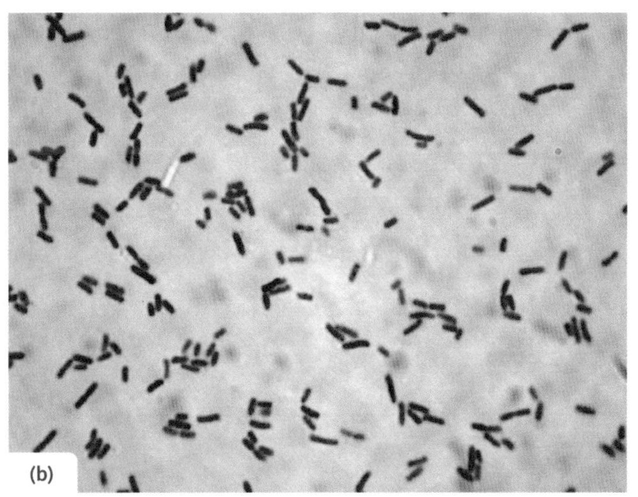

Figure 1.4 Photomicrographs of: (a) gram-positive cocci in clusters; (b) gram-negative rods.

take up the neutral red or safranine stain and appear pale red (**Figures 1.4a,b**). This simple technique is still central to diagnostic bacteriology, and in certain circumstances has not been superseded by modern molecular methods.

Within minutes of a specimen being received in the laboratory, a Gram stain result can be obtained. A Gram stain of a specimen of cerebrospinal fluid (CSF) of a previously healthy young child admitted with meningitis shows numerous gram-positive diplococci scattered around a neutrophil (**Figure 1.5**). In this setting the organism is *Streptococcus pneumoniae*, and the child has pneumococcal meningitis. This result is immediately used to confirm the correct antibiotic treatment.

Other Gram stain features that are used are gram-positive cocci in clusters (staphylococci), gram-positive cocci in chains (streptococci) or in pairs and short chains (enterococci). *Neisseria gonorrhoeae*, *Neisseria meningitidis*, which are gram-negative, and *Streptococcus pneumoniae*, gram-positive, are characteristically found in pairs (diplococci).

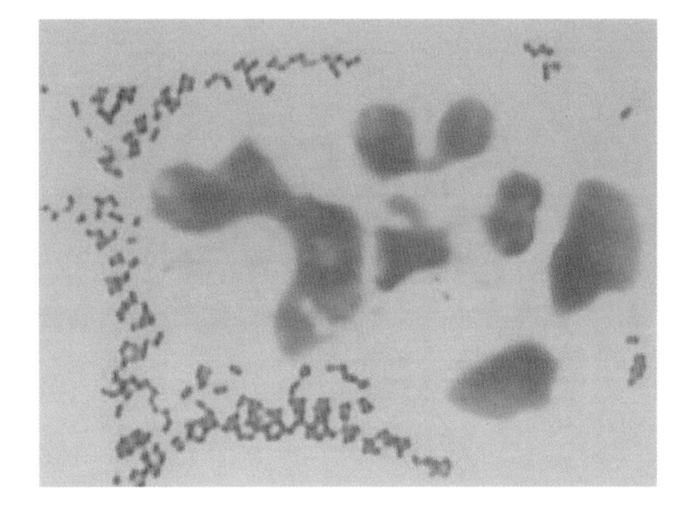

Figure 1.5 Photomicrograph of the cerebrospinal fluid of a young child with the symptoms and signs of meningitis. Numerous gram-positive diplococci are seen around the red-stained nucleus of a neutrophil. This is *Streptococcus pneumoniae*.

CELL WALL

The bacterial cell wall has many functions. The most important is to protect the inner cell structures from osmotic and other physical forces that a bacterium can encounter in a changing environment. This protection is provided by a mesh of peptidoglycan surrounding the cytoplasmic membrane. The cell wall of gram-positive bacteria is largely made up of peptidoglycan. Gram-negative bacteria have an internal layer of peptidoglycan; external to this is an outer lipid (bilayer) membrane, to which lipopolysaccharide (endotoxin) is attached. The outline structure of the gram-positive and gram-negative cell wall is shown in **Figure 1.6**.

The peptidoglycan polymer is cross-linked by short peptide side chains that are essential for the stability of the peptidoglycan and cell wall. Cross-linking is carried out by trans- and carboxy-peptidases, enzymes anchored in the cytoplasmic membrane; these are also known as the penicillin-binding proteins (PBP) (**Figure 1.7**). Note that while the number of different PBP may be five or more, three are used for illustrative purposes in this book. The amino acid serine is the key component in the active site of these enzymes. It is also the target of the β-lactam antibiotics (penicillins, cephalosporins and carbapenems) whose activity resides in the β-lactam ring (**Figure 1.8**). Covalent binding of a β-lactam antibiotic to

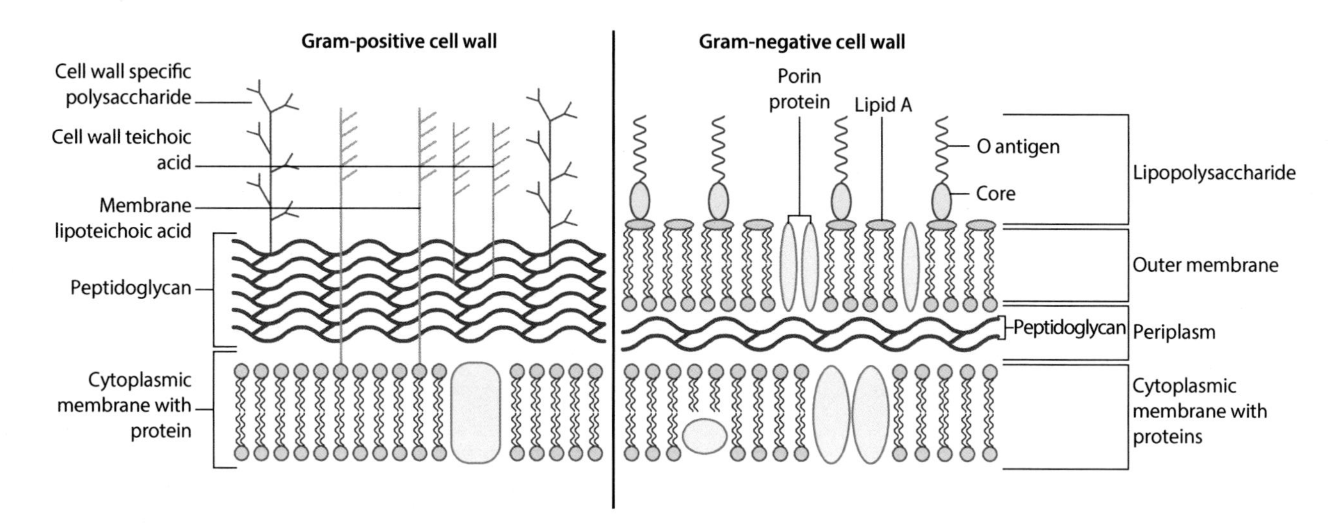

Figure 1.6 The structure of the cell wall of gram-positive and gram-negative bacteria.

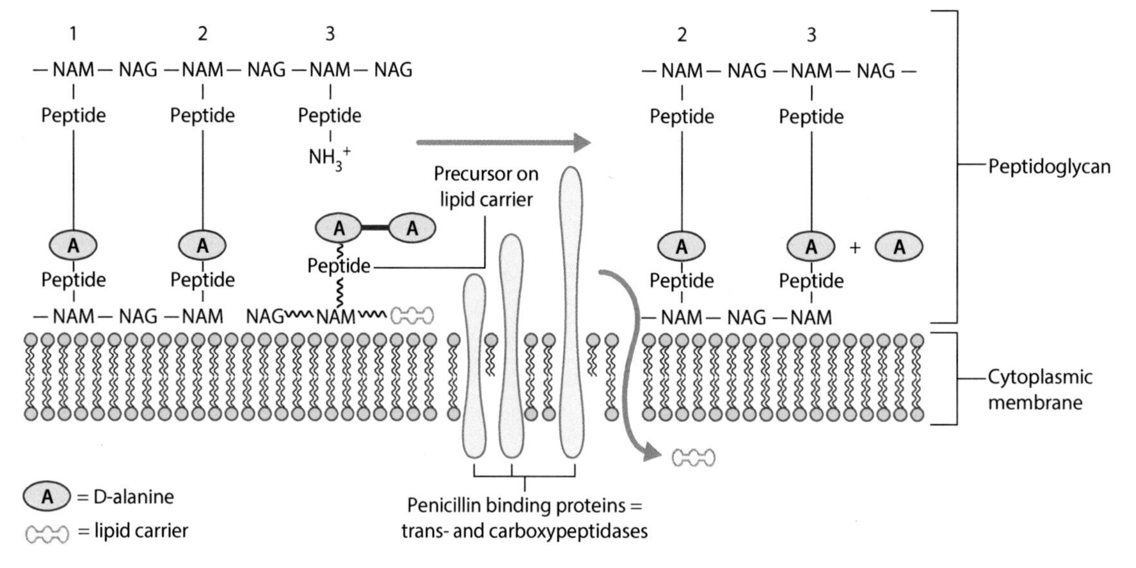

Figure 1.7 Peptidoglycan consists of repeating units of N-acetylglucosamine (NAG) and N-acetylmuramic acid (NAM) cross-linked by peptide side chains. The penicillin-binding proteins (PBP) are responsible for cross-linking these peptide side chains.

the serine residue of the PBP inactivates the enzyme, preventing cross-linking (**Figures 1.9a,b**). Without the protective peptidoglycan mesh, the unprotected cytoplasmic membrane and cell contents bulge through defects in the mesh and the cell bursts.

Penicillins Cephalosporins Carbapenems

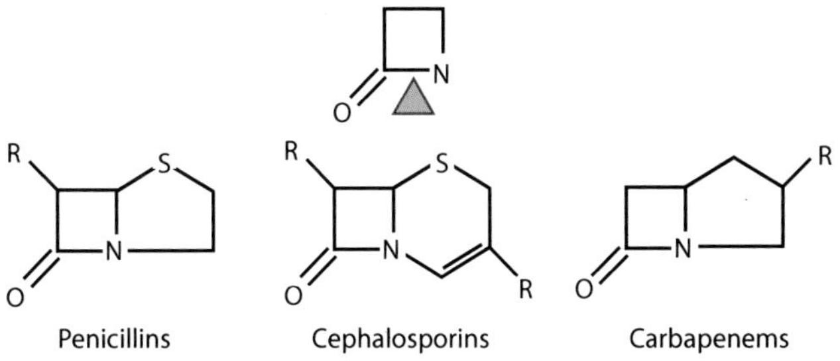

Figure 1.8 An outline of the structure of β-lactam antibiotics. The arrow shows the bond in the β-lactam ring that accounts for the antimicrobial activity of the β-lactam antibiotics.

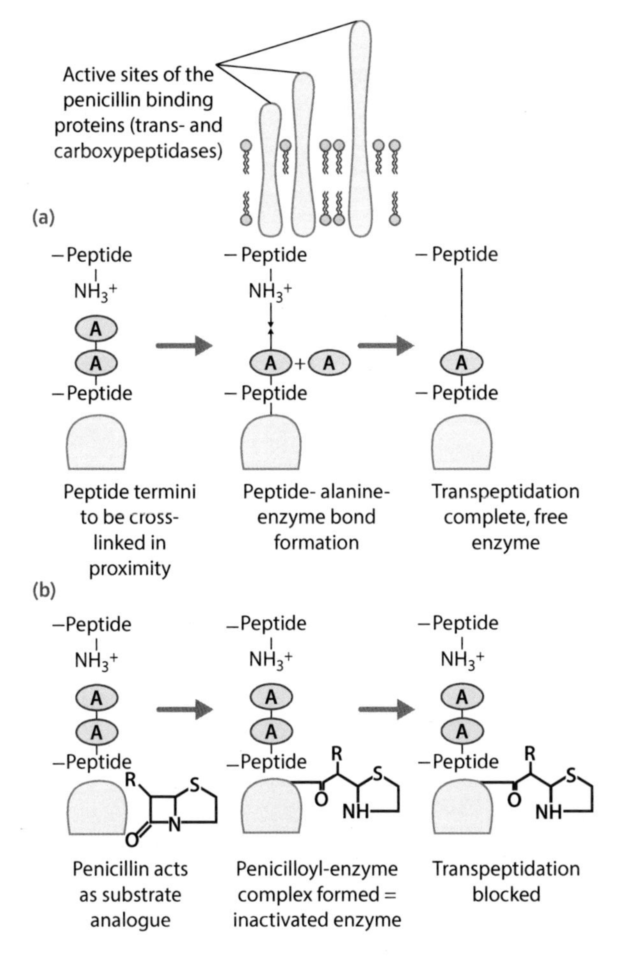

Figure 1.9 The action of a β-lactam antibiotic. (a) The steps in the formation of the peptide cross-link of the peptidoglycan chain. (b) β-lactam antibiotics bind covalently to the active site (the amino acid serine) of the penicillin-binding proteins preventing the transpeptidation step, and the cross-link is not formed.

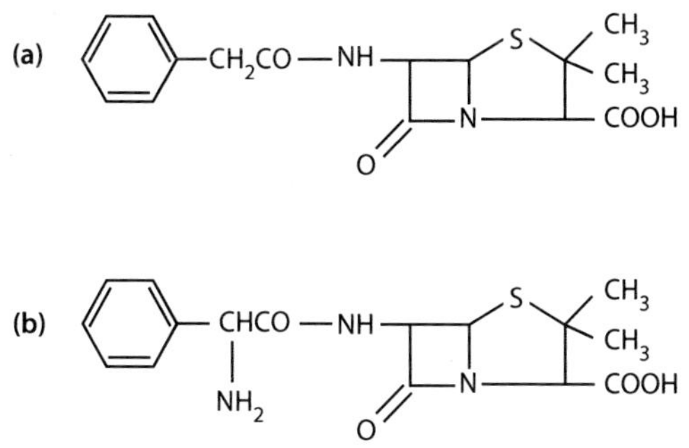

Figure 1.10 The structure of: (a) benzylpenicillin; (b) ampicillin.

The cell wall of gram-negative bacteria is more complex than that of gram-positive bacteria (**Figure 1.6**). The outer lipid bilayer has proteins, such as adhesins, and flagella traversing it. Porins act as channels that allow hydrated molecules to pass through the membrane. From the periplasmic space, molecules can be transported across the cytoplasmic membrane into the cell. It should be noted that porins enable antibiotics such as the β-lactams to reach their site of action. Benzylpenicillin is not effective against most gram-negative organisms because it is not sufficiently polar to pass through a porin channel. Ampicillin, a derivative of benzylpenicillin, differs in the addition of an amino group on the side chain (**Figure 1.10**). The polar ampicillin passes through the hydrated porin channel into the periplasmic space where it can act on the PBP.

BACTERIAL PHYSIOLOGY

Bacteria function by many complex and interacting biochemical pathways. Energy to drive these pathways needs to be provided by a carbon source such as glucose. Physiologically, bacteria are classed as aerobic, where oxygen is essential for growth (e.g. *Pseudomonas aeruginosa*), facultative, where the organism can grow in the presence or absence of oxygen (e.g. the gram-negative 'coliforms') and anaerobic, where the bacteria have to grow in the absence of oxygen (e.g. clostridia and bacteroides).

When aerobic bacteria such as *Pseudomonas aeruginosa* and facultative 'coliforms' grow in oxygen, glucose is completely metabolized by aerobic respiration, using oxygen as the final electron acceptor:

$$\text{Glucose} + 6O_2 \rightarrow 6H_2O + 6CO_2 \qquad \Delta G_0 = -686 \text{ kcal/mole}$$

When 'coliforms' grow in the absence of oxygen, they metabolize glucose by the less efficient process of fermentation, where mixed acids are the end-products. This reaction is as follows:

$$2\text{Glucose} + H_2O \rightarrow 2\text{Lactate} + \text{Acetate} + \text{Ethanol} + 2CO_2 + 2H_2$$
$$\Delta G_0 = -47 \text{ kcal/mole}$$

The mode of metabolism that facultative organisms such as 'coliforms' are in at a particular time influences the action of some antibiotics. The aminoglycosides, such as gentamicin, act on 'coliforms' that are growing in the presence of oxygen, because these antibiotics probably enter the cell by an energy-dependent process that is part of aerobic respiration.

Organisms such as streptococci and enterococci can grow in the presence or absence of oxygen, but they always use fermentation. For anaerobic bacteria, molecular oxygen derivatives such as superoxide are toxic. These organisms do not have the necessary enzyme systems to inactivate these toxic radicals, hence they grow only in the absence of oxygen.

From a practical laboratory aspect, the different requirements of bacteria for oxygen are important. The correct gaseous conditions must be available to ensure that obligate aerobes, facultative organisms or obligate anaerobes are isolated from clinical specimens. For the routine culture of anaerobic bacteria from clinical specimens, all laboratories have anaerobic cabinets, or similar systems, from which oxygen is excluded.

SYNTHESIS OF DNA, RNA AND PROTEINS

The bacterial genome consists of double-stranded deoxyribonucleic acid (DNA) and semi-conservative replication produces two genomes, and the cell divides to produce two cells; continuing reproduction gives rise to exponential growth (**Figures 1.11a,b**). Most bacteria divide every 20 minutes or so under optimum growth conditions, hence a single organism inoculated onto an agar plate will have reproduced to form a visible colony the next day.

In contrast, *Mycobacterium tuberculosis* divides every 18 hours or so, and under standard laboratory conditions it can take several weeks for a colony to be seen. With semi-conservative replication the two parent strands separate and daughter DNA is laid down (**Figures 1.12a,b**); a DNA polymerase enzyme complex is responsible for this (**Figure 1.12c**). Details of the action of this enzyme complex at the molecular level are shown in **Figure 1.12d.**

The genome of *Escherichia coli* has a length of about 1000 microns and has to fit into a cell 3 × 1 microns, and to accomplish this is supercoiled (**Figure 1.13a**). This involves the topoisomerase enzymes, which include DNA gyrase whose mode of action is shown in **Figure 1.13b**. An important group of antibiotics, the fluorinated quinolones (e.g. ciprofloxacin and levofloxacin), bind to the α-subunit and inactivate the enzyme.

DNA is divided into sequences of nucleotides that code for proteins via messenger ribonucleic acid (mRNA). These sequences are arranged into transcription units termed operons. An operon consists of promotor/operator, protein coding and 'termination' sequences (**Figure 1.14a**). The RNA polymerase complex carries out the transcription of mRNA from the DNA chromosome (**Figure 1.14b**). The action of this enzyme is inhibited by the antibiotic rifampicin.

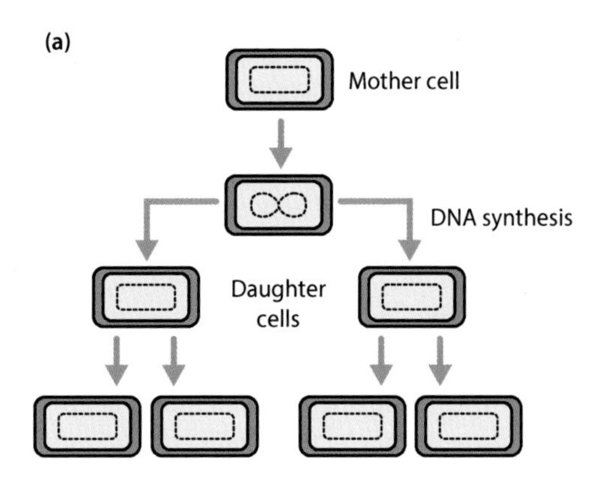

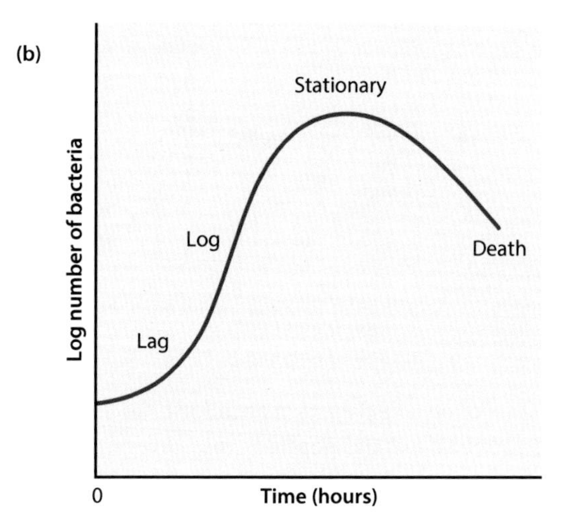

Figure 1.11 (a) Replication and cell division give rise to logarithmic growth, where bacteria divide into 2, 4, 8 organisms and so on. (b) A growth curve; when nutrients and other factors become self-limiting, the bacterial population enters a stationary and then death phase.

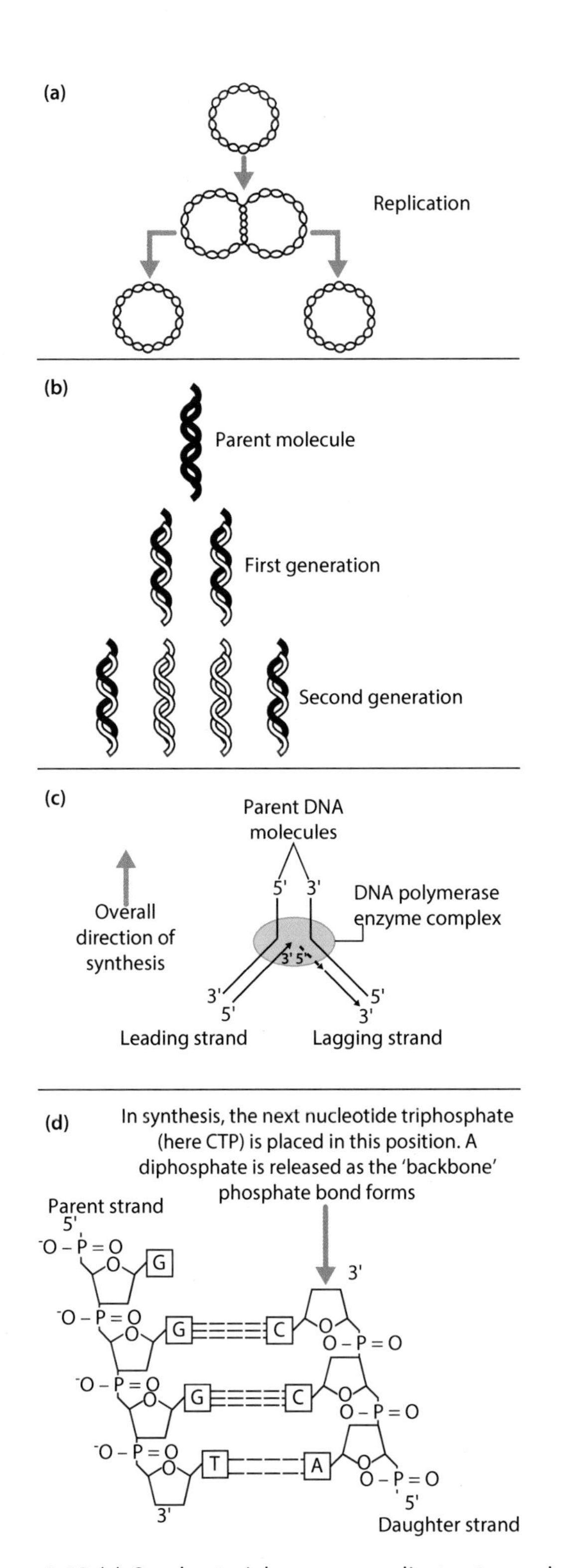

Figure 1.12 (a) One bacterial genome replicates to produce two 'daughters'. (b) The 'mother' deoxyribonucleic acid (DNA) strands separate and a 'daughter' strand is laid down. (c) An outline of the DNA polymerase complex. (d) Synthesis of a daughter strand relies on specific 'base pairing'. (CTP: cytidine triphosphate.)

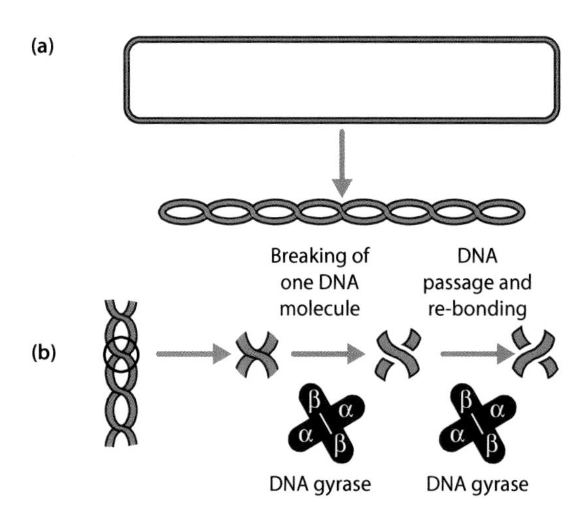

Figure 1.13 (a) The deoxyribonucleic acid (DNA) chromosome has to be supercoiled to fit into a cell. (b) Strand breakage and cross-over are essential in this process, which is performed by enzymes such as DNA gyrase.

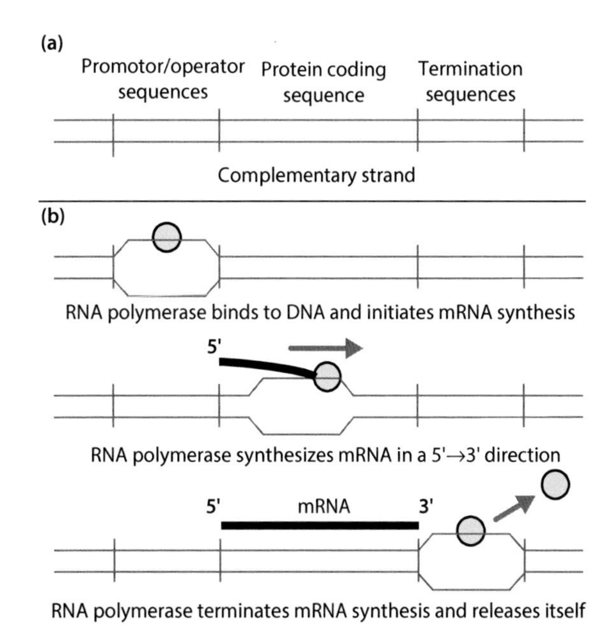

Figure 1.14 (a) DNA is divided into sequences that code for particular proteins. There are 'promotor/operator' and 'termination' sequences at the beginning and end of every transcription complex. (b) An outline of the process of transcription.

The regulation of gene expression at the transcription level is central in coordinating the metabolic activity of the cell. In prokaryotic organisms such as bacteria, it is usual for all the enzymes necessary for a particular metabolic pathway to be expressed by means of one polycistronic mRNA molecule. The organization of the lactose operon of *Escherichia coli*, which codes for three enzymes, β-galactosidase, permease and transacetylase, is shown in **Figure 1.15**. All three enzymes are needed for uptake into the cell and initial processing of the carbohydrate lactose.

PROTEIN SYNTHESIS

Ribosomes translate mRNA molecules to produce proteins. Each three nucleotide 'codon' of the mRNA specifies a particular amino acid. All protein synthesis starts with the amino acid formylmethionine, coded by the sequence AUG, the initiation codon. An outline of protein synthesis is shown in **Figure 1.16**. Two ribosomal subunits bind specifically to the 5′ end of the mRNA. Individual ribosomes move down the mRNA molecule, and as each three base codon is 'read', an amino acid is inserted into the growing peptide chain. At the end of each coding sequence on the mRNA, 'stop' codons such as UAA specify termination, and the completed peptide chain is released. After completing synthesis of the last protein on the mRNA, the ribosomal subunits recycle to form new initiation complexes. The process of translation is the target of the aminoglycosides (gentamicin) and macrolides (erythromycin, clarithromycin, azithromycin).

CYTOPLASMIC MEMBRANE AND SOME OF ITS FUNCTIONS

While eukaryotic cells have several lipid bilayer membrane systems where they can organize metabolic and synthetic functions, bacteria have only the cytoplasmic membrane, which delimits the cytoplasm from the cell wall. This membrane is essential for transport of a wide range of compounds both in and out of the cell. Metabolic and structural entities reside in the cytoplasmic membrane, which has the basic structure of all lipid bilayers (**Figure 1.6**).

β

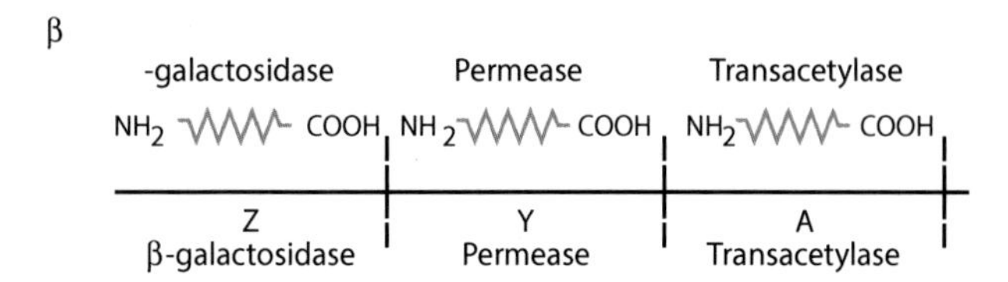

Figure 1.15 The organization of the lactose operon of *Escherichia coli*.

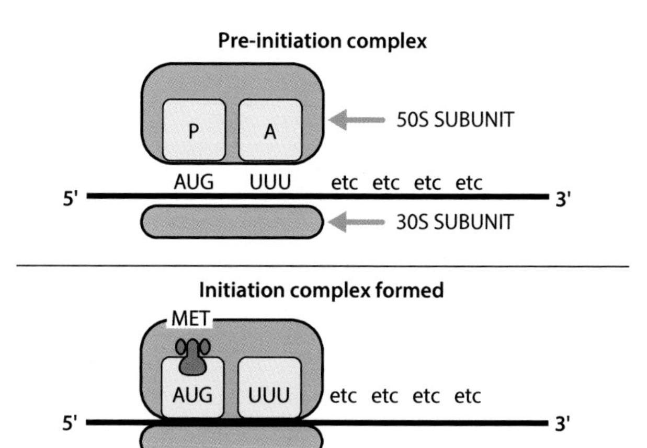

Aminoglycosides (e.g. gentamicin) bind to the 30S subunit and prevent peptide chain initiation

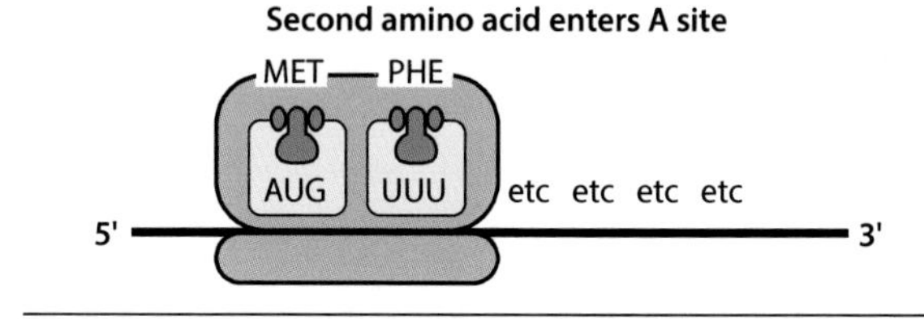

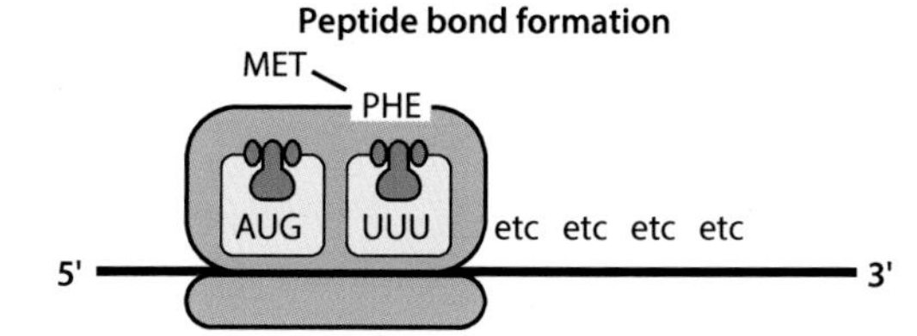

Chloramphenicol binds to the 50S subunit preventing peptide bond formation

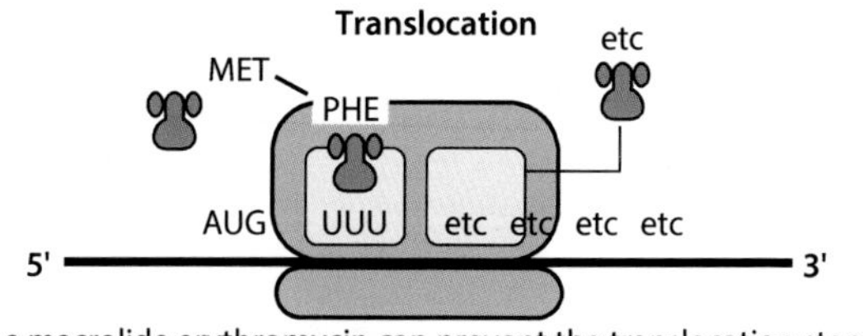

The macrolide erythromycin can prevent the translocation step by interfering with release of the 'free' tRNA from the P site

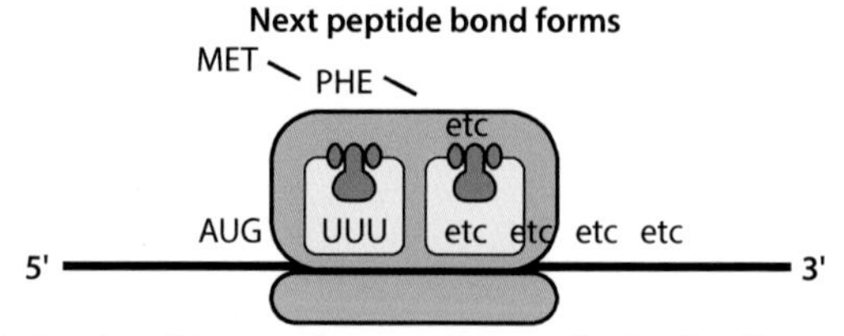

Aminoglycosides can also cause 'mis-reading' as the ribosomes move down the mRNA, causing incorporation of wrong amino acids into the growing peptide to produce malfunctioning proteins

Figure 1.16 An outline of protein synthesis. Many antibiotics interfere with protein synthesis. (P: peptidyl; A: acceptor site of the 50S subunit; MET: formylmethionine; PHE: phenylalanine, the codon for which is UUU.)

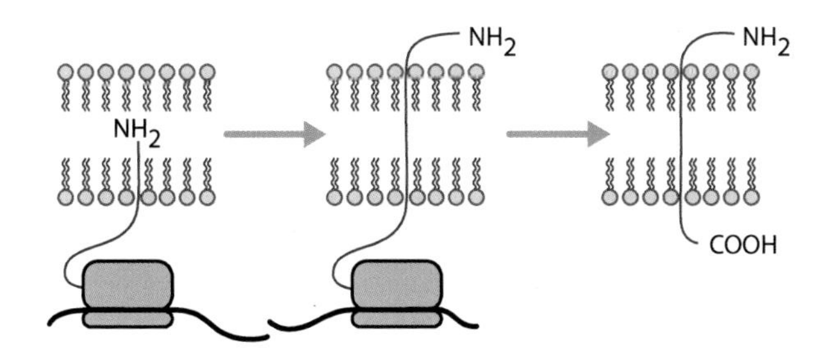

Figure 1.17 The cytoplasmic membrane has the typical lipid bilayer structure. Proteins such as the penicillin-binding proteins are synthesized on ribosomes adjacent to the membrane.

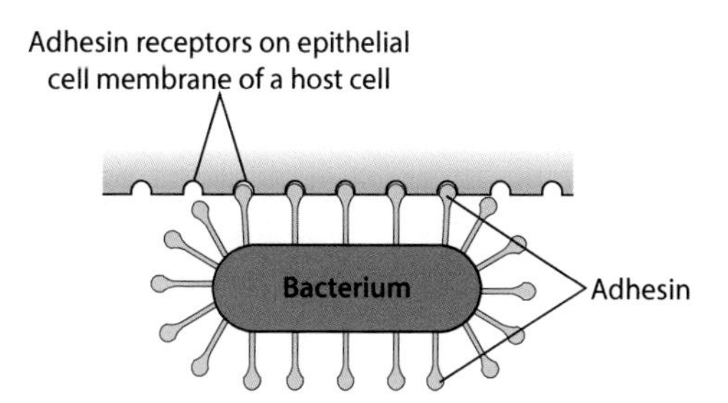

Figure 1.18 Adhesin proteins enable bacteria to adhere to specific receptors on the surface of host cells.

PROTEIN SYNTHESIS

Proteins that are destined to reside in the cytoplasmic membrane, or that are to be secreted out of the cell, are synthesized in the proximity of the cytoplasmic membrane (**Figure 1.17**). Specific sequences at the amino terminal end of the growing peptide chain enable the protein to enter the cytoplasmic membrane. A protein may be completely secreted, as occurs with the toxins of *Clostridium difficile*, or it can be anchored in the cytoplasmic membrane where it will have a specific function. The proteins making up the electron transport chain of 'oxidative' gram-negative bacteria and the PBP are examples of anchored proteins. Other important protein structures resident in the cytoplasmic membrane are adhesins and flagella.

ADHESIN PROTEINS

An essential pathogenic property of many bacteria is the ability to adhere to epithelial and endothelial surfaces. Adhesin proteins enable this to occur. Uropathogenic strains of *Escherichia coli*, commonly associated with urinary tract infections (UTI), colonize the periurethral area of susceptible females by specific adhesins, which recognize receptors on the host cell surface (**Figure 1.18**). From here the bacteria gain access to the bladder via the urethra and initiate cystitis.

FLAGELLUM

Examples of flagellated bacteria include *Escherichia coli*, *Vibrio cholerae* and *Clostridium tetani*. Flagella are complex protein structures anchored in the cell membrane. In conjunction with chemical signalling systems, bacteria can use their flagella to move towards a source of nutrients or away from an unfavourable environment. The basic structure of a flagellum is shown in **Figure 1.19**. They can occur around the cell, as in coliforms (peritrichous), or they can be restricted to one end as found with *Pseudomonas aeruginosa*.

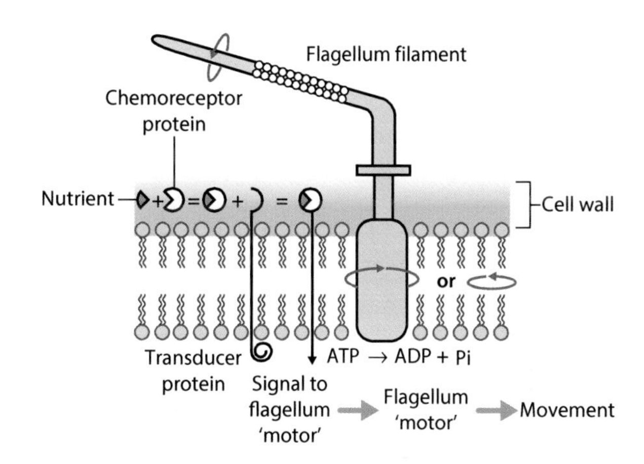

Figure 1.19 A flagellum. Interaction with chemical signals and transducer proteins determines which direction the flagellum and the bacterium moves. (ADP: adenosine diphosphate; ATP: adenosine triphosphate.)

CAPSULES

A number of gram-positive and gram-negative bacteria have capsules, which are structures exterior to the cell wall. They usually consist of polysaccharide, enabling the bacterium to resist phagocytosis by macrophages and neutrophils (**Figure 1.20**). Injection of millions of unencapsulated pneumococci into the peritoneum of a mouse is not lethal, whereas injecting a few hundred encapsulated organisms is.

For reasons of pathogenicity, polysaccharides are not particularly good antigens on their own. Their antigenicity is improved by conjugation with a protein carrier. Clinically, the most important serotype of *Haemophilus influenzae* is serotype b (Hib), which can cause invasive disease such as bacteraemia, meningitis and epiglottitis in children usually less than 5 years old. Disease due to serotype b is now rare

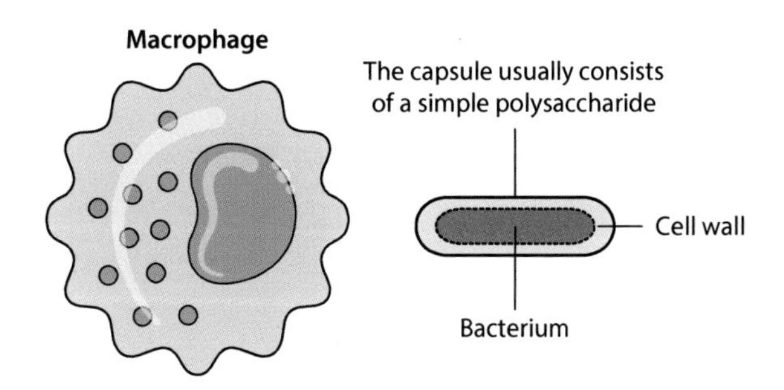

Figure 1.20 Certain bacteria are encapsulated. This capsule inhibits phagocytosis and is an important pathogenic property of bacteria such as *Haemophilus influenzae*, pneumococcus and meningococcus.

Organism	Capsule type	Vaccine
Haemophilus influenzae	6: a,b,c,d,e,f	Hib (polysaccharide of capsule serotype b conjugated to a carrier protein)
Neisseria meningitidis	8: A,B,C,E,X,Y, Z,W-135	A,C,B,Y,W-135
Streptococcus pneumoniae	>90: e.g. 3,10, 19	23 common serotypes in the polysaccharide vaccine and 13 in one of the conjugated vaccines

Figure 1.21 Three important encapsulated bacteria, with the number and classification of the serotypes and examples of vaccines indicated. (Hib: *Haemophilus influenzae* type b.)

in countries where Hib vaccination is practised. The Hib vaccine consists of the polyribose ribitol phosphate (PrP) capsule polysaccharide bound to tetanus toxoid. Examples of important encapsulated bacteria, their various capsular serotypes and available vaccines are shown in **Figure 1.21**.

The *Streptococcus pneumoniae* 23-valent vaccine is an unconjugated vaccine of the commoner 23 serotypes. The 13-valent vaccine has the capsular polysaccharides conjugated to the diphtheria toxoid as the carrier, which improves the immunological response.

SPORULATION

Medically important bacteria including *Clostridium botulinum, Clostridium difficile, Clostridium tetani* and *Bacillus cereus* produce spores. Under unfavourable growth conditions the vegetative cell produces a heat stable spore (**Figure 1.22**). These spores can survive for years. When growth conditions are favourable, the spores germinate and vegetative growth is re-established with production of exotoxin.

Bacillus cereus is found in dry grain foods such as rice. Its spores can survive the cooking process, and when this rice is stored at room temperature, the spores germinate and secrete a heat-stable emetic exotoxin. When food is heated and eaten, nausea, abdominal cramps and vomiting occur within 1–4 hours.

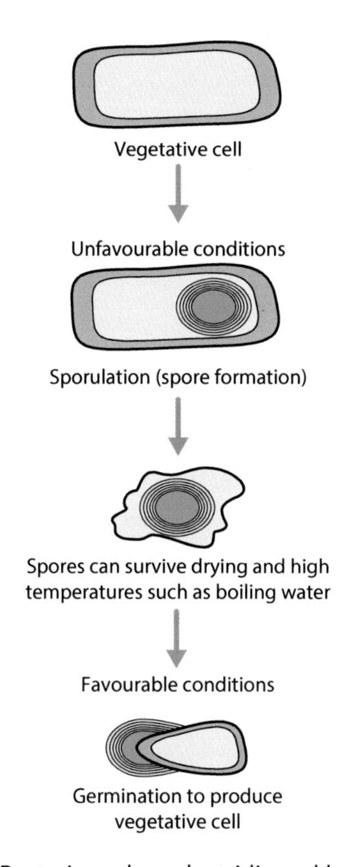

Figure 1.22 Bacteria such as clostridia and bacillus sporulate under unfavourable growth conditions.

GENETIC EXCHANGE IN BACTERIA

DNA can be transferred between bacteria by bacterial viruses (bacteriophages), by transformation or by conjugation. Many bacteria contain extra chromosomal plasmids that can occur as one or more copies per cell (**Figure 1.23a**). If there are two or more copies per cell, each daughter cell will usually inherit a plasmid after cell division. Plasmids can also transfer from a 'male' F+ cell to a 'female' F− cell by the process of conjugation (**Figure 1.23b**). In this process the 'male' cell remains 'male'.

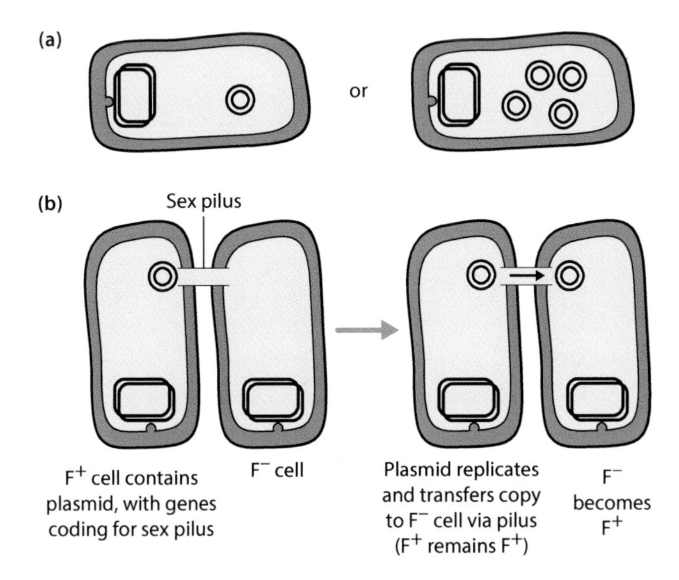

Figure 1.23 Many bacteria contain plasmids. (a) Plasmids can exist as one or more copies per cell. (b) Plasmids can transfer between bacteria by the process of conjugation. A sex pilus, a simple protein tube, is necessary for this process.

As many plasmids carry genes coding for antibiotic resistance, the spread of plasmids is central to the problem of antibiotic resistance. There are various types of mobile genetic elements that can move between the bacterial chromosome and a plasmid. Insertion sequences (IS) are one example. They are about 1000 base pairs in length and consist of short inverted repeat sequences on either side of a gene that enables the element to move to different sites in chromosomal or plasmid DNA. Bacteria contain many copies of one or more IS structures; *Escherichia coli* has more than 40 scattered throughout its chromosome. When two IS domains combine at either end of an antibiotic resistance gene, this forms a transposon, and the resistance gene gains mobility. Integrons add a further level to the mechanisms of antibiotic resistance. These are sequences of DNA where different antibiotic resistance genes are linked together in an operon, under the control of a single promotor. Integrons can be integrated into transposons. These mobile genetic elements can be transferred via plasmids between members of the same species, or different species.

A common form of resistance in the gram-negative Enterobacteriaceae, or 'coliforms', is the production of a β-lactamase enzyme that hydrolyses the β-lactam ring of antibiotics including benzylpenicillin, ampicillin and amoxicillin. Examples are the TEM-1 β-lactamase found in *Escherichia coli* and SHV1 of *Klebsiella pneumoniae*. Other examples are the extended-spectrum β-lactamases (ESBL) and carbapenemases, which are also found in these two bacteria.

High-level vancomycin resistance in enterococci is due to the presence of a composite transposon that contains the genes necessary for the resistance phenotype.

VIRUSES

A classification of the major virus groups that infect humans is shown in **Figure 1.24**. The genome of viruses is either single- or double-stranded DNA or RNA. This is surrounded by a protein capsid, consisting of one or more protein subunits that are used repeatedly to make a protective shell. Many viruses are enveloped, obtaining a lipid membrane from the infected cell. This is often the plasma membrane, but certain viruses 'bud' into the Golgi apparatus or endoplasmic reticulum. Virus-specific proteins, such as gp120/41 of human immunodeficiency virus (HIV) or the surface antigen (HBsAg) of hepatitis B virus (HBV), are inserted into these and are responsible for recognizing and binding to the receptors on the surface of susceptible cells. These envelope proteins are glycosylated, the addition of specific sugar residues by a post-translation cellular function.

The structure of HBV and electron micrographs of several viruses are shown in **Figure 1.25**. The HBV 2-DNA genome is surrounded by the capsid made up of core antigen (HBcAg). This nucleocapsid is surrounded by the envelope containing the large (L), medium (M) and small (S) surface antigen proteins (HBsAg). The ruptured envelope of herpes simplex virus (HSV) reveals the capsid, while the enveloped coronavirus clearly shows the glycoprotein extending out from the surface of the virus. Norovirus and adenovirus are examples of naked (unenveloped) icosahedral (spherical) viruses.

Apart from the pox viruses, DNA viruses replicate in the nucleus where they have full access to the DNA synthetic, RNA transcription and post-transcription RNA splicing machinery of the eukaryotic nucleus. The parvoviruses have a 1-DNA genome and, like the adenoviruses, polyomaviruses and papilloma viruses, are naked. The members of the herpesviridae and hepadnaviridae (HBV) are enveloped.

The majority of RNA viruses replicate in the cytoplasm and must provide the enzymes for replicating their genomes and synthesizing mRNA. The classification of RNA viruses has several themes. The reoviruses, which include rotavirus, have a 2-RNA segmented genome. The 1-RNA viruses are grouped on the basis of having either a positive- or negative-sense genome. Positive-sense genomes act directly as a mRNA and are translated

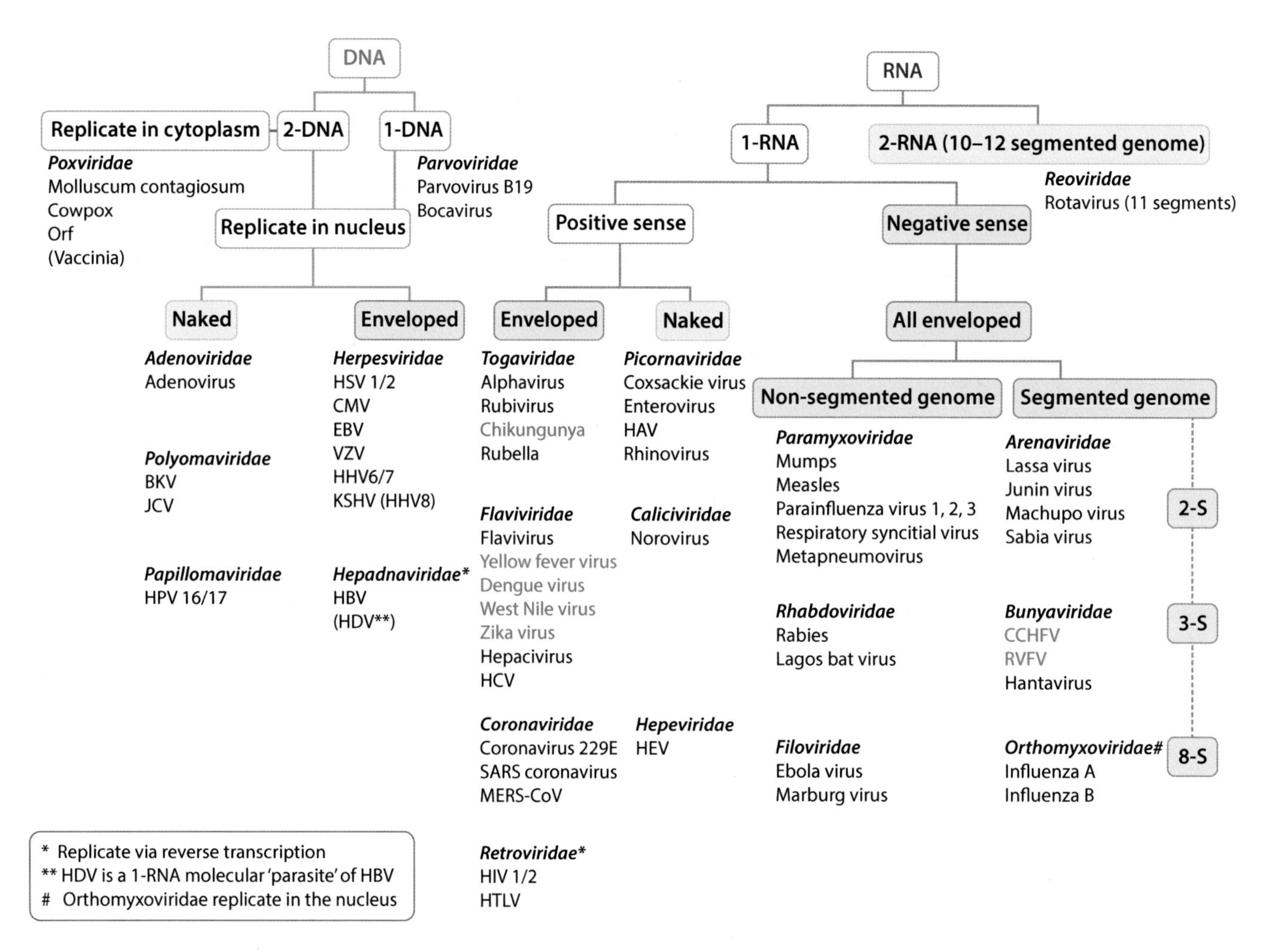

Figure 1.24 A classification of the major groups of viruses that infect humans. Yellow boxes: naked viruses, pink boxes: enveloped viruses. The number of segments (S) is indicated. Viruses highlighted in green are transmitted by mosquitoes, with the exception of tick-borne Congo Crimean haemorrhagic fever. (MERS-CoV: Middle East respiratory syndrome coronavirus.)

into the viral proteins necessary to initiate replication of the incoming genome. The genome of negative-sense 1-RNA viruses is the complementary sense to their mRNA. These viruses must include their own RNA polymerase when virus is assembled, as this enzyme converts the incoming negative-sense genome into a double-stranded replicative form in the newly infected cell. All the negative-sense 1-RNA viruses are enveloped.

The 1-RNA viruses have a particular challenge to overcome in the cytoplasm of the cell they infect. Eukaryotic mRNA is monocistronic, coding for one protein, and RNA viruses have evolved mechanisms to overcome this restriction. The genome of picornaviruses is translated into a 'poly-protein', which is cleaved into the individual structural and enzymatic functions during translation by viral and host proteases. Reoviruses and orthomyxoviruses have a genome divided into segments that code for a single functional protein, having 10–12 and 7–8 segments respectively, with each segment coding for one protein. The bunyaviruses use a combination of these mechanisms. The Rift Valley fever virus (RVFV) genome has three segments: L codes for the RNA polymerase, M for the envelope glycoproteins G1 and G2 and S codes for the capsid protein and a non-structural protein.

Influenza viruses A and B, as orthomyxoviruses, and HIV, a retrovirus, are exceptions within the RNA viruses as they must use the host nucleus for reproduction. Within the nucleus, influenza steals the 5′ leader sequence of the host's mRNA molecules to prime synthesis of its own mRNA. By having this unusual dependence, influenza virus has parasitized the eukaryotic nuclear splicing machinery, enabling it to code for an additional protein in the second reading frame of segments 7 and 8.

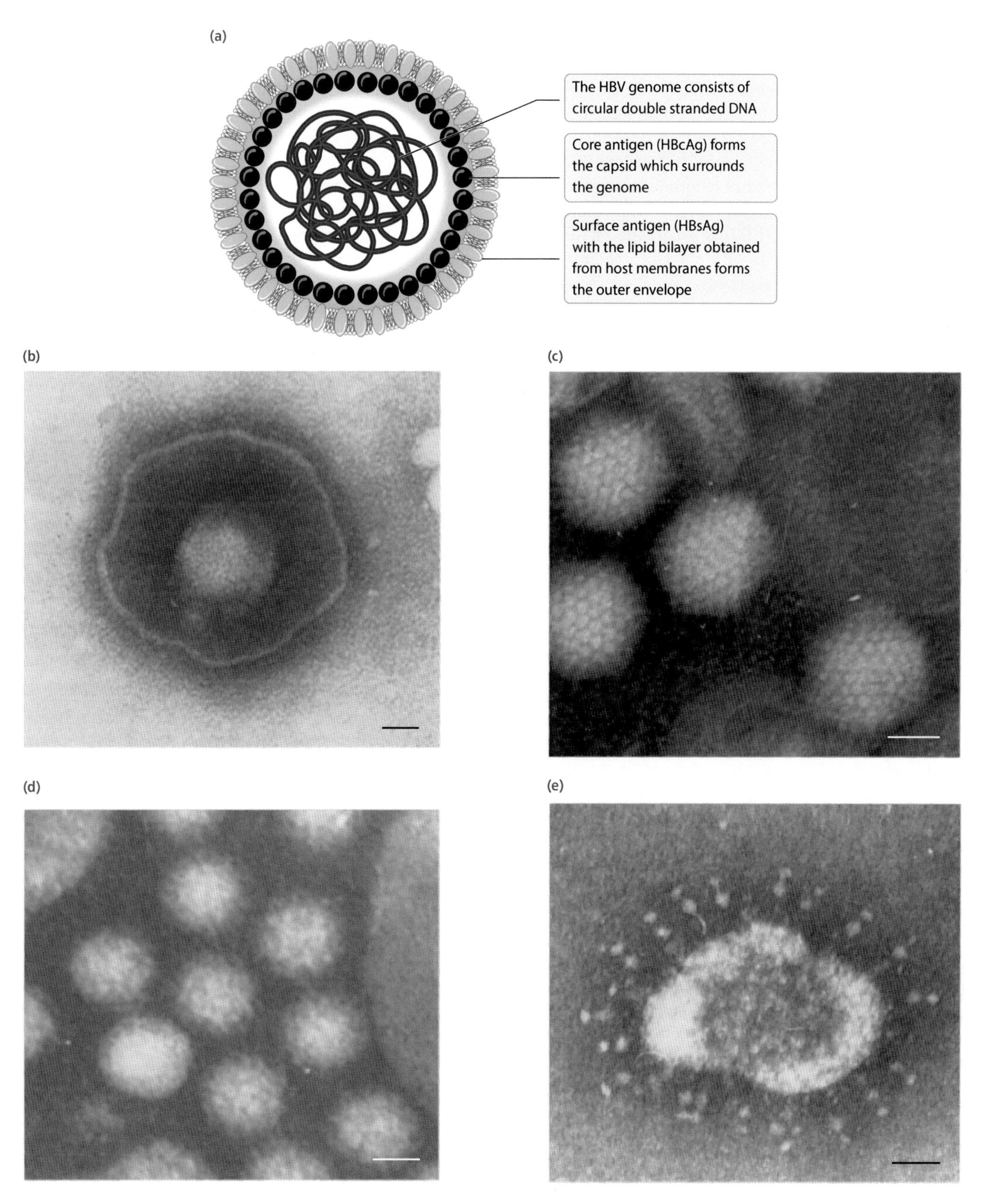

Figure 1.25 (a) A cross-sectional model of HBV, showing the DNA surrounded by the capsid made up of HBcAg capsomeres (nucleocapsid), enclosed in the lipid bilayer with the HBsAg. (b) A ruptured envelope of HSV reveals the icosahedral nucleocapsid. (c) Adenovirus. (d) Norovirus. (e) The surface glycoproteins of a coronavirus are clearly visible. (Bar: 20 nm.)

As a retrovirus, HIV synthesizes a DNA copy of the RNA genome by reverse transcription, and a 2-DNA form is integrated into the host cell DNA. Transcription of this provirus produces progeny genomes, as well as all the required mRNAs. This process is entirely dependent on the RNA splicing machinery of the host cell nucleus. HBV is an unusual DNA virus, using a full-length RNA transcript as a template for DNA synthesis via reverse transcription.

Hepatitis delta virus (HDV) is a parasite of HBV, and has a circular 1-RNA virus with one gene, coding for its RNA polymerase. It only reproduces in cells infected with HBV, having an absolute requirement to steal the HBsAg for its own capsid. Massive numbers of HDV, in excess of 10^{11}, can be produced by each hepatocyte.

THE REPLICATION OF VIRUSES

The diseases that viruses cause are linked to their tissue tropism, replication strategy and the effect that they have on the target cell. The immune response, including the role of interferons and apoptosis, is closely entwined. Viruses can be lytic and kill the infected cell, and influenza is an example. Norovirus leads to apoptosis of infected cells causing to transient atrophy and blunting of the villi of the small intestine, as well as loss of functional microvillae.

Parvovirus B19, a relatively simple 1-DNA virus, requires the nuclear functions of an actively dividing host cell, and the erythrocyte precursor in the bone marrow is its main target. In otherwise healthy individuals, the reserve of these cells results in only a minor drop in red blood cell (RBC) numbers, whereas in the individual with a disease of RBC production, such as sickle cell anaemia, severe anaemia results in a transient aplastic crisis. In the first 20 weeks of gestation parvovirus B19 infection gives rise to non-immune hydrops fetalis, characterized by severe anaemia and high output cardiac failure, with death of 2–10% of affected fetuses.

Latency is found in DNA viruses (and retroviruses via the DNA provirus), and HSV and varicella zoster virus (VZV) are examples. Following initial replication in the respiratory mucosa, a VZV viraemia enables the virus to reach its target cell, the basal keratocyte of the epidermis, where replication gives rise to the typical 'chickenpox' rash. Virus can then ascend sensory fibres to the ganglia, residing in a latent form as episomal DNA within the nucleus, and viral replication is prevented. With declining immune function of older age, or induced immunosuppression, the episomal DNA can switch to replication. Virus descends to the epidermis, where replication produces the typical dermatomal distribution of herpes zoster. As a retrovirus, HIV has a period of prolonged latency, via the integrated proviral DNA. With the gradual depletion of functioning T CD4 lymphocytes as a result of virus replication, the stage is reached when this depletion allows both uncontrolled virus replication and the opportunistic infections that define the acquired immunodeficiency syndrome (AIDS).

Certain DNA viruses produce cancer by specifically interfering with the regulation of the host's DNA synthesis, overriding control mechanisms within the nucleus, for example human papillomavirus (HPV 16, 18) and cervical carcinoma. Epstein–Barr virus (EBV) is associated with infectious mononucleosis, but can also cause lymphoproliferative disease in patients who have decreased T-cell function due to immunosuppressive agents. This complex DNA virus is able to suppress programmed cell death (apoptosis) and drive the infected B cell from the G1/G0 cycle into the synthesis stage. Although integrated HBV DNA is found in patients with hepatocellular carcinoma (HCC), it is likely that hepatocellular cancer is due to repeated cycles of inflammation, cell death, regeneration and fibrosis, which over a period of years give rise to heptocytes with uncontrolled cell division. A similar pathology takes place with chronic infection caused by HCV, a 1-RNA virus. A summary of several virus replication strategies, and how disease arises is shown in **Figure 1.26**.

To introduce virus replication in more detail at the cellular level, HBV and HAV are discussed here, and HIV in Chapter 4.

THE REPLICATION OF HBV

The genome organization of this virus, based on the linear RNA template, is shown in **Figure 1.27**. A 5′ cap is a feature of eukaryotic mRNA and there are terminal redundancy (repeat) elements, R at each end. Full length RNA is translated into the reverse transcriptase (RT) enzyme complex polyprotein (terminal protein [TP], spacer, the reverse transcriptase and RNase H). RNase H is responsible for degrading RNA on the RNA/DNA duplex, enabling synthesis of the 2-DNA genome. The core antigen (cAg) is also synthesized from genome-length RNA, but in a different reading frame, as is X, a protein that regulates transcription of the viral genome.

The e antigen (HBeAg) is an unusual protein containing the first 70% of the amino acids that make up the HBcAg protein; however, its synthesis starts upstream from the start codon of the cAg, and this amino acid sequence directs the growing protein into the endoplasmic reticulum membrane system. The e antigen is then cut by a host protease, removing the last 30% of amino acids found in the cAg, releasing it into the lumen of the membrane and when membrane fusion occurs at the cell surface, the HBeAg enters the blood. While sharing amino acids sequences of HBcAg, the eAg antigen has distinct

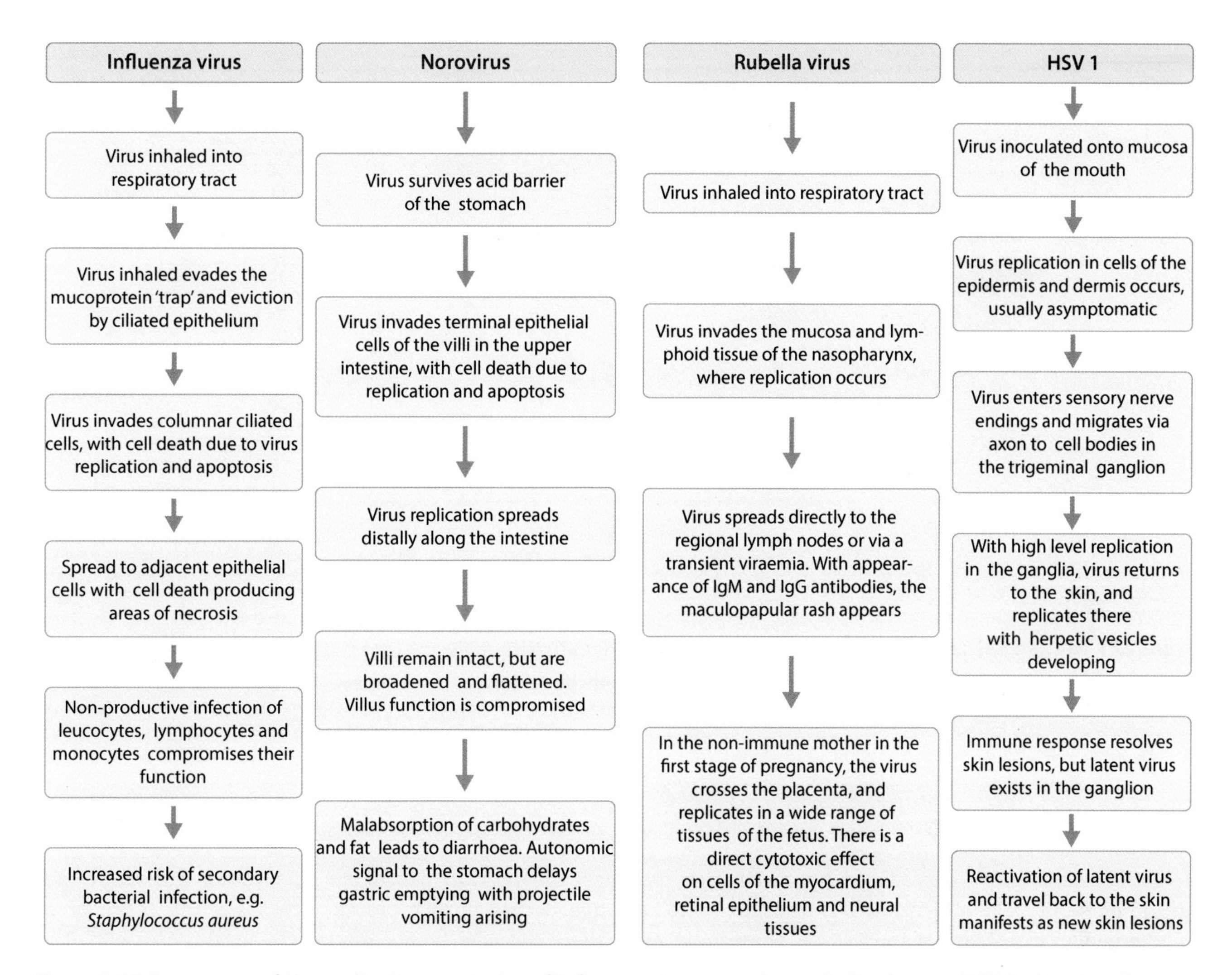

Figure 1.26 A summary of the replication strategies of influenza virus, norovirus, rubella virus and HSV1, showing the outcomes of their replication in relation to the tissues they can access.

epitopes. It acts to suppress the immune response to the HBcAg transiently, increasing the likelihood of chronic infection being established. Once the virus achieves that status, eAg becomes redundant, and antibodies (HBeAb) appear (see Chapter 12).

The replication of HBV is outlined in **Figure 1.28**. Binding to the hepatocyte cell membrane is initiated by the N-terminus of L surface antigen; the hydrophilic loops of all three sAg proteins are then involved in full attachment and internalization of the nucleocapsid (1, 2). The nucleocapsid is then transported to, and enters, the nuclear pore, where viral DNA is released (2, 3). The viral DNA is only double stranded for 70% of its length, and the next step is its conversion to a complete circular 2-DNA molecule by host DNA polymerase (4, 5). This is then transcribed into full length single-stranded RNA, and two shorter RNA molecules (6). The full length RNA acts as the mRNA for the core protein (cAg) and the RT enzyme complex (7). Core protein forms a capsid around full length RNA molecules, and within this the RT enzyme complex forms the RNA/DNA hybrid, which is then converted to 2-DNA, to form the mature nucleocapsid (8, 9). The DNA-containing nucleocapsids have two routes to follow. They can either enter the nucleus to amplify the cycle there (10), or be transported to the cytoplasmic membrane where the capsid 'buds' to form mature virus with the sAg-containing envelope (17).

Essential to the life cycle of HBV is production of the eAg. As discussed above, this is a form of the cAg that is exported out of the infected cell into the blood (11–13).

HBsAg is the envelope protein. The large (L), medium (M) and small (S) surface antigen proteins have the same sequence of amino acids at the carboxy end, which contain the determinant that enables cell binding, to which

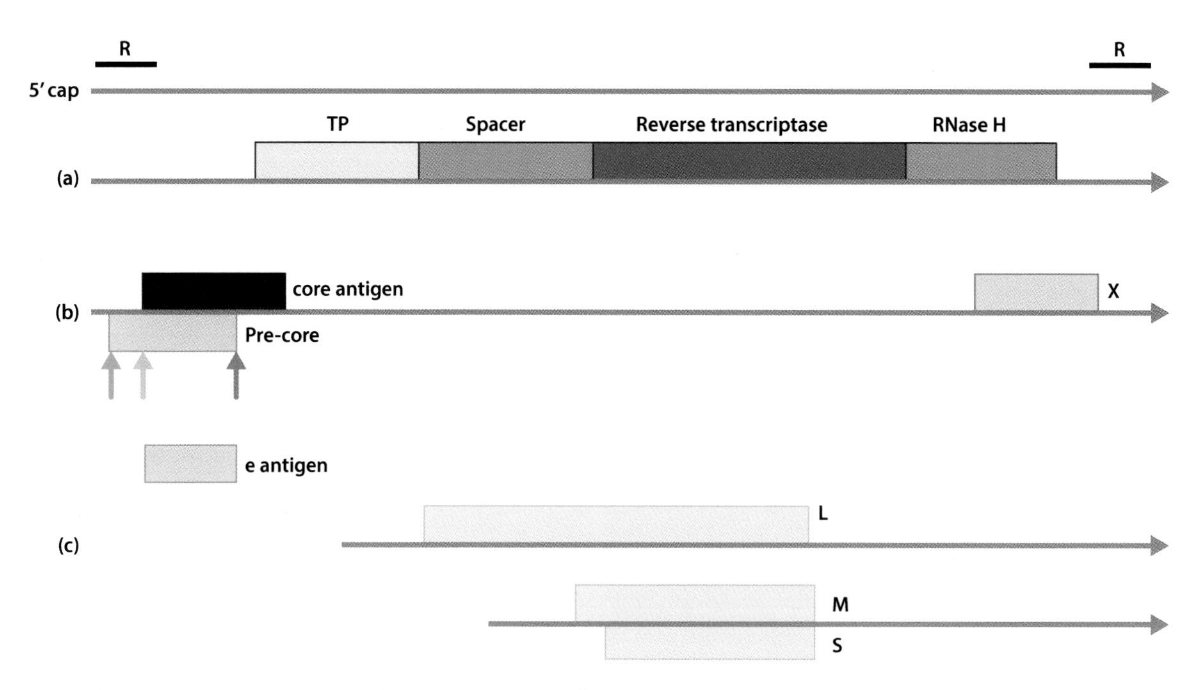

Figure 1.27 The genome organization of HBV based on the full length RNA molecule that is used as the template for DNA synthesis by reverse transcription. Full length RNA is the mRNA for (a) the RT complex, and (b) HbcAg, HBeAg and X protein. Green to yellow arrows shows the leader sequence, and the red arrow the translation termination sites of the HBeAg. (c) The L and M/S forms of the HBsAg are translated from different mRNA molecules.

protective antibodies are produced by vaccination, or natural immunity. The larger mRNA is translated into L, and the smaller mRNA into M and S (7, 14). These proteins are synthesized and anchored in the endoplasmic reticulum and then move to the cytoplasmic membrane where 'budding' occurs. There are two roles for the HBsAg. It is released into the blood in massive amounts as rod or filament shaped subviral protein/lipid particles, absorbing virus-neutralizing antibodies, thus facilitating spread and maintenance of the virus in the liver (15, 16). DNA-containing nucleocapsids (9) bud into sAg-containing membranes, forming mature enveloped virus (17, 18). Following acute infection, the ongoing presence of HBsAg in the blood is the marker of chronic infection.

THE REPLICATION OF HAV

The genome organization of HAV with the functions of the various proteins is shown in **Figure 1.29**.

HAV is a naked, icosahedral virus with a 1-RNA positive-sense genome, acting directly as mRNA within the infected cell. The viral protein Vpg is covalently attached at the 5′ end and is the primer for RNA synthesis. Acting as one long mRNA, all the individual proteins are theoretically part of one polyprotein. However, cleavage occurs during protein synthesis; the majority of these cleavage events are done by viral protease 3C. The domains P1/2A, P2 and P3 are primary cleavage sites.

Following uptake into enterocytes it is likely that the virus initiates replication here, and then, via the portal vein, is transported to the liver. The replication cycle in the hepatocyte is summarized in **Figure 1.30**. Attaching to receptors on the surface, termed cr7 and TIM-1, it enters the cytoplasm (1–3).

The RNA genome is released into the cytoplasm (4) and acts directly as mRNA. The host ribosomes bind the internal ribosome entry site (IRES) near the 5′ end and move down the RNA, synthesizing protein. During this process, proteases cleave the translation product into the individual virus proteins (5) (the cleavage of 1A from 1B takes places as a final step in virus assembly).

The viral RNA/protein synthesis complex induces tubular vesicular structures in association with the endoplasmic reticulum, in close proximity with the nucleus, and this is where virus replication takes place. The viral RNA polymerase (3D) and NTPase/helicase (2C) are responsible for RNA synthesis. The viral RNA is converted to a 2-RNA

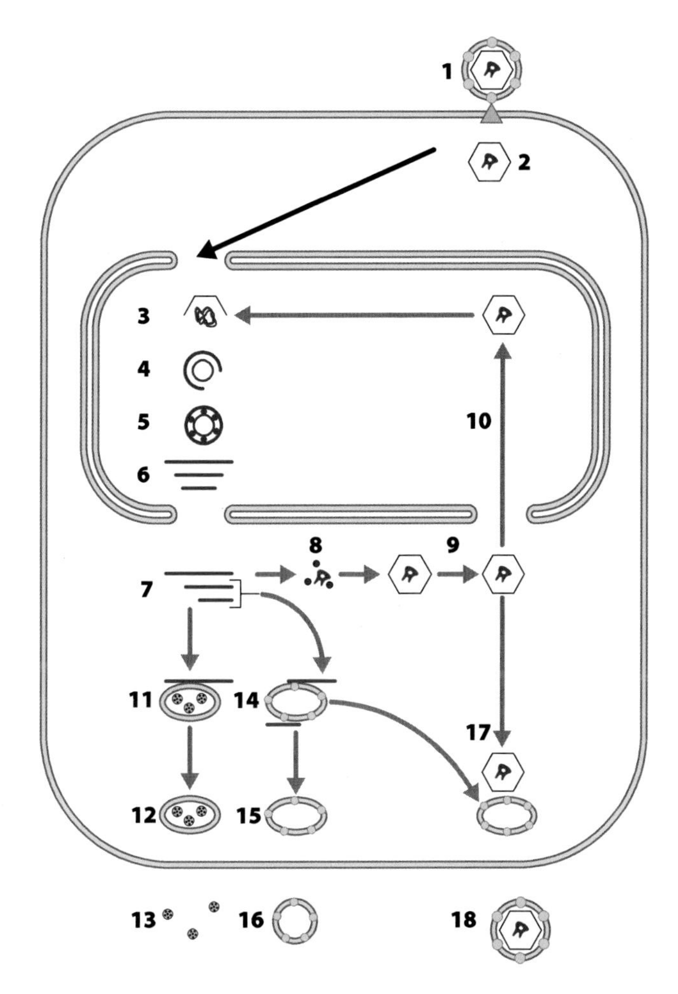

Figure 1.28 An outline of the replication strategy of HBV in the hepatocyte. The numbers refer to the key stages discussed in the text.

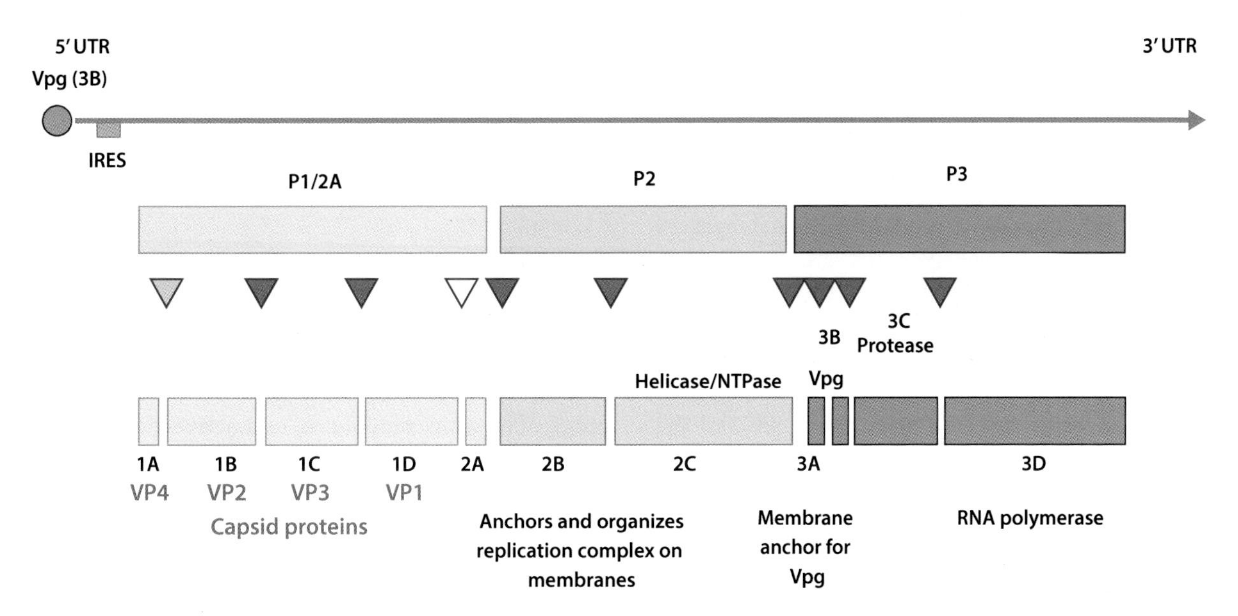

Figure 1.29 The genome organization of HAV. (UTR: untranslated region; IRES: internal ribosome entry site. Arrows show protease cleavage sites: 3C: ▼, cellular protease: ▽, during virus maturation: ▼. Vpg: virus protein genome.)

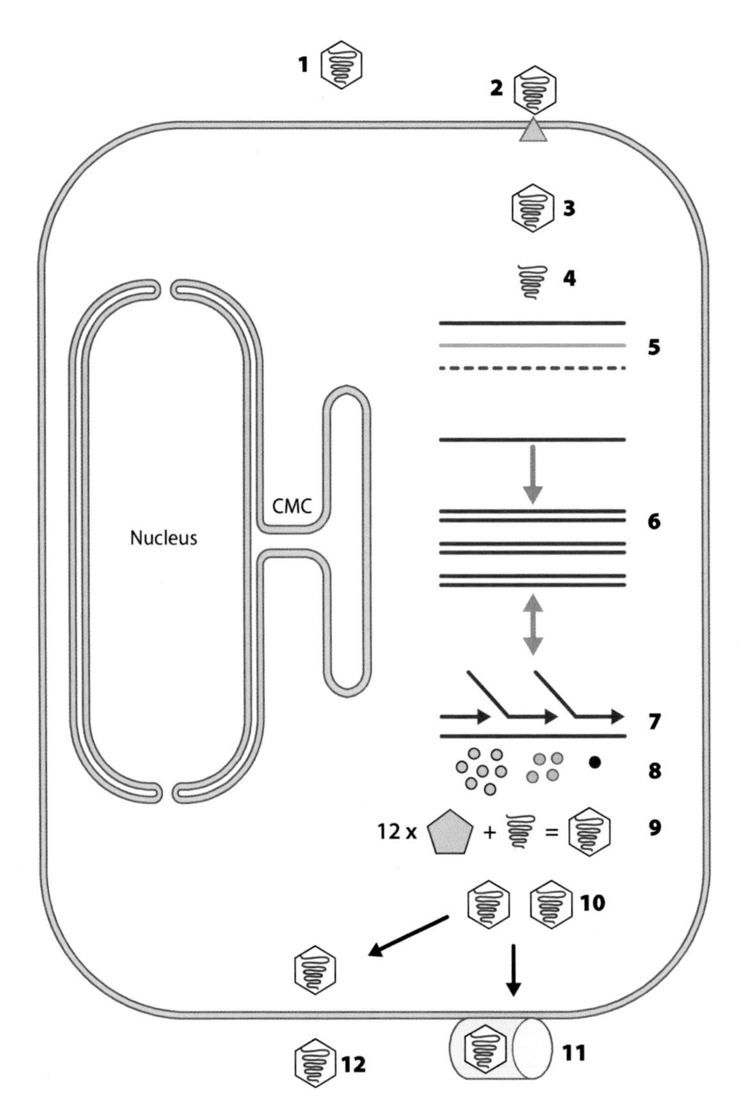

Figure 1.30 The replication strategy of HAV in the hepatocyte. The numbers refer to the key stages discussed in the text. (CMC: cytoplasmic membrane complex.)

replicative form (RF), and by cycling through an integrated process, mRNA and progeny genomes are synthesized, while the pool of 2-RNA intermediates (RI) is amplified (6, 7). The necessary pools of the RFs and intermediates of the RNA are also controlled by cell proteins that enhance or inhibit ribosome entry into the IRES site. The end result is the maximum production of progeny genomes and the structural proteins to encapsulate them (8).

Under the direction of 2A, proteins 1AB, IC, 1D assemble into pentamer structures; 12 of these assemble with a single RNA genome to form the mature virus particle (9, 10). It is at this stage that 1A and 1B are cleaved. The virus is released from the hepatocyte into the biliary canaliculi of the liver, and is excreted in the stool (11). Virus also enters the blood, and via this viraemia, is also excreted in urine (12).

FUNGI

Fungi are eukaryotes, and the majority grow on dead and decaying material in the environment and are essential in the recycling of nutrients in nature. As chemotrophic organisms they secrete enzymes that degrade a wide range of organic compounds, and actively absorb soluble nutrients. Yeasts of the genus *Candida* are minor members of the flora of the body, and live in moist areas such as the groin, perineum or mouth. The cell wall of fungi contains 1,3 β-glucan and chitin, and small quantities of other carbohydrates. Ergosterol is the steroid in the plasma membrane of fungi, which is the target of agents such as amphotericin B and the azole antifungal agents. *Pneumocystis jirovecii* is unicellular and is probably an early colonizer of the human lung; it uses cholesterol and not ergosterol as its cell membrane steroid.

A general classification of fungi is shown in **Figure 1.31**, and includes diseases associated with the various groups. The separation into Ascomycota, Basidiomycota and Zygomycota is based on structures that are found in sexual reproduction, the ascus, basidium and zygospore, respectively.

There are two morphological forms of fungi, yeasts or moulds (**Figure 1.32**). *Candida* is a yeast and reproduces by budding, and when these buds remain attached, pseudohyphae form. When *Candida* is incubated in human plasma, a primitive germ tube (hypha) forms. Moulds include organisms such as *Aspergillus*, *Penicillium* and *Mucor*, and these grow as hyphae (mycelia). Asexual reproduction relies on the production of spores. These are termed phialoconidia (*Aspergillus*, *Penicillium*), thallic macroconidia (*Trichophyton*, *Microsporum*) and sporangiospores (*Mucor*).

Histoplasma capsulatum is a dimorphic fungus, having a hyphal form in the environment and yeast form at 37°C in the human host. It is widespread in tropical and subtropical parts of the world, growing in soil with a high nitrogen content. Bird and bat droppings are an ideal nutrient source for this organism. When inhaled it can cause acute or chronic pulmonary disease, as well as disseminated infection. These are of particular severity in the immunocompromised individual. *Coccidioides immitis* grows in hyphal form in the environment, reproducing by arthroconidia that are derived from the hyphae. These are easily disseminated in the air and survive extremes of temperature, remaining viable for years. In the human host the fungus enters its 'yeast-like' cycle, producing characteristic structures termed spherules (**Figure 1.33**). Diseases are similar to those caused by *Histoplasma*, and this fungus is found in the south west USA and Mexico.

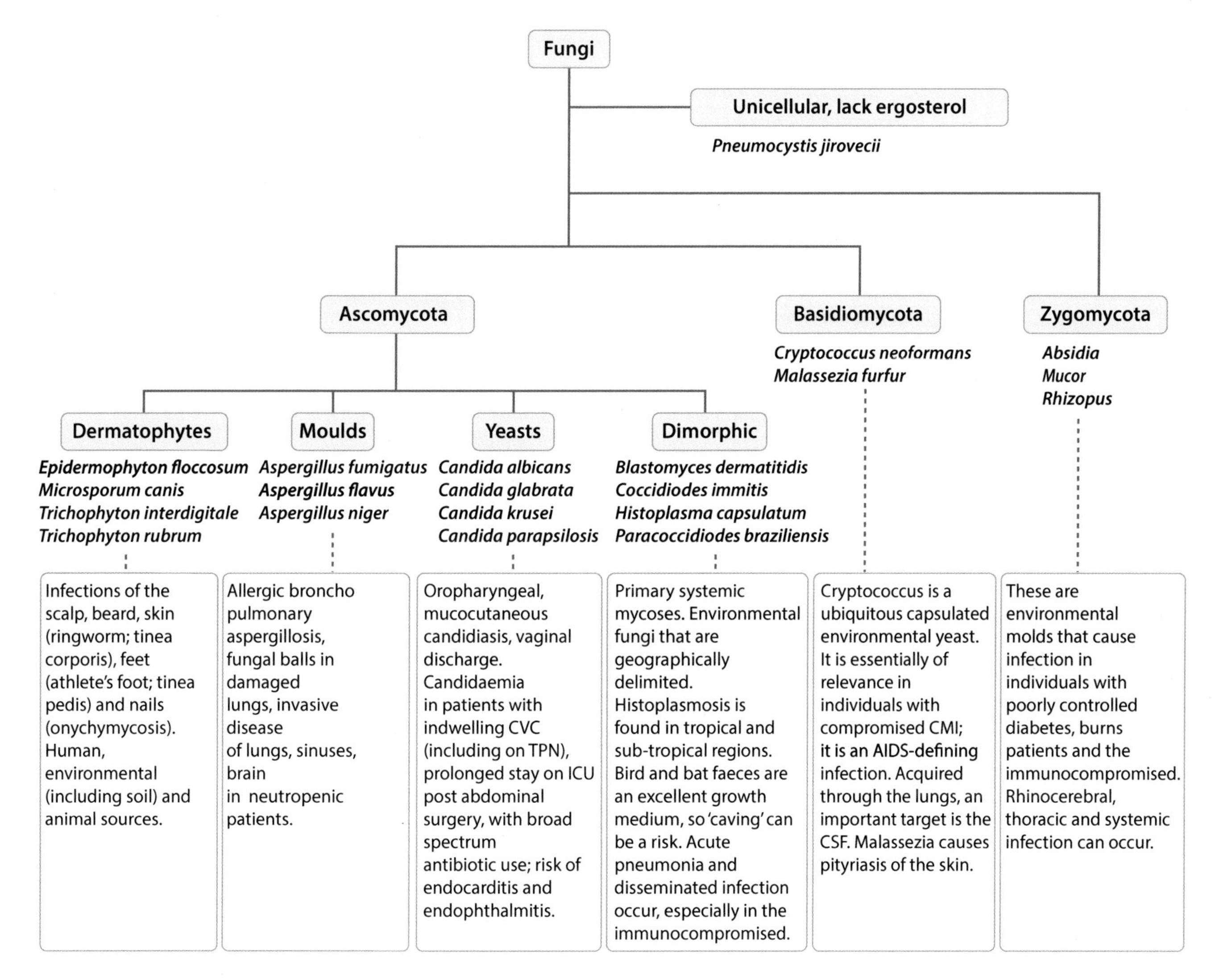

Figure 1.31 A classification of the main groups of fungi that infect humans.

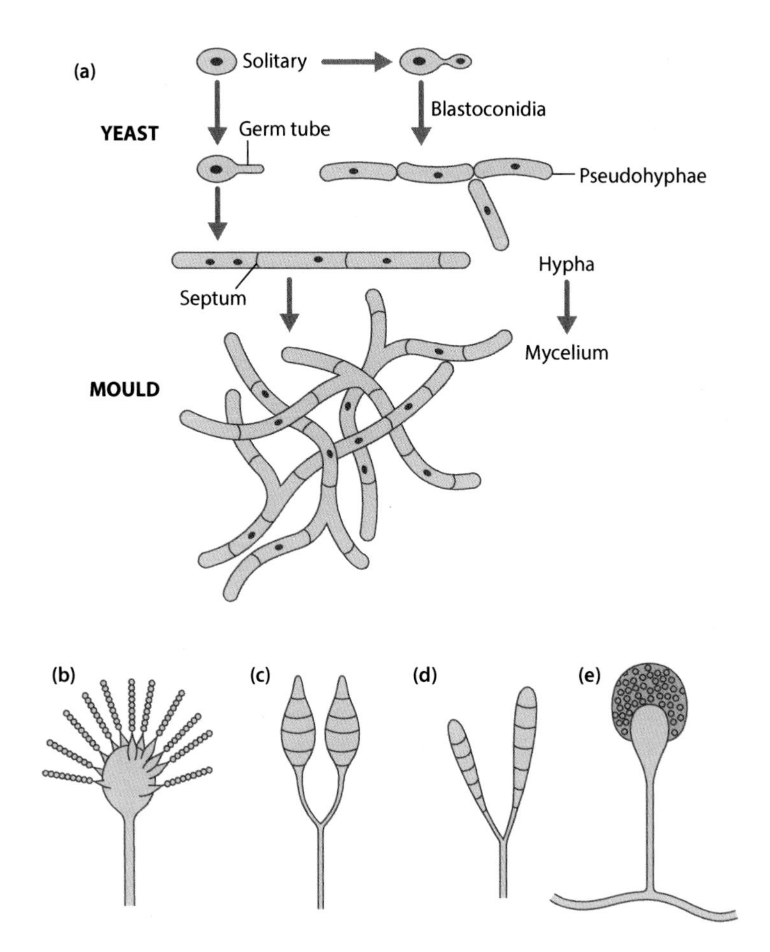

Figure 1.32 (a) Fungi can grow as yeasts or moulds, while dimorphic fungi use both methods of reproduction. (b) Phialoconidia spores (*Aspergillus*, *Penicillium*); (c, d) thallic macroconidia of *Trichophyton* and *Microsporum*; (e) sporangiospores (*Mucor*).

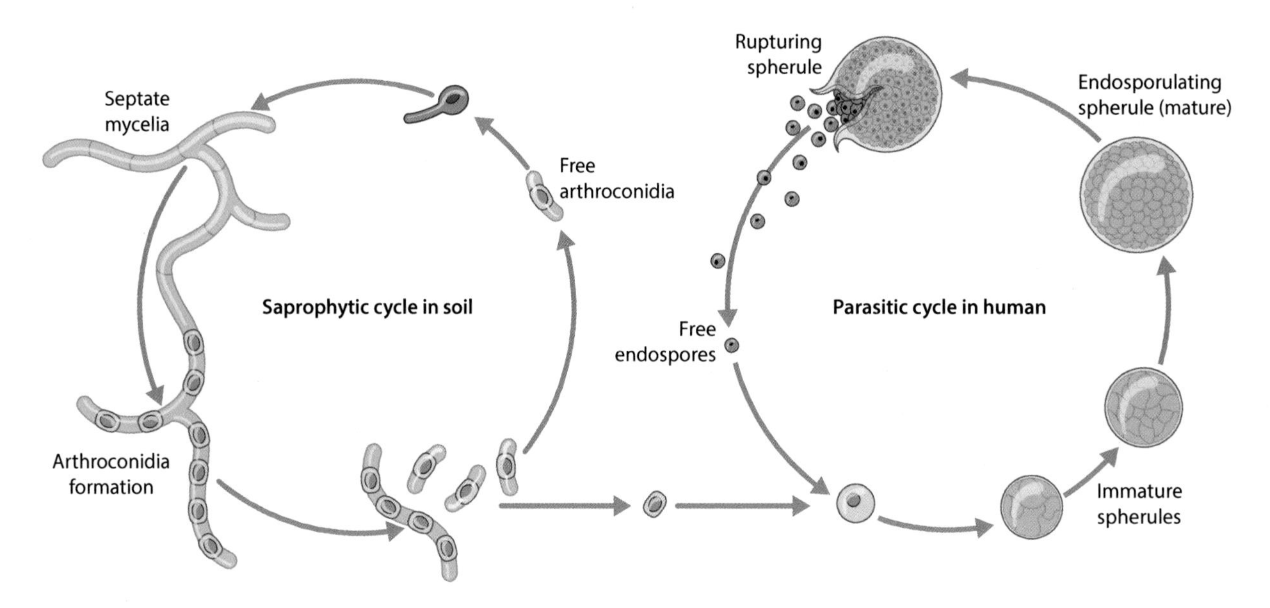

Figure 1.33 Life cycle of *Coccidioides* showing the mycelial form found in the environment and the yeast-like form that occurs in the human host.

DISEASES CAUSED BY FUNGI

While certain dermatophytes have environmental and animal sources, these organisms also have a predilection for human skin, nails and hair, and invade the stratum corneum. Athlete's foot is one example, with a susceptible individual acquiring the organism in a communal swimming pool changing room. Colonizing wet and macerated skin folds of the little toe in particular, their quest for nutrients enables them to grow into the outer skin layers, giving rise to significant local irritation. If unchecked, this invasion can erode to produce a painful bleeding fissure. (This can be the route whereby *Streptococcus pyogenes* accesses the blood.) Treatment is usually effected by keeping the feet and toes dry after bathing, along with the application of a topical antifungal such as miconazole.

Aspergillus spores are ubiquitous in the air and are inhaled into the lung, where they usually cause no harm, as the spores are taken up and destroyed by the alveolar macrophages and neutrophils. High concentrations of inhaled spores can precipitate allergic bronchospasm, an IgE-mediated reaction. *Aspergillus* causes two other lung diseases. When spores settle in an old tuberculous cavity, they can obtain sufficient nutrients to grow within the cavity to produce a 'fungal ball'. Movement of this ball in the cavity, along with degradative effects of secreted enzymes, leads to weakening of the wall and rupture of adjacent blood vessels, with haemoptysis.

Invasive aspergillosis is a most important disease, and is a frequent consideration in the neutropenic immunosuppressed patient. Spores germinate in the alveoli and local invasive disease occurs. The organism can enter the blood to reach other organs, including the brain.

Candida are usually minor members of the mucosal surfaces of the mouth and vagina, and can also colonize moist skin areas of the groin and perineum. However, if conditions change to its advantage, it will reproduce to numbers that cause infection. The oral contraceptive causes changes in vaginal epithelial cells that allow *Candida* to overgrow and an unpleasant curdy discharge results. *Candida* readily colonizes the urine of the patient with a long-term catheter.

When broad-spectrum antibiotics are used in the patient with postoperative complications following abdominal surgery, these have a major effect on the normal bacterial flora of the bowel, and the loss of colonization resistance leads to overgrowth of *Candida*. The yeast can then become a member of a polymicrobial abdominal collection, and it can colonize long-term central venous catheters (CVC) such as total parenteral nutrition (TPN) lines. From these sites it can then invade the blood. Diagnosis of a candidaemia by positive blood culture identifies a serious complication. Once in the blood, *Candida* can settle in other organs; the ophthalmologist must be called promptly to exclude *Candida* endophthalmitis.

The basidiomycota include *Malassezia furfur* and *Cryptococcus neoformans*. *Cryptococcus* is an environmental yeast which was of limited medical importance until it became a key AIDS-defining illness in the HIV epidemic. Inhaled into the lungs of those with depressed cell-mediated immunity, it can cause not only pneumonia, but invades the blood. A key target is the CSF, with *Cryptococcus* meningitis resulting. The organism has a thick polysaccharide capsule, readily visualized on microscopy against the black background of the India ink stain. The yeast can also settle in other organs such as the prostate and skin; new skin lesions of the immunocompromised patient should always be biopsied and cultured for bacteria and fungi; *Cryptococcus* can be the causative organism.

The zygomycota are environmental fungi that usually cause no problem, despite the fact that their spores are ubiquitous in the air. However, in the patient with uncontrolled diabetes or the immunosuppressed, invasive disease of the nasal sinuses can occur. The proximity of the sinuses to the brain means that this difficult to treat infection can have devastating consequences once the brain is invaded.

PARASITES

There are two broad groups, the protozoa and the helminths/flatworms, and their classification is shown in **Figures 1.34a,b**. The helminths comprise the roundworms (nematodes), tapeworms (cestodes) and blood flukes (trematodes). In both groups the life cycles have differing degrees of complexity, often involving insect or animal hosts. African trypanosomiasis is transmitted by the tsetse fly, *Glossina morsitans*, while the South American disease is transmitted by the reduvid bug, a hemipteran. Tissue nematodes such as *Wucheria bancrofti* are transmitted by *Anopheles*, *Culex* and *Aedes* mosquitoes.

THE LIFE CYCLE OF MALARIA

The life cycle of malaria is shown in **Figure 1.35**, with the human (a) and mosquito (b) sections of the cycle shown. The infected female *Anopheles* mosquito injects sporozoites into the blood (B), and within 30–40 minutes these disappear from the blood and enter the hepatocytes (L). Cell division produces schizonts (L 1–3), which divide

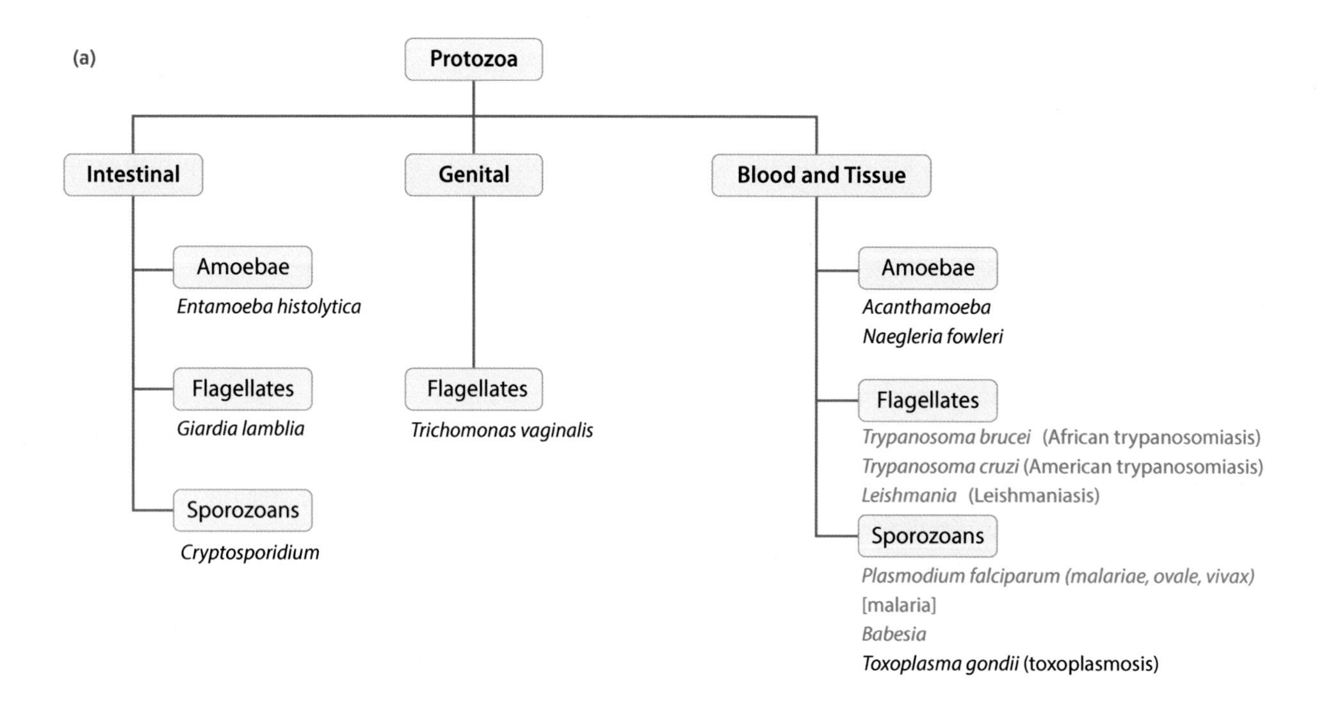

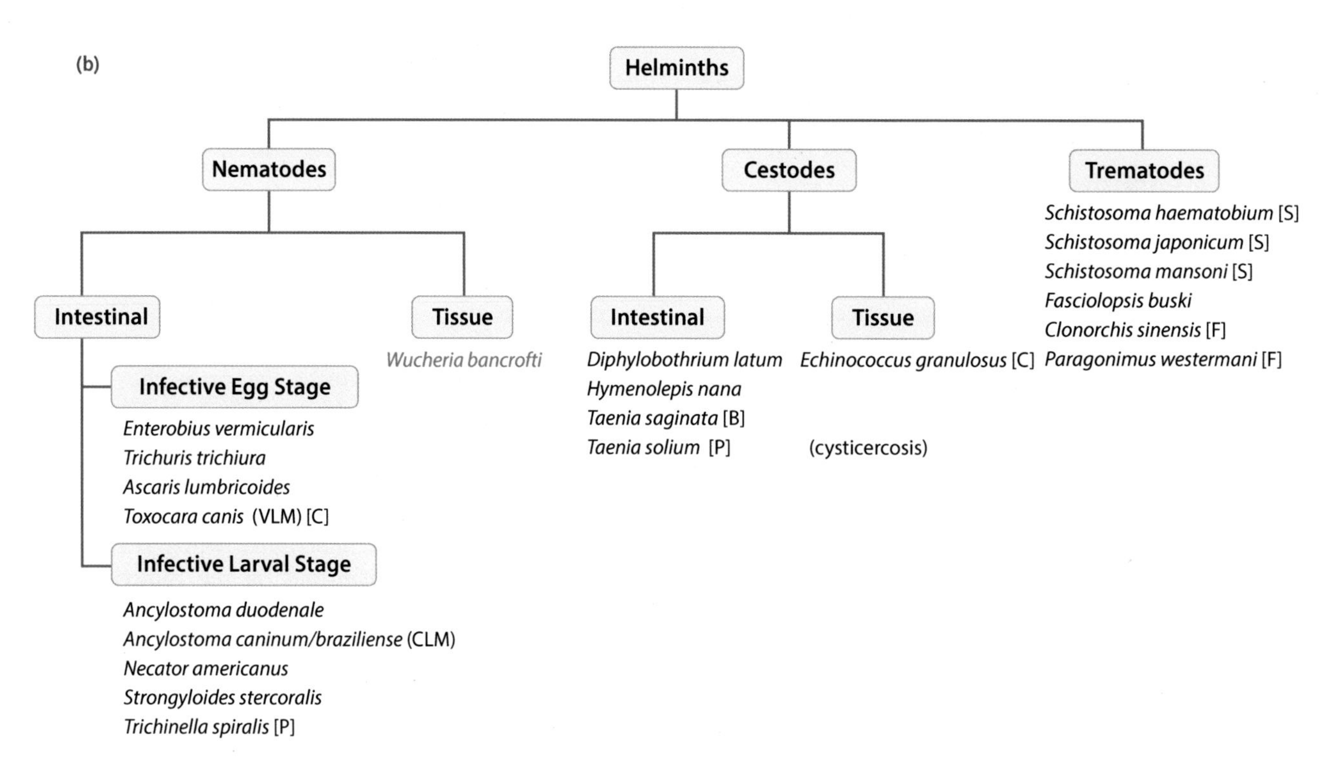

Figure 1.34 A classification of (a) protozoal and (b) helminth parasites. Organisms highlighted in green are transmitted by insect vectors, with the exception of *Babesia*, which is a tick-borne zoonosis. For the helminths, the other hosts are B: bovine; C: canine; F: piscine; P: porcine; S: snail. (CLM: cutaneous larval migrans; VLM: visceral larval migrans.)

further to form up to 50,000 merozoites in each infected hepatocyte (L 4), which are released into the blood. The liver stage takes about 6 days. (The liver forms of *Plasmodium vivax* and *Plasmodium ovale* remain viable as hyponoites after the blood forms have been eliminated, accounting for episodes of relapse and parasitaemia weeks or months later.)

Merozoites in the blood attach to and enter RBC (R 1, 2) and reproduce through ring, trophozoite, schizont and merozoite stages. Up to 30 daughter merozoites are released from each infected RBC (R 3–6), and then infect new RBC, initiating the next cycle. This RBC stage takes about 2 days, so the minimum incubation period after mosquito bites is about 8 days.

It is the massive release of merozoites from RBC within a short period that activates the immune response, with release of cytokines that result in fever, chills and rigors. The repeated cycles of RBC infection require large numbers of damaged RBC to be cleared by the spleen, and splenomegaly occurs. When abnormal RBC become trapped in capillaries, with resulting ischaemia, manifestations such as cerebral malaria can occur (see Chapter 14).

After 1 week or so, developing male and female gametocytes appear in the blood (G 1, 2). Mature female macrogametocytes and male microgametocytes are taken up by the blood-feeding female mosquito (G 3). In the mosquito midgut (**Figure 1.35b**), the male gametocyte undergoes exflagellation to produce sperm-like bodies, which mate with the haploid female gametocyte to form the oocyte (M 1,2). This develops into a motile ookinete that burrows through the midgut epithelium to the haemocoele membrane, outside the midgut wall where it encysts (M 3–5). This matures and releases sporozoites into the insect's haemolymph, that migrate to the salivary glands. After a few days they mature, await *for* the next blood feed the mosquito takes, and enter the human host.

THE LIFE CYCLE OF *SCHISTOSOMA HAEMATOBIUM*

The life cycle of *Schistosoma haematobium* is shown in **Figure 1.36**. The adult worms live for years in the venules of the urogenital venous system, and chronic egg production by the female means that eggs are being

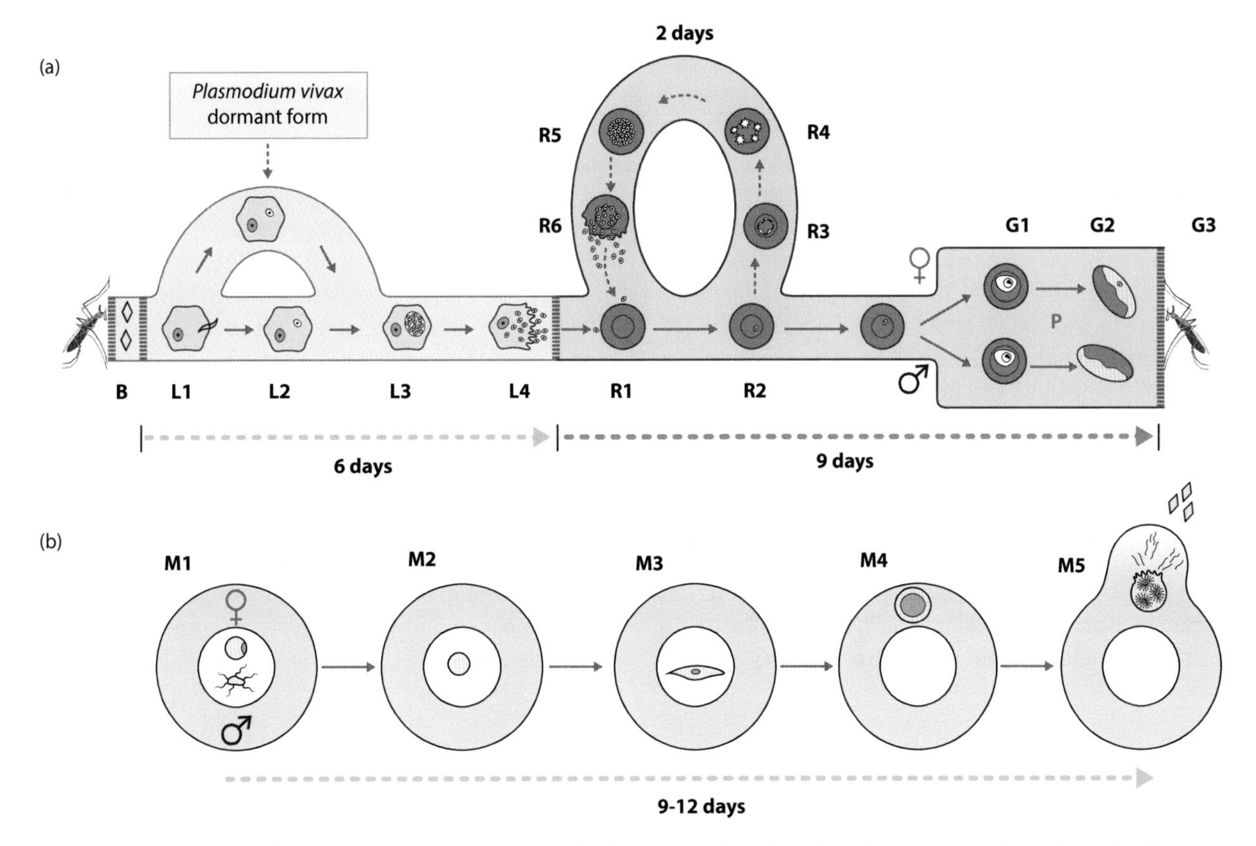

Figure 1.35 The life cycle of the malaria parasite. (a) In the human: B: blood; L: liver hepatocyte; R: red blood cell; G: gametocyte stages; (b) in the mosquito (M) midgut, and on the midgut epithelium.

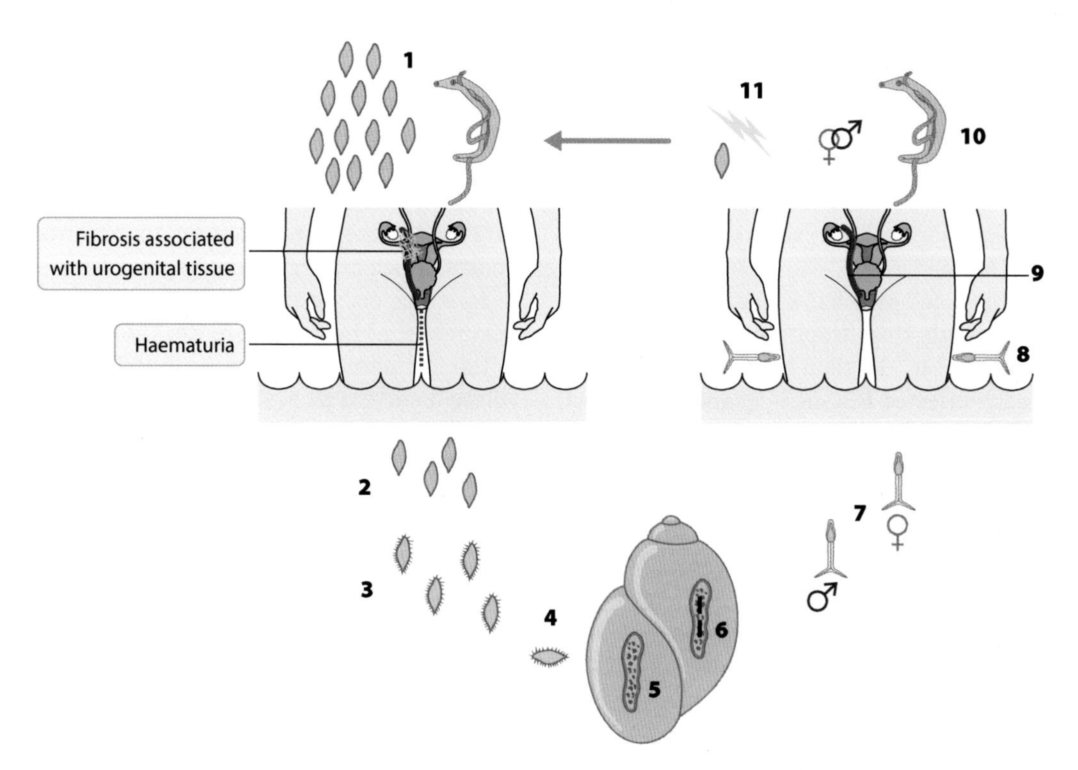

Figure 1.36 The life cycle of *Schistosoma haematobium*. Numbers are as described in the text. Antigen from the initial egg production can induce the allergic reaction of Katayama syndrome.

pushed out of the venules and through the bladder wall, and haematuria can arise (1). Fibrosis occurs in the bladder wall and around adjacent organs. Eggs enter the bladder, with the highest numbers being pushed out at the hottest time of the day. The parasite 'knows' that its host is more likely to seek the coolness of fresh water at this time, and when urination occurs, the eggs have reached the environment of the intermediate host (2). They hatch into ciliated miracidia (3) that search out a specific *Bulinus* snail as the intermediate host (4). Once inside the tissue of the snail, two larval stages develop into motile (tailed) cercaria that are released into the water (5–7).

Upon attaching to a human, the cercaria burrow into the skin, releasing the tail (8). If there is exposure to large numbers of cercaria, an urticarial reaction can occur. By some mechanism the parasite, maturing to adulthood, reaches the urogenital venous system where it searches out a mate (9). The slender female lies in the gynaecophoric canal of the male, and mating is initiated (10), with egg production starting 2–6 weeks later. The terminal spine of the egg enables them to lodge in the walls of small vessels, and venous pressure and pressure changes in the pelvis around the bladder force eggs into the bladder wall, and into the lumen. Eggs 'leak' antigens that incite an acute inflammatory response, with chills and fever; Katayama syndrome is usually mild in *Schistosoma haematobium* in comparison to *Schistosoma mansoni* and *Schistosoma japonicum* (11).

The fibrosis that arises from chronic release of eggs can lead to bleeding and haematuria, and is the main sign of chronic infection. Fibrosis adjacent to the fallopian tube can give rise to ectopic pregnancy many years after exposure and can be the first manifestation of the infection.

Chapter 2

How Bacteria Cause Disease

INTRODUCTION

The body has a complex bacterial population that is an essential part of normal physiology. These microbiota are associated with the skin, upper respiratory tract, throat, gastrointestinal system and vagina. Intact epithelial surfaces as well as anatomical and physiological defences separate these bacteria from the sterile tissues. While small numbers of bacteria can ascend the relatively short urethra and reach the bladder of the premenopausal female, defences on the bladder epithelium and regular micturition with complete emptying usually remove them. It is reasonable to assume that a few bacteria can occasionally cross the bowel epithelium and enter the blood, to be transported by the portal vein to the liver. In most circumstances they are taken up and destroyed in the sinusoids by macrophages, the Kupffer cells.

The microbiota of the colon is the most complex, being made up of hundreds of species of bacteria, mainly anaerobes that comprise over 99.9% of this population; 50% of the dry weight of faeces is bacteria. They reside in a biofilm in the zone adjacent to the epithelium, and provide a physiological environment in coexistence with the epithelium. This is not a casual residential symbiosis, but a closely coordinated ecosystem. In mammalian studies it has been shown that bowel epithelial cells secrete micro ribonucleic acid (RNA) molecules that enter these bacteria to control their gene expression. Key bacteria are responsible for production of secondary bile acids, an essential part of the enterohepatic cycle.

The bacterial population creates a zone of 'colonization resistance' that thwarts the ability of pathogens, in low numbers at least, to establish themselves and cause disease. When antibiotics are used to treat a respiratory tract infection, they will exact collateral damage on the microbiota of the body, and that of the bowel in particular, compromising its normal protective function. If exotoxin-producing *Clostridium difficile* is resident in the colonic contents, it will take advantage of this, reproduce and establish itself adjacent to the epithelium to precipitate antibiotic-associated diarrhoea (AAD).

Bacteria can be broadly grouped as being commensals, colonizers or exogenous; and 15 are named in **Figure 2.1**.

Commensals are members of a specific microbiota that is part of the normal functioning of the body, and should be considered symbionts. The first four commensals named are, albeit minor, members of the normal bowel flora. These four can intermittently colonize the groin and perianal area. Uropathogenic strains of *Escherichia coli* colonize the periurethral area of the female introitus. *Staphylococcus epidermidis* is a member of the coagulase-negative staphylococci (CNS), which, along with *Corynebacterium* and *Propionibacterium*, are dominant in the bacterial microbiota of the skin.

Two colonizers, *Staphylococcus aureus* and *Streptococcus pneumoniae*, are named. This staphylococcus is commonly found in the anterior nares, throat, axillae and groin of over 30% of the population, while pneumococcus colonizes the upper respiratory tract of about 10%.

Eight exogenous bacteria are named, but they have different relationships with the body. *Pasteurella multocida* is a member of the endogenous bacterial flora in the mouth of cats, dogs and other animals. It can cause local, invasive disease and sepsis in the person who sustains a bite from an animal. Apart from *Legionella*, derived from water sources, and *Pasteurella*, the others can 'colonize' a site of the body. *Mycobacterium tuberculosis* can be considered an 'immunologically controlled' exogenous organism, considering that one-third of the world's population is infected, but in the vast majority, its reproduction is prevented by cell-mediated immunity. When this immunity is compromised, for example by the acquired immunosuppression of human immunodeficiency virus (HIV) infection, the organism will reproduce to cause disease.

While the three intestinal pathogens are clearly exogenous, asymptomatic colonization (carriage) occurs; in the case of *Clostridium difficile*, this ranges from 3% in the community to 30% in the hospital setting. *Streptococcus pyogenes* and *Neisseria meningitidis* colonize the upper respiratory tract of a small proportion of individuals as that is how they are maintained in the population.

Figure 2.1 demonstrates that every human is under microbiological threat from several directions. The five colonizer/exogenous bacteria highlighted essentially have a human source, and are shown as having a narrow range of distribution across two groupings. Disease arises

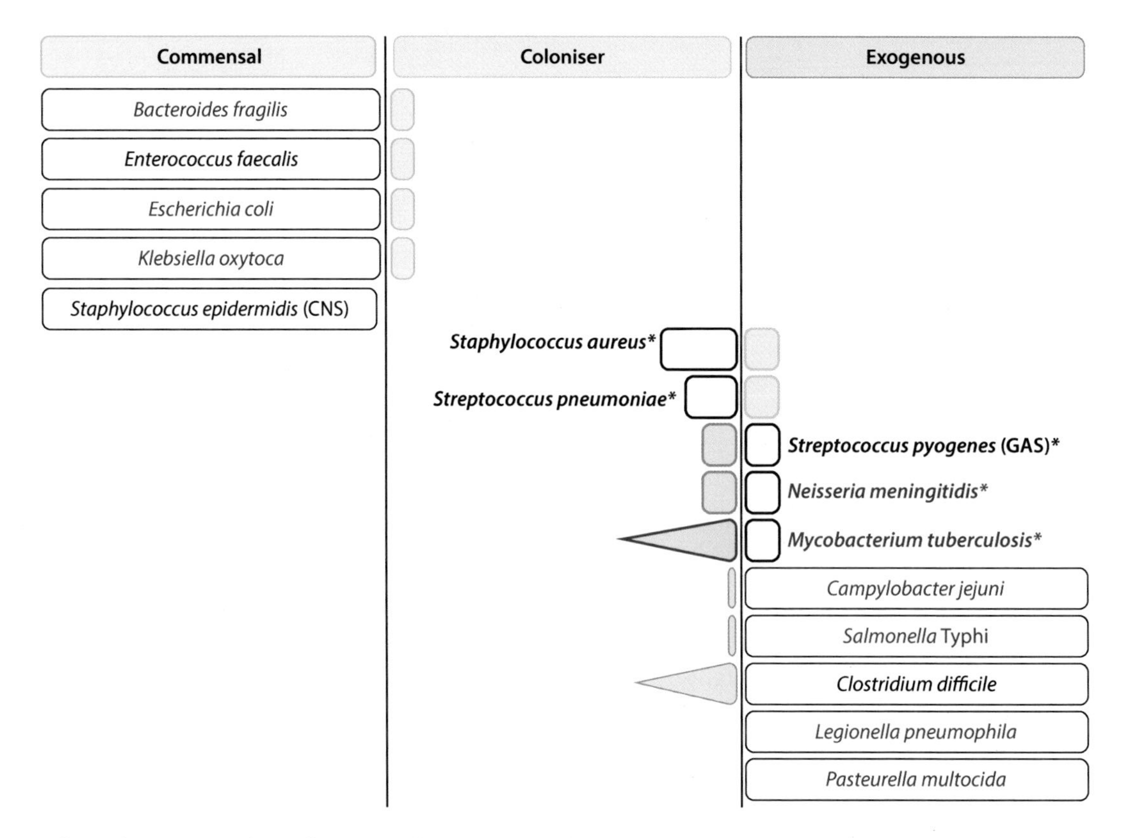

Figure 2.1 Fifteen bacteria are broadly grouped as commensal, colonizer or exogenous in their relationship with the human body. For the purposes outlined in the text, their usual environment is defined by the open box. *Five organisms having a human source. (Gram-positive bacteria: black; gram-negative: red; mycobacteria: purple).

from an organism derived from either source, which is directly linked to the organism's pathogenic properties. If the individual person is not exposed to an exogenous organism, and they have no risk that a colonizer can take advantage of, they will usually be healthy.

The four commensal members of the bowel flora would not normally cause a problem. However, if there is an anatomical abnormality, disease can occur. A diverticulum of the large bowel is one example. When the normal bowel flora is retained within a diverticulum as colonizers, it is likely that the ecological balance can shift, and certain bacteria, such as these four, gain an advantage. Their unsupervised reproduction leads to inflammation, and diverticulitis. If inflammation closes off the diverticulum opening, the situation progresses to a polymicrobial abscess. Without intervention, peritonitis, sepsis and death can occur.

The members of the endogenous flora of the skin can be considered to have limited pathogenic potential. However, if they are provided with a situation that gives a growth advantage, it will be exploited. A long-term central venous cannula (CVC) in the intensive care unit (ICU) patient is one example. Once *Staphylococcus epidermidis* establishes a biofilm on the cannula, it will enter the blood in large numbers to cause line-associated disease (usually a fever that does not settle despite antibiotics); line removal can promptly resolve this.

The important point is that if there is an anatomical, physiological or immunological breach in the body's defences, it is reasonable to expect that there is a bacterial (or fungal) species that will exploit that weakness to cause disease. This is enlarged upon in the rest of this chapter, with emphasis on the pathogenic properties of bacteria that gives them the advantage over defences, enabling them to precipitate a specific disease.

NON-SPECIFIC DEFENCES

The non-specific defences of the body thus rely on anatomical, physiological and bacteriological/microbiome components, and key examples are summarized in **Figure 2.2**.

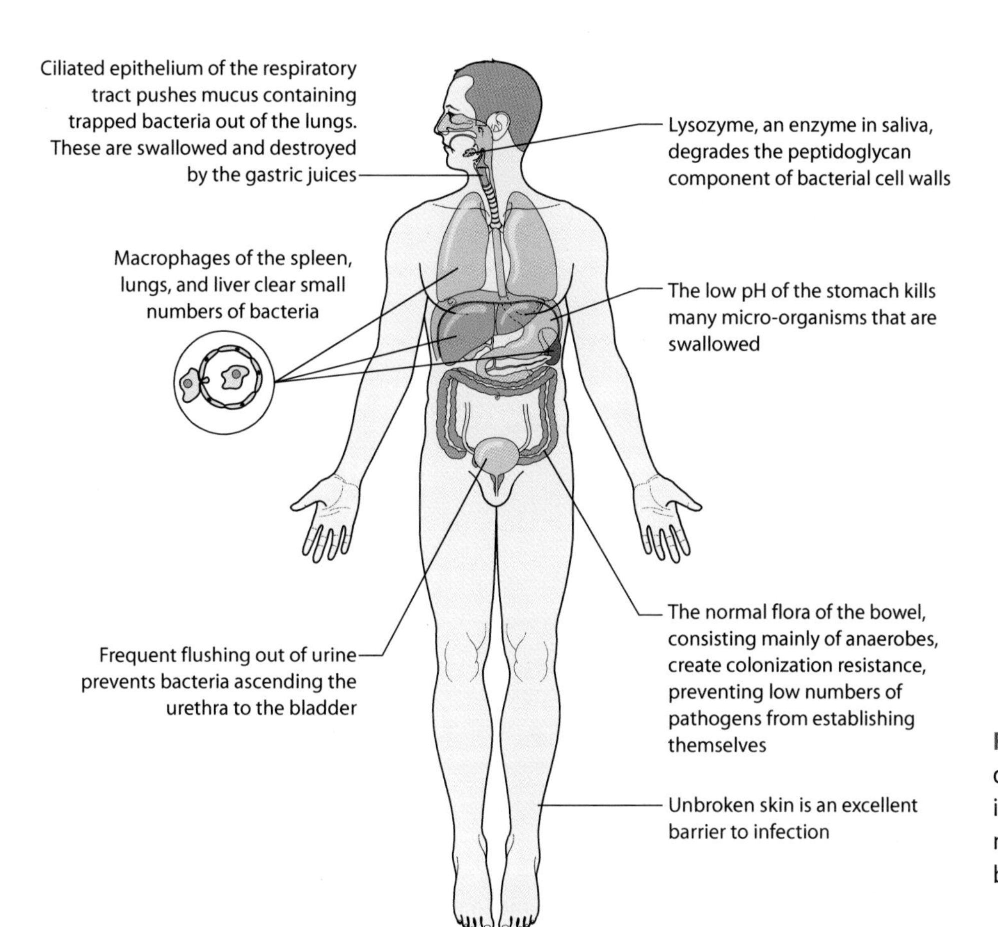

Figure 2.2 Some examples of anatomical, physiological, immune-based and microbial defences of the body.

ACTIVE DEFENCES

The active defence system is based on the immune response and its ability to differentiate self from non-self, and identify and destroy invading micro-organisms. The key cells at the start of the process are the phagocytic cells such as the tissue macrophages, which have the ability to take up micro-organisms and destroy them. The first step is uptake of the organism into a phagosome. The latter fuses with another cytoplasmic vesicle, the lysosome, containing oxygen radicals, acid hydrolases, peroxidases and lysozymes responsible for destruction of the organism. The end-products are a toxic combination of bacterial and host cell chemicals, which must not be allowed to pollute the extracellular environment (**Figure 2.3**).

Such a stressed macrophage has to undergo programmed cell death, or apoptosis. This is the process by which the body removes cells that have either done their work, or are under a form of external or internal stress, so that it is best to remove them before they cause damage. Apoptosis is highly regulated and controlled, and directs synthesis of cellular proteins whose function is to

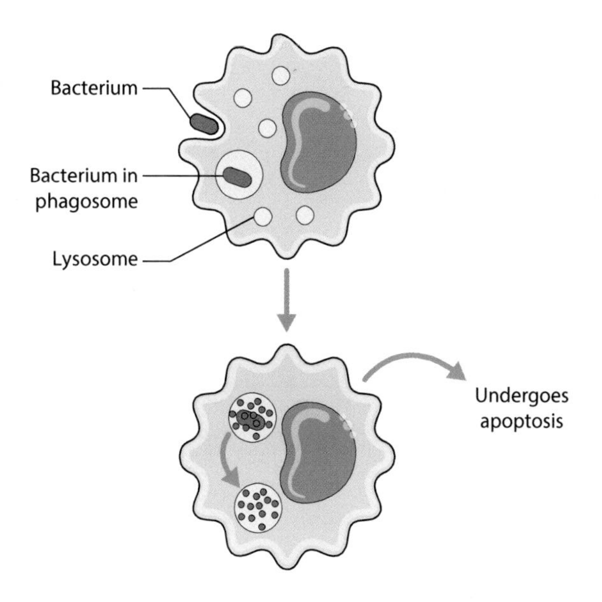

Figure 2.3 A phagocytic cell takes up a bacterium into a phagosome; fusion with the lysosome enables the organism to be killed. Resulting toxic intracellular products direct that the cell is safely eliminated by the process of apoptosis.

puncture the outer membrane of mitochondria releasing cytochrome c. This triggers activation of caspases that degrade all types of proteins within the cell. An aptosome of the redundant cell is formed that enables safe compartmentalization of all its cellular debris into apoptotic bodies that are taken up and safely disposed of by neighbouring macrophages.

If a few pneumococci enter an alveolus, the resident macrophages will take them up and destroy them, and as directed, undergo apoptosis. If large numbers of the organism enter the same site the resident macrophages may be overwhelmed (**Figures 2.4a,b**). Uncontrolled release of toxic waste products from overloaded phagocytic cells causes local tissue necrosis, compromising defences and giving the pneumococcus an advantage. Reproduction of the bacteria will incite an amplifying immune response, with recruitment of large numbers of neutrophils that attempt to control the infection. It is here that antibiotics are critical in halting a deteriorating situation.

Other macrophages will also be taking up bacterial antigens, and as antigen-presenting cells, will relay this information to other cells of the immune system, and the T lymphocyte is key in this process. Some features of the main cells involved are shown in **Figure 2.5**.

Macrophages take up foreign material (an antigen) with the aim of destroying it. They are also antigen-presenting cells (APC) and present a small part of an antigen, the antigenic determinant, on their cell surface. It is here that the specificity of the immune response is initiated. The antigenic determinant, with its unique structure, is presented in the arms of the major histocompatibility type II (MHC-II) proteins on the surface of the APC. (MHC-II proteins are restricted to cells of the immune system, in contrast to the MHC-I proteins that are widely distributed amongst all other cells of the body.)

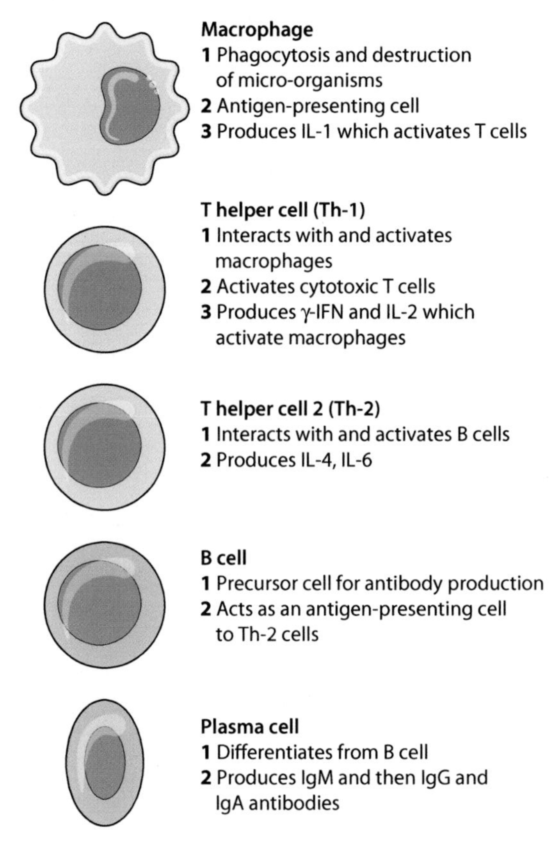

Figure 2.5 Some important cells of the immune response. (IFN: interferon; Ig: immunoglobulin; IL: interleukin.)

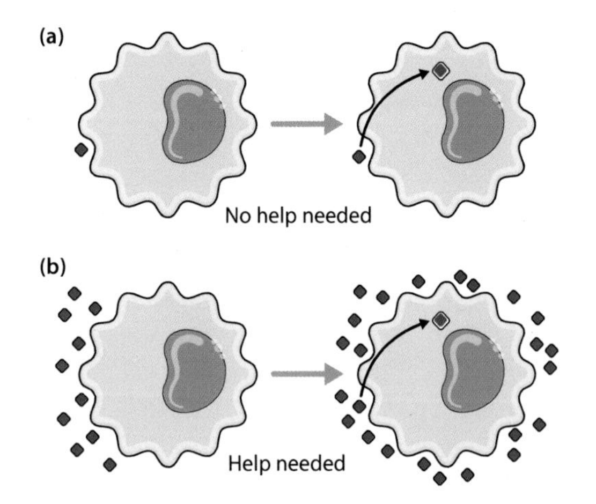

Figure 2.4 (a) A macrophage may be able to take up one pathogen and destroy it. (b) Large numbers of the pathogen will overwhelm the macrophage and it has to recruit help.

There are small populations of T lymphocytes in every individual, specific for every conceivable foreign antigenic determinant that an individual may come in contact with during his or her life. An APC such as a macrophage with a unique antigenic determinant on its surface will be recognized by the population of T cells that have T-cell receptors (TCRs) specific for that complex (**Figure 2.6**). If an organism has a number of antigenic determinants, for example □♠♦♥, these four different determinants would each be recognized by four different sets of T cells, each waiting for that specific determinant to be presented to them by an APC.

For the purposes of this section consider that the 'organism' is an antigen consisting of a single antigenic determinant ♦. As shown in **Figure 2.7**, macrophages initiate the response, and after taking up ♦ and attempting

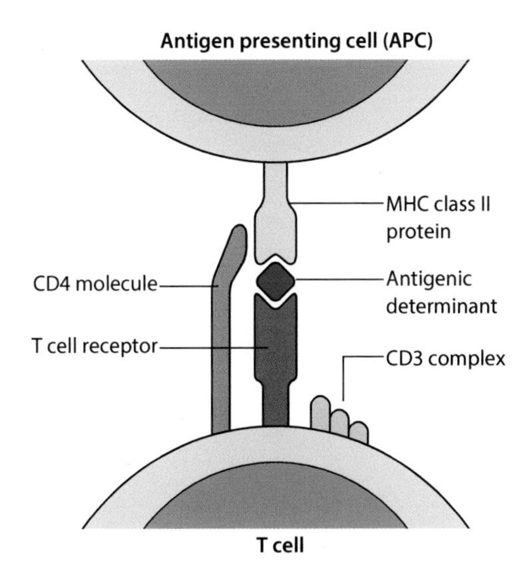

Figure 2.6 The macrophage presents antigen ♦ with the MHC-II protein. This specific structure is recognized by one set of T cells only. Proteins such as CD4 and CD3 are essential in enhancing this specificity. (MHC: major histocompatibility.)

to destroy it, present ♦ to T-CD4 cells and T-CD8 lymphocytes. The T-CD4 cells are the key coordinators of the entire process, and via cytokines, stimulate macrophages to take up and destroy ♦. A set of T-CD4 memory cells is also established for future use. The T-CD8 cells are stimulated by the T-CD4 cells, proliferate and differentiate, via gamma interferon (γ-IFN) and interleukin-2 (IL-2), into cytolytic cells that search out and destroy cells harbouring ♦. This applies in particular to virus-infected cells that usually express viral antigens on their surfaces. These cells are separate from the immune system, and antigen appears on their surface with the MHC-I protein. It is here that the immune system differentiates these cells from cells such as macrophages that express the antigen with the MHC-II protein. This T-cell response is the basis of cell-mediated immunity (CMI).

The other arm of the immune response results in antibody production, and is initiated by the B lymphocyte. In each individual there is a small population of B cells specific for every conceivable antigenic determinant,

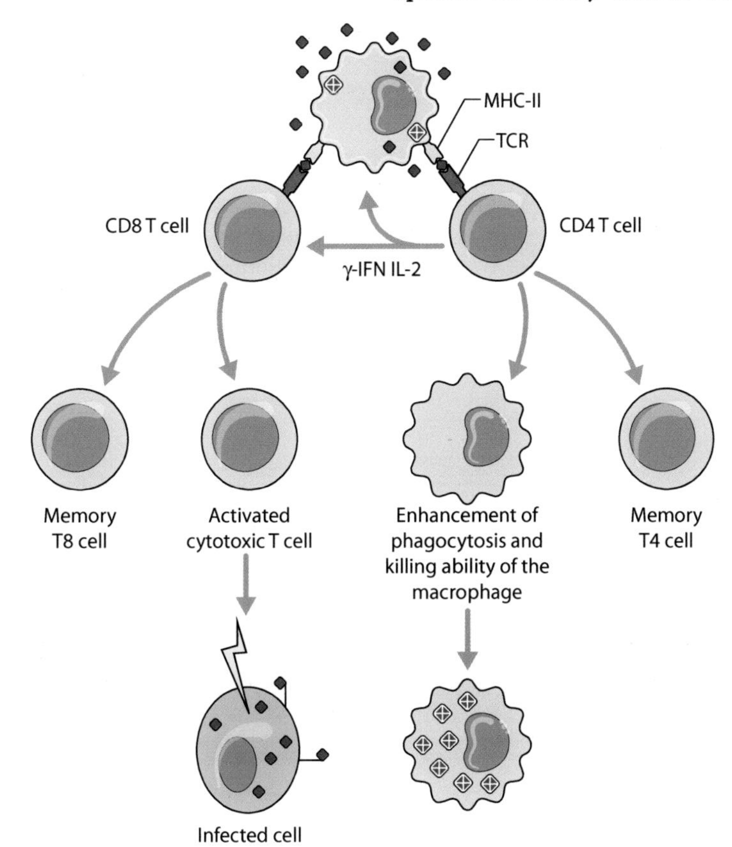

Figure 2.7 A macrophage takes up ♦ and presents it with the MHC-II protein. This is recognized by the T-cell receptor of a specific population of T-CD4 and T-CD8 cells. The T-CD4 cell is the key coordinator. Via cytokines, the macrophage takes up more ♦ and destroys it. CD8-derived cytotoxic T cells recognize ♦ with the MHC-I protein on cells that are not part of the immune system and destroy them too. Memory cells are retained for future use. TCR: T-cell receptor.

such as ♦, the example used here. These B cells have on their surfaces immunoglobulin (Ig) M and IgD antibodies, which act as receptors specific for that determinant. Acting as APC, they recognize ♦ and bind it via the Ig receptors, internalize and then present ♦ to CD4 Th-2 cells in the presence of the MHC-II proteins. Cytokines, including IL-4 and IL-6, activate the B cell, which matures into a plasma cell that synthesizes antibodies that are specific for ♦ (**Figure 2.8**). After several days, IgM and then IgG and IgA antibodies are released by the plasma cells (**Figure 2.9a**). They will bind to antigen ♦ and in this bound form the antigen is 'neutralized' (**Figure 2.9b**). Antigen bound to antibody is more readily phagocytosed by macrophages and neutrophils.

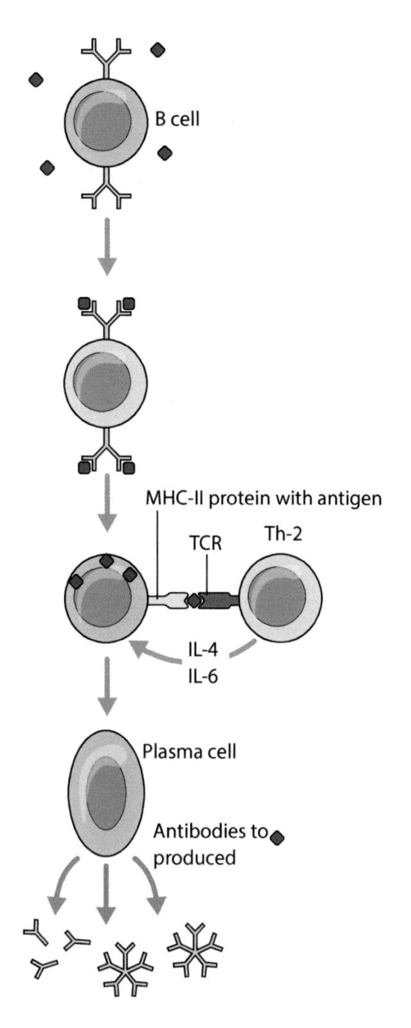

Figure 2.8 B cells with immunoglobulin receptors bind ♦ that is presented on the surface with the MHC-II proteins. This is recognized by the T-cell receptor of a specific population of Th-2 cells. IL-4 and IL-6 activate the B cell to mature into an antibody-producing plasma cell.

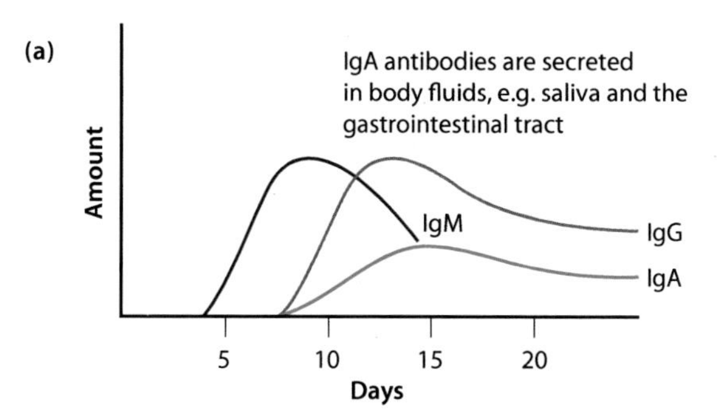

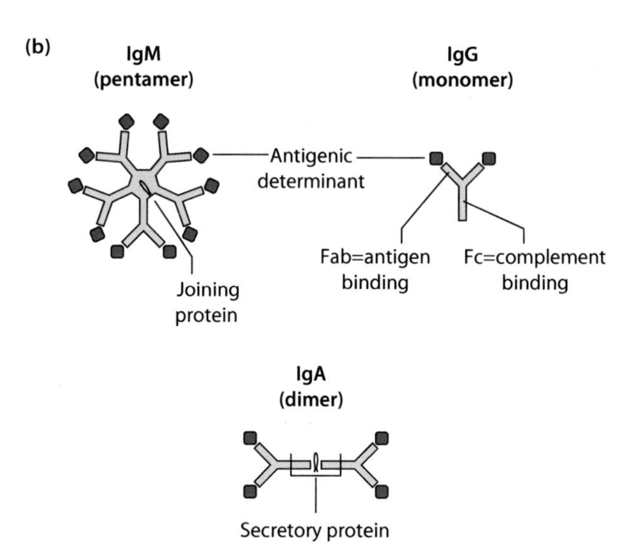

Figure 2.9 (a) Antibodies are released from plasma cells after a number of days. Pentameric IgM is produced first, followed by a switch to IgG and IgA. (b) The specific antigenic determinant (♦) binds to the Fab portion of the antibodies. The Fc portion of IgG antibodies binds certain complement proteins.

Following a successful response to a foreign antigen, populations of T and B memory cells for that antigen are maintained. They can mount an immediate response when the individual is exposed to the antigen in the future. This is the basis of vaccination.

THE ACUTE REACTION TO INFECTION

The problem with the antibody response is that antibodies start to appear 4–5 days after the immune system recognizes an organism as foreign. In the case of meningococcal or pneumococcal infection this antibody response will be too late in helping to overcome the acute infection. Within a matter of hours, the infected person may either recover with no medical intervention, recover because antibiotics are given promptly or die. Other arms of the immune defence are of critical importance in fighting an infection in the early stages. The interaction of the macrophage with the T cell initiates the cytokine response. Cytokines involved include tumour necrosis factor (TNF), γ-IFN and IL-1. Being distributed throughout the body, cytokines coordinate the acute response to infection and stimulate many organs (**Figure 2.10**).

In the hypothalamus cytokines elevate body temperature, manifesting as fever. Certain cells involved in the immune response work more effectively at a higher temperature and certain organisms, especially viruses, are labile at these temperatures. Cytokines mobilize mature and immature neutrophils from the bone marrow. The phagocytic and killing ability of these cells is enhanced. Changes occur locally in the vascular endothelium of the capillary bed, and neutrophils are directed to leave the blood to reach the site of infection, as occurs in bacterial meningitis. When these vascular changes occur in an uncontrolled manner throughout the body, for example in gram-negative sepsis, hypotension and septic shock arise. The acute phase proteins, discussed below, are released from the liver.

The effects of prolonged cytokine stimulation, which occurs in chronic infections such as tuberculosis (TB) or bacterial endocarditis, are characterized by weight loss, representing the outcome of the cytokine-stimulated catabolic process. A low serum albumin is another marker of this catabolism.

MARKERS OF THE CYTOKINE RESPONSE

The acutely ill septic patient can have a high temperature (e.g. 40°C), low blood pressure (e.g. 60/40 mmHg) and laboratory blood tests showing a raised white cell count (WCC), with a neutrophilia in the WCC differential. The clinical parameters and laboratory tests are used to monitor the response of a patient to medical intervention and are an essential adjunct to management of these patients on the ICU. The longer the acutely unwell patient has a high fever and low blood pressure (septic shock) in the setting of maximum cardiac and ventilatory support, the less likely the outcome will be favourable.

Markers such as temperature, C-reactive protein (CRP) and erythrocyte sedimentation rate (ESR) are useful in monitoring the effect of treating chronic infections such as endocarditis. The raised ESR is due to the fact that fibrinogen is released from the liver as an acute phase protein. Fibrinogen causes red blood cells (RBCs) to stick to each other, with a faster sedimentation rate. Resolution of a fever after several days of appropriate antibiotics followed by the progressive return of the CRP and ESR to normality are reassuring, and mean that the antibiotic regime instituted is likely to be effective in eliminating the infection from the infected heart valve. Some of the parameters commonly used in the management of the infected patient are shown in **Figure 2.11**. Nowadays ESR is less frequently used as an inflammatory marker.

ACUTE PHASE PROTEINS

The liver produces the acute phase proteins, which are released under the influence of cytokines, and IL-6 in particular. Important acute phase proteins are CRP, mannose-binding lectin (MBL) and endotoxin binding protein. Mannose is a constituent of the cell wall of gram-positive and gram-negative bacteria. MBL binds to the mannose residues and as phagocytic cells have receptors for this bound MBL, the bacteria are more readily taken up; this is termed opsonization (**Figure 2.12**).

COMPLEMENT

Complement is a group of proteins produced by the liver and cells such as monocytes and macrophages. When IgG antibodies bind to an antigen, such as that on the surface of a virus-infected cell, the classic complement cascade is activated. In a specific sequence, complement proteins are cleaved and components of proteins 5–9 form a 'membrane attack complex' that inserts into the cytoplasmic membrane of the infected cell (**Figure 2.13**). The channels produced by these complexes puncture the cell, which dies, undergoing apoptosis, thus aborting the replication of the virus.

In the acute stages of infection before any antibodies are present, the alternative complement pathway can also be activated. Here bacterial cell wall structures can bind component C3, which is cleaved to C3b and C3a, and subsequently C5 is cleaved into C5b and C5a. Bound C3b acts as an opsonin, as phagocytic cells have receptors for C3b (**Figure 2.14**). The C5a produced acts as a chemoattractant, recruiting neutrophils to the site of infection (**Figure 2.15**).

Lipopolysaccharide of the outer membrane of gram-negative bacteria can also activate the alternative pathway; the 'membrane attack complex' that is formed results in the release of endotoxin.

Opsonins such as MBL, C3b and specific antibodies enhance phagocytosis, as phagocytes have receptors for these components when they are bound to bacteria (**Figure 2.16**). Cytokine stimulation of macrophages by the T-CD4 cell enhances the expression of the cell surface receptors for C3b and the Fc portion of IgG, increasing the phagocytic properties of these cells.

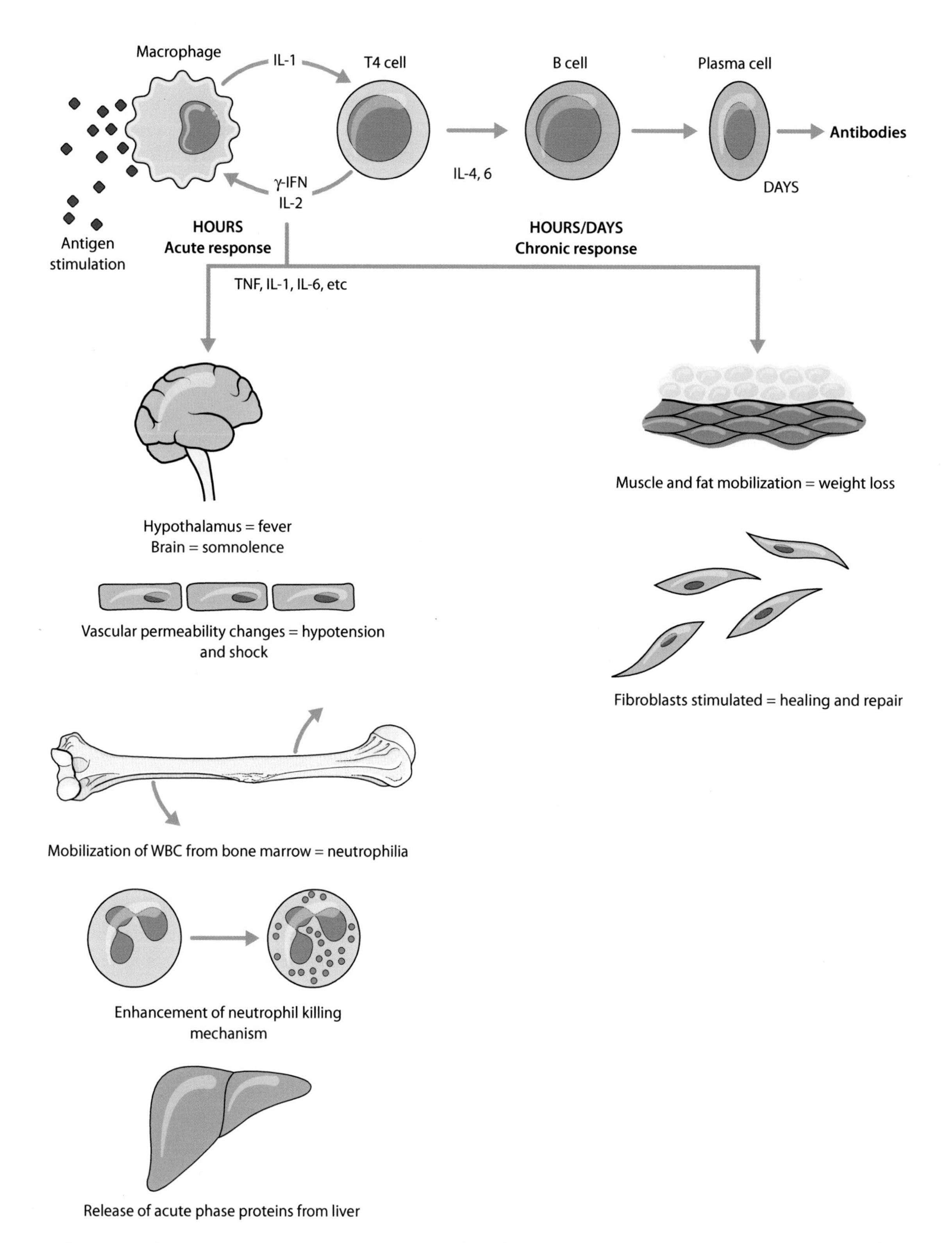

Figure 2.10 The reaction between the macrophage and T cell initiates the cytokine response. This stimulates the acute response of many organ systems. Chronic stimulation of the cytokine response also has an effect on the body. (γ-IFN: gamma interferon; IL: interleukin; TNF: tumour necrosis factor.)

	Normal	Acute infection	Chronic infection
Temperature	37°C	40°C	37.8°C
Blood pressure	120/80 mmHg	60/40 mmHg	120/80 mmHg
WCC	4–11 x 10^9/L	26.4 x 10^9/L	12.4 x 10^9/L
Differential count (% neutrophils)	70	95	70
CRP	<10 mg/L	160 mg/L	160 mg/L
ESR	<20 mm/hour	120 mm/hour	120 mm/hour
Albumin	34–48 g/L	32 g/L	25 g/L

Figure 2.11 Some of the clinical and laboratory markers of the inflammatory response, with normal values and examples of those found in acute and chronic infection. (CRP: C-reactive protein; ESR: erythrocyte sedimentation rate; WCC: white cell count.)

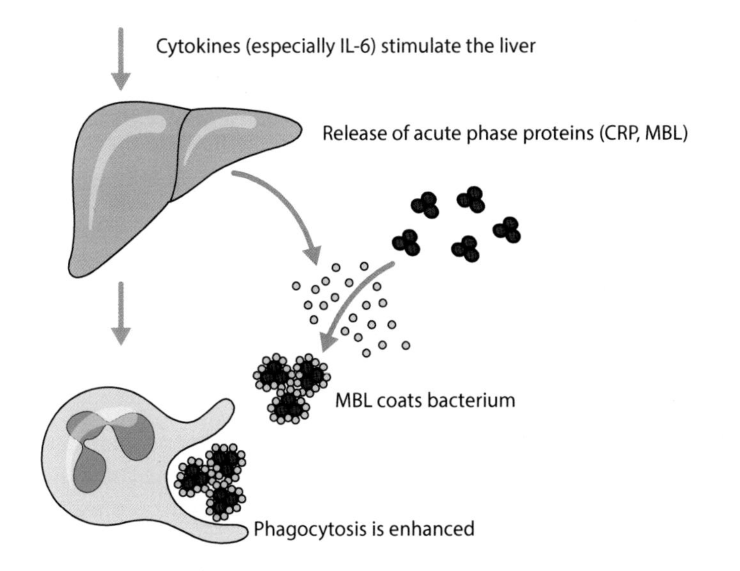

Figure 2.12 Cytokines such as IL-6 stimulate the liver to release the acute phase proteins. Here mannose binding lectin (MBL) coats the surface of the bacteria, which are then more easily phagocytosed.

PATHOGENIC PROPERTIES OF BACTERIA

Perhaps the most remarkable feature of bacteria is the diversity of their pathogenic characteristics. *Staphylococcus aureus* has an impressive array of weaponry, which includes cell wall proteins, extracellular enzymes and toxins. Some of the pathogenic features of bacteria are shown in **Figure 2.17**, and are discussed in more detail below.

STRUCTURAL FEATURES

A number of important bacteria have an extracellular capsule, which is central to their ability to survive within the host. Capsules have antiphagocytic properties (**Figure 2.18**). If thousands of unencapsulated pneumococci are injected into the peritoneum of a mouse they will have no effect, as the bacteria are rapidly phagocytosed and killed. When a few encapsulated bacteria are injected into the same site they resist phagocytosis, reproduce, overwhelm defences and kill the host.

Staphylococcus aureus is able to resist phagocytosis by a specific protein termed protein A. This protein, anchored in the cell membrane, extends through the cell wall to the outside of the cell. The terminal part of the protein is able to bind the Fc portion of IgG antibodies.

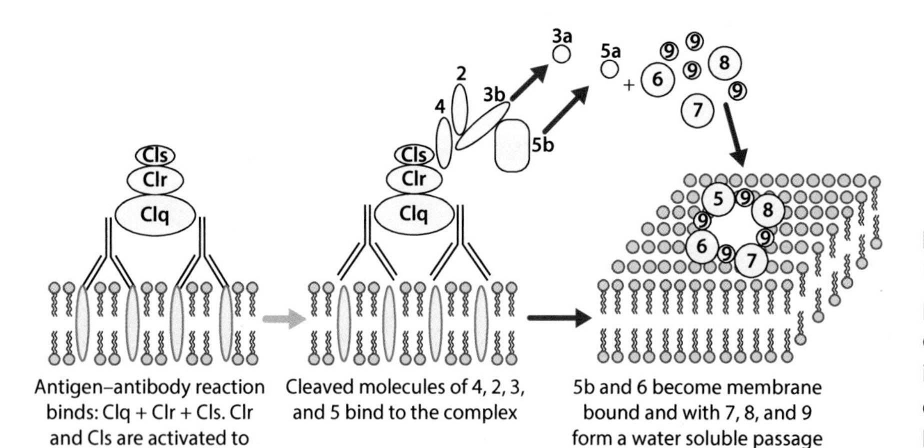

Figure 2.13 When antibodies bind to a cell membrane protein, the classic complement cascade is initiated. The resulting channel allows contents to leak out and the cell dies.

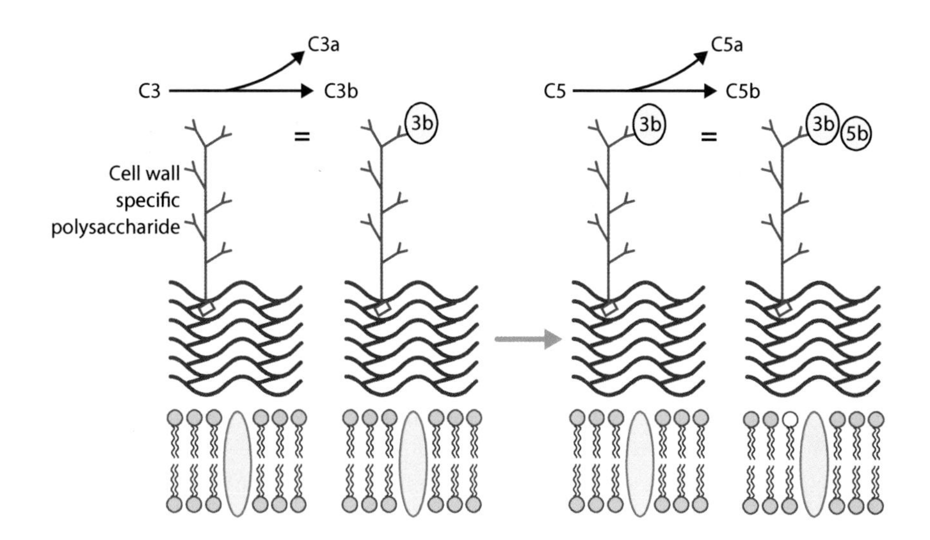

Figure 2.14 Bacterial cell wall components can stimulate the complement cascade via the alternative pathway. C3 is split into C3a and C3b, the latter protein then cleaves C5. Bound C3b is an opsonin and enhances the phagocytosis of bacteria.

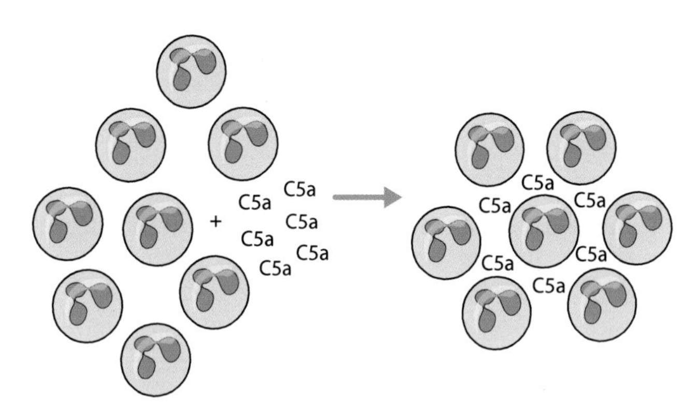

Figure 2.15 Complement component C5a is a powerful chemoattractant and attracts neutrophils and macrophages to the site of infection.

By coating itself with antibodies in this manner, the organism is protected from the phagocytic actions of the host (**Figure 2.19**).

It is recognized that many bacteria grow in biofilms adherent to an inert or living surface. CNS of the skin form biofilms on and in long-term CVCs, multiplying in the extracellular material produced by the organism, and seed the blood from there (**Figure 2.20**). It is difficult to eradicate these bacteria, as penetration of antibiotics into biofilm is generally poor. The patient with persistent fever, positive central line and peripheral blood cultures, despite being given appropriate antibiotics, needs to have the central line removed. The formation of biofilms by members of the commensal skin flora is of major importance in infections of implantable cardiac electronic devices (ICED) and prosthetic joint replacements.

ADHESION

An important step for many bacteria in establishing a nidus of infection is the ability to adhere to a surface. In urinary tract infections, organisms such as uropathogenic *Escherichia coli* bind to specific receptors on the

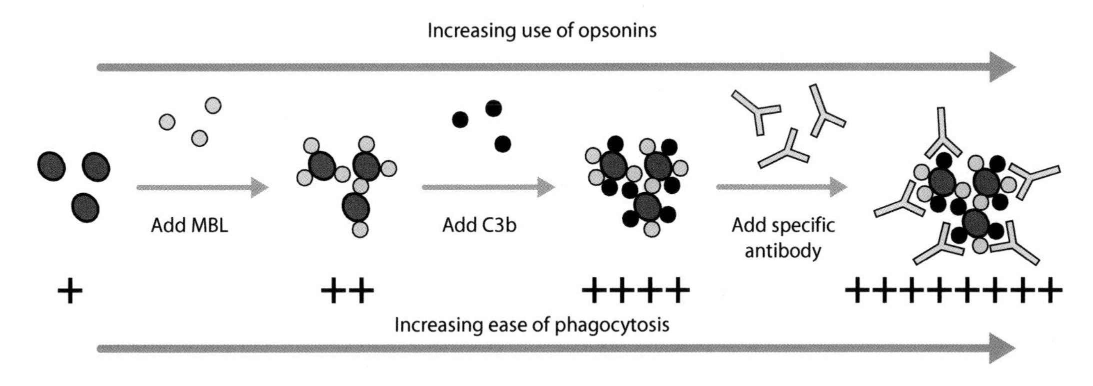

Figure 2.16 Phagocytic cells have surface receptors for mannose-binding lectin, C3b, and the Fc portion of IgG antibodies bound to an antigen. The binding of the opsonin to these receptors enhances phagocytosis. (MBL: mannose-binding lectin.)

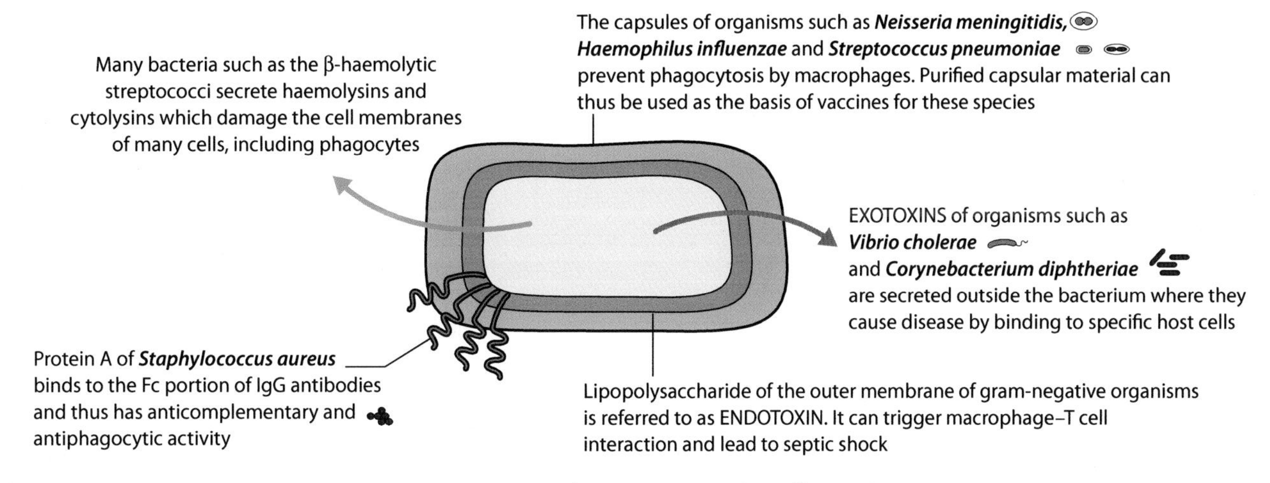

Figure 2.17 Diagram summarizing some important pathogenic properties of bacteria.

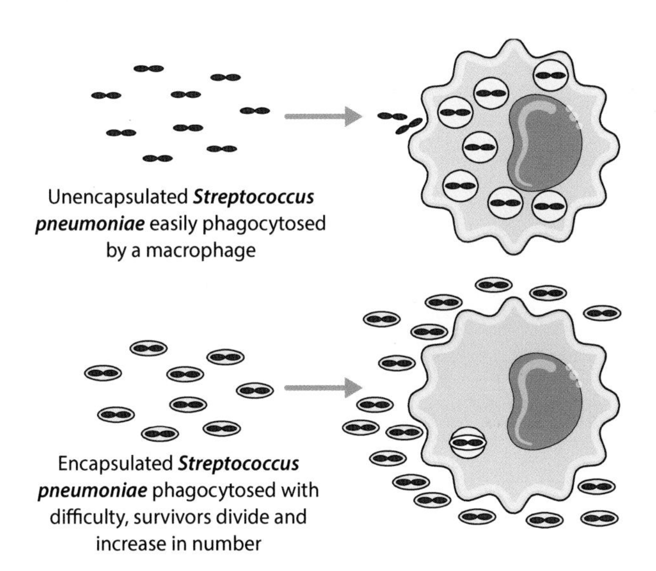

Figure 2.18 While unencapsulated pneumococci are easily phagocytosed by a macrophage, encapsulated organisms can resist phagocytosis and are able to multiply; they then overwhelm the local defences.

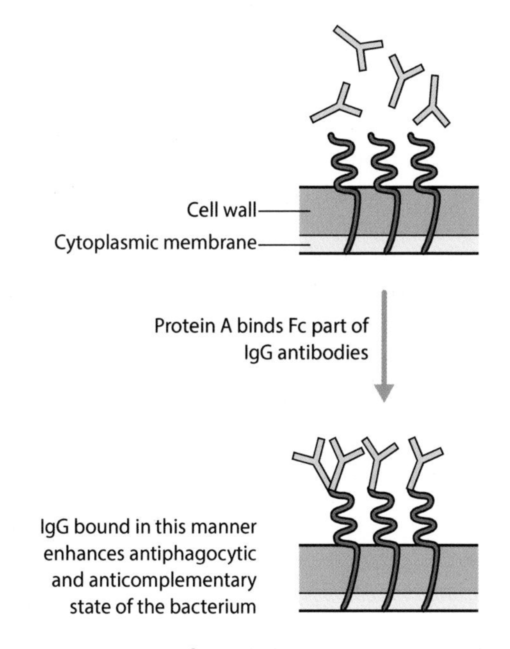

Figure 2.19 Protein A of *Staphylococcus aureus* can bind the Fc portion of IgG antibodies.

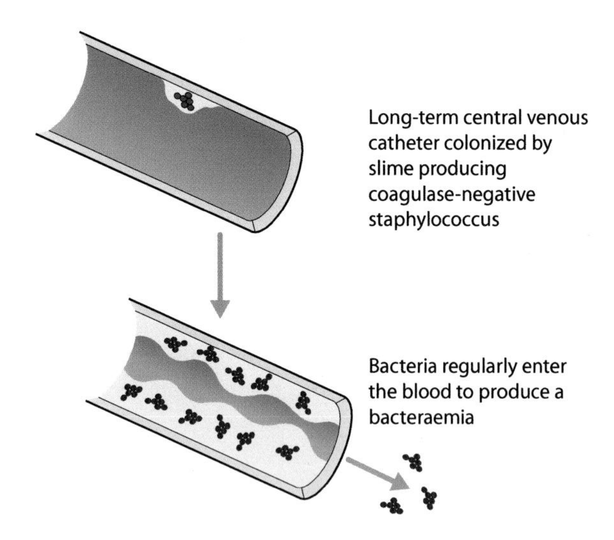

Figure 2.20 Organisms such as the coagulase-negative staphylococci can colonize foreign bodies such as central venous catheters. They exist here surrounded by an extracellular biofilm.

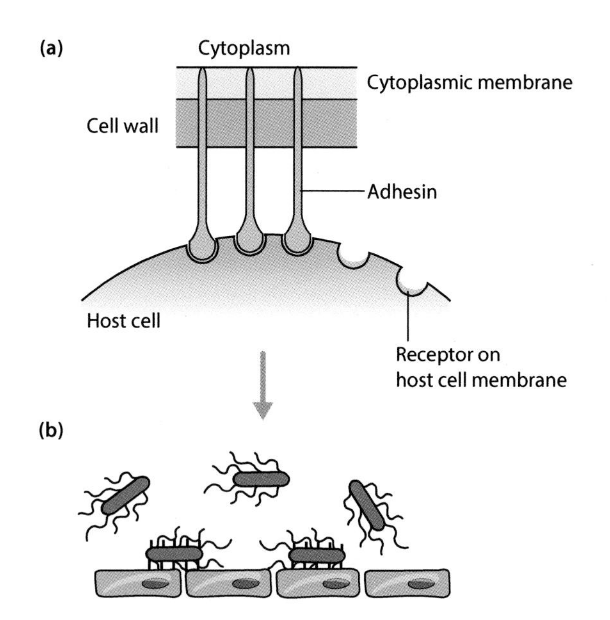

Figure 2.21 (a) Many bacteria adhere to the surface of cells by specific adhesins. (b) Adherent bacteria have an advantage over non-adherent bacteria in a place such as the bladder, as they are not washed out at micturition.

bladder epithelial cell using adhesins (**Figures 2.21a,b**). Adherent bacteria in the bladder have an advantage over non-adherent bacteria.

SECRETED PROTEINS

The secretion of extracellular proteins is a significant pathogenic feature of bacteria. *Staphylococcus aureus* secretes the enzyme hyaluronidase. A scratch of the skin may allow resident organisms to enter into the superficial layer of the skin. Hyaluronidase breaks down the hyaluronic acid matrix between the cells, allowing the bacteria to penetrate into deeper layers of the skin (**Figure 2.22**). The Panton Valentine leucocidin (PVL) is a secreted protein produced by PVL-positive strains of *Staphylococcus aureus*. This protein inserts into the cell membrane of cells such as neutrophils, and results in leakage of cell contents that the bacterium can use for its own reproduction. The function of neutrophils is compromised and uncontrolled release of toxic neutrophil contents exacerbates the necrotic process.

Bacteria including the β-haemolytic streptococci produce extracellular proteins termed haemolysins. These proteins are able to disrupt host cell membranes by enzymatic or detergent action, resulting in cell death by lysis (**Figure 2.23**). This haemolytic activity is demonstrated in the laboratory by the lysis (clearing) around colonies of bacteria when they are grown on blood agar. Important to the organism is the ability of the haemolysin to disrupt the cytoplasmic membrane of phagocytes, thus compromising the action of these cells. *Listeria monocytogenes* is a small gram-positive bacillus that can cause invasive disease.

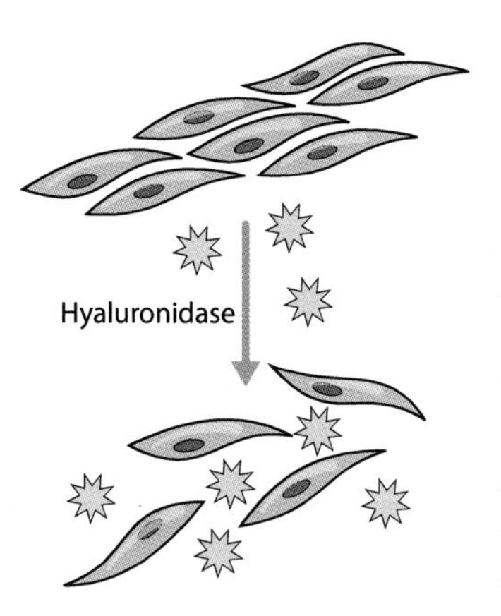

Figure 2.22 In connective tissue, the production of hyaluronidase by *Staphylococcus aureus* allows the organism to spread through the tissue.

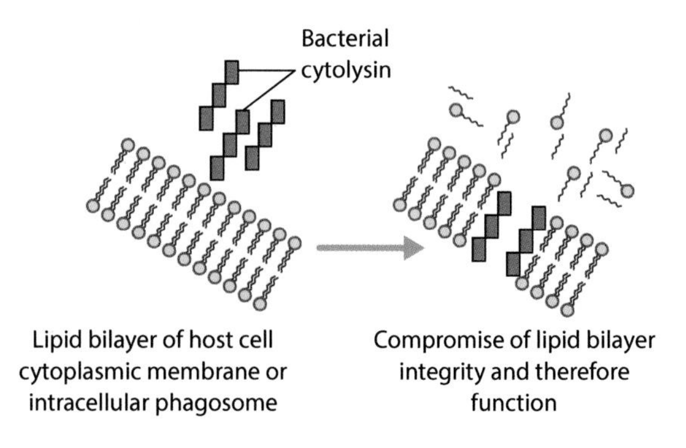

Figure 2.23 Bacterial cytolysins such as the ß-haemolysin of the streptococci can degrade host cell membranes by acting as enzymes or detergents.

This organism produces a haemolysin, which degrades the membrane of the phagosome. When phagocytosed, *Listeria monocytogenes* is able to escape into the cytoplasm where it cannot be destroyed (**Figure 2.24**).

Once in the cytoplasm, *Listeria monocytogenes* manipulates polymerization of cellular actin for its own benefit to propel itself to the plasma membrane of the infected cell. Here it produces pseudopod-like projections that are recognized by cells such as macrophages, which it then enters. In this way the organism not only bypasses the intracellular killing machinery, but then spreads to other cells without being recognized by antibodies, complement or neutrophils.

Many bacteria secrete proteinaceous exotoxins, which usually consist of an A and B component. The A component is the active part of the toxin, while the B component is responsible for the binding of the toxin to specific receptors on the cell's surface (**Figure 2.25**). *Vibrio cholerae* is an important cause of diarrhoea in lower-income parts of the world. The organism is spread via the faecal–oral route through contaminated water. The exotoxin produced by *Vibrio cholerae* binds via component B to receptors on the surface of enterocytes of the bowel. Internalization of the A component results in increased cyclic adenosine monophosphate (cAMP) levels in the cell, which gives rise to water and salt loss; a profuse watery and life-threatening diarrhoea results. Botulism is another exotoxin-mediated disease. *Clostridium botulinum* produces spores that can survive cooking. If contaminated food is stored for a long period, spores germinate and the vegetative bacteria secrete the neurotoxin responsible for the clinical manifestation of botulism. Both the A and B proteins produced by *Clostridium difficile* are toxins.

ENDOTOXIN

Bacteria such as *Escherichia coli* and *Neisseria meningitidis* can initiate endotoxic shock. Within the outer membrane of gram-negative bacteria is the lipopolysaccharide

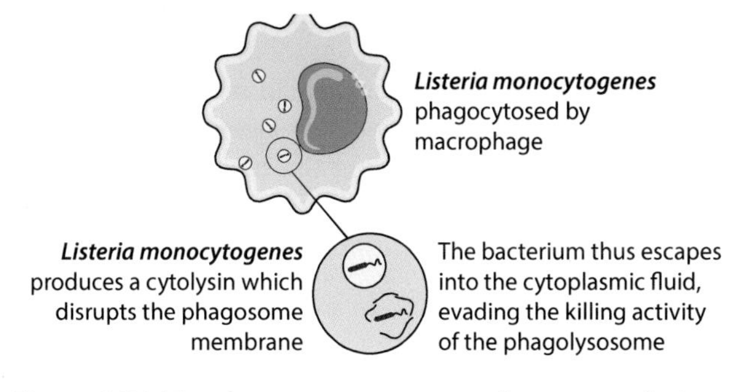

Figure 2.24 *Listeria monocytogenes* produces a cytolysin that enables the organism to escape from the phagosome into the cytoplasm of the macrophage.

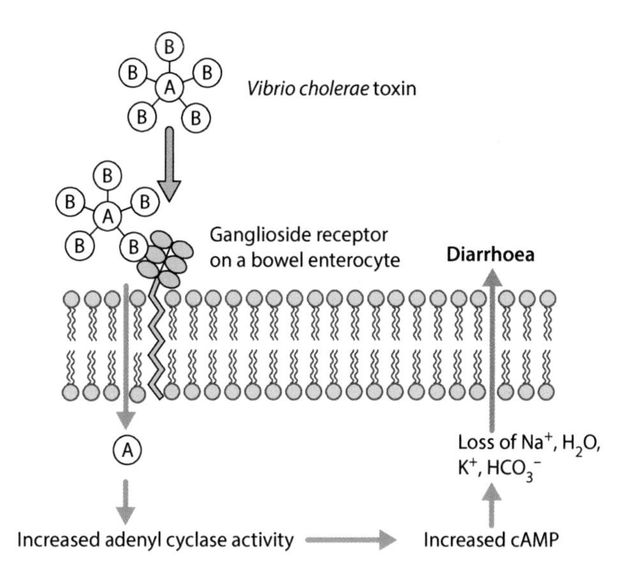

Figure 2.25 The toxin produced by *Vibrio cholerae* binds to specific receptors on the surface of the enterocyte by its B component. The internalized A component increases adenyl cyclase activity. (cAMP: cyclic adenosine monophosphate.)

endotoxin, which can activate the alternative complement pathway (**Figure 2.26a**). The resulting 'membrane attack complex' releases endotoxin, which is bound by endotoxin-binding protein, an acute phase protein released from the liver. This complex is then taken up by macrophages. When large numbers of macrophages are activated in this manner, uncontrolled release of cytokines gives rise to endotoxic or septic shock (**Figure 2.26b**).

SUPERANTIGENS

Under normal circumstances when an APC and T cell interact, relatively few cells are involved. Toxic shock syndrome (TSS) is caused by toxins such as the pyrogenic toxin of *Streptococcus pyogenes* and the enterotoxins of *Staphylococcus aureus*. These proteins act as superantigens, which cause macrophages and T cells to interact in a non-specific manner (**Figure 2.27**). This results in massive cytokine release, which gives rise to TSS, with fever, hypotension and multi-organ failure.

IMMUNE-MEDIATED DISEASES

Rheumatic heart disease is relatively uncommon in higher-income parts of the world but is still important in lower-income countries. The responsible organism is a group A streptococcus, *Streptococcus pyogenes*. Recurrent untreated pharyngitis results in activation of an immune response to the organism.

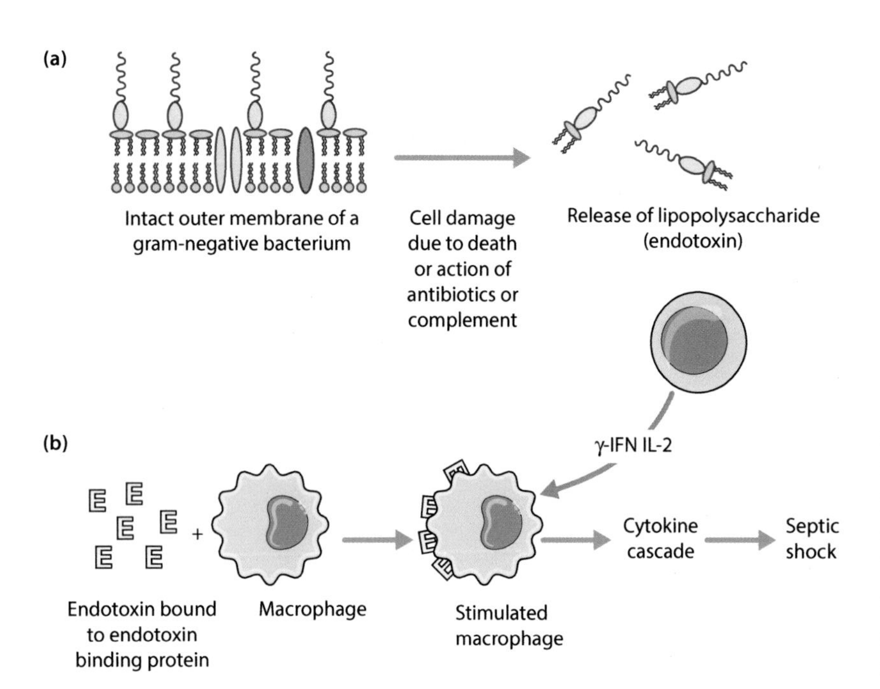

Figure 2.26 (a) Damage to the outer membrane of gram-negative bacteria results in release of lipopolysaccharide (endotoxin). (b) Endotoxin/endotoxin binding protein complex (E) binds to macrophages, producing an uncontrolled cytokine response with endotoxic or septic shock.

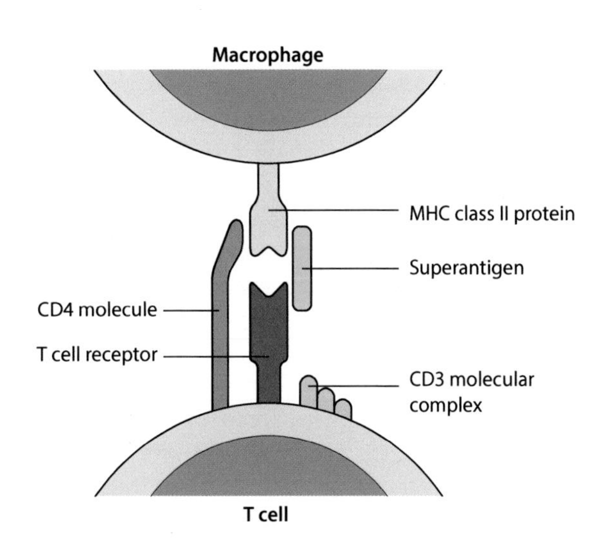

Figure 2.27 Superantigens such as pyrogenic toxin bypass the standard antigen presentation mechanism and uncontrolled cytokine release results in toxic shock syndrome (TSS).

The antibodies produced to the cell wall M protein of the streptococcus cross-react with antigenic determinants on the vascular endothelium of the host. The heart valves are the most important anatomical sites affected, as the high flow rate and turbulence around the valves means the complement deposition and the resulting inflammatory response is likely to cause structural damage to the valve. The healing process results in a thickened and abnormal valve (**Figure 2.28**). Any subsequent damage to the endothelium by turbulence will result in the deposition of platelets and fibrin. Oral streptococci such as *Streptococcus salivarius* entering the blood following, for example, manipulation of the teeth by dentistry, may settle in these deposits and initiate the process of infective endocarditis.

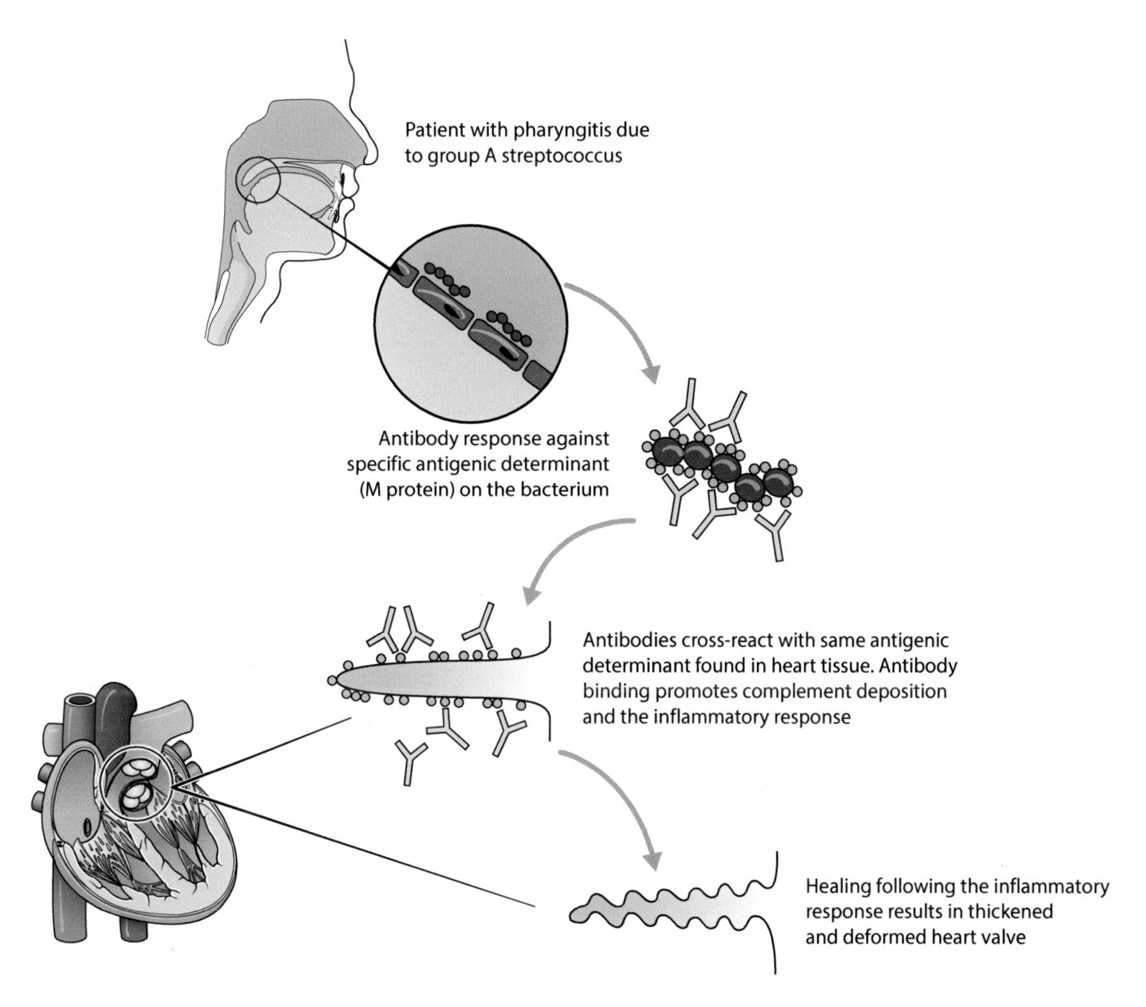

Figure 2.28 Antibodies to the M protein of *Streptococcus pyogenes* cross-react with antigenic determinants of the host. By binding to the endothelium of the heart valves these antibodies initiate an inflammatory response. The resulting healing with fibrosis leads to abnormal valves.

Chapter 3

Epidemiology and Assessment

INTRODUCTION

The bacteria introduced in Chapter 2 were categorized as endogenous, colonizers or exogenous, in order to classify them on the basis of their likely source. The list below is expanded here to include a number of viruses, fungi and parasites to broaden this theme. They are shown alphabetically, and then in their major groups.

Aspergillus fumigatus
Bacteroides fragilis
Candida albicans
Campylobacter jejuni
Clostridium difficile
Cytomegalovirus
Enterococcus faecalis
Epstein–Barr virus
Escherichia coli
Hepatitis B virus
Hepatitis C virus
Herpes simplex virus
Human immunodeficiency virus
Influenza virus
Klebsiella pneumoniae
Legionella pneumophila
Mycobacterium tuberculosis
Neisseria meningitidis
Norovirus
Pasteurella multocida
Plasmodium falciparum
Respiratory syncytial virus
Salmonella Typhi
Schistosoma haematobium
Staphylococcus aureus
Staphylococcus epidermidis
Streptococcus pneumoniae
Streptococcus pyogenes
Varicella zoster virus
Wucheria bancrofti

Plasmodium falciparum
Schistosoma haematobium
Wucheria bancrofti

Aspergillus fumigatus
Candida albicans

Cytomegalovirus (CMV)
Epstein–Barr virus (EBV)
Hepatitis B virus (HBV)
Hepatitis C virus (HCV)
Herpes simplex virus (HSV)
HIV
Influenza virus
Norovirus
Respiratory syncytial virus
Varicella zoster virus (VZV)

Bacteroides fragilis
Campylobacter jejuni
Clostridium difficile
Enterococcus faecalis
Escherichia coli
Klebsiella pneumoniae
Legionella pneumophila
Mycobacterium tuberculosis
Neisseria meningitidis
Pasteurella multocida
Salmonella Typhi
Staphylococcus aureus
Staphylococcus epidermidis
Streptococcus pneumoniae
Streptococcus pyogenes

The following colour code key is used throughout the book:
Gram-positive bacteria
Gram-negative bacteria
Mycobacteria
DNA viruses,
RNA viruses
Parasites
Fungi

Based on the characteristics of an organism, its pathogenic properties, usual site of residence (source) and mode of transmission, several questions can be asked in the clinical assessment of the patient. These are:

- What is/are the likely organism(s) causing the infection?
- Is there an identifiable source in the patient?
- Has the organism gone to other sites in the patient's body?
- What other conditions in the patient need to be considered (e.g. chronic obstructive pulmonary disease [COPD], injecting drug user [IDU], immunosuppression, no spleen)?
- Have all the necessary steps been taken to make a microbiological diagnosis?
- Has the appropriate anti-infective treatment been given?
- Have all the relevant infection control and public health issues been identified?

In wider terms, the following are also part of the assessment:

- If the organism is exogenous, was it acquired by person-to-person spread?
- If the organism is exogenous and acquired from the environment what is the source (food, water, air, animal [a zoonosis])?
- If there is a source, who else is at risk?
- If there is an identifiable risk in the population or the environment, how can it be controlled?
- Is this a vector-borne organism? If so, what is the nature of the cycle between the human, vector and any other another animal host?

EPIDEMIOLOGY

In its broadest sense, an appreciation of the epidemiology of organisms and the diseases they cause is an important backdrop to making the diagnosis. This is examined here using examples of world-wide, regional and local information.

WORLD-WIDE

The distribution of malaria is directly related to where the *Anophyles* mosquito lives, and exposure of the human population to infected blood-feeding female mosquitoes (**Figure 3.1**). There are many other arthropod-borne diseases that rely on this vector–human interaction, including the mosquito-borne dengue, West Nile and Zika viruses. For South American and African trypanosomiasis, the vector is the reduvid bug and the tsetse fly respectively. In the UK, Lyme disease, caused by *Borrelia burgdorferi*, is transmitted by the *Ixodes* tick, as part of a zoonosis with deer; there are several thousand cases in humans each year. The distribution of these arthropod vectors is not static, as climate change, human conflict, migration and incompetent government can affect their distribution. If authorities do not provide a safe, reliable water supply in urban areas, residents will use other means to store water, which is ideal for mosquitoes to breed in.

In addition to having an appreciation of the general epidemiology of diseases and their vectors, it is important to consider the likelihood that an individual has come in contact with the organism or the vector. The prevalence of schistosomiasis in Africa by predicative distribution at local level to 5 × 5 km spatial resolution and by country

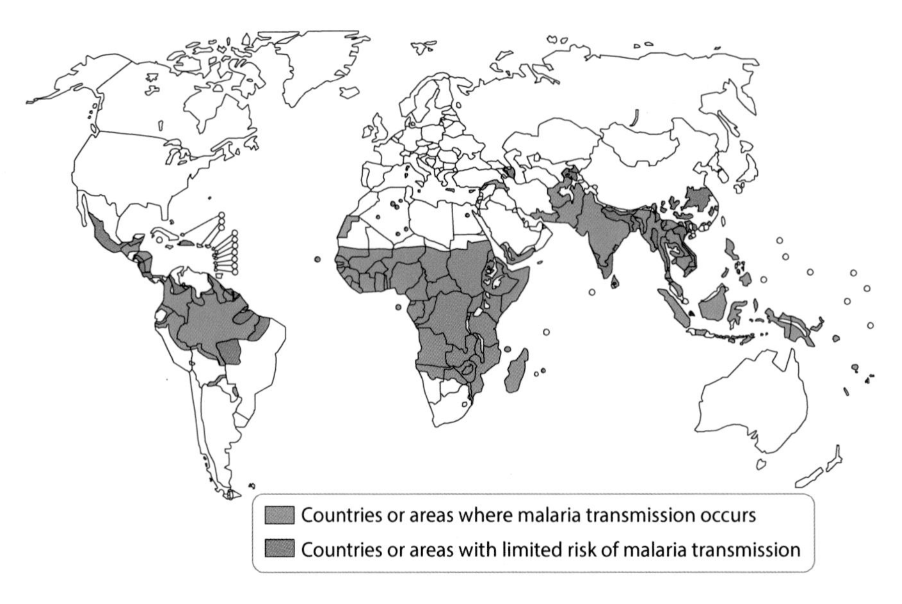

Figure 3.1 The World Health Organization's (WHO) countries and areas at risk of malaria transmission (2011). (Reproduced with permission from the WHO Program: Information for Travellers.)

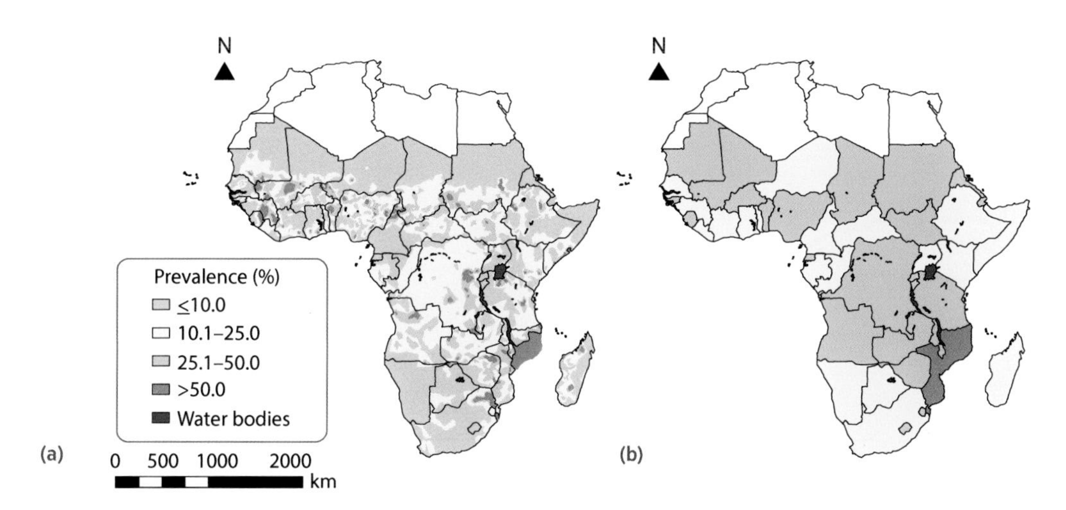

Figure 3.2 Predicted prevalence of schistosomiasis in the WHO Africa region. (a) Based on the median of the posterior predictive distribution at pixel level (5 × 5 km spatial resolution); (b) based on population adjusted estimation at country level for school-aged children (5–14 years old) from 2000 onwards. (Reproduced with permission, Xu J *et al.* [2016] Enhanced collaboration between China and African countries for schistosomiasis control. *Lancet Infectious Diseases* **16**(3):376–83.)

is shown in **Figure 3.2**. The point prevalence information at local level is more useful than the country data in determining the likelihood of exposure to the parasite. Mozambique at country level has a prevalence of >50%, but there are parts where the prevalence is in fact <10%. Similar conclusions can be drawn for the distribution in neighbouring South Africa. The high prevalence areas shown in **Figure 3.2a** reflect the environmental and human factors that enable the parasite to circulate continually through the definitive human and intermediate snail host. Areas where there is intensive farming, with irrigation from dams and rivers, are ideal for this parasite. When considering a specific organism and its geographical distribution, the likelihood of exposure needs to be determined too. Even in high prevalence areas, if the individual has not come in contact with a natural water source, they will not acquire the schistosome parasite.

REGIONAL

NHS laboratories in the UK are required each week to submit the numbers of key organisms identified, which are collated by the Regional Epidemiologist of Public Health England (PHE). This provides useful information concerning trends in individual organisms over time. It should be appreciated that the numbers recorded do not reflect the total cases, but only those patients who submitted a specimen for examination, and in which the organism was identified.

Several examples are shown in **Figure 3.3** for 2015–2016. Gastrointestinal illness, based on the number of weekly consultations in general practice over a year, are reasonably constant, as are the number of cases of laboratory-confirmed *Campylobacter* infection. Norovirus shows a peak in late winter/early spring, generally reflecting the usual period of activity of this virus, although exceptions to this do occur (**Figure 3.3a**). It is estimated that for every case of laboratory identified *Campylobacter*, there are ten other affected individuals. This means that every week in the West Midlands there are about 1000 cases of campylobacter infection. The source of this organism is its natural colonization of the intestines of chickens, and contamination of fresh chicken meat during processing. There is an ongoing public health message relating to the preparation, cooking and consumption of chicken.

The reported numbers of norovirus appear unimpressive. However, most positive results reflect the initial diagnosis of infection on a hospital ward or nursing home in the first (index) case(s). Outbreaks in these institutions usually reflect activity in the community. For every case of laboratory-confirmed norovirus, it is generally accepted that there are 1500 others, so that in a week with 40 laboratory diagnosed cases, there are an estimated 60,000 cases in the West Midlands that week. The economic impact of *Campylobacter* and norovirus infection in the community is considerable.

Viral respiratory infections have a distinct lull in activity in the summer months, based on the number of consultations for influenza-like illness, and the higher rates of consultation in the winter periods reflect the activity of these viruses, and of respiratory syncytial virus (RSV) and influenza in particular (**Figure 3.3b**).

GP In Hours Syndromic Surveillance System Weekly Incidence Rate (per 100,000 population) for Gastroenteritis in the West Midlands

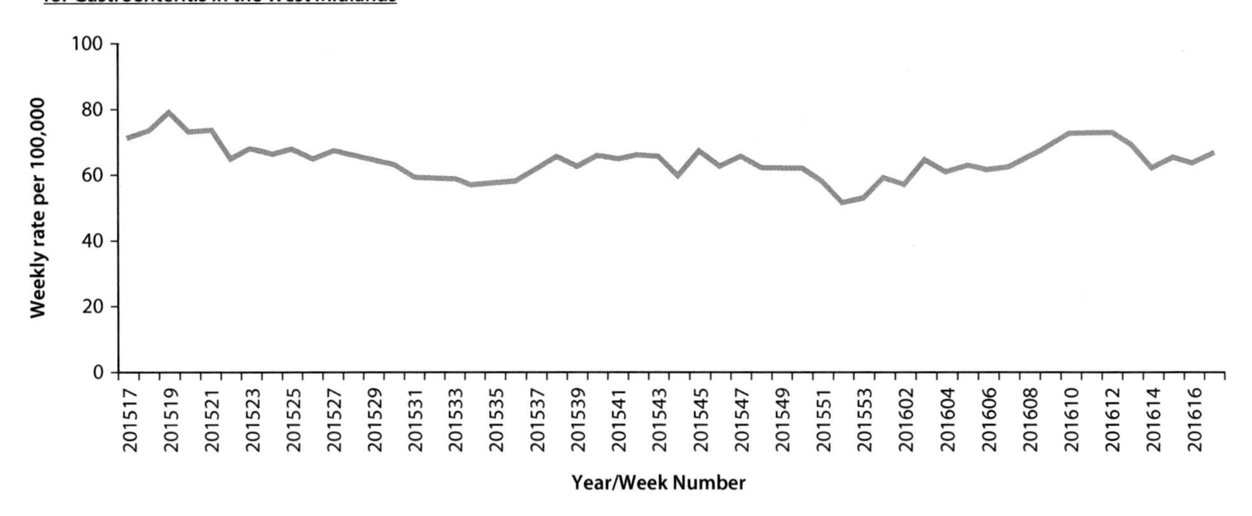

Laboratory confirmations of *Campylobacter* in the West Midlands

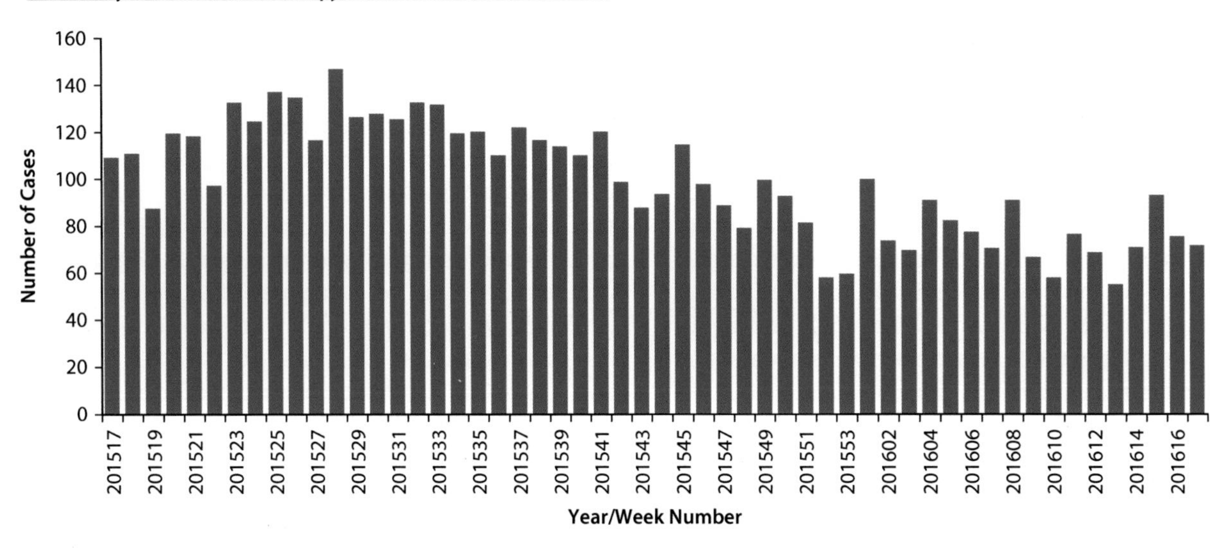

Laboratory confirmations of norovirus in the West Midlands

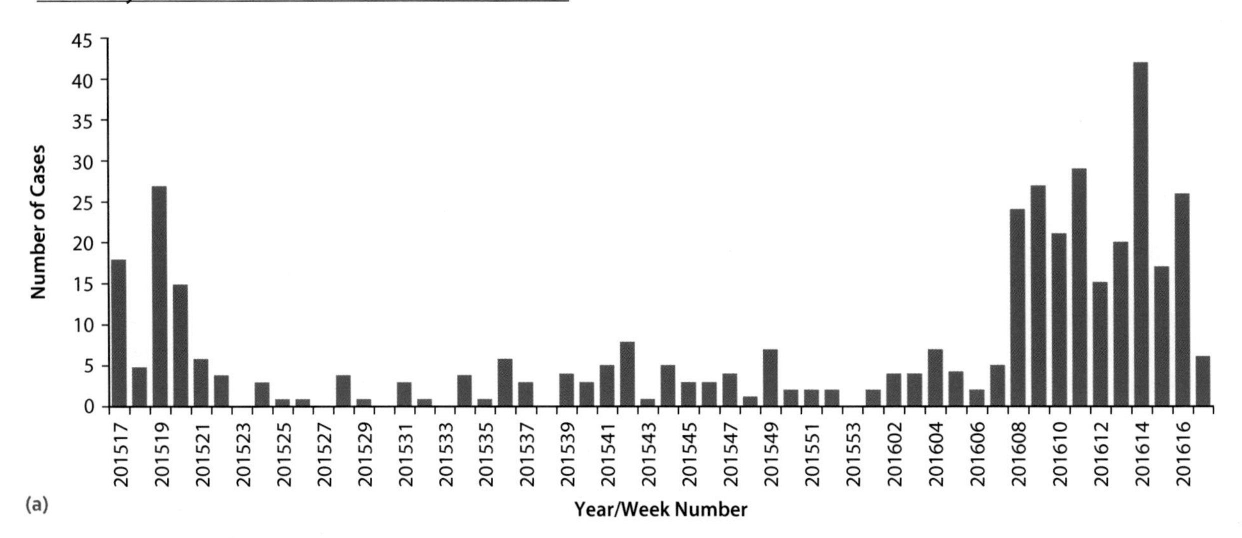

(a)

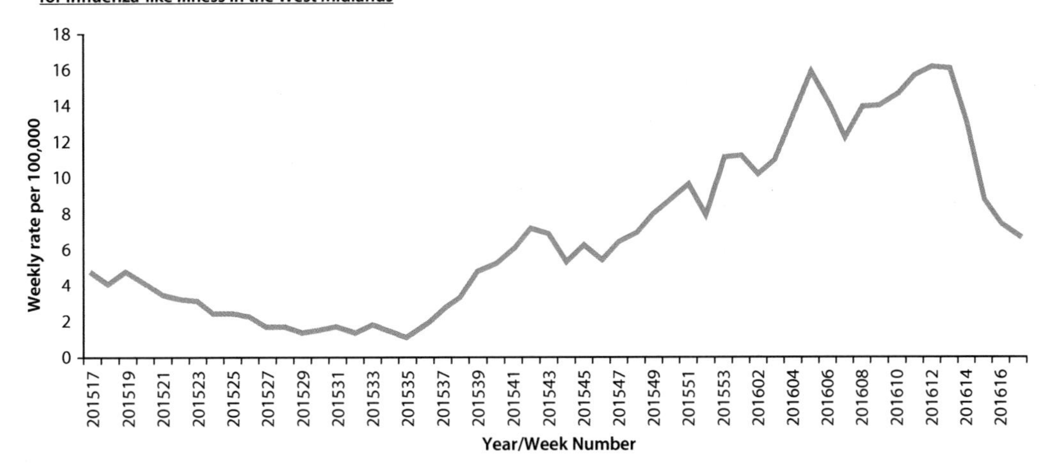

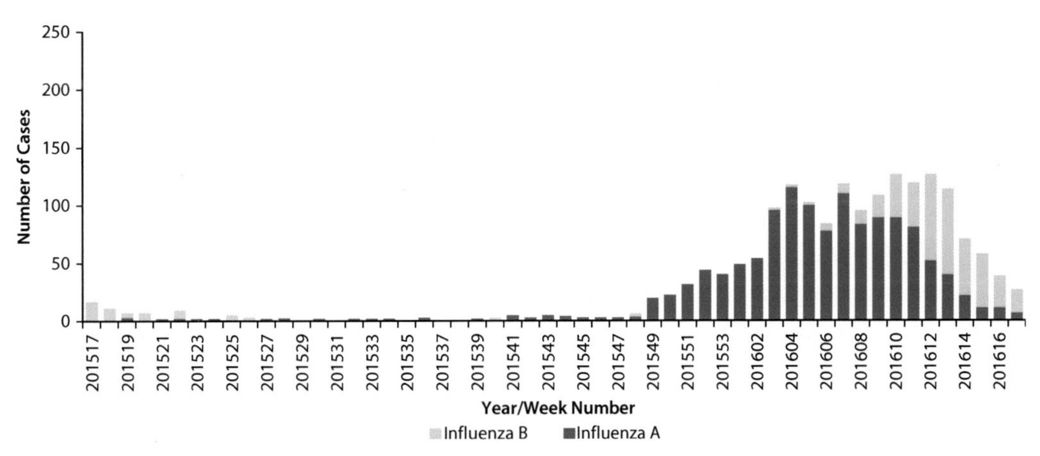

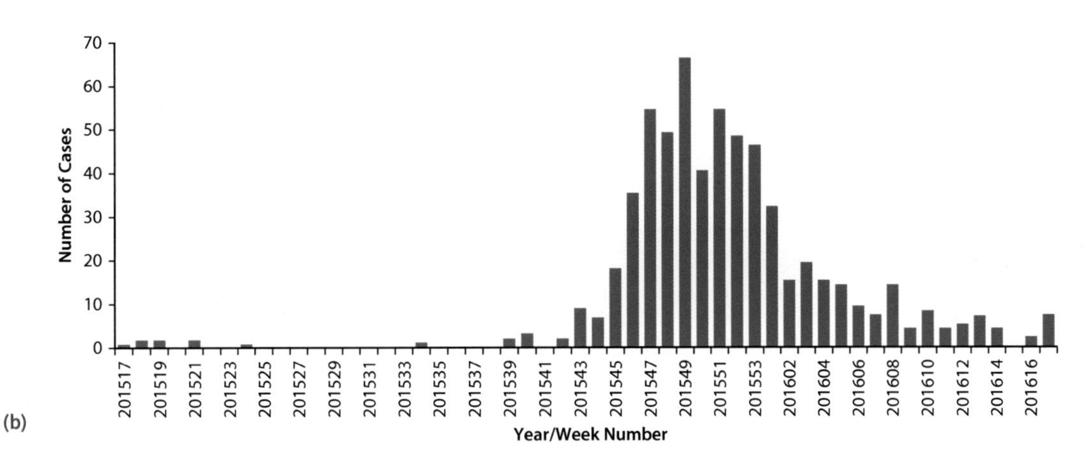

(b)

Figure 3.3 Surveillance data for selected infections and organisms in the West Midlands region 2015–2016 based on weekly data collated by the Regional Epidemiologist, Public Health England. (a) Gastroenteritis incidence rate in primary care, with numbers of *Campylobacter* and norovirus infections; (b) influenza-like illness incidence rate in primary care, with numbers of influenza A and B virus and respiratory syncytial virus infections. (Modified with the permission of the Office of the Regional Epidemiologist for Public Health England, West Midlands, UK.)

The RSV season usually precedes that of influenza, and influenza B becomes dominant over influenza A later in the influenza virus season. All respiratory viruses will be maintained by a low level of activity in the human population at other times of the year.

The increase in viral respiratory tract infections over winter reflects changes in the atmosphere and increased crowding in warmed spaces, which enhances transmission both in the family and in social, work and institutional settings. The influenza season is reasonably predictable, and for this reason persons in at-risk groups and health care professionals (HCPs) need to vaccinated.

LOCAL

Following a patient's admission to hospital, the reason(s) for admission are coded and recorded. A summary of information for a whole year of adult patients admitted to a large hospital in the West Midlands shows the age and sex distribution (**Figure 3.4**). Not surprisingly, the majority are over the age of 60 years, reflecting the general effect of disease on the older population. When this information is grouped according to broad categories of disease, infection accounts for six out of the top 20. Lower respiratory tract infections (LRTI) are top in both sexes, while urinary tract infections (UTI) are second in females and fourth in males (**Figure 3.5**). This information is important in the appreciation of clinical microbiology, as it provides a background knowledge of local epidemiology, which should be used to consider which organism(s) is the culprit, what the diagnostic and treatment options are, and what infection control and public health alerts need to be acted upon from the outset. Some examples of signs and symptoms and more common organisms are summarized in **Figure 3.6**. This goes back to the theme introduced in Chapter 2.

This local knowledge should be combined with a background appreciation of regional and national information, as well as that of the wider world. While 'common infections are commonest', it is the background

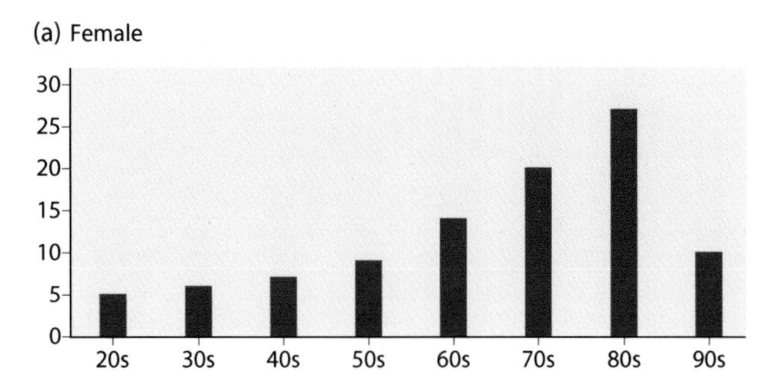

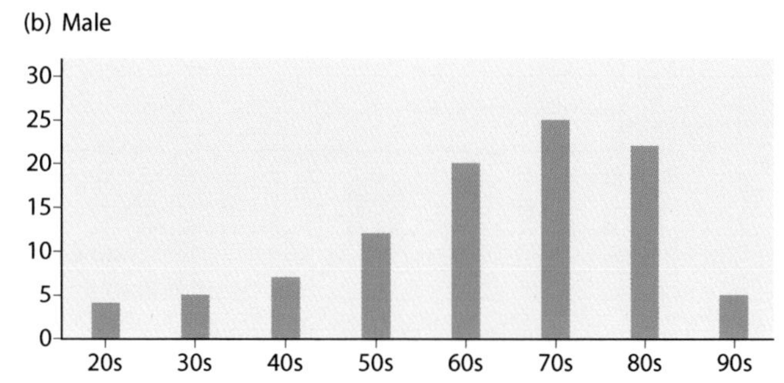

Figure 3.4 The age distribution as a percentage of (a) female and (b) male patients, in decades, admitted to a large teaching hospital over 1 year.

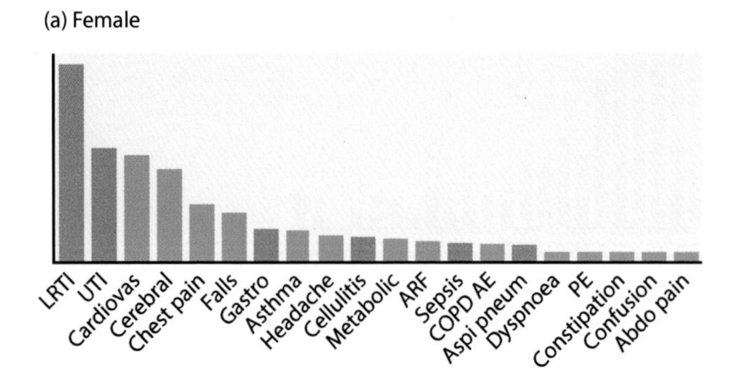

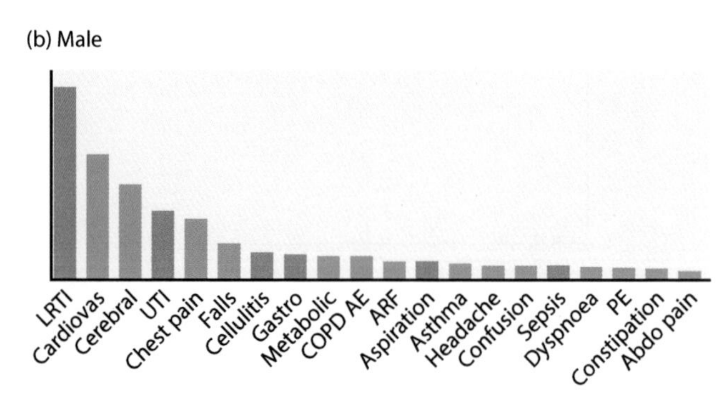

Figure 3.5 The 'top 20' reasons for admission for the cohort of female and male patients in **Figure 3.4**, with the six infection categories highlighted in red. (ARF: acute renal failure; Aspi: aspiration pneumonia; Gastro: gastroenteritis; PE: pulmonary embolus.)

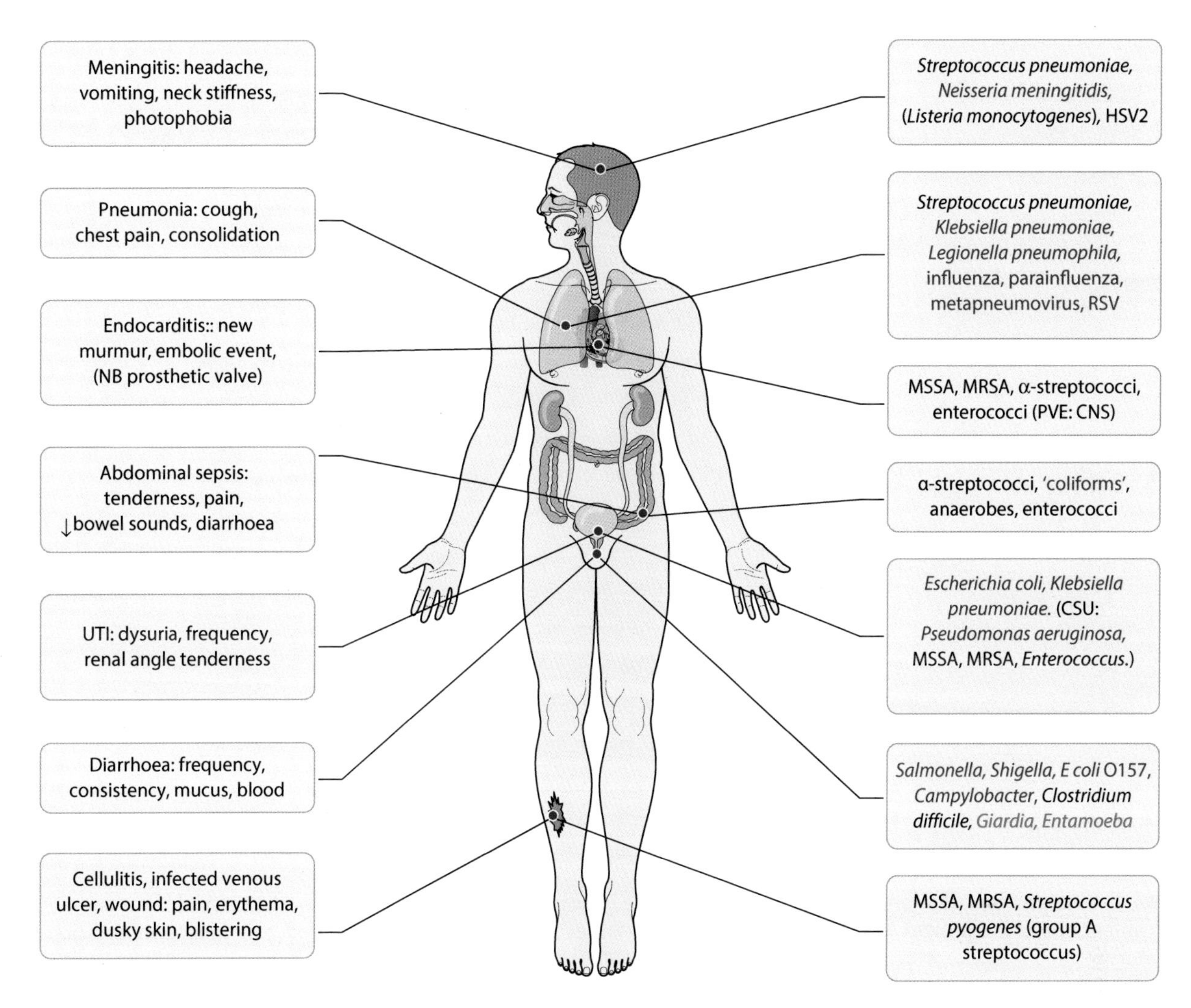

Figure 3.6 Some examples of signs and symptoms of common infections, and the more likely organisms that need to be considered.

knowledge that alerts one to the exception at the earliest opportunity.

THE INCUBATION PERIOD

Following infection with an organism, there is an incubation period, usually of days, until the symptoms and signs of that disease manifest. Some examples are shown in **Figure 3.7**. *Plasmodium falciparum* malaria has an incubation period of 8–30 days, but that for *Plasmodium vivax* and *Plasmodium ovale* can be months or years, with the liver hypnozoite remaining dormant for long periods.

Many infections can be asymptomatic, including hepatitis A, B and C (HAV, HBV, HCV), and this is especially so in children; in the case of HBV and HCV these may only be revealed as asymptomatic chronic infection in later years.

THE HIV TEST

Human immunodeficiency virus (HIV) infection is now diagnosed in a wide age range of patients, including those in their 70s and 80s. There should be a low threshold for discussing the test with the patient. This can include the previously well older adult with a first episode of lobar

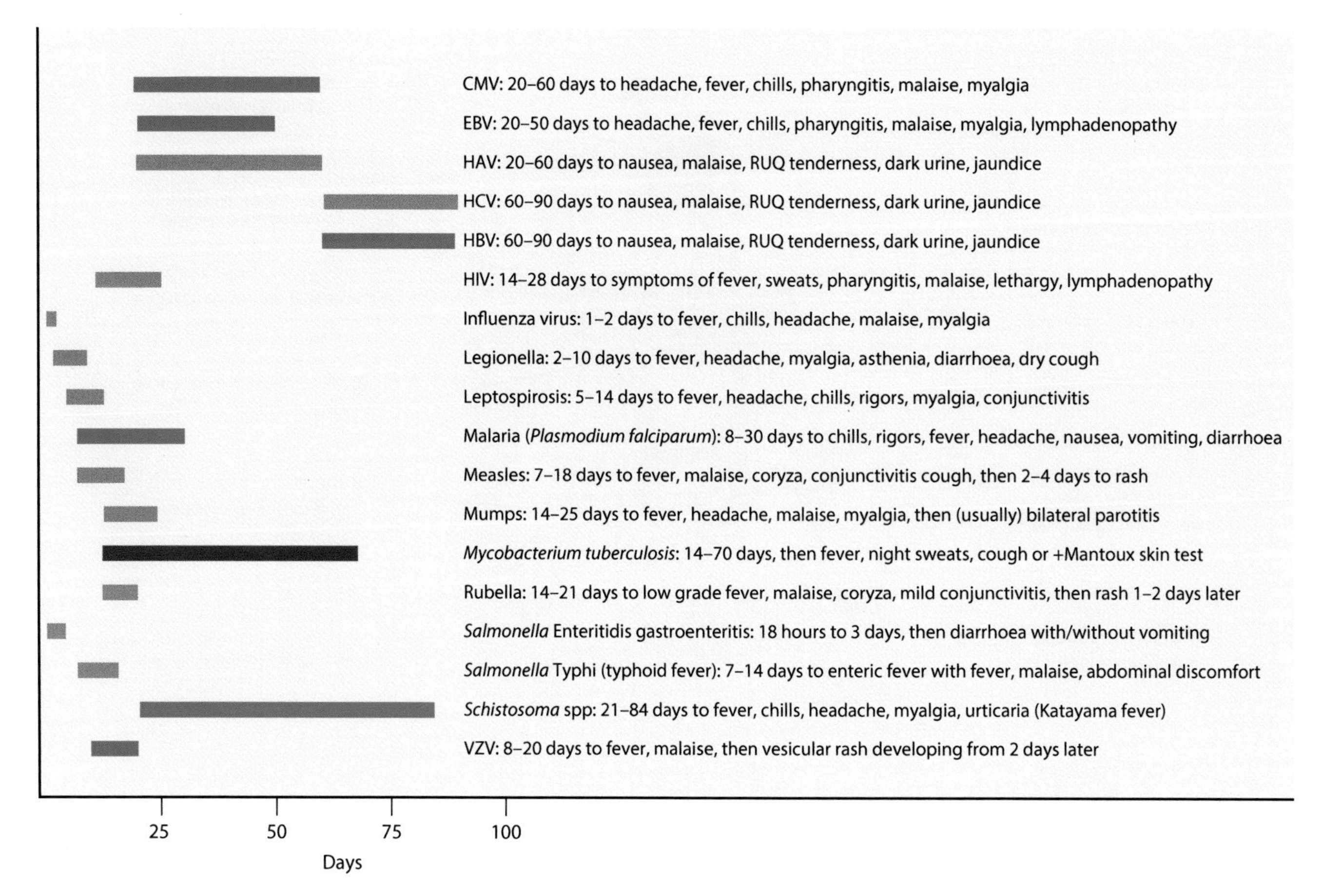

Figure 3.7 The range of incubation period in days (bar) to the onset of symptoms of a number of bacterial, viral and parasitic infections. If the patient with *Plasmodium falciparum* infection is only partially treated, recrudescence can occur weeks later.

pneumonia. UTIs are uncommon in young men, and a first episode may reveal a previously silent anatomical defect. There is also an increased risk of UTI in the setting of men-who-have-sex-with-men (MSM), and here, an HIV test should be discussed with the patient. Fever, headache, sore throat, malaise and lymphadenopathy are features of an infectious mononucleosis-like illness. If the monospot test for Epstein Barr virus (EBV) and serology tests for both cytomegalovirus (CMV) and EBV are negative, HIV needs to be considered, particularly if lymphadenopathy persists.

THE STEPS IN MANAGEMENT OF THE PATIENT

The various stages of assessment of the patient admitted to the A&E department, or who is being reviewed on the ward, are shown in **Figure 3.8**. Here the clinical assessment, history and examination are complemented by the collection of the correct specimens and initiating the necessary anti-infective treatment. Infection control and public health alerts are also addressed.

The patient with the signs and symptoms of sepsis requires immediate attention. Manifesting as the systemic inflammatory response syndrome (SIRS), it is the earliest delivery of the appropriate antibiotic(s) into the patient that reverses further deterioration.

MEWS AND THE 'SEPSIS SIX'

The modified early warning score (MEWS) provides a rapid way of recording the severity of infection, as shown by the example in **Table 3.1**. Most hospitals use a score of 3 (or more) to trigger their sepsis pathway, but the MEWS score is an aid to good clinical judgement, not a substitute for it.

Sepsis results in systemic vasodilatation, increased capillary permeability and increased oxygen demand, resulting in reduced cardiac and circulatory function.

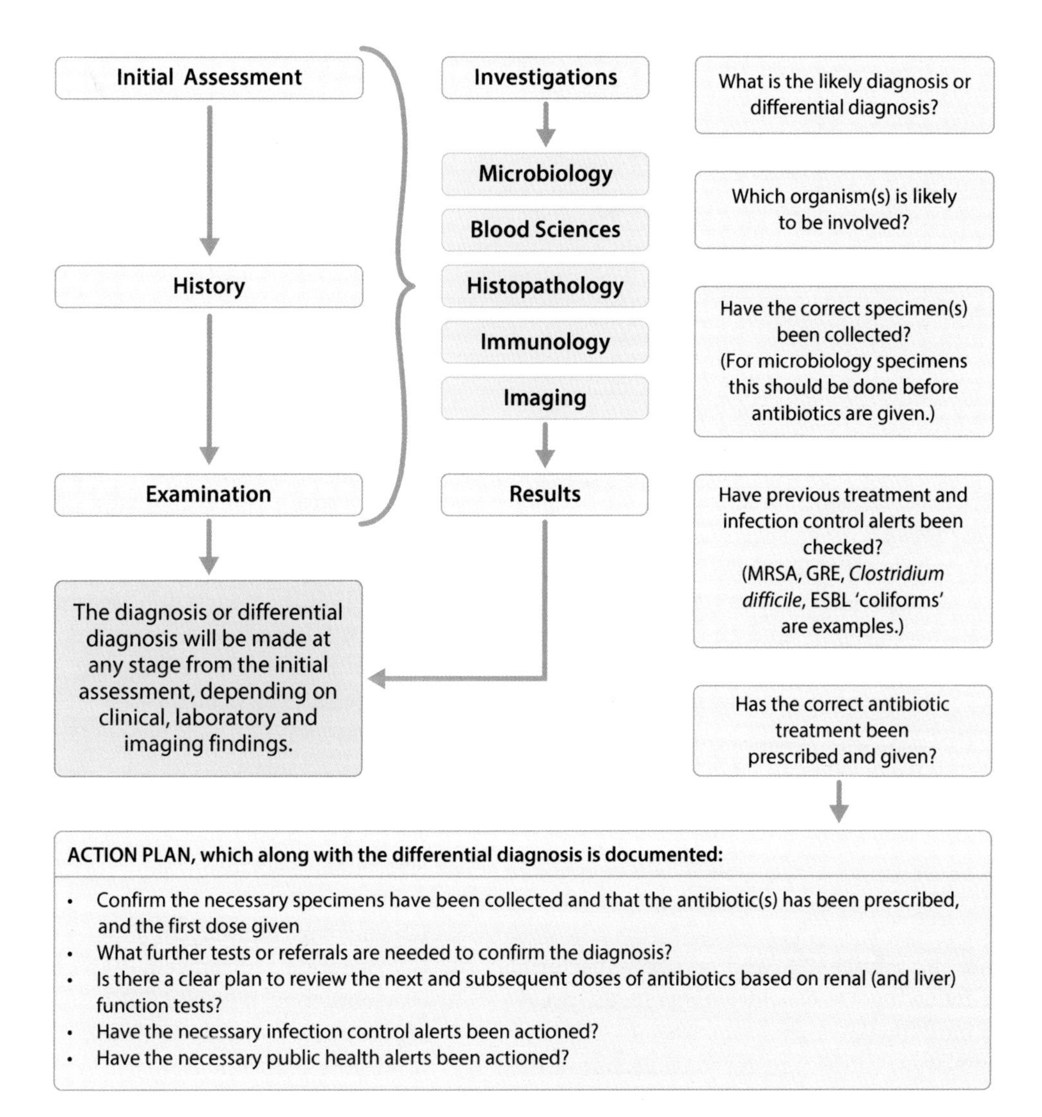

Figure 3.8 The key stages to consider when managing the patient with infection. Clearly the severity of the situation will determine the sequence of events. Treatment/infection control alerts need to be identified (see **Figure 3.10**).

Deranged clotting and an imbalance between oxygen supply and demand give rise to ischaemia and organ dysfunction. An abnormal white cell count (WCC) of <4 or >12 × 10^9/L is frequently observed, as is a blood glucose of >7 mmol/L in the non-diabetic patient.

This sepsis pathway sets in train the actions of the 'Sepsis Six', guided by the necessary clinical assessment:

- Give high-flow oxygen to maximize availability to the tissues (in the patient with severe COPD, seek advice from the respiratory team).
- Collect blood cultures (and other necessary specimens that can be obtained) in order to make a microbiological diagnosis.
- Then give intravenous antibiotic(s) immediately.
- Give a fluid challenge to improve cardiac output and organ perfusion pressure, and to maximize oxygen delivery to the tissues.
- Measure serum lactate regularly.
- Measure urine output (the patient may need to be catheterized).

Additional markers of severity in sepsis include:

- Serum lactate >4 mmol/L.
- International normalized ratio (INR) >1.5.
- Activated partial thromboplastin time (aPTT) >60 s.
- Bilirubin >34 µmol/L.
- Oxygen to keep SpO_2 >90%.
- Platelets <100 × 10^9/L.

Table 3.1

Score	3	2	1	0	1	2	3
Systolic BP	<71	71–80	81–100	101–199		≥200	
Heart rate (bpm)	<30	<40	41–50	51–100	101–110	111–129	≥130
Respiratory rate (RPM)		<9		10–16	17–20	21–29	
Temperature (°C)		≤35	35.1–36	36.1–37.5	37.6–38.1	≥38.2	
Mental state			New confusion	Alert	To voice	To pain	Unresponsive

- Creatinine >177 µmol/L.
- Urine output <0.5 mL/kg/h for 2 h.
- Neutropenic sepsis.

The clinical assessment of the patient, the MEWS score and degree of derangement of other markers identify the severity of the situation, and the need for urgent action. Delivery of the first dose of the antibiotic(s), and ensuring that subsequent doses are given on time is essential. Theintensive care unit (ICU) team must be contacted, so that intensive care can be provided as soon as possible.

WHICH ORGANISMS AND WHICH ANTIBIOTICS?

With reference to **Figure 3.6** and a likely bacterial infection, the following are considerations in the previously healthy individual:

- Usually one organism will be responsible, but in the setting of an abdominal source the infection can be polymicrobial, including anaerobes.
- The clinical diagnosis may be straightforward and the septic patient with a typical non-blanching meningococcal rash is likely to have *Neisseria meningitidis* sepsis.
- When a UTI is the source, a 'coliform' will be the causative agent in most circumstances.
- A community-acquired pneumonia (CAP) is most likely to be caused by *Streptococcus pneumoniae*, but other organisms including *Legionella pneumophila* and *Mycobacterium tuberculosis* must be considered.
- Cellulitis of the limbs is most likely caused by *Staphylococcus aureus* or *Streptococcus pyogenes*. If the cellulitis is progressing to necrotizing fasciitis, the latter is the more likely culprit.
- Infective endocarditis is usually caused by gram-positive bacteria and, in the septic patient, *Staphylococcus aureus* in particular.

Antibiotics are discussed in more detail in Chapter 4, but are introduced here to underline the principles that are used in treating a bacterial infection, especially when they are given empirically. The major classes of antibiotics include:

1. β-lactam: penicillins (benzylpenicillin [PEN], amoxicillin [AMO]); β-lactam/β-lactamase inhibitor combinations (amoxicillin/clavulanic acid, co-amoxiclav [COA]), piperacillin/tazobactam [P/T]); cephalosporins (cefalexin, ceftriaxone [CTX]); carbapenems (ertapenem [ERT], meropenem [MER]).
2. Aminoglycoside: gentamicin (GEN), amikacin.
3. Fluorinated quinolone: ciprofloxacin (CIP), levofloxacin (LEV).
4. Macrolide: erythromycin (ERY), clarithromycin, azithromycin.
5. Lincomycin: clindamycin (CLI).
6. Glycopeptide: vancomycin (VAN), teicoplanin.
7. Oxazolidinone: linezolid (LIN).
8. Lipopeptide: daptomycin (DAP).
9. Imidazole: metronidazole (MET).

Based on usual *in vitro* antibiotic susceptibility testing profiles, the general range of activity for 'well known' bacteria is presented in **Figure 3.9**; the above abbreviations ('ABC') refer to those in this diagram.

There are several themes that can be recognized. MET is only effective against anaerobes, and the glycopeptides, oxazolidinones and lipopeptides have activity against gram-positive bacteria.

Several other features are of note:

- The broadening of activity of the β-lactams, from penicillin and amoxicillin, to β-lactam/β-lactamase inhibitor combinations, cephalosporins and the carbapenems.
- Enterococci are not only resistant to CTX but in general to all the cephalosporins (as is *Listeria monocytogenes*). Note that *Enterococcus faecium* is resistant to the β-lactams.
- The reduced treatment options for the extended-spectrum β-lactamase (ESβL) 'coliforms'.
- No treatment options for the carbapenemase-producing Enterobacteriaceae (CPE); certain

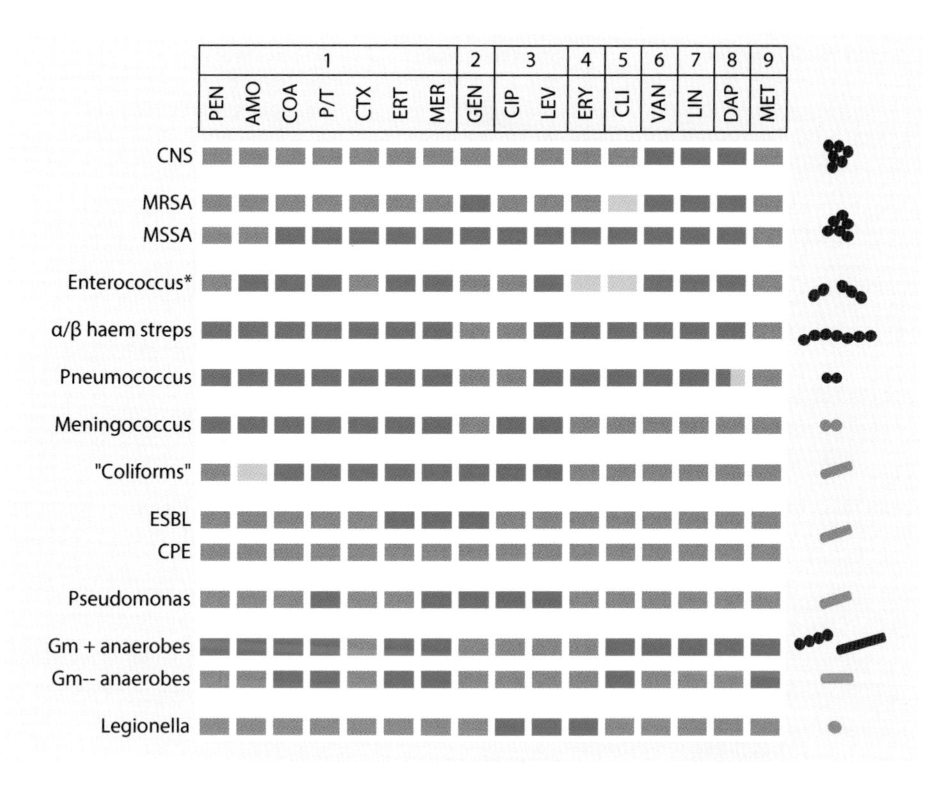

Figure 3.9 The general range of activity of antibiotics (based on *in vitro* testing) used against a selection of bacteria encountered in clinical practice. Green: usually consider susceptible, red: consider resistant; amber: caution. The shaded caution with daptomycin and pneumococcus is due to inhibition of this antibiotic by lung surfactant. *This is *Enterococcus faecalis* (*Enterococcus faecium* is β-lactam resistant). (Gm: Gram; other abbreviations see text.)

isolates are sensitive to the polymyxin colistin or the glycylcycline tigecycline.

Several amber cautions are indicated:

- There can be increased resistance of methicillin-resistant *Staphylococcus aureus* (MRSA) to CLI.
- DAP is inactivated in the lung by surfactant, and should not be used to treat pneumonia caused by gram-positive bacteria such as pneumococcus.

The spectrum of activity of these antibiotics is used to determine the antibiotics recommended in hospital guidelines (Chapter 4), and this theme is used further in the systems chapters.

ANTIBIOTIC TREATMENT IN THE SEPTIC PATIENT WITH AN UNKNOWN SOURCE

Bacteria to consider in the septic patient where the source and organism(s) are unknown are shown in **Figure 3.10.** *Clostridium septicum*, a member of the bowel flora, is included here, to emphasize the need to consider anaerobes if a bowel source is a possibility.

Usually antibiotic(s) with a broad spectrum of activity are administered, until microbiology results and/or clinical assessment direct the appropriate narrow-spectrum agent(s). Using the information in **Figure 3.9**, AMO/GEN/MET is one combination, and VAN/CIP/MET could be used in the penicillin-allergic patient. The carbapenem MER has broad cover on its own. Hospitals vary in their choice.

A treatment alert (**Figure 3.10**) identifies the need to ensure that the empirical treatment includes an antibiotic active against the resistant organism.

FEVER OF UNKNOWN ORIGIN

Fever, or pyrexia, of unknown origin (FUO/PUO) is where the temperature has regularly breached 38.4°C for a period of 3 weeks or more. Infection, connective tissue diseases and neoplasia account for nearly 70% of cases, but the diagnosis is not confirmed in about 25% of cases (**Figure 3.11**). A summary of the history and clinical parameters to consider in 'classic' FUO in the patient admitted from the community, as well as in hospital-acquired and neutropenic FUO, is shown in **Figure 3.12.**

Organisms to consider:	Gram stain feature	Treatment alert
Clostridium septicum		Anaerobes, acute abdomen
Enterobacteriaceae ('coliforms')		ESBL 'coliform', CPE
Neisseria meningitidis		Meningitis
Staphylococcus aureus		MRSA, GISA, PVL producer, endocarditis
Streptococcus pneumoniae		Penicillin resistance, meningitis
Streptococcus pyogenes (group A streptococcus)		Necrotizing fasciitis

Figure 3.10 The key bacteria to consider in the patient with sepsis with an unclear source that determines the empirical antibiotics used. Other relevant history, such as occupational, recreational, travel, will influence the range of organisms to consider. (GISA: glycopeptide-intermediate *Staphylococcus aureus*.)

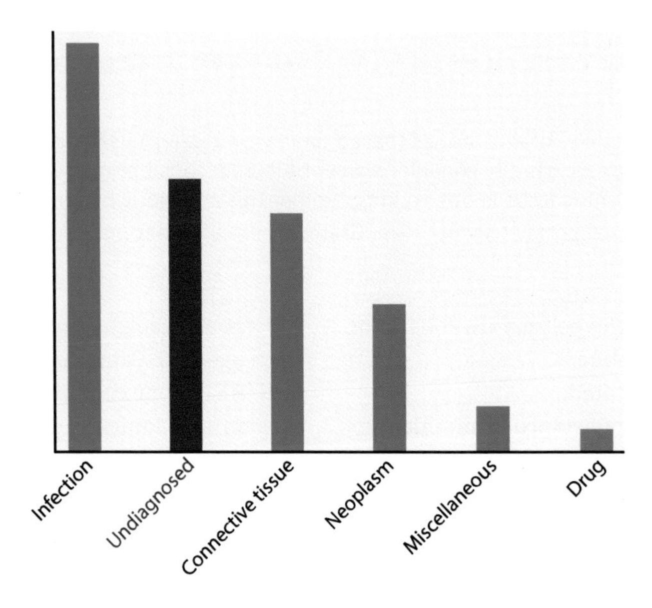

Figure 3.11 A bar diagram showing the reasons for a fever of unknown origin (FUO). Approximate values: infection 35%, undiagnosed 25%, connective tissue disease 22%, neoplasm 12%, miscellaneous 4%, drug 2%.

Taking into account the information presented in **Figure 3.11**, and the fact that about one-quarter of cases do not obtain a diagnosis, the patient with a FUO warrants a multi-disciplinary team review, with the patient's participation. Organisms and pathologies that need to be considered include *Mycobacterium tuberculosis*, culture-negative infective endocarditis and intra-abdominal abscess formation. If not considered previously, the need for the HIV test must be confirmed.

It is essential to review occupational, family, social, rural and travel-associated activities in discussion with the patient. They may have migrated through a number of countries under conditions of extreme deprivation, and in each country there will potentially be organisms that could have been acquired. For example, *Brucella* could have been acquired by drinking unpasteurized goat's milk weeks previously in the country where the journey began.

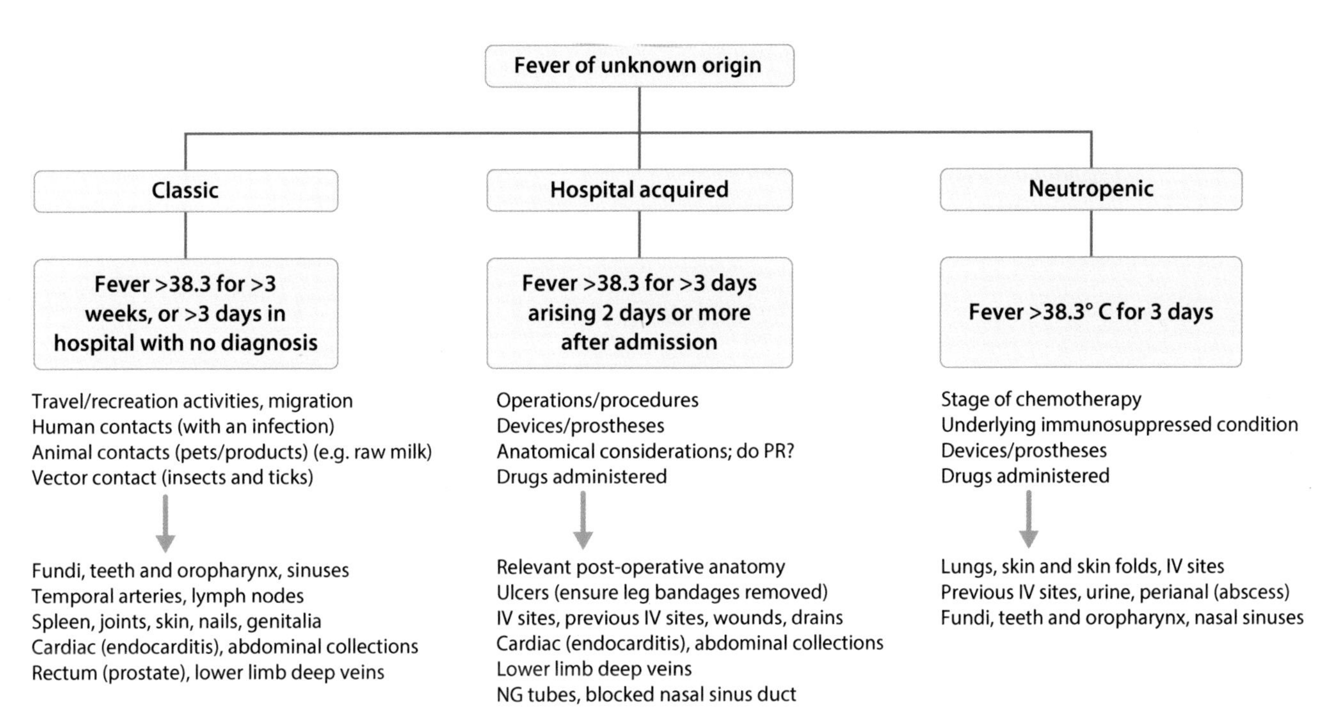

Figure 3.12 Some parameters to consider in the patient with a FUO, which can be grouped as classic, nosocomial and neutropenic. Each creates particular alerts to be investigated.

INFECTION CONTROL

With the clinical management process, there must be consideration of the infection control interventions needed to keep the patient, other patients, HCP, visitors and, as relevant, members of the community safe. These alerts should be the prompt, where necessary, to contact the infectious diseases physician, microbiologist or infection control team.

Infection control actions include promptly nursing the patient in a single room that contains only the necessary equipment. Staff access is restricted and where appropriate infection control practice is put in place, including the correct personal protective equipment (PPE). Negative pressure single rooms are the safest.

The antibiotic resistance profile of an organism, its method of spread and the group that either the patient, or fellow patients, fall into, are risks that need to be considered (**Figure 3.13**). They emphasize the range of situations encountered and direct how each needs to be addressed.

PUBLIC HEALTH

The list of organisms that require notification under the Notifiable Diseases Act of 2010 is shown in **Figure 3.14**. This is in two parts, Schedule 1 and Schedule 2. It is the legal responsibility of the clinical team to notify clinical diagnoses listed under Schedule 1 to the local public health team. Specific organisms listed in Schedule 2 are the responsibility of the laboratory to notify.

This is done as soon as the diagnosis is made, in order that other individuals who may have had the same exposure, or who have been in contact with the index patient, are identified and managed correctly. The reasons for prompt notification are clear. One example is the patient with food poisoning and infectious bloody diarrhoea; *Escherichia coli* O157 must be considered. There are likely to be other individuals who have had the same exposure risk and need to be identified, as a matter of urgency. This also enables public health to review the food risk promptly, and take appropriate action to eliminate them.

LABORATORY TESTS

Details on the role of the microbiology laboratory and the processes used to identify organisms are given in Chapter 5, and further information on specimen collection and processing is discussed in the systems chapters. Pathology departments will have on-line guidance about

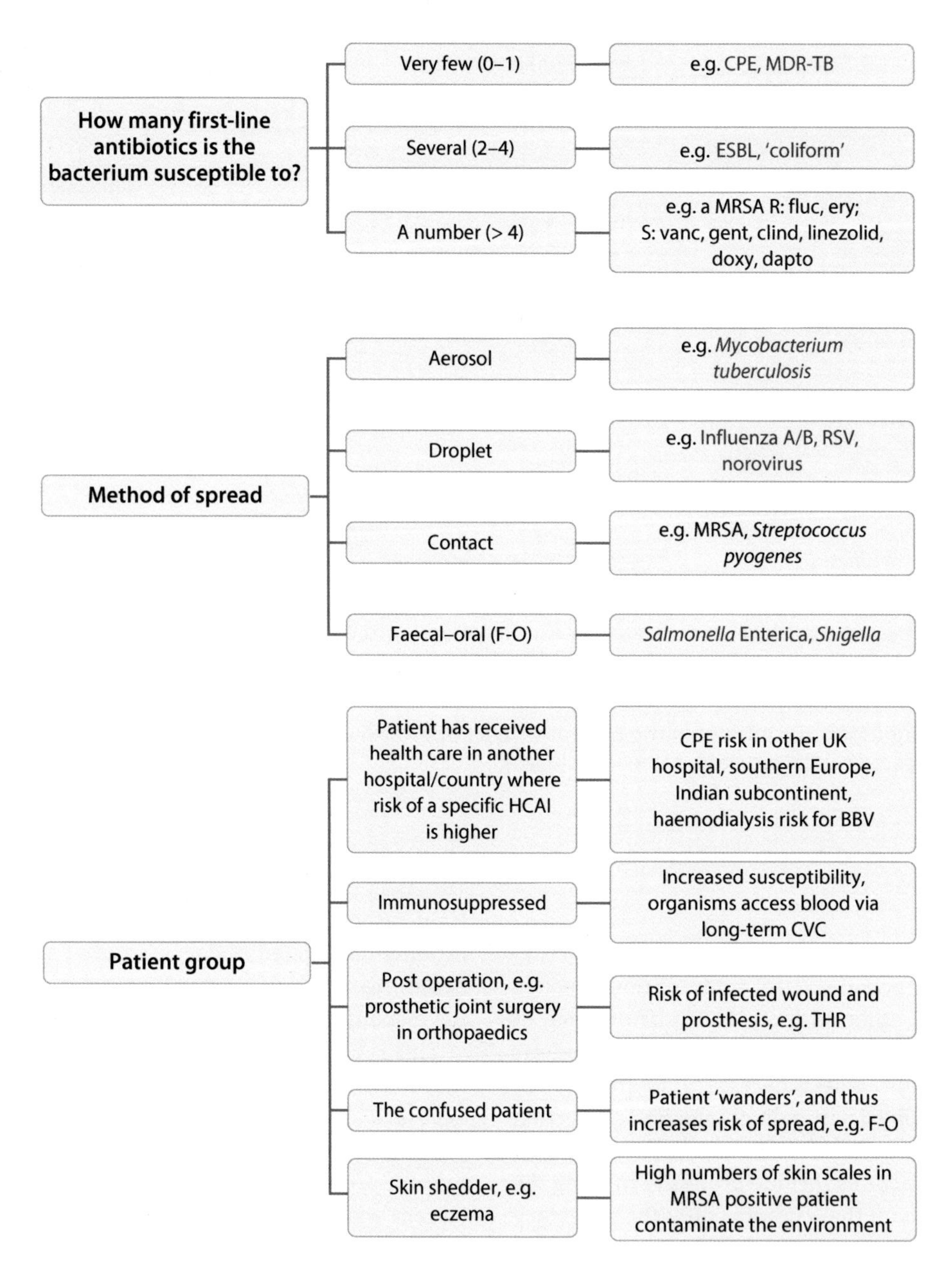

Figure 3.13 A number of alerts highlight treatment, infection control and additional risk groups that can be identified in the individual patient.

the range of tests that are available, with advice on collection of the various specimen types.

Examples are shown in **Figure 3.15**. This highlights the central role that blood cultures have in diagnostic microbiology. Specific containers are used to collect blood for tests such as meningococcal and pneumococcal polymerase chain reaction (PCR), as well as HIV viral load (ethylenediamine tetra-acetic acid [EDTA]), serology tests for immunoglobulin (Ig) M and IgG antibodies (clotted blood for serum) and citrated tubes for the culture of mycobacteria from blood (**Figure 3.16**). Thick and thin blood films (EDTA blood) are used for the diagnosis of malaria by microscopy; the antigen test for the parasite is done on this specimen too.

Good quality specimens optimize the isolation of organisms and direct appropriate antibiotic treatment.

Notification Regulations Schedule 1	
Notifiable Diseases	
Acute encephalitis	Malaria
Acute meningitis	Measles
Acute infectious hepatitis	Meningococcal septicaemia
Acute poliomyelitis	Mumps
Anthrax	Plague
Botulism	Rabies
Cholera	Rubella
Diphtheria	SARS
Enteric fever (typhoid or paratyphoid fever)	Smallpox
Food poisoning	Tetanus
Haemolytic uraemic syndrome (HUS)	Tuberculosis
Infectious bloody diarrhoea	Typhus
Invasive group A streptococcal disease and scarlet fever	Viral haemorrhagic fever (VHF)
Legionnaires' disease	Whooping cough
Leprosy	Yellow fever

Notification Regulations Schedule 2	
Causative Agents	
Bacillus anthracis	*Corynebacterium diphtheriae*
Bacillus cereus (only if associated with food poisoning)	*Corynebacterium ulcerans*
Bordetella pertussis	*Coxiella burnetii*
Borrelia spp.	Crimean–Congo haemorrhagic fever virus
Brucella spp.	*Cryptosporidium* spp.
Burkholderia mallei	Dengue virus
Burkholderia pseudomallei	Ebola virus
Campylobacter spp.	*Entamoeba histolytica*
Chikungunya virus	*Francisella tularensis*
Chlamydophila psittaci	*Giardia lamblia*
Clostridium botulinum	Guanarito virus (Venezuela haemorrhagic fever virus)
Clostridium perfringens (only if associated with food poisoning)	*Haemophilus influenzae* (invasive)
Clostridium tetani	Hanta virus

Figure 3.14 (*Continued*)

Hepatitis A, B, C, Delta and E viruses (All acute and chronic cases, as relevant)

Influenza virus

Junin virus (Argentina haemorrhagic fever virus)

Kyasanur Forest disease virus

Lassa virus

***Legionella* spp.**

Leptospira interrogans

Listeria monocytogenes

Machupo virus (Bolivia haemorrhagic fever virus)

Marburg virus

Measles virus

Mumps virus

***Mycobacterium tuberculosis* complex**

Neisseria meningitidis

Omsk haemorrhagic fever virus

Plasmodium falciparum, vivax, ovale, malariae, knowlesi*

Polio virus wild or vaccine types

Rabies virus (classical rabies and rabies-related lyssaviruses)

***Rickettsia* spp.**

Rift Valley fever virus

Rubella virus

Sabia virus (a South American haemorrhagic fever virus)

***Salmonella* spp. (including *S.* Typhi and *S.* Paratyphi)**

SARS coronavirus

***Shigella* spp.**

***Streptococcus pneumoniae* (invasive)**

***Streptococcus pyogenes* (invasive)**

Varicella zoster virus

Variola virus

Verotoxigenic *Escherichia coli* (including *Escherichia coli* O157)

Vibrio cholerae

West Nile virus

Yellow fever virus

Yersinia pestis

Figure 3.14 The list of notifiable diseases (for which the attending clinician is responsible) and causative agents (for which the laboratory is responsible) listed in the Health Regulations (Notifications) Act of 2010 for England and Wales. These are to be conveyed to the local HPU at the earliest opportunity. Note: **Plasmodium* spp. are usually screened for in blood films in haematology laboratories in the United Kingdom.

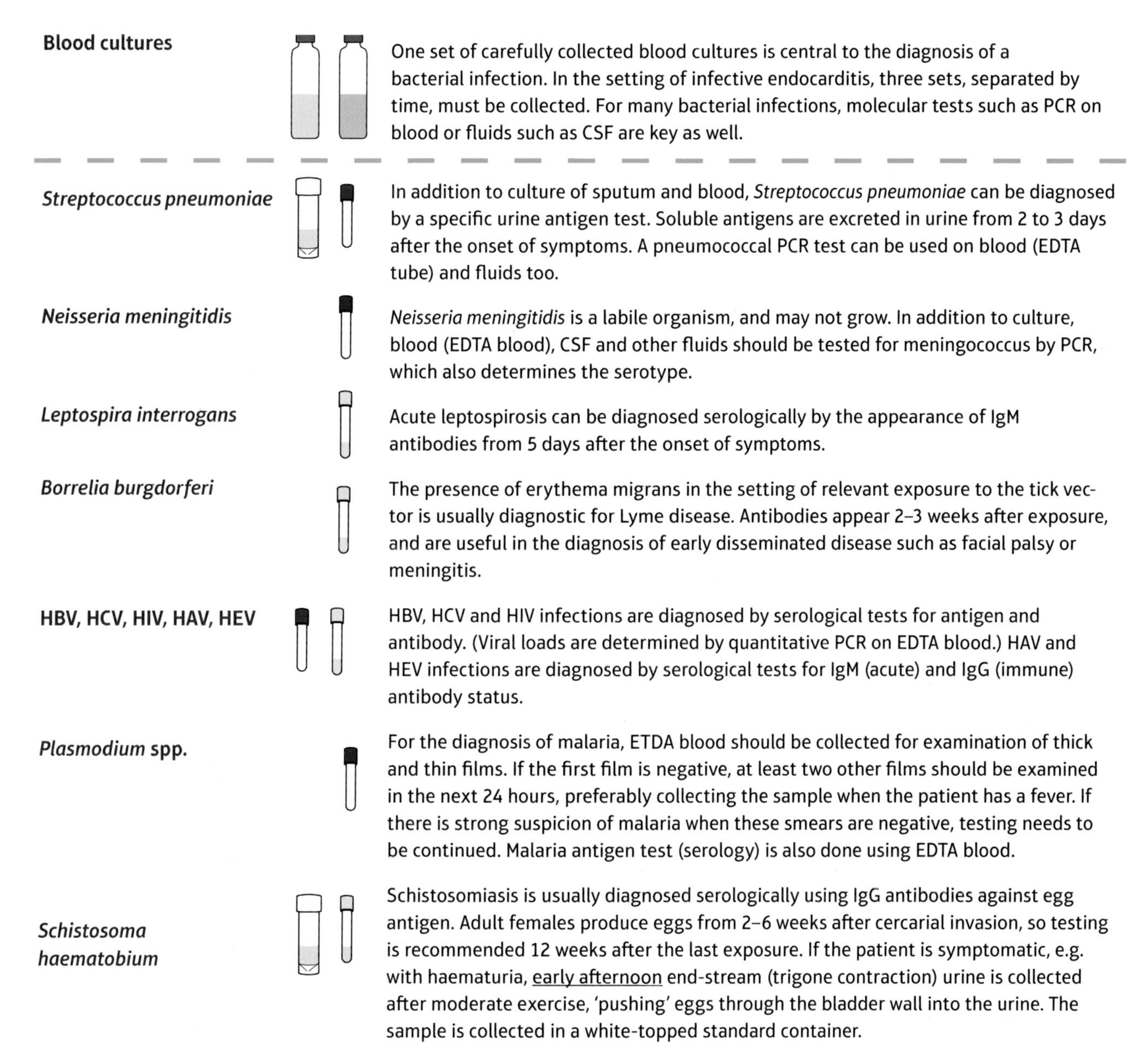

Figure 3.15 Some examples of the range of tests available in diagnostic microbiology.

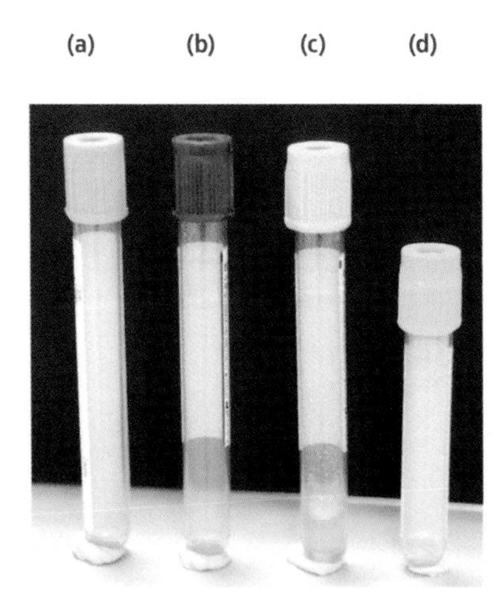

Figure 3.16 Blood collection tubes for microbiology specimens. (a) Ethylenediamine tetra-acetic acid (polymerase chain reaction, malaria films and malaria antigen); (b) clotted blood (antibiotic assays/serology); (c) clotted blood with clot activator: serology; (d) citrate: mycobacteria culture from blood. Reference to local hospital guides is essential.

Figure 3.17 A good quality specimen needs to be collected for the diagnosis of bacteria in pneumonia. A specimen of saliva (left) will be discarded; purulent sputum represents the inflammatory process in the lung.

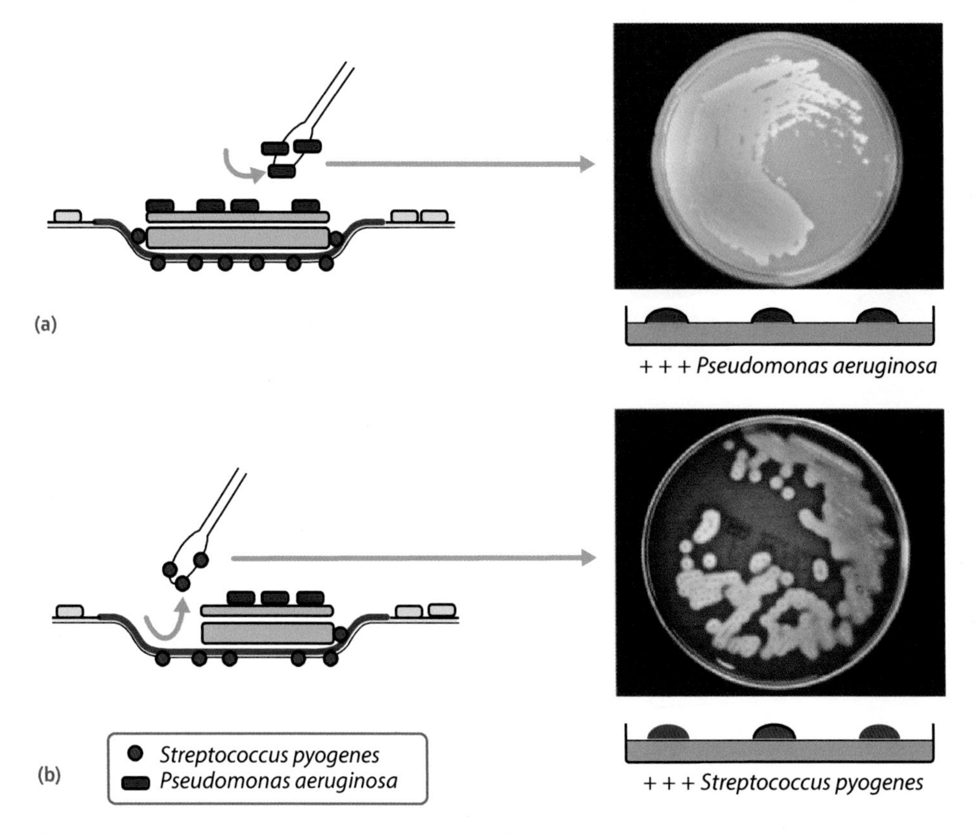

Figure 3.18 (a) Taking a swab from the surface slough of an infected chronic leg ulcer grows a colonizer such as *Pseudomonas aeruginosa*. (b) Removal of slough and sampling the deeper material collects the pathogen.

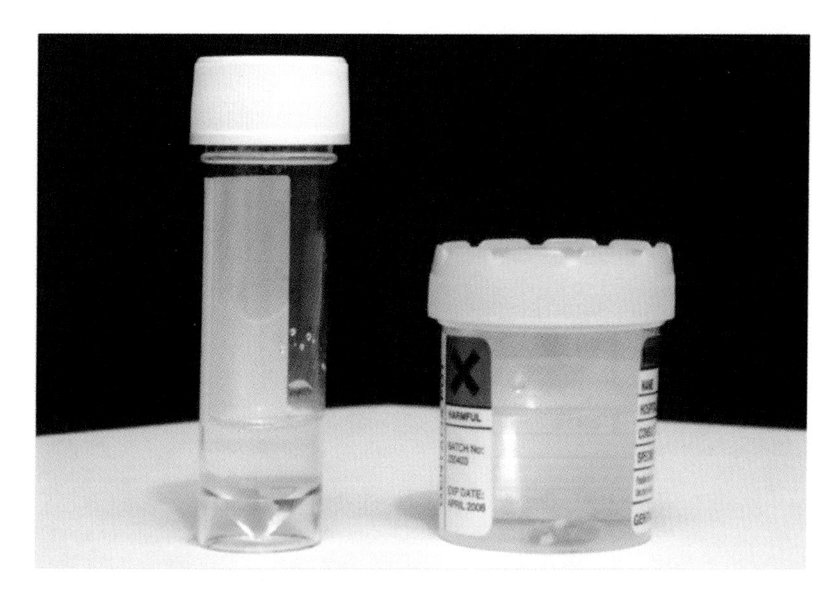

Figure 3.19 When a microbiological test is required, a biopsy specimen must be divided, with one half placed in sterile saline (left) and the other half placed in the formalin fixative for histology (right).

In the case of the patient with pneumonia, the purulent secretions from the lung are required. Salivary specimens are of no use, as they do not represent the infective process, and will be discarded by the laboratory, with a diagnostic opportunity being lost (**Figure 3.17**).

A situation that frequently leads to misleading results, with subsequent inappropriate and unnecessary antibiotic treatment, is the chronic leg ulcer. A swab should be collected when there is evidence of infection including pain, increasing pain, surrounding cellulitis and fever. The ulcer that is 'sloughy' or is a 'bit smelly' is generally not infected; it requires appropriate tissue viability assessment, with slough removal.

When infection is identified, a swab specimen should be collected from an area likely to contain the probable pathogen. A swab collected from the superficial layer of the ulcer, where slough or exudate is present, is of little use, as the bacteria identified will often not represent the organism responsible for the infection. *Pseudomonas aeruginosa* is a common example of this, as are 'coliforms'. Where possible, a swab should be used to remove slough, in order to expose material as close to viable tissue as possible. A fresh swab is used to collect the sample from an infected area, rotating it to maximize loading. In this way, the bacteria isolated are more likely to represent the infective process. The incorrect and correct procedures are outlined in **Figure 3.18**.

When a biopsy is taken during surgery and an infection such as tuberculosis is a consideration, this biopsy must be divided, with half placed in formalin for histopathology and half into sterile saline for microbiological examination (**Figure 3.19**). Not infrequently, the whole specimen is placed in formalin, and a critical diagnostic opportunity is lost.

Chapter 4

Anti-Infective Agents

INTRODUCTION

In Chapter 3, important antibiotics and their ranges of activity against selected bacteria were introduced (see **Figure 3.9**). This subject is expanded here to include antifungal, antiviral and antiparasite agents (**Figure 4.1**).

In addition to the range of activity of an anti-infective, it is important to appreciate the pharmacokinetics, route of excretion, penetration into various body compartments (volume of distribution) and toxicity profile (**Figure 4.2**). Renal and liver function are monitored as anti-infective agents are primarily excreted or metabolized by these

ANTIBACTERIALS

β-lactam antibiotics	**Aminoglycosides**	**Macrolides**	**Quinolones**
Stop cell wall synthesis	**Stop protein synthesis**	**Stop protein synthesis**	**Stop DNA supercoiling**
Penicillins	Gentamicin	Erythromycin	Ciprofloxacin
Penicillin	Tobramycin	Clarithromycin	Levofloxacin
Amoxicillin	Amikacin	Azithromycin	Moxifloxacin
Penicillin/β-lactamase inhibitor combinations			
Amoxicillin/clavulanic acid	**Lincosamide**	**Glycopeptides**	**Oxazolidinone**
Piperacillin/tazobactam	**Stop protein synthesis**	**Stop cell wall synthesis**	**Stop protein synthesis**
Cephalosporins	Clindamycin	Vancomycin	Linezolid
Cefalexin (1st generation)		Teicoplanin	
Cefuroxime (2nd generation)			
Ceftriaxone (3rd generation)	**Lipopeptide**	**Nitroimidazoles**	**Tetracyclines**
Carbapenems	**Inserts into plasma membrane**	**Metabolite shears DNA**	**Stop protein synthesis**
Ertapenem	Daptomycin	Metronidazole	Tetracycline
Meropenem		Tinidazole	Doxycycline
			Tigecycline (a glycylcycline)

ANTIFUNGALS

Amphotericin B/lipid formulations	**Azoles**	**Echinocandins**	**Agents against *Pneumocystis jirovecii***
Insert into plasma membrane, cell leaks	**Stop ergosterol synthesis**	**Stop 1-3 β glucan synthesis**	**Stop folate synthesis**
Amphotericin B	Fluconazole	Anidulafungin	Trimethoprim/sulphamethoxazole
Ambisome®	Itraconazole	Caspofungin	Pentamidine
	Posaconazole	Micafungin	
	Voriconazole		

Figure 4.1 A list of anti-infective agents used to treat bacterial, viral, fungal and parasite infections, with a comment on their mode of action.

(Continued)

ANTIVIRAL AGENTS

Influenza viruses	**HSV and VZV**	**CMV and EBV**
Oseltamivir **Inhibits neuraminidase**	Aciclovir **Inhibits DNA synthesis**	Gancilovir **Inhibits DNA synthesis**
Zanamivir **Inhibits neuraminidase**	Valaciclovir **Inhibits DNA synthesis**	Cidofovir **Inhibits DNA synthesis**
		Foscarnet **Inhibits DNA synthesis**

Hepatitis B virus	**Hepatitis C virus**	**HIV**
PEG-IFN **Blocks mRNA translation**	PEG-IFN **Blocks mRNA translation**	Lamivudine **Nucleoside RT inhibitor**
Lamivudine **Nucleoside RT inhibitor**	Boceprivir **Protease (NS3) inhibitor**	Tenofovir **Nucleotide RT inhibitor**
Adefovir **Nucleotide RT inhibitor**	Telaprivir **Protease (NS4A) inhibitor**	Zidovudine **Nucleoside RT inhibitor**
Entecavir **Nucleoside RT inhibitor**	Ledipasvir **RNA regulator inhibitor (NS5A)**	Efavirenz **Non-nucleoside RT inhibitor**
Telbivudine **Nucleoside RT inhibitor**	Sofosbuvir **Polymerase inhibitor (NS5B)**	Nevirapine **Non-nucleoside RT inhibitor**
Tenofovir **Nucleotide RT inhibitor**	Ribavirin **Nucleoside analogue**	Indinavir **Protease inhibitor**
		Ritonavir **Protease inhibitor**
Anti RNA virus agent		Saquinavir **Protease inhibitor**
General activity against RNA viruses		Enfurvitide **Prevents fusion of virus to cell**
Ribavirin **Nucleoside inhibitor of RNA polymerase**		Elvitegravir **Integrase inhibitor**
		Raltegravir **Integrase inhibitor**

ANTIPARASITE AGENTS

Antimalaria		**Antihelminth**	
Quinine	**Stops (toxic) haem(ozoin) crystallization**	Albendazole	**Causes loss of microtubules, especially in intestinal cells of worm**
Doxycycline	**Stops expression of apicoplast genes**	Mebendazole	**Causes loss of microtubules, especially in intestinal cells of worm**
Artesunate	**Stops glutathione transferase**	Ivermectin	**? inhibits parasite neurotransmission**
		Praziquantal	**? increases cellular permeability to calcium**

Figure 4.1 A list of anti-infective agents used to treat bacterial, viral, fungal and parasite infections, with a comment on their mode of action.

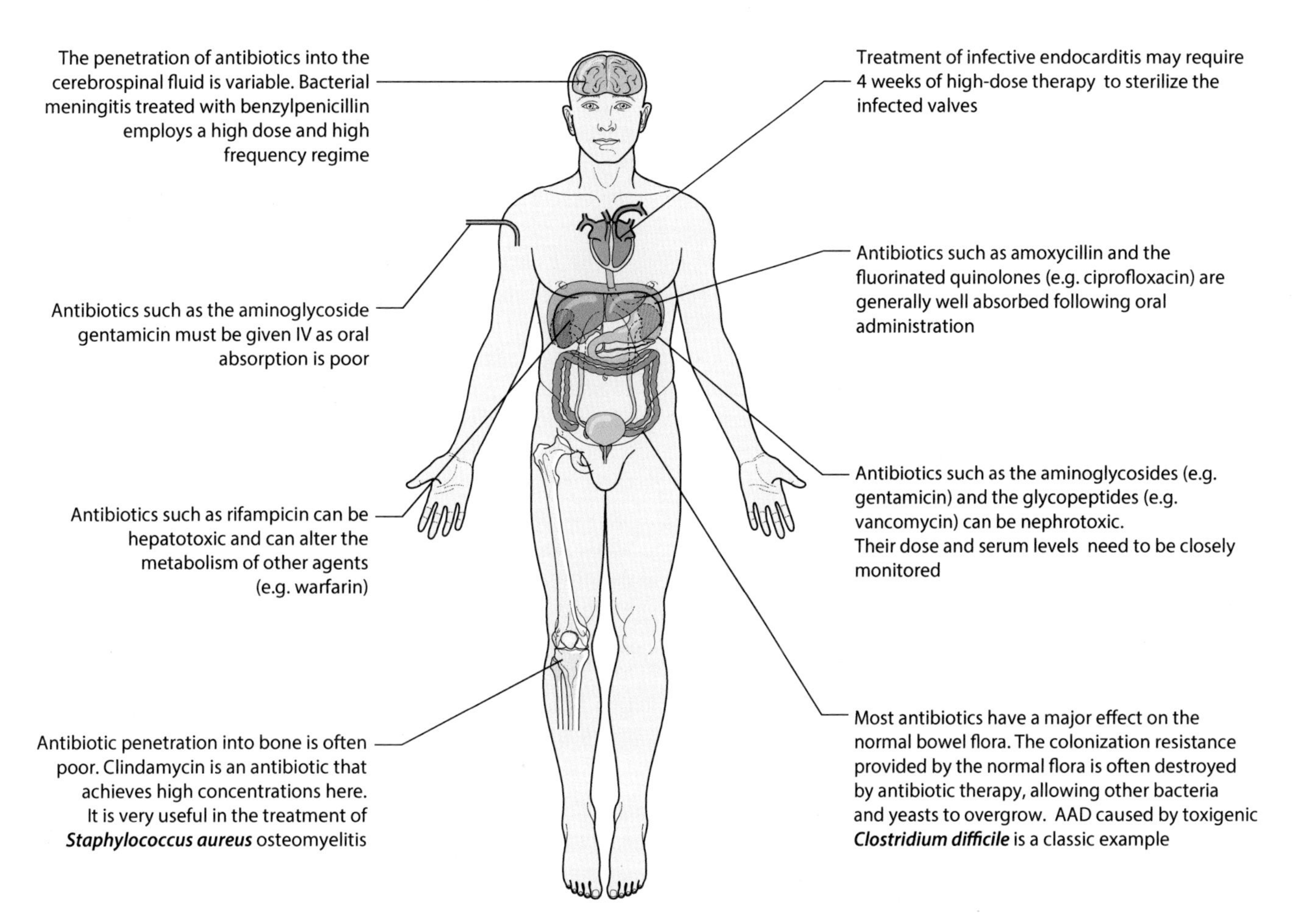

Figure 4.2 The 'body' outline shows a number of important principles to consider when prescribing antibiotics. (AAD: antibiotic-associated diarrhoea.)

routes. An estimation of renal function should be part of standard patient care, and the urea and electrolyte (U+E) results used to calculate creatinine clearance. Rifampicin, isoniazid and pyrazinamide, used in the treatment of tuberculosis, can be hepatotoxic. Liver function tests (LFTs) are monitored in patients taking these agents, and treatment is reviewed promptly when LFTs deteriorate. By inducing liver enzymes, rifampicin alters the metabolism of other agents, resulting in reduced levels of warfarin and the oral contraceptive.

Allergies to antibiotics can significantly reduce treatment options, and this is especially relevant to the β-lactam class. While safety is paramount, every effort must be made to confirm the basis of an allergy history, to ensure that the patient is not deprived of the best treatment option.

The duration of anti-infective therapy depends on the infection being treated. It is appropriate to treat meningococcal sepsis for 5 days with ceftriaxone. Prosthetic heart valve endocarditis caused by a coagulase-negative staphylococcus usually requires 6 weeks of therapy, while tuberculous meningitis is treated for 12 months. In the case of human immunodeficiency virus (HIV) infection, treatment with antiretroviral agents is lifelong, as there is no way of removing the integrated viral deoxyribonucleic acid (DNA) provirus from the host cell genome.

Whatever the infection and the duration of treatment, it is essential to bear in mind that most, if not all, anti-infective agents will have side-effects. The collateral damage that antibiotics exact on the endogenous bacterial flora of the bowel has already been highlighted (Chapter 2), and is a side-effect that often is not at the forefront of clinical practice. In addition, the unnecessary and prolonged use of antibiotics is central to the international problem of antibiotic resistance. The policy of antibiotic use must be to keep any regimen as narrow spectrum as possible and as short as possible.

ANTIBACTERIAL AGENTS

As shown in **Figure 3.9**, Chapter 3, the β-lactam antibiotics include benzylpenicillin, amoxicillin, flucloxacillin, β-lactam/β-lactamase inhibitor combinations, cephalosporins and carbapenems. The β-lactams are the antibacterial agents of choice. At the correct dose and frequency, they are rapidly bactericidal, and apart from allergy that occurs in the minority of individuals, side-effects are relatively few.

The macrolides such as erythromycin, clarithromycin and azithromycin are active against gram-positive bacteria, and other organisms that cause community-acquired pneumonia (CAP), including *Legionella pneumophila, Chlamydophila pneumoniae* and *Coxiella burnetii*. Erythromycin and azithromycin are used to treat genital chlamydial infection. Clarithromycin is used to treat *Campylobacter* gastroenteritis, and azithromycin is active against *Salmonella*, including *Salmonella* Typhi/Paratyphi.

The fluorinated quinolones include ciprofloxacin, levofloxacin and moxifloxacin. These have activity against a wide range of gram-negative bacteria and methicillin-sensitive *Staphylococcus aureus* (MSSA), with levofloxacin having additional activity against streptococci. They are used in the treatment of *Legionella pneumophila, Mycoplasma, Chlamydophila pneumoniae* and *Coxiella burnetii* infections.

Anaerobic infections are often treated with metronidazole. β-lactam/β-lactamase inhibitor combinations, such as co-amoxiclav and piperacillin/tazobactam, are active against anaerobes, as are the carbapenems.

THE β-LACTAM ANTIBIOTICS

The structure of the penicillins, cephalosporins and carbapemens, and numbering of atoms from the N residue is shown in **Figure 4.3**. All have the β-lactam ring, which is responsible for covalent binding of the antibiotic to the serine residue in the active site of the penicillin-binding protein (PBP). The penicillins have a sulphur atom at position 4 of the thiazolidine ring, cephalosporins have a sulphur at the 5 position of the dihydrothiazine ring. The carbapenems have a carbon at the 4 position of a 5-membered ring; the carbon 6 and short hydroxyethyl R1 side chain account for their stability and broad range of activity.

For comparative purposes, the structure of representative β-lactam antibiotics is shown in **Figure 4.4**. The difference between benzyl penicillin and amoxicillin is the NH_2 residue on the R1 side chain, which gives the aminopenicillin its activity against β-lactamase-negative 'coliforms' (Chapter 1, **Figure 1.10**). The R1 side chain of flucloxacillin sterically hinders the β-lactamase of MSSA. Piperacillin has a R1 polar side chain that allows it to traverse a porin of the outer membrane of *Pseudomonas aeruginosa*.

The β-lactamase inhibitors clavulanic acid and tazobactam contain the β-lactam ring. While they have little affinity for the PBP involved in cell wall synthesis, they have high affinity for a range of β-lactamase enzymes, covalently binding to the serine residue in the active site of these enzyme, inactivating them.

The R1 side chain of the cephalosporins determines their range of activity. In general the first generation agents (cefalexin) have greater gram-positive activity, while second generation agents (cefuroxime) and third generation agents (cefotaxime, ceftriaxone) have a relative increase in gram-negative activity. The side chain of cefalexin is essentially the same as that of amoxicillin, and can account for the cross-reactivity between these two antibiotics in the allergic patient, as discussed in more detail later in this chapter. The R2 side chain of

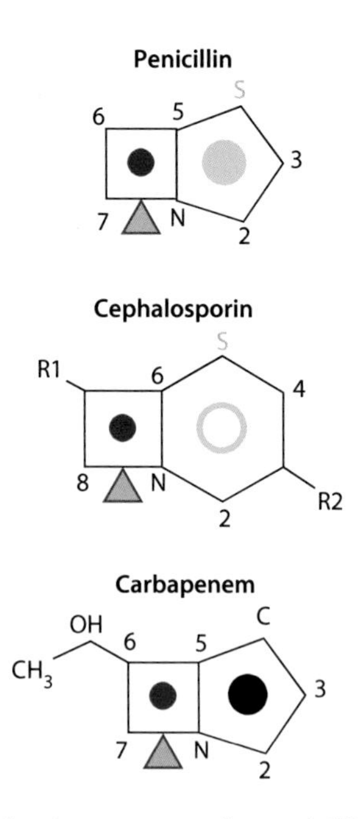

Figure 4.3 The basic structure of a penicillin, cephalosporin and carbapenem, with the atoms numbered anticlockwise from the N-atom. The β-lactam (red), thiazolidine (●) dihydrothiazine (○) and carbapenem (●) rings are marked. The arrow shows the CO–N bond involved in the interaction with penicillin-binding protein (PBP). This bond is cleaved by β-lactamase enzymes.

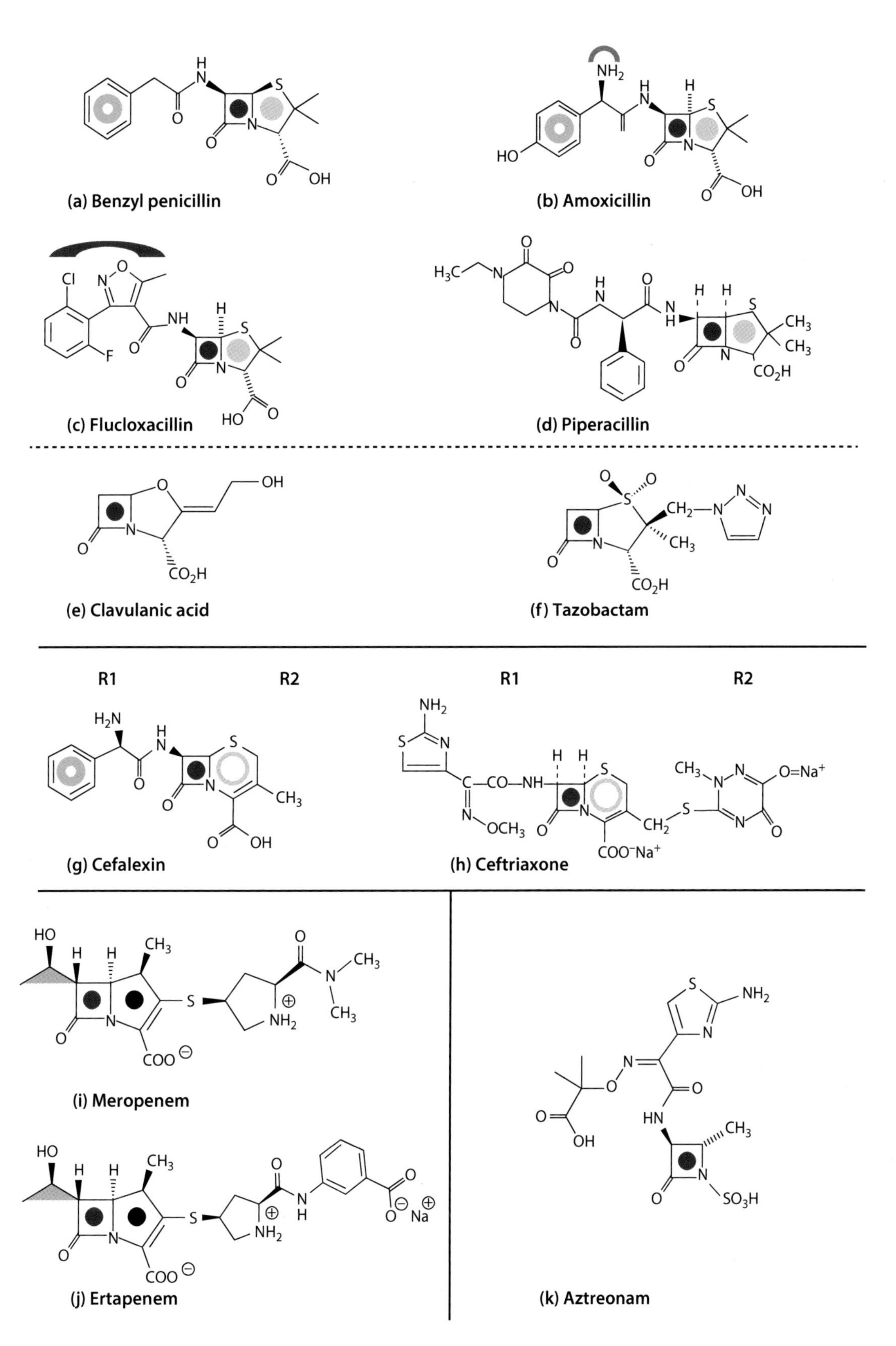

Figure 4.4 The structure of representative β-lactams, with the rings identified as in **Figure 4.3**. Key features identified are discussed in the text. The side chain (bracket) of flucloxacillin sterically inhibits the β-lactamase of *Staphylococcus aureus*.

the cephalosporins determines their pharmacokinetics. Ceftriaxone has a half-life of 8 hours, which is considerably longer than the 1–2 hours of most other β-lactams. It can be given as a regime of 2 g q24h, which makes it particularly useful for intravenous outpatient parenteral antibiotic therapy (OPAT).

Two carbapenems shown are meropenem and ertapenem; doripenem and imipenem/cilastin are other members of this group. (Cilastin is included with imipenem to prevent inactivation of the antibiotic by a kidney dipeptidase.) The carbapenems have the broadest range of activity of the β-lactam antibiotics, as they readily cross the outer membrane of many gram-negative bacteria, have high affinity for a wide range of PBP and are resistant to β-lactamase enzymes, with the exception of the carbapenemases. Ertapenem does not have activity against *Pseudomonas aeruginosa*, due to a porin protein restriction. Meropenem is prescribed q8h, ertapenem has a long half-life, and is prescribed q24h, making it useful for OPAT. There are no oral carbapenems available.

Aztreonam is a monobactam, with a β-lactam ring, and is active against gram-negative bacteria, including *Pseudomonas aeruginosa*. Allergic cross-reactivity with penicillins and cephalosporins is recognized as rare, so aztreonam is a useful option to consider in the patient with an allergy to antibiotics in these groups, especially in the intensive care unit (ICU) setting.

ANTIBIOTIC RESISTANCE IN BACTERIA

An outline of mechanisms of resistance of gram-positive and gram-negative bacteria to antibiotics is summarized in **Figure 4.5**:

a. Because of their size and/or charge, a number of antibiotics that are active against gram-positive bacteria are unable to cross the outer lipopolysaccharide outer membrane (OM) of gram-negative bacteria. These include the glycopeptides vancomycin and teicoplanin, the lipopeptide daptomycin and the oxazolidinone linezolid. Gram-negative bacteria are thus 'structurally' resistant to these agents.

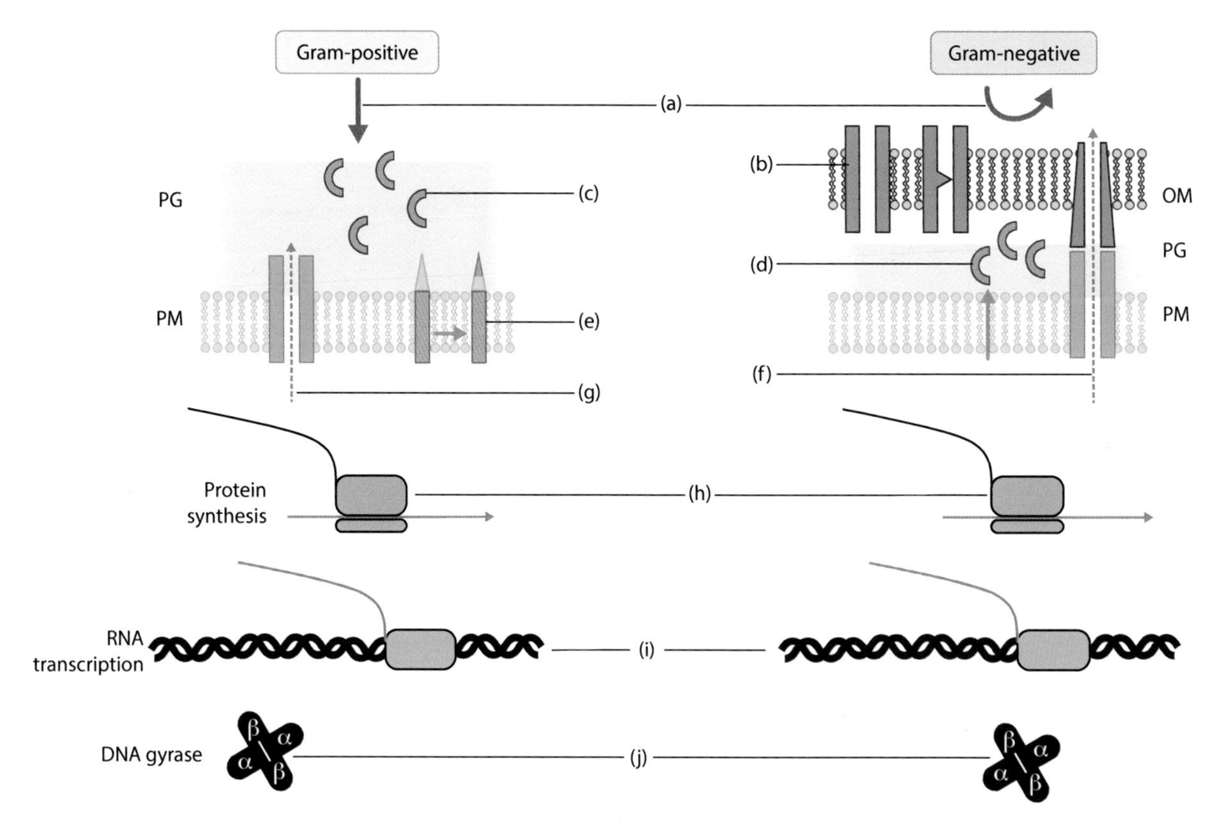

Figure 4.5 Examples of inherent antibiotic resistance (**a**) and acquired resistance (**b–j**), as discussed in the text. (OM; outer membrane, PG: peptidoglycan layer; PM: plasma membrane.)

Acquired resistance can arise by a number of mechanisms:

b. *Pseudomonas aeruginosa* has a porin D2 in the OM, allowing passage of imipenem across the membrane. A single point mutation in the D2 gene produces an amino acid substitution that blocks the passage of imipenem.
c. The β-lactamase of *Staphylococcus aureus* diffuses into the peptidoglycan (PG) cell wall layer, and can degrade penicillin and amoxicillin. This enzyme is inhibited by clavulanic acid, but due to steric hindrance cannot inactivate flucloxacillin.
d. The β-lactamase synthesized by gram-negative bacteria is retained in the periplasmic space of the cell wall.
e. Penicillin resistance in *Streptococcus pneumoniae* is due to synthesis of PBPs that have decreased affinity for the antibiotic. In the case of methicillin-resistant *Staphylococcus aureus* (MRSA), the mecA gene codes for a different PBP that has no affinity for flucloxacillin (or methicillin).
f. Efflux pumps are driven by a H^+ proton motive force and actively transport molecules across the cytoplasmic membrane and out of the cell. Antibiotics such as the quinolones and tetracyclines are removed from cells in this way. In *Pseudomonas aeruginosa* an efflux pump is linked to a porin, enabling complete removal of the agent from the cell.
g. Efflux pumps are also found in gram-positive bacteria.
h. Protein synthesis is a target of many antibiotics, including tetracyclines, macrolides and aminoglycosides such as gentamicin. Point mutations give rise to amino acid changes in the specific ribosomal proteins that prevent binding of the antibiotic. (Macrolides are also inactivated by methylase enzymes. With aminoglycosides the most common form of resistance is due to acyltransferase enzymes that inactivate this class of antibiotic.)
i. Rifampicin binds to the bacterial ribonucleic acid (RNA) polymerase. A point mutation produces a protein to which the antibiotic cannot bind. This is also the mechanism of rifampicin resistance in *Mycobacterium tuberculosis*.
j. The fluorinated quinolones bind to the α-subunit of the DNA gyrase enzyme responsible for supercoiling of the bacterial DNA within the cell. A point mutation gives rise to a protein that prevents binding of the antibiotic to the enzyme.

The mechanism of resistance of enterococci to glycopeptides (e.g. vancomycin-resistant enterococci; VRE), is due to the ability to add a terminal lactose residue at the end of the peptide side chain, replacing the terminal alanine. This prevents the glycopeptide binding to the 'building block', and PG synthesis can continue (**Figure 4.6**). Glycopeptide tolerance/resistance is an increasing problem in coagulase-positive and coagulase-negative staphylococci.

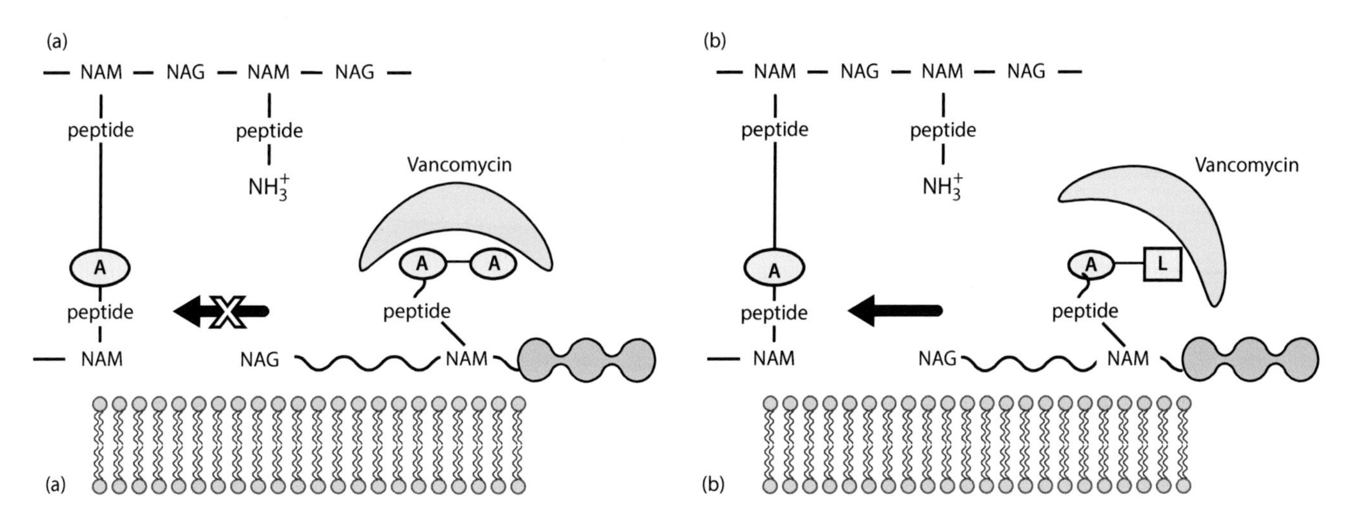

Figure 4.6 The mechanism of vancomycin resistance in enterococci. (**a**) With sensitive enterococci, the glycopeptide embraces the incoming precursor, preventing its incorporation. (**b**) The lactose (L) prevents the embrace, and peptidoglycan synthesis continues.

β-LACTAMASE RESISTANCE

There are a number of detailed classification systems for β-lactamase enzymes; however, for the purposes of this book these are broadly grouped and simplified as follows:

The penicillinase β-lactamases:

- Inactivate penicillin, amoxicillin and piperacillin. Examples include TEM1 of *Escherichia coli* and SHV1 of *Klebsiella pneumoniae*.
- Are inhibited by the β-lactamase inhibitors clavulanic acid and tazobactam.
- Bacteria that synthesize very high levels of enzyme can compromise the effect of inhibitors, and also inactivate first generation cephalosporins such as cefalexin.

The extended-spectrum β-lactamases:

- Inactivate penicillin, amoxicillin, piperacillin and the majority of cephalosporins.
- Are inhibited by the β-lactamase inhibitors *in vitro*.

The carbapenemases: these inactivate the carbapenems and are divided into two groups:

- Those that have a serine residue in the active site, the serine carbapenemases.
- Those that have one (or two) zinc atoms at the active site of the enzyme, and are termed the metallo-β-lactamases (MBL).

HOW β-LACTAMASES WORK

The primary protein structure of a TEM1 β-lactamase is shown in **Figure 4.7**. The six numbered amino acids make up the active site of the enzyme, as shown in **Figure 4.8**, with the key catalytic activity being provided by the serine residue (2), while glutamine (4) directs the water molecule needed to hydrolyze the β-lactam ring. This 1–6 numbering is used 'clockwise' in **Figure 4.8**; note how these widely, randomly spaced amino acids come together to form the active site in the tertiary folded structure of the enzyme. In the example of a TEM8 extended-spectrum β-lactamase (ESBL), the point mutations change four amino acids that are responsible for enlarging the active site of the ESBL enzyme, so that it is able to accommodate the R1 side chain of the second and third generation cephalosporins.

Using a simplified model enzyme, the process of inactivation of a β-lactam such as amoxicillin by a (TEM1) penicillinase is shown in **Figure 4.8a** (i–iii). The antibiotic enters the active site where the amino acids work in concert to stabilize the antibiotic over the serine (2), while glutamate (4) drives in a water molecule with cleavage of the CO–N bond.

As shown in **Figure 4.8b** (i), the active site of this β–lactamase cannot accommodate the long R1 side chain of the second and third generation cephalosporins, and the β-lactam ring does not align over the serine. An ESBL enzyme such as TEM8 provides an enlarged active site that accommodates the cephalosporin (**Figure 4.8b** [ii]). The hydrolysis reaction can take place (**Figure 8b** [ii–iii]).

The carbapemens have a very stable structure provided by the critical strength of carbon 6 and the short hydroxyacyl side chain at R1. This cannot be

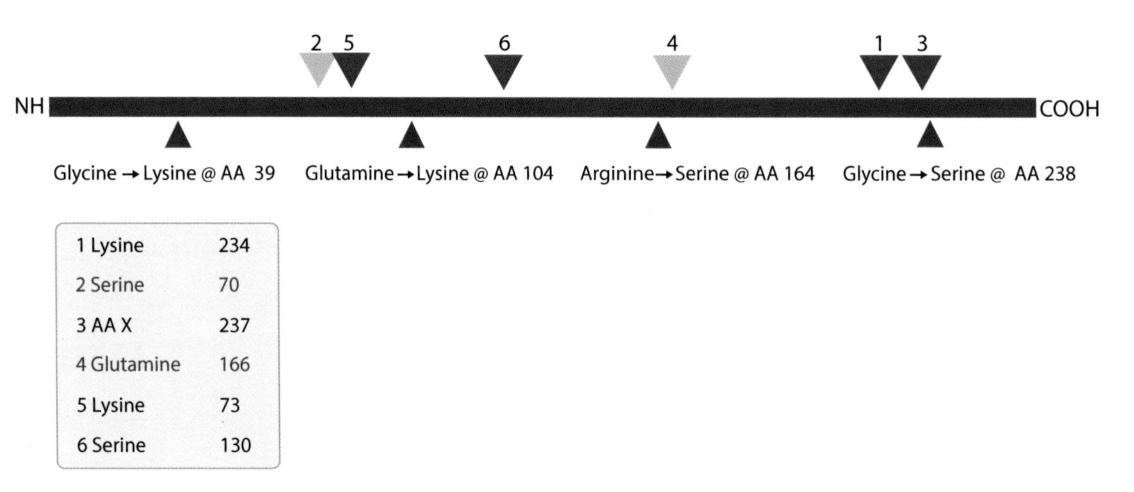

Figure 4.7 The position and names of the amino acids (1–6) that make up the active site of a TEM1 β-lactamase (see **Figure 4.8**). The four mutations that give rise to the altered active site of the TEM8 β-lactamase are marked.

'pulled' by enzymes such as a TEM1 (**Figure 4.8c** [i]). However, the serine carbapenemases have amino acid changes that create the necessary 'pull', allowing the hydrolysis reaction to occur. A disulphide bond between two cysteine residues, 69 and 238, is key to this action (**Figure 4.8c** [ii–iii]).

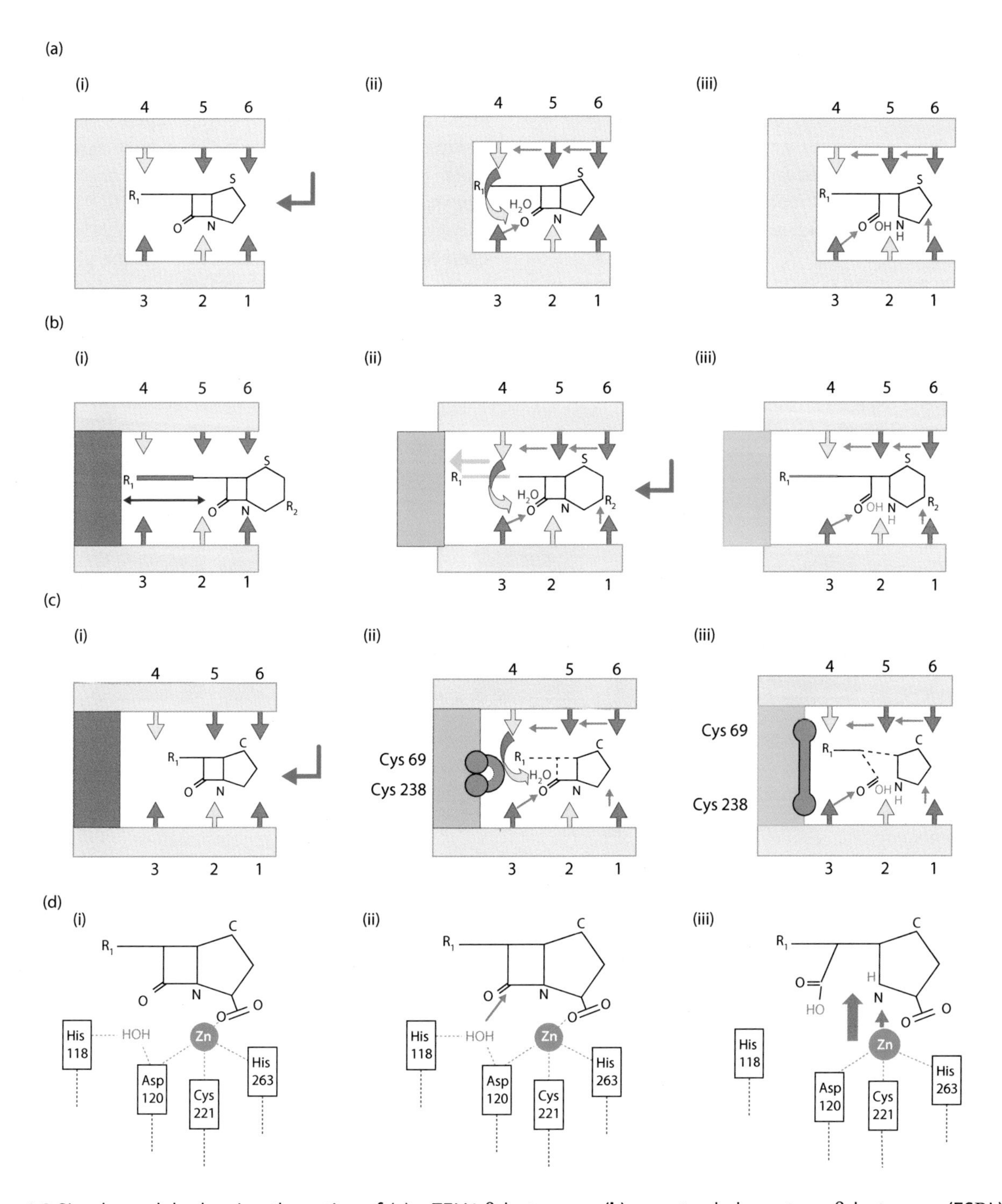

Figure 4.8 Simple models showing the action of (**a**) a TEM1 β-lactamase, (**b**) an extended-spectrum β-lactamase (ESBL), (**c**) a serine carbapenemase, (**d**) a metallo β-lactamase, as described in the text.

The MBL have one or two zinc atoms as the active moiety and not the amino acid serine. With four adjacent amino acids at the active site stabilizing the antibiotic/enzyme complex, hydrolytic cleavage of the CO–N bond occurs (**Figure 4.8d** [i–iii]).

THE ANTIFUNGALS

There are three important groups of antifungal agents, amphotericin B, the azoles and echinocandins. The general range of activity of representative agents against the more common fungi identified in clinical practice is shown in **Figure 4.9**.

There are two main targets of these agents, ergosterol, which is an essential steroid of the plasma membrane of fungi (with the exception of *Pneumocystis jirovecii*), and 1-3 β-D glucan, the building block of the cell wall glucan, which along with chitin makes up this protective structure. The site of action of these antifungals is shown in **Figure 4.10**.

The azoles, such as fluconazole, itraconazole, posaconazole and voriconazole, inhibit the 4-α demethylase enzyme involved in the synthesis of ergosterol. The echinocandins include anidulafungin, caspofungin and micafungin. These semisynthetic agents have a cyclic hexapeptide covalently linked to a fatty acid chain and block 1-3 β-D glucan synthesis by the transmembrane glucan synthetase complex.

Amphotericin is a polyene and is able to orientate itself between ergosterol molecules in the plasma membrane, creating a channel so that cell contents leak out. As amphotericin has toxic renal side-effects, it is usually combined with a lipid such as cholesterol sulphate to

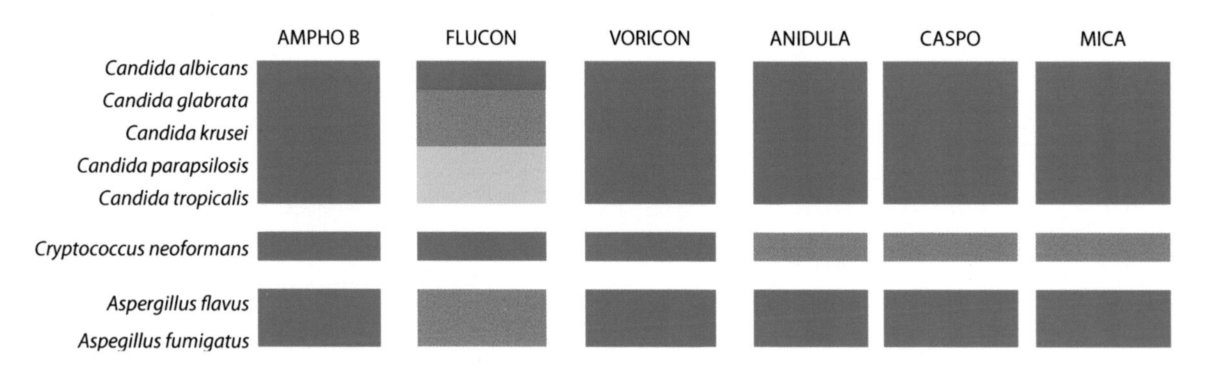

Figure 4.9 The general *in vitro* susceptibility of selected fungi to amphotericin B, the azoles fluconazole and voriconazole and the echinocandins, anidulafungin, caspofungin, micafungin. (Green: usually susceptible, orange: confirm by susceptibility testing, red: consider resistant.)

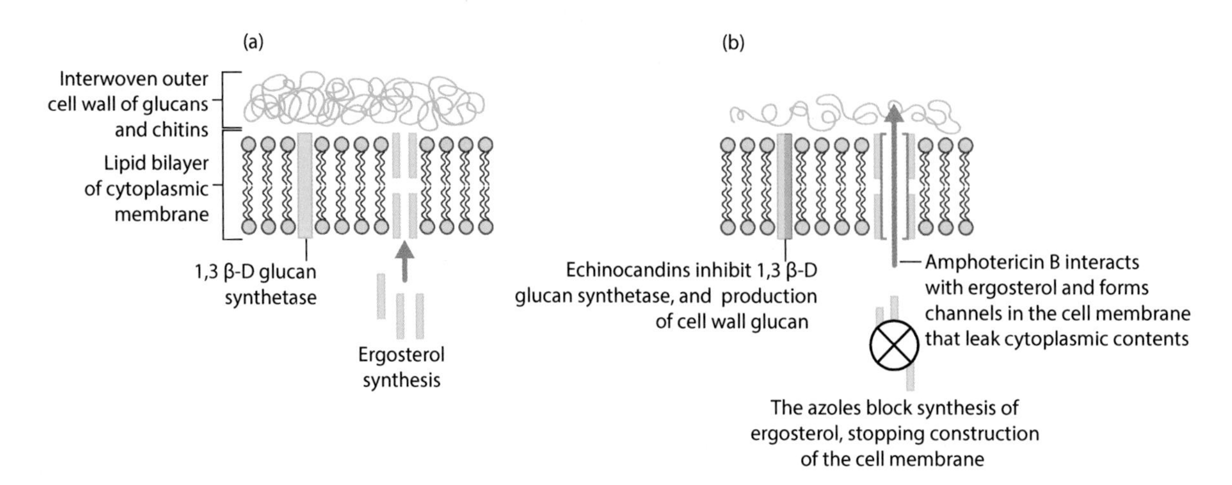

Figure 4.10 (**a**) The structure of the membrane and cell wall of fungi; (**b**) the site of action of important antifungal agents.

reduce this toxicity. A liposomal preparation of the drug is Ambisome®.

RESISTANCE TO ANTIFUNGALS

Resistance to amphotericin is due to synthesis of an alternative steroid to ergosterol, bypassing the site of action of the drug. Resistance in azoles such as fluconazole is due to drug efflux mechanisms, or a point mutation that gives rise to a 4-α demethylase enzyme with decreased drug binding. Increase of enzyme production levels is another mechanism, so that ergosterol levels exceed the inhibitory concentration that the azole can achieve.

THE ANTIVIRALS

INFLUENZA VIRUS

The neuraminidase envelope protein (N) cleaves terminal sialic acid residues from host plasma membrane glycoproteins, enabling free virus to leave the cell at the end of the replication cycle. Oseltamivir and zanamivir are neuraminidase inhibitors that bind to and block the action of neuraminidase. They reduce the release of free virus from the infected cell, and are in effect 'last chance' antiviral agents.

THE HERPESVIRUSES

Aciclovir is an analogue of guanosine, with activity against herpes simplex virus (HSV) 1 and 2 and varicella zoster virus (VZV). This is dependent on conversion of aciclovir to a monophosphate by the viral thymidine kinase, and subsequent phosphorylation by host enzymes to a triphosphate. Incorporation of this molecule into the replicating viral DNA by the viral DNA polymerase blocks chain elongation (**Figure 4.11**). Ganciclovir is used to treat cytomegalovirus (CMV) and Epstein–Barr virus (EBV) infection.

Foscarnet is a pyrophosphate that acts as an analogue, blocking the pyrophosphate site of the DNA polymerase of CMV and EBV. Cidofovir is an analogue of deoxycytidine monophosphate, which is converted to the diphosphate form by a cellular enzyme. In this form it competes with the cellular triphosphate deoxycytidine for the active site of the DNA polymerase of HSV, CMV and EBV. When two consecutive cidofovir molecules are incorporated, viral

(a)

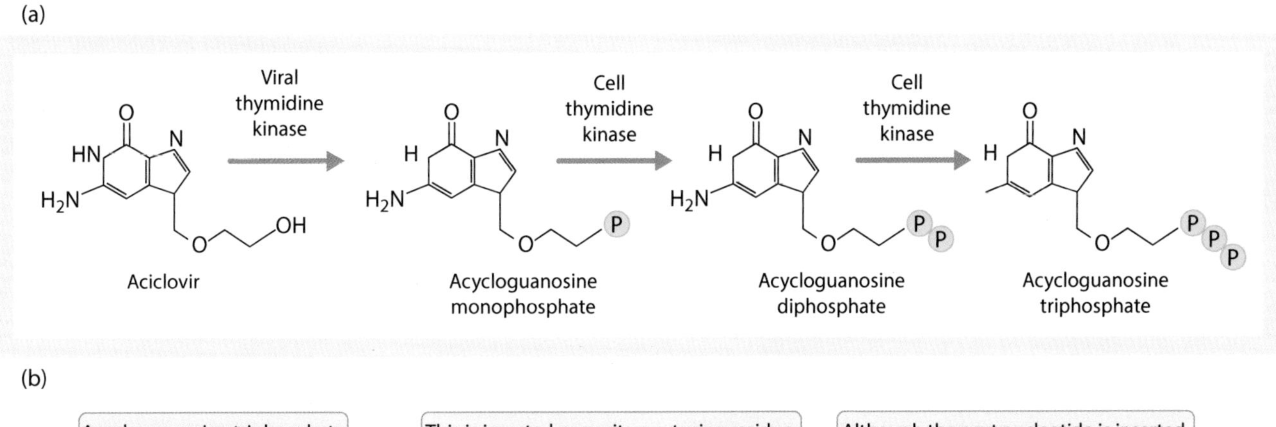

(b)

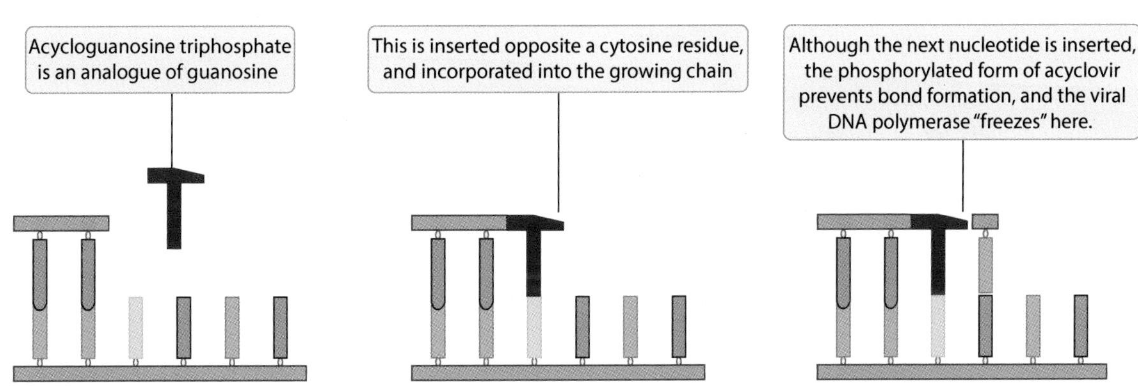

Figure 4.11 (**a**) The activation of aciclovir requires it to be phosphorylated by the viral thymidine kinase, and converted to a triphosphate by cellular enzymes. (**b**) Incorporation of the guanosine analogue terminates viral deoxyribonucleic acid synthesis.

DNA synthesis is abrogated. Cidofovir also has activity against adenovirus.

HEPATITIS B VIRUS AND HEPATITIS C VIRUS

Agents included in the treatment of chronic HBV and HCV infection are the non-specific antiviral agent interferon (IFN), as well as specific agents that target enzymatic functions of these viruses.

INTERFERON, RIBAVIRIN AND SPECIFIC ANTIVIRAL AGENTS

IFNs are a group of cytokines that activate antiviral defences within a cell, but are also released by cells to activate these defences in neighbouring cells. They play a critical role in the ability of the body to combat virus infections. The trigger that activates the IFN pathway is double-stranded RNA (2-RNA) within the cytoplasm of the infected cell. Apart from the 22 nucleotide microRNA, 2-RNA is not a feature of the eukaryotic cell, but RNA viruses that replicate in the cytoplasm must use a 2-RNA intermediate as an essential component of semi-conservative replication (Chapter 1, **Figure 1.30**). DNA viruses transcribe RNA from overlapping sequences of both strands of their genomes, and so 2-RNA structures accumulate in the cytoplasm of these infected cells.

2-RNA activates a pathway that results in the synthesis of two host enzymes. The first is an RNA degrading enzyme, RNaseL, which indiscriminately cuts up both host cell and viral mRNA. The other enzyme induced is a protein kinase that phosphorylates host translation initiation factor eIF2α, and this phosphorylation step prevents translation of mRNA by the ribosomes. The individual infected host cell will in effect switch off protein synthesis, while attempting to abort virus replication, or it undergoes apoptosis. The infected cell, via cytokines, alerts neighbouring cells to activate their IFN pathways, so that they are in a state of preparedness, should the virus attempt invasion.

Treatment of both HBV and HCV uses IFN covalently attached to polyethylene glycol (PEG-IFN), which prolongs the half-life of IFN, so that it is given once a week in treatment. Because it is a cytokine, IFN treatment is associated with significant side-effects.

The replication cycle of HBV has been discussed in Chapter 1 (**Figure 1.28**). Antiviral agents target the reverse transcription (RT) process, and a number of nucleoside and nucleotide RT inhibitors are listed in **Figure 4.1**. Unlike HIV, the RT process is the specific target for current antivirals used to treat HBV infection.

Options for HCV treatment include use of PEG-IFN, ribavirin, as well as virus-specific agents. Phosphorylated ribavirin is considered to be active against the RNA polymerase of a wide range of RNA viruses, and induces 'error catastrophic reading' by the enzyme. It also decreases cellular guanosine triphosphate (GTP) pools, which has a detrimental effect on progression of RNA virus replication. Virus-specific agents target the NS3 and NS4A proteases, the regulator of RNA synthesis, NS5A and the RNA polymerase (NS5B), as shown in the genome organization of HCV (**Figure 4.12**). Further information on the treatment of chronic HBV and HCV infections is given in Chapter 12.

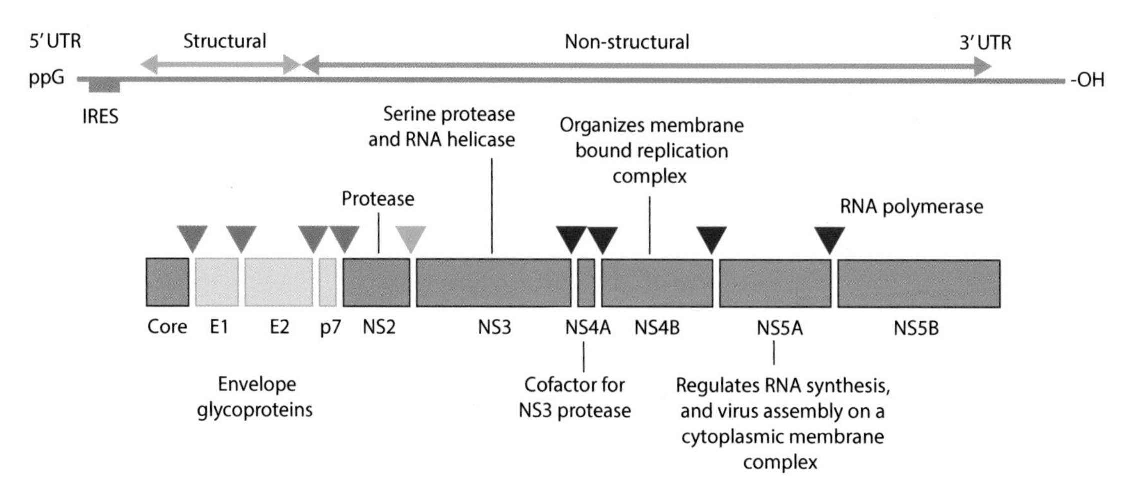

Figure 4.12 The genome organization of hepatitis C virus (HCV). The overall arrangement is similar to HAV, but there are key differences in the 5′ and 3′ ends. (UTR: untranslated region; IRES: internal ribosome entry site. Cleavage by host signalase [blue], NS2 [green], NS3/NS4A [red].)

HIV

The structure of HIV is shown in **Figure 4.13**. The virus has two copies of the 1-RNA genome, but only one is needed for replication. Although retroviruses have a 'positive sense' 1-RNA genome, this cannot be expressed in the same way as viruses such as HAV and HCV, as retrovirus replication must go through an integrated DNA proviral form. The virus has an absolute dependency on the RNA transcription and splicing machinery of the host cell nucleus to initiate and continue its replication.

An outline of HIV replication is shown in **Figure 4.14**. Virus binds to CR5 receptors of susceptible cells, via gp120/41 (1), the nucleocapsid is internalized (2) and the genome is released (3). RT occurs via the RT/RNaseH complex, and the RNA genome is converted to a 2-DNA form through the RNA/DNA intermediate (4). The 2-DNA structure is then transported through the nuclear pore, and viral integrase is then responsible for integration of the 2-DNA to form the provirus (5, 6).

Transcription of the whole genome is done by the host RNA polymerase apparatus. Initially the viral RNA is highly spliced by the post-transcription splicing machinery of the nucleus, and amongst others, the mRNA for two 'early' proteins are produced, translated in the cytoplasm (7) and return to the nucleus (8). Tat binds to the host RNA polymerase complex, increasing the rate of transcription of the provirus by 200-fold or more, while Rev binds to specific Rev binding segments on the viral RNA, and actively moves viral RNA out to the cytoplasm, reducing their 'splicing time' in the nucleus (9). Rev supplies a shuttle service, ensuring that full length viral RNA (the genome for new virus), mRNA for the capsid (gag)/polymerase (pol) proteins, as well as the subgenomic mRNA for the envelope (env) proteins reach the cytoplasm (10).

Whole genomes accumulate in the cytoplasm (11), gag/pol proteins are synthesized in the cytoplasm (12), while env proteins are synthesized on the endoplasmic reticulum (13). Immature nucleocapsids form (14) and bud through the plasma membrane to the exterior obtaining the envelope with the gp120/41 (15/16). The virus at this stage is not infectious. In addition to containing RT and integrase enzymes needed to infect the next cell, viral proteases are also included to complete cleavage of the viral structural and non-structural proteins, and mature and infective virus is formed outside the cell (17).

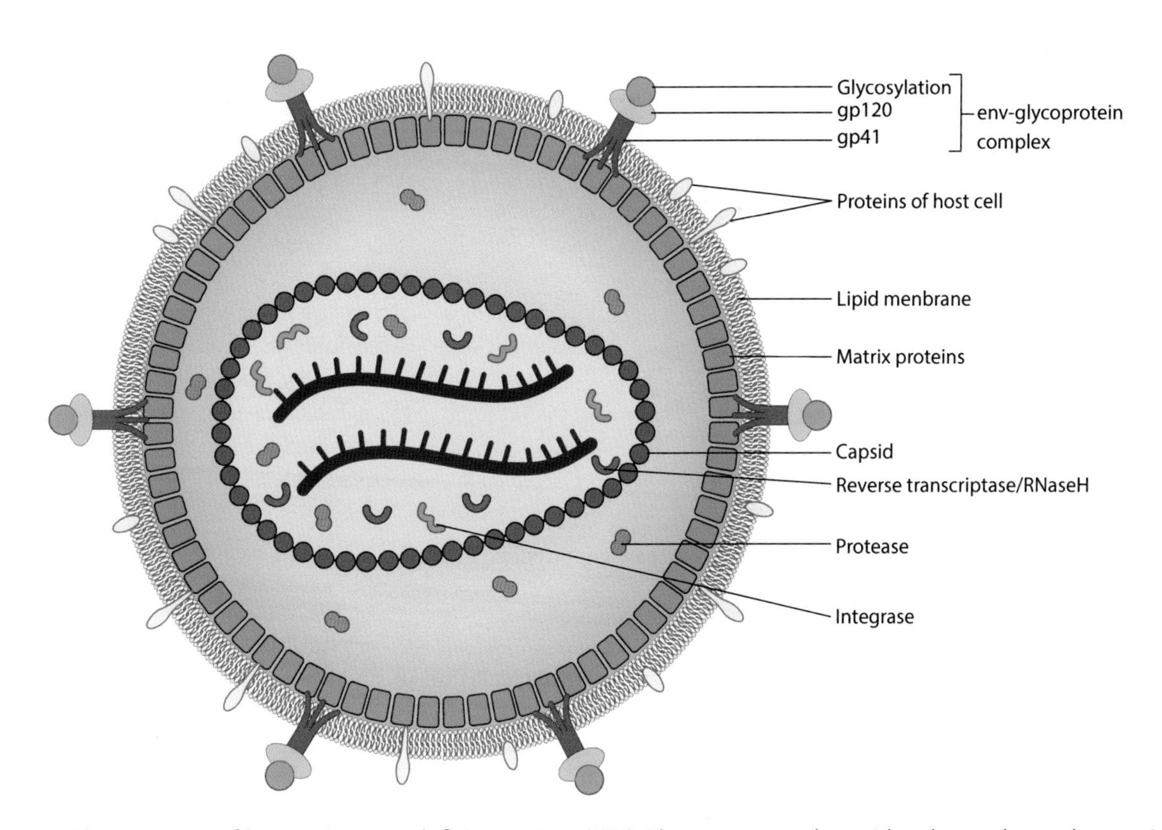

Figure 4.13 The structure of human immunodeficiency virus (HIV). The genome and capsid make up the nucleocapsid. Note the presence of enzymes that are essential for producing infective virus (protease), replication (RT/RNaseH complex) and integration of the provirus (integrase).

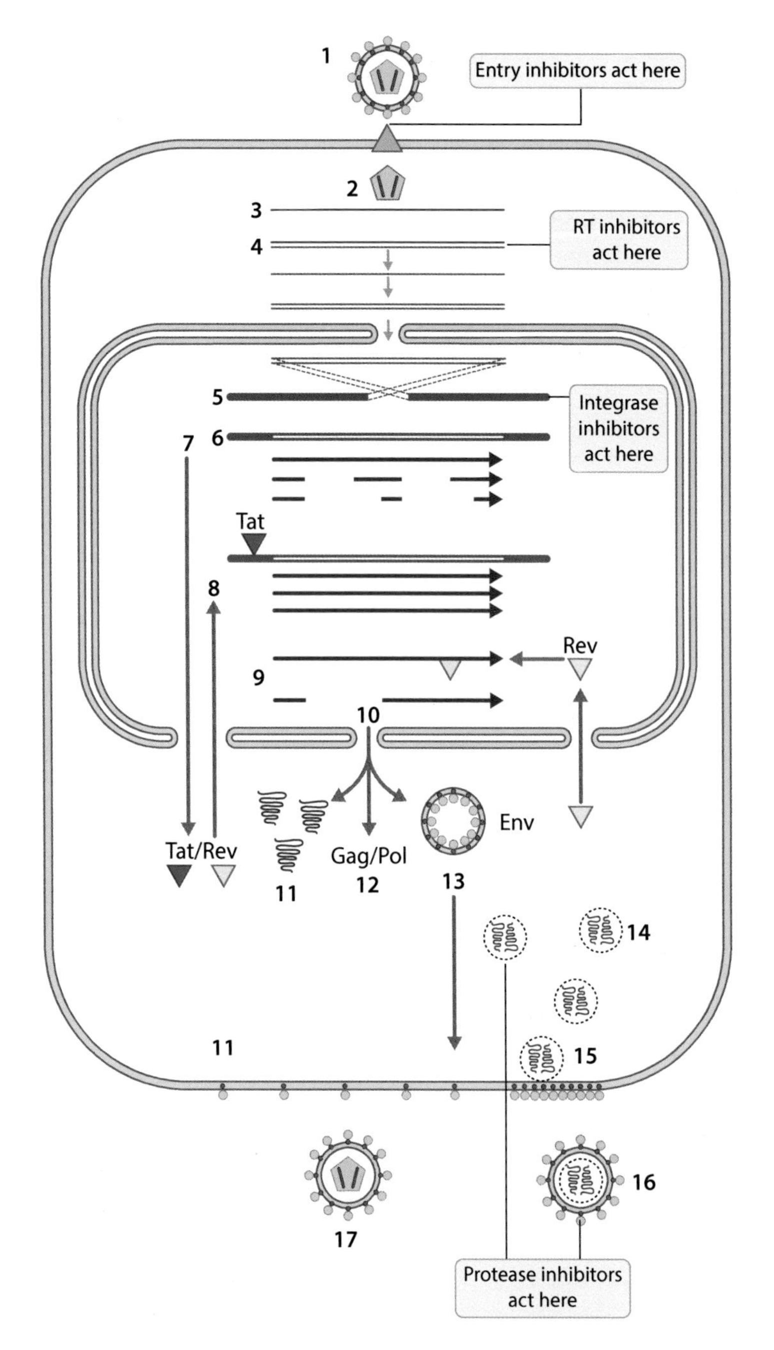

Figure 4.14 The replication cycle of HIV, with specific sites that antiviral agents target. Numbers refer to those used in the text.

Agents that target key stages of the HIV replication cycle include nucleoside and nucleotide RT inhibitors, such as lamivudine and tenofovir, which are competitive inhibitors at the active site of the RT enzyme. Non-nucleoside inhibitors such as nevirapine bind outside the active site of the enzyme, and sterically alter the active site so that it can no longer incorporate nucleotides. Agents that inhibit the attachment of the virus to the cell, integration of the provirus, as well as the protease inhibitors are listed in **Figure 4.1**.

RESISTANCE TO ANTIVIRAL AGENTS

There are various mechanisms whereby viruses develop resistance to antiviral agents, and examples are given here.

Resistance frequently arises by point mutations giving rise to amino acid changes in enzymes that decrease drug binding.

Resistance in herpesviruses to aciclovir can arise by mutations that reduce or stop synthesis of the viral thymidine kinase (TK) protein, and thus conversion of the aciclovir nucleoside to the monophosphate form does not occur.

HIV resistance to zidovudine (AZT) can arise in mutant virus that has depleted levels of functional RNaseH. This slows down the process of RT, enabling the RT polymerase to identify the abnormal drug insert and edit it out. DNA synthesis then continues until the next drug insert is found and evicted.

ANTIMALARIAL AGENTS

ARTESUNATE

Artesunate, a derivative of artemisinin, is the drug of choice to treat complicated malaria, and it is active against all *Plasmodium* species. It acts by inhibiting glutathione S-transferase, an essential export enzyme in the parasite's plasma membrane.

4-AMINOQUINOLINE

Quinine and chloroquine inhibit the ability of the parasite to convert toxic haem to crystalline and non-toxic haemzoin. Accumulation of haem kills the parasite within the red blood cell (RBC). Resistance to these agents is due to a drug transporter that removes the agent from the cell.

8-AMINOQUINOLINES

Primaquine is an example. It disrupts the mitochondrial electron transport chain of the parasite, and has wide activity, and is effective against liver hypnozoite and schizont forms.

TETRACYCLINES

Doxycycline is used as a second agent with quinine to treat *Plasmodium falciparum* infection, and is active against the blood schizonts. It targets RNA translation in the apicoplast, a plant chloroplast-like organelle that resides in the cytoplasm. Non-functional apicoplasts accumulate in daughter merozoites, resulting in termination of parasite reproduction.

THE SCIENCE OF ANTIBIOTIC USE

This section centres on antibacterial agents, but the same principles generally apply to all anti-infective drugs.

MINIMUM INHIBITORY CONCENTRATION

The minimum inhibitory concentration (MIC) is the fundamental test in determining the sensitivity of a bacterium to an antibiotic. Although essentially replaced by the strip Etest (Chapter 5, **Figures 5.29, 5.30**), the 'tube MIC' explains the principle (**Figure 4.15**). A dilute suspension of the test organism is inoculated into tubes containing a growth fluid, each with a different concentration of the antibiotic being tested. A concentration range from 32 mg to 0.25 mg/L is used in this example, which is broadly within the range where antibiotics are generally effective in the blood, fluids and tissues, and where their toxic effects are usually minimal. After overnight incubation, the MIC is determined by examining for growth of bacteria in each tube, indicated by turbidity or 'cloudiness'. The lowest concentration of the antibiotic that inhibits growth is the MIC; in the example in **Figure 4.15**, the MIC is 2 mg/L.

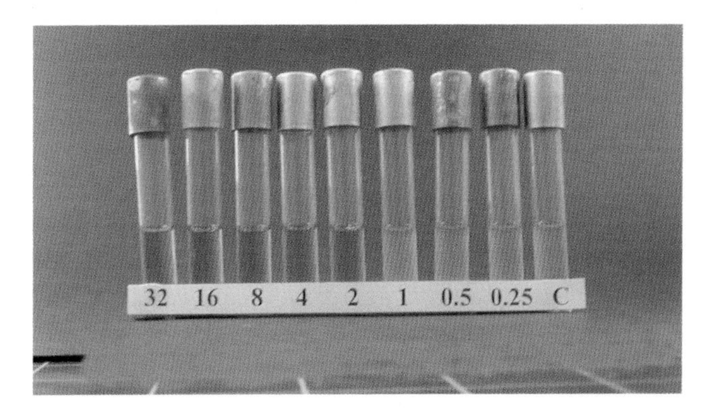

Figure 4.15 Determination of the minimum inhibitory concentration (MIC) by the tube broth dilution test. The MIC is 2.0 mg/L. C: control tube, no antibiotics.

MINIMUM BACTERICIDAL CONCENTRATION

Although seldom done, the minimum bactericidal concentration (MBC) is a further test that determines the *in vitro* killing effect of an antibiotic, as inhibition does not mean death of the bacteria. In this example, a loop-full of medium from each of the tubes from the MIC test that show no turbidity is inoculated onto agar plates that are incubated 'overnight'. Any growth on the agar plates shows that organisms from the initial inoculum have been inhibited but not killed by the antibiotic; in the example in **Figure 4.16**, the MBC is 8.0 mg/L. If bacteria survive in the tubes that are more than two dilutions above the MIC, i.e. 16.0 mg/L, the bacteria are considered 'tolerant' to the antibiotic.

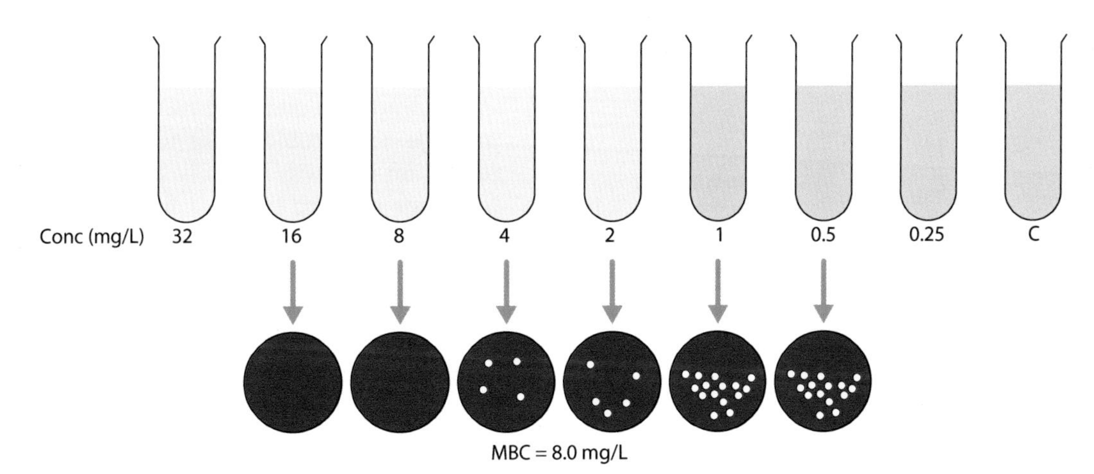

Figure 4.16 The minimum bactericidal concentration (MBC) is determined by plating out a loop-full of broth from the tubes above the MIC. In the example shown here, the MBC is 8 mg/L.

This is an important concept in treating bacterial endocarditis caused by streptococci and enterococci, which can be tolerant to penicillin or amoxycillin alone. In order to be killed, they are treated with a synergistic combination of agents, for example amoxycillin and gentamicin. The rationale is that the β-lactam agent damages the cell wall sufficiently to allow gentamicin, at the therapeutic concentration used, to reach and cross the cytoplasm membrane, which it could not do if the organism's cell wall was intact. (The streptococci and enterococci are not oxidative organisms and do not have the mechanisms necessary to pump aminoglycosides across the cell membrane.)

ANTIBIOTIC KINETICS

In order to help the natural defences of the body control and defeat a bacterial infection, an antibiotic must be present at an effective concentration and for sufficient time. The concentration should usually exceed the MIC by some factor. However, the concentration of the antibiotic must not be so high that it is toxic. The kinetics of a single dose of an antibiotic is shown in **Figure 4.17**. Most antibiotics have a half-life of 1–2 hours and, following distribution into the various body compartments, are eliminated, usually by the kidneys or liver. As shown in **Figure 4.17**, when the MIC value of an antibiotic for a particular organism is superimposed on the graph, the antibiotic concentration falls below the MIC value after a certain period, and in most circumstances the next dose must be given by that time.

There are three principles that determine the ability of an antibiotic to be effective, as shown in **Figure 4.18**. These are:

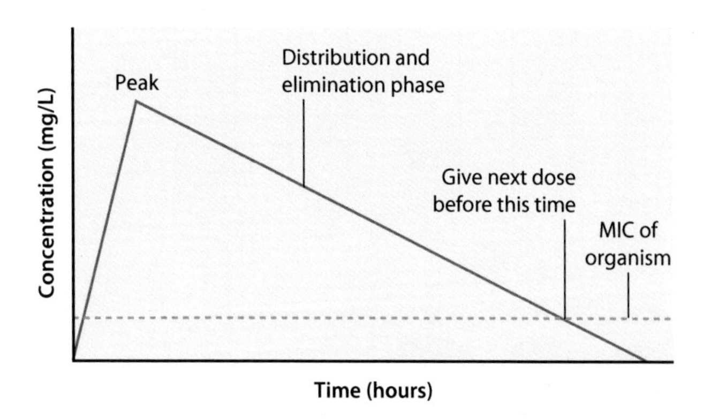

Figure 4.17 The kinetics of a single dose of an antibiotic. Superimposing the MIC of the antibiotic for a particular organism shows that the timing of doses is essential to maintain antibiotic levels that will inhibit, or preferably kill, the organism.

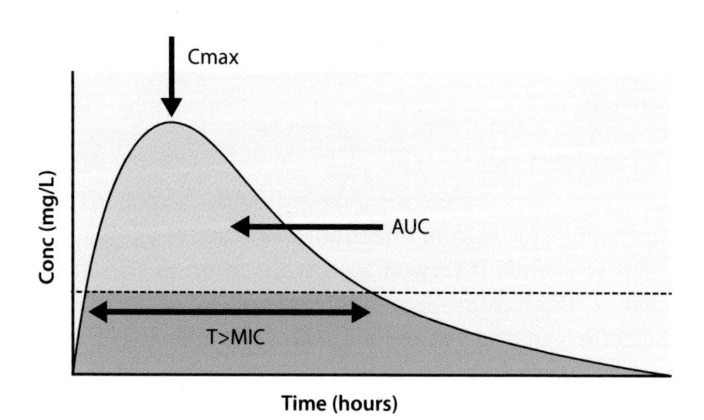

Figure 4.18 The kinetics diagram with the MIC superimposed. The maximum concentration (Cmax), area under the curve (AUC) and time above MIC are relevant to the action of various antibiotic groups.

The maximum concentration (Cmax):

This applies to the aminoglycosides when used against gram-negative organisms that are in the aerobic state of metabolism. Under these circumstances, the antibiotic is actively transported into the cell by the proton-motive force generated by the electron transport chain in the plasma membrane.

If the bacterial cell is exposed to high concentrations of gentamicin, e.g. a dose of 5–7 mg/kg lean body weight, the amount of antibiotic pumped into the cell will be sufficient to overwhelm and bind to the active site of all the ribosomes in the cell, and protein synthesis will be switched off and the cell dies. Under these circumstances the aminoglycosides are rapidly bactericidal.

Time above the minimum inhibitory concentration (T >MIC):

This applies to the β-lactam antibiotics that bind to the trans and carboxy peptidases (PBP) that carry out their function on the surface of the cell membrane. As these enzymes are continually synthesized and inserted into the plasma membrane of rapidly dividing cells, the antibiotic has to be present at a concentration that will not only bind covalently to and inactivate resident enzymes, but it must also be present to await the PBPs that the cell is still synthesizing and inserting into the plasma membrane.

In this situation the antibiotic has to be present above the MIC for long enough to disable all the PBP, depriving the cell of its PG mesh so it bursts. The β-lactam antibiotics are rapidly bactericidal, but adherence to the correct dosing frequency is essential to achieve this.

Area under the curve (AUC):

This applies to the glycopeptides that bind to the two alanine residues of the PG precursor, preventing the correct alignment required for incorporation (**Figure 4.6a**). Unlike time above MIC, the AUC profile is required to ensure that all precursors are bound and prevented from entering the correct site by a concentration-dependent process. When using vancomycin it is essential that the predose antibiotic level is regularly monitored and is between 10 and 20 mg/L. It is within this range that the necessary inhibitory concentration is maintained at the site of PG synthesis, and renal toxicity is minimized.

THE DISTRIBUTION OF ANTIBIOTICS IN THE BODY

A number of other important factors influence the effectiveness of an antibiotic, and include the volume of distribution in the body, degree of protein binding to plasma proteins (e.g. albumin) and whether they are hydrophilic or lipophilic.

Drugs with a high lipid solubility that are non-polar with low plasma protein binding have higher volumes of distribution than those which are polar, more highly ionized and have high plasma protein binding. Examples of antibiotics that are hydrophilic and with a smaller volume of distribution include the β-lactams, aminoglycosides and glycopeptides. Lipophilic agents with a larger volume of distribution include the fluorinated quinolones and macrolides. Volume of distribution is also increased in renal failure, due to fluid retention, and liver failure due to decreased plasma protein synthesis.

IMPORTANCE OF ESTIMATING RENAL FUNCTION

Creatinine clearance is an estimate of the glomerular filtration rate (GFR). Hospitals will provide formulas to calculate these, or will provide a GFR from the patient's renal function tests.

With gentamicin the 'once-daily' (in normal renal function) or 'pulse dose' regimes are often used. This must prompt the prescribing doctor to determine the patient's renal function, as the aminoglycosides are excreted via the kidneys. The formula below gives a reasonable estimate of creatinine clearance. It takes into account important factors such as sex, age, lean body mass, and current serum creatinine (**Figure 4.19**). The use of such a formula should be routine. After calculating the clearance, the interval between doses is determined:

$$\text{Creatinine clearance (mL/min)} = \frac{170 - \text{age in years} \times \text{wt (kg)}}{\text{serum creatinine}}$$

The value of 170 applies to male patients below 70 years of age. For males above this age and all female patients, use 160 and 150 respectively (see Cronberg S [1994]. Simplified monitoring of aminoglycosides. *Journal of Antimicrobial Chemotherapy* **34**:819–27.)

Thus, if clearance is:

100–60 mL/min the interval between doses is 24 hours.
59–40 mL/min the interval between doses is 36 hours.
39–20 mL/min the interval between doses is 48 hours.

When using gentamicin a dose of 5 mg/kg body weight is often used, with a maximum dose of 500 mg. The calculated dose should be rounded to the nearest multiple of 40, which is the amount of the antibiotic in a single vial. The dose is infused over 30 minutes. In the setting of prescribing an aminoglycoside to treat gram-negative sepsis, the length of course is usually up to 5 days.

This formula can also be used to determine the dose when the glycopeptide vancomycin is used. The usual adult dose is 1 g q12h. If the creatinine clearance is, for example, 50 mL/min, the dose is decreased to 500 mg.

ANTIBIOTIC ASSAYS

When a patient is prescribed an aminoglycoside or the glycopeptide vancomycin, the mimimum serum concentration has to be determined at regular intervals throughout the course of treatment. This is to ensure that the antibiotic is not accumulating, as renal toxicity can occur (the aminoglycosides can also cause ototoxicity). A blood sample, often referred to as a 'predose level', is obtained in the hour before the next dose is given. Hospitals have

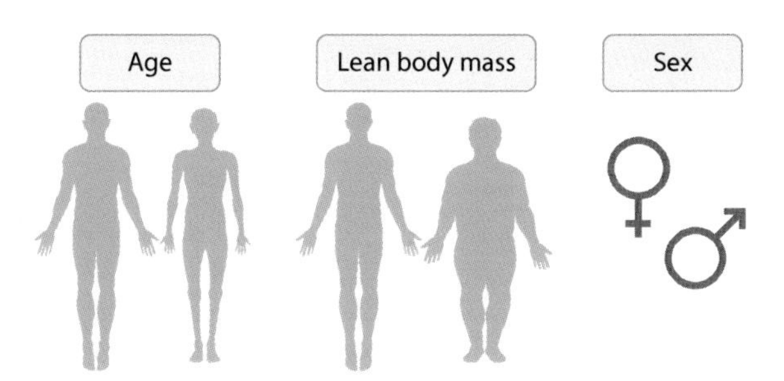

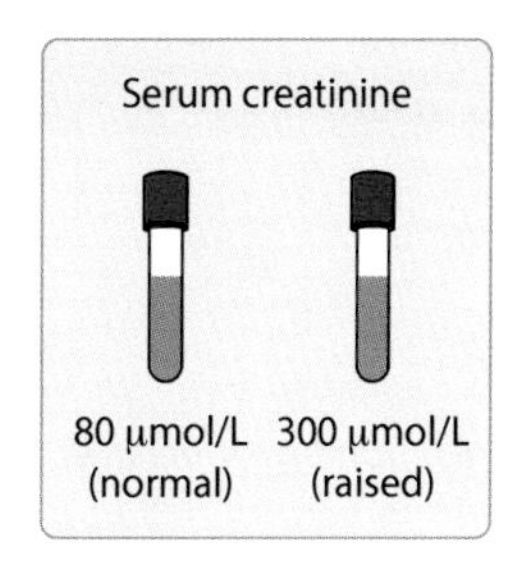

Figure 4.19 Four important parameters to remember when calculating renal function.

guidelines for interpreting these assays, and for gentamicin a predose of <1.0 mg/L is acceptable. For vancomycin, predose value should be maintained between 10 and 20 mg/L. Any predose level above those recommended requires a prompt reassessment of the patient's renal function, and adjustment of the dosing regimen of the antibiotic.

Daptomycin levels should be monitored when courses are longer than 1 week, to ensure that therapeutic levels are obtained. Creatinine kinase levels should also be determined weekly at least when using this agent for long-term treatment.

ANTIBIOTIC GUIDELINES

Every hospital and community health care practice will have antibiotic guidelines that cover the key infections, and direct the agents that should be used. The selection of particular antibiotics in each patient must take account of previous microbiology results, allergy to β-lactam and other antibiotics as well as the patient's renal and liver function. When bacteria are subsequently isolated from specimens, their identity and susceptibility profile must be used to determine further management.

An outline of a set of simple antibiotic guidelines for a hospital is shown in **Figure 4.20**. It is essential to access the guidelines at each hospital and be fully aware of them, as there are differences.

PRUDENT ANTIBIOTIC PRESCRIBING

The science of antibiotic prescribing centres on recognizing an infection, what the likely source is, which organism(s) are involved and which antibiotic(s) must be given,

Figure 4.20 An example of a set of anti-infective guidelines for educational information only. It is important to be familiar with the document at each hospital, as they do differ. Apart from infections such as community-acquired pneumonia and hospital-acquired pneumonia, the length of a course depends on the organism(s) isolated from clinical specimens, for example bacterial meningitis.

SEPSIS OF UNKNOWN ORIGIN

Organisms to consider:

Neisseria meningitidis, Staphylococcus aureus, Streptococcus pyogenes, Streptococcus pneumoniae, Enterobacteriaceae, *Clostridium* spp.

Antibiotics

First line

Amoxicillin 2 g q8h IV WITH
Gentamicin 5 mg/kg IV (max 500 mg)* WITH

Metronidazole 500 mg q8h**

*** Gentamicin interval q24h with GFR > 60 mL/min**
**** Change to oral when clinically safe to do so**

Second line

Meropenem 1 g IV q8h OR
Vancomycin 1 g IV q12h + ciprofloxacin 400 mg q12h IV +**
metronidazole 500 mg q8h IV**
FOR BOTH OPTIONS CONSIDER
Gentamicin 5 mg/kg IV (max 500 mg) once

COMMUNITY-ACQUIRED PNEUMONIA

Organisms to consider:

Streptococcus pneumoniae, Klebsiella pneumoniae, Legionella pneumophila, Chlamydophila pneumoniae

Other organisms:

Staphylococcus aureus, Haemophilus influenzae, Mycoplasma pneumoniae

Influenza viruses (note seasonal activity)

Influenza A and B

Antibiotics (if aspiration is a concern, add metronidazole [PO or IV as appropriate])

First line	Second line (e.g. penicillin allergy)
Low severity (CURB-65: 0-1)	
Amoxicillin 500 mg q8h PO for 7 days	**Doxycycline 100 mg q12h PO for 7 days**
Moderate severity (CURB-65: 2)	
Amoxicillin 500 mg q8h PO for 7 days, WITH Clarithromycin 500 mg q12h PO for 7 days	**Levofloxacin 500 mg q24h PO for 7 days**
High severity (CURB-65 3-5)	
Co-amoxiclav 1.2g q8h IV * for 7–10 days, WITH Clarithromycin 500 mg q12h IV* for 7–10 days	**Levofloxacin 500 mg q12h IV* for 7–10 days, WITH Clarithromycin 500 mg q12h IV * for 7–10 days**

*** Change to oral when clinically safe to do so**

Antiviral (influenza virus)

Oseltamivir 75 mg q12h PO for 5 days

Hospital-acquired pneumonia

Organisms to consider:

Streptococcus pneumoniae, Staphylococcus aureus, Enterobacter, Klebsiella

Other organisms:

It is important to review microbiology results (NB: MRSA and ESBL status).

Antibiotics

First line	Second line (e.g. penicillin allergy)
Co-amoxiclav 1.2 g q8h IV/625 mg q8h PO for 5 days	**Levofloxacin 500 mg q12h IV/PO for 5 days**

Meningitis

Organisms to consider:

Streptococcus pneumoniae, Neisseria meningitidis, (Haemophilus influenzae)

Is the patient immunosuppressed? (*Listeria monocytogenes* is inherently resistant to cephalosporins)

Listeria monocytogenes

Antibiotics

First line	Second line (e.g. penicillin allergy; non-anaphylaxis: meropenem)
Ceftriaxone 2 g q12h IV for 10 days	**Meropenem 2 g q8h IV for 10 days**
	Chloramphenicol 1 g q6h IV for 10 days

WITH (for listeria, as needed, 2–3 weeks treatment):

Amoxicillin 2 g q4h IV

If there is evidence that viral encephalitis is possible, consider HSV and VZV:

Aciclovir 10 mg/kg q8h, and review with CSF PCR result

(Continued)

ABDOMINAL INFECTIONS (COMMUNITY)

Infection in general–cover for usual bowel organisms to be considered:

Streptococci, enterococci, 'coliforms', anaerobes

First line

Co-amoxiclav 1.2 g q8h IV for 7 days

Second line (e.g. penicillin allergy; non-anaphylaxis)

Ertapenem 1 g q24h IV for 7 days

Acute cholecystitis/cholangitis

First line

Co-amoxiclav 1.2 g q8h IV for 7 days OR

Piperacillin/tazobactam 4.5 g q8h IV for 7 days

Second line (e.g. penicillin allergy; non-anaphylaxis)

Meropenem 1 g q8h IV for 7 days

THE PATIENT WITH DIARRHOEA ADMITTED FROM THE COMMUNITY

Organisms to consider in diarrhoea NOT associated with current/recent antibiotic use:

Campylobacter spp., *Escherichia coli* O157, *Salmonella*, *Shigella*

It is not accepted practice to empirically prescribe antibiotics for 'community' diarrhoea. Antibiotics are contraindicated in the setting of bloody diarrhoea/likely haemolytic uraemic syndrome caused by *Escherichia coli* O157. Most diarrhoeal illnesses will settle without the need for antibiotics.

Protracted diarrhoea due to *Campylobacter*:
Clarithromycin 500 mg q12h PO for 5 days

Protracted diarrhoea due to non-typhoid *Salmonella* or *Shigella*:
Azithromycin 500 mg q24h PO for 5 days

THE PATIENT WITH *CLOSTRIDIUM DIFFICILE* TOXIN POSITIVE DIARRHOEA

Mild disease:

<4 stools/day of type 5–7 Bristol stool chart, WCC within normal range, creatinine unchanged. Ensure laxatives, and antibiotics stopped (where possible), observe, as symptoms can settle without treatment.
Metronidazole 400 mg q8h PO for 14 days or Fidaxomycin 200 mg q12h for 10 days

Moderate disease:

3–5 stools/day of type 5–7 Bristol stool chart, WCC <15 x 10^9/L:
Metronidazole 400 mg q8h PO for 14 days or Fidaxomycin 200 mg q12h for 10 days

Severe disease:

5 or more stools/day of type 5–7 Bristol stool chart, WCC >15 x 10^9/L, creatinine 50% or more above recent baseline:
Vancomycin 125 mg q6h PO for 14 days or Fidaxomycin 200 mg q12h for 10 days

URINARY TRACT INFECTIONS

NB: The patient with a long-term catheter will likely have a bacteruria (or candiduria). Usually only consider treatment if systemically unwell; has a CSU (and blood culture) been collected?

Organisms to consider:

Escherichia coli, *Klebsiella* spp., *Enterococcus* spp., group B streptococcus

Cystitis (lower urinary tract infection)

First line	Second line
Trimethoprim 200 mg q12h (3 days female, 7 days male)	**Nitrofurantoin 100 mg q12h (3 days female, 7 days male). If GFR is <45 renal excretion reduces significantly; use alternative agent.**

Pyelonephritis

First line	Second line (e.g. penicillin allergy; non anaphylaxis [ertapenem])
Co-amoxiclav 1.2 g q8h IV for 7 days*/**	**Ertapenem 1 g q12h IV for 7 days****
	Ciprofloxacin 500 mg q12h PO for 7 days**

*** Change to oral as soon as clinically safe to do so**

**** A single dose of gentamicin, 5 mg/kg, maximum 500 mg, should also be considered.**

CELLULITIS AND WOUND INFECTIONS

Organisms to consider (For skin and soft tissue infection, always review the patient's MRSA status)

Staphylococcus aureus, group A streptococcus, group C/G streptococcus

First line	Second line
Mild (no systemic illness, superficial skin only)	
Flucloxacillin 1 g q6h PO for 7 days	**Doxycycline 100 mg q12h for 7 days**
Moderate (no systemic illness, deeper tissues involved, but no lymphangitis)	
Flucloxacillin 2 g q6h IV for 7 days	**Clindamycin 600 mg q6h IV* for 7 days**
Severe (systemic illness, deeper tissues involved, with lymphangitis)	
Flucloxacillin 2 g q6h IV for 7 days WITH	**Clindamycin 600 mg q6h IV* for 7 days WITH**
Clindamycin 600 mg q6h IV* for 7 days	**Vancomycin 1 g q12h IV for 7 days**

*** Change to oral as soon as clinically safe to do so**

preferably after specimen collection. Not infrequently, broad-spectrum agents are given initially, but these must be changed to narrower-spectrum agents when appropriate to reduce selection of antibiotic resistant organisms, and minimize side-effects, especially the collateral damage they exact on the bowel microbiota.

Clear prescription of antibiotics, with regular review is required and this is highlighted in the key steps outlined in **Figure 4.21**. Some important points in antibiotic prescription are given in **Figure 4.22** (these prescriptions relate to patients who do not have a penicillin allergy):

a. This female patient has been diagnosed with cystitis, and a 3-day course of treatment is appropriate. It was noted that a recent midstream urine (MSU) grew an ESBL-producing *Escherichia coli*. A repeat MSU is collected before the first dose to confirm the microbiological diagnosis. This patient's GFR is 65 mL/min (nitrofurantoin's renal excretion significantly reduces when the GFR falls below 45 mL/min, and an alternative antibiotic has to be used).
b. This patient has been diagnosed with a hospital-acquired pneumonia (HAP) and oral co-amoxiclav is considered appropriate by the on-call doctor who has reviewed the patient at 0500 hours. The sputum specimen result must be reviewed by day 2. If the likely culprit is amoxicillin sensitive, the co-amoxiclav can be changed to the narrower-spectrum agent.
c. This patient has been admitted with cellulitis around a chronic venous leg ulcer and intravenous flucloxacillin is considered appropriate. The admitting doctor has documented the need for review with the microbiology results at day 2.
d. In relation to the patient in (**c**), on the evening 3/3, the microbiologist communicated that the blood culture set taken on admission was growing gram-positive cocci in clusters. The patient is improving on flucloxacillin; the microbiologist advises continuing this intravenously until the organism is identified. This is confirmed as a MSSA on the morning of 4/3. Prescription (c) is crossed out and re-written (d). At least 2 weeks of intravenous

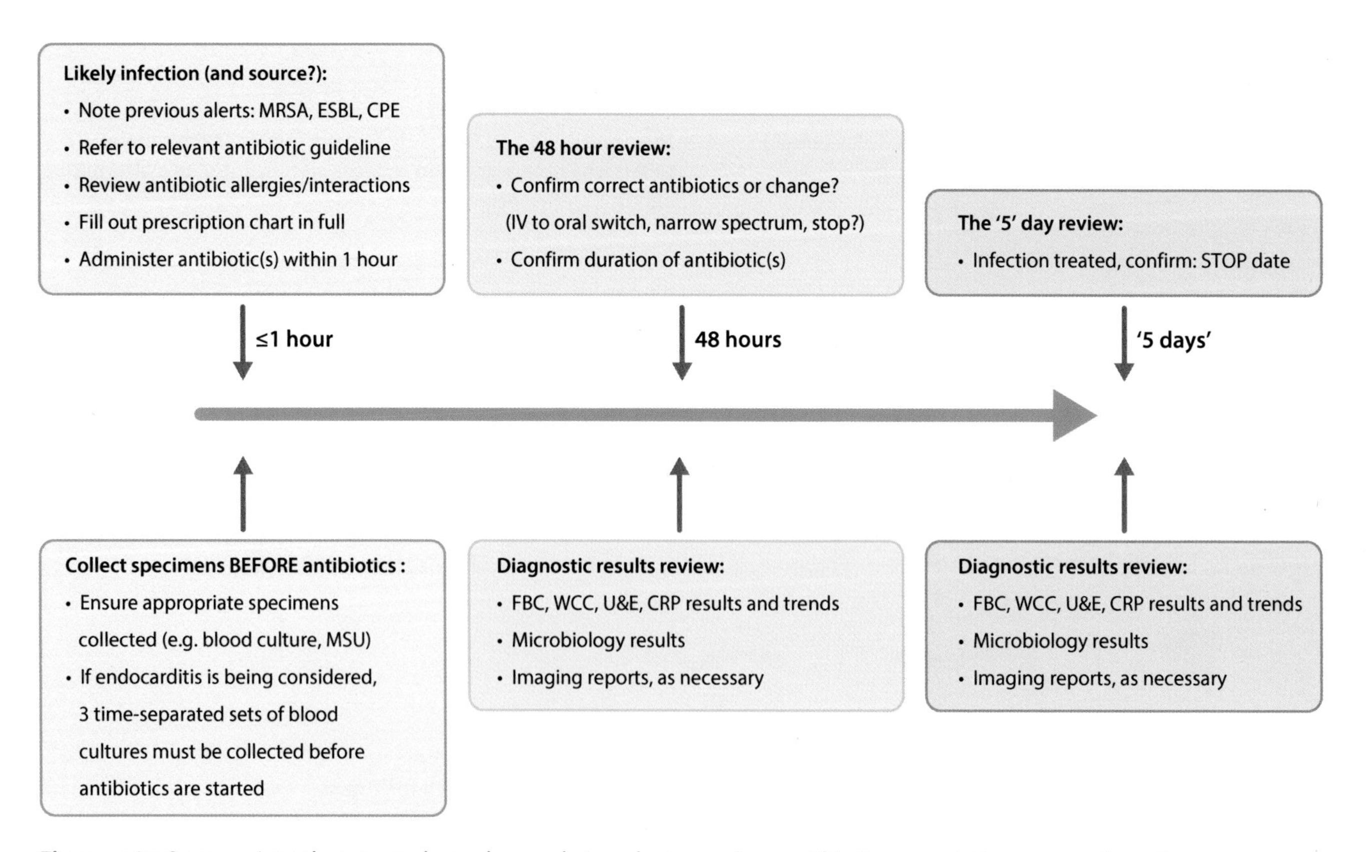

Figure 4.21 Some points that must always be used at each stage when antibiotic prescriptions are reviewed.

antibiotics are needed, and the continuation box of the prescription is signed (see Chapter 7).

Based on good clinical judgement, antibiotics should be given by mouth when appropriate. This clearly depends on the severity of the illness, the organism(s) involved, antibiotic susceptibility profile, site of infection and the ability of the patient to swallow and absorb drugs. When clinically appropriate, the intravenous to oral switch is important because it:

- Removes the risk of venflon-associated phlebitis, cellulitis and bacteraemia.
- Reduces discomfort and inconvenience for the patient, and enables increased mobility, facilitating early discharge from hospital.
- Saves medical and nursing time.
- Can reduce treatment costs.

PROPHYLACTIC ANTIBIOTICS

Postoperative surgical site infections (SSI) cause morbidity (and mortality), and add significantly to the length and thus cost of the hospital stay. There are many factors that influence risk of these infections, and an appreciation of them is important. These include:

- Increasing age (> 65 years).
- Poor nutritional state.
- Obesity (>20% above ideal body weight).
- Diabetes mellitus.
- Smoking.
- Immunosuppression (steroid or other immunosuppressive drugs).
- Malignancy.
- Prolonged preoperative stay.
- Foreign material at the surgical site.
- Surgical drains.
- Surgical technique including quality of haemostasis, poor closure, tissue trauma and ischaemia.
- Length of surgical scrub and the standard of preoperative skin preparation.
- Length of operation.
- The standard of theatre practice and the separation of sterile from contaminated processes (the one-way flow of equipment from 'clean to dirty').
- Inadequate instrument sterilization.
- Operating theatre ventilation.

These risks link to:

- A coexisting infection at other site(s) in the patient.
- Bacterial colonization (e.g. colonization of nostrils with *Staphylococcus aureus*).
- The effectiveness of antimicrobial prophylaxis.

(a)

Approved Name	Dose	Route	DATE → TIME ↓ (circle)	1/2	2/2	3/2	4/2				
NITROFURANTOIN	100mg	PO	(06.00)						Prescriber's Signature for continuation		
Date 1/2 1700; Course length 3 DAYS; Review date	Indication UTI-CYSTITIS		08.00								
Signature, Print Name and Bleep number: a.n.other A.N.OTHER 1234		Pharmacy	12.00								
ESBL IN RECENT MSU NITROFURANTOIN SENSITIVE			(18.00)				X				
			22.00								

(b)

Approved Name	Dose	Route	DATE → TIME ↓ (circle)	1/2	2/2	3/2	4/2	5/2			
CO-AMOXICLAV	625mg	PO	(06.00)						Prescriber's Signature for continuation		
Date 1/2 0500; Course length 5 DAYS; Review date 3/2	Indication HAP		08.00								
Signature, Print Name and Bleep number: a.n.other A.N.OTHER 1234		Pharmacy	12.00								
			(14.00)								
PURULENT SPUTUM COLLECTED RESULT REVIEW 3/2 PLEASE			18.00								
			(22.00)								

(c)

Approved Name	Dose	Route	DATE → TIME ↓ (circle)	2/3	3/3	4/3					
FLUCLOXACILLIN	2g	IV	(06.00)						Prescriber's Signature for continuation		
Date 2/3 0500; Course length R/V DAY 2; Review date 4/3	Indication L LEG ULCER:CELLULITIS		08.00								
Signature, Print Name and Bleep number: a.n.other A.N.OTHER 1234		Pharmacy	(12.00)			R/V					
BLOOD CULTURE, ULCER SWAB COLLECTED. PLEASE REVIEW 4/3			(18.00)								
			(24.00)								

(d)

Approved Name	Dose	Route	DATE → TIME ↓ (circle)	4/3	5/3	6/3	7/3	8/3		9/3	10/3
FLUCLOXACILLIN	2g	IV	(06.00)	X					Prescriber's Signature for continuation: a.n.other		
Date 4/3 1100; Course length 14 DAYS (+); Review date ONGOING	Indication L LEG CELLULITIS, MSSA BACTERAEMIA		08.00								
Signature, Print Name and Bleep number: a.n.other A.N.OTHER 1234		Pharmacy	(12.00)								
ULCER SWAB AND BLOOD CULTURE GROWING MSSA. SEE NOTES			(18.00)								
			(24.00)								

Figure 4.22 Four examples (**a–d**) of clear antibiotic prescribing on a manual chart, as described in the text.

THE ROLE OF ANTIBIOTIC PROPHYLAXIS

In order to reduce the chance of a postoperative infection, an antibiotic is used during the procedure, and when appropriate levels of the antibiotic are present throughout the operation, introduced bacteria are unlikely to survive. The relative effectiveness of a single dose of an antibiotic such as the cephalosporin cefuroxime, in preventing infection in relation to the time of operation is shown in **Figure 4.23**. This shows two important points; the antibiotic levels should be high during the procedure, and repeated doses of antibiotic after operation are of decreasing value. If an operation is prolonged, e.g. >4 hours, a further dose can be given, but prophylaxis should not exceed 12 hours. Continuing prophylaxis for days is not good practice, and only adds to the selection of resistant bacteria and yeasts. Collateral damage to the bowel flora must be considered.

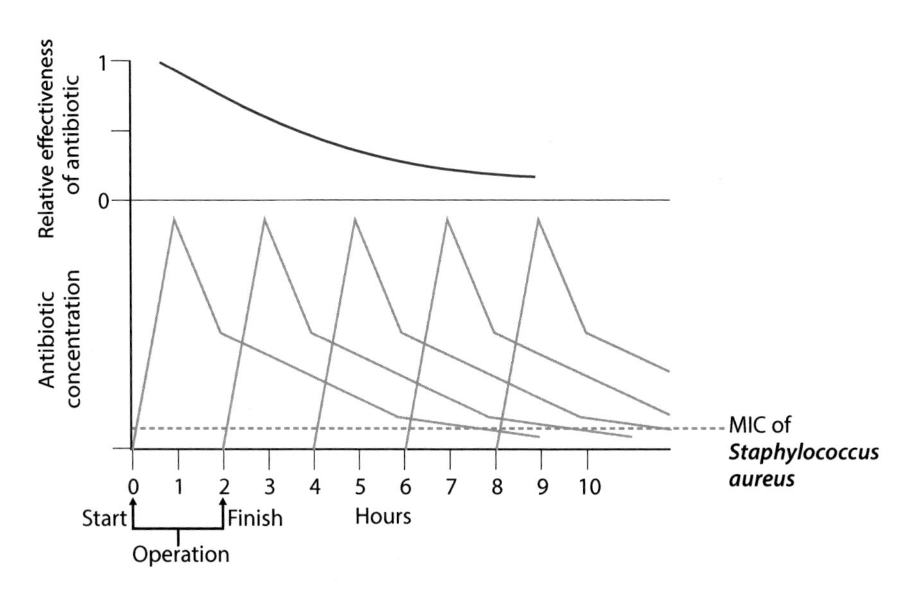

Figure 4.23 The effectiveness of a single dose of an antibiotic in an operation without complications is progressively diminished when it is given after the procedure.

THE CONTAMINATION CLASSIFICATION OF SURGICAL PROCEDURES

All hospitals will have antibiotic prophylaxis guidelines for each surgical specialty. It is important to be familiar with these. For the purposes of general antibiotic prophylaxis, surgical procedures are classified below; these definitions are used to broadly define the risk of infection where the skin is being cut to gain access to the relevant anatomy.

CLEAN OPERATION

Here no inflammation is encountered, the respiratory, alimentary or genitourinary tracts are not entered and prophylactic antibiotics are not usually required. The critical exception to this is when foreign material is inserted through this route. These include implantable cardiac electronic devices (ICED), prosthetic heart valves, prosthetic joints and other situations where implants are used, for example in breast surgery.

CLEAN CONTAMINATED

Here the respiratory, alimentary or genitourinary tracts are entered but without significant spillage. These operations require prophylactic antibiotics.

CONTAMINATED OPERATION

Here acute inflammation (without pus) is encountered, or there is visible contamination of the wound. Examples include gross spillage from a hollow viscus during the operation or compound fracture or open injuries operated on within 4 hours. These operations require prophylactic antibiotics.

DIRTY OPERATIONS

Here pus is present, or where there is a previously perforated hollow viscus, or a compound fracture or open injuries more than 4 hours old. These situations require a treatment course.

WHICH PROPHYLACTIC ANTIBIOTIC TO USE

In general, the antibiotics used are similar to those used to treat an infection at the relevant site. For example, in abdominal surgery involving the large bowel, an agent such as co-amoxiclav can be used, as this has the necessary cover for the usual bowel flora, and in addition has good activity against MSSA, which is relevant in superficial wound infection.

The most common organisms causing infection of implants such as ICED are gram-positive (skin) bacteria. These include *Staphylococcus aureus* and coagulase-negative staphylococci in particular. For these procedures, the glycopeptide teicoplanin is often recommended.

MRSA carriage is a risk factor for SSI. Patients known to carry MRSA should have a course of eradication therapy prior to admission for the surgery. This involves the application of mupirocin to the nostrils three time a day, and a daily wash with a chlorhexidine body wash, both for 5 days. Where antibiotic prophylaxis is required, this would include MRSA cover, and teicoplanin is appropriate.

ALLERGY TO PENICILLINS

Patients can report an allergy to one or more antibiotics, and this significantly restricts treatment options. It is an allergy to penicillins that is most frequently noted. Every attempt should be made to confirm this allergy, and identify whether a β-lactam can be used. The reason for using a β-lactam where possible is that at the correct dose and frequency they are rapidly bactericidal. Within the penicillin group are four antibiotics that provide a graded spectrum of activity against streptococci (penicillin), β-lactamase-negative 'coliforms' and *Haemophilus* (amoxicillin) and β-lactamase-producing TEM1, SHV1 'coliforms' as well as MSSA (co-amoxiclav). Piperacillin/tazobactam is active against *Pseudomonas aeruginosa* as well, and as with co-amoxiclav, is active against anaerobes. (It is worthwhile to review Chapter 3, **Figure 3.9**.) In addition, flucloxacillin is key to treating infections caused by MSSA; it also has reasonable activity against a number of streptococci, including *Streptococcus pyogenes*.

At least 10% of hospitalized patients report an allergy to penicillin, often as a result of a rash appearing during a course of penicillin in childhood. It is recognized that up to 95% of patients with a history of penicillin allergy are negative on further investigation. This means that penicillin allergy can potentially be excluded in 99% of the population, particularly when it is recognized that infections themselves can elicit a rash.

The frequent outcome when a patient is 'labelled' as penicillin allergic is that alternative antibiotics are used, including broad-spectrum quinolones, along with agents such as vancomycin, where close monitoring of serum levels is required. Unnecessary side-effects, suboptimal treatment levels with selection of resistant bacteria, and side-effects such as antibiotic-associated diarrhoea (AAD) can result. The important step should be to identify which patients are allergic and determine which β-lactam agents can be safely used.

THE NATURE OF THE IMMUNOLOGICAL REACTION TO PENICILLIN

Penicillin itself is not antigenic, and does not initiate an immune response. However, degradation products of penicillin such as benzylpenicilloyl can covalently bind to human proteins. The penicilloyl residue is then converted to an antigenic determinant or hapten, and antibodies specific to this can be produced (**Figure 4.24**). On subsequent exposure to penicillin and related agents, an immunologically-mediated reaction can occur.

THE TYPES OF DRUG REACTIONS TO CONSIDER

IMMEDIATE IgE-MEDIATED DRUG REACTION

This usually arises within 1 hour of administration, and is the result of interaction of the antibiotic antigen with preformed specific IgE antibodies bound to the surface of mast cells and eosinophils. This results in release of histamine, proteases, prostaglandins, leukotrienes and platelet activating factor. The clinical manifestations are:

- Urticarial rash.
- Angioedema (swelling of the face, hands and or feet).
- Rhinitis.
- Bronchospasm.
- Anaphylaxis.
- Swelling of tongue and/or throat.
- Difficulty swallowing and/or breathing.
- Hoarse voice, cough, wheeze.
- Hypotension.

NON-IMMEDIATE DRUG REACTIONS

These IgG-mediated reactions occur between 1 hour and 7 days after administration and can be divided into two groups:

Accelerated, occurs within 1–72 hours.

Delayed, occurs after 72 hours.

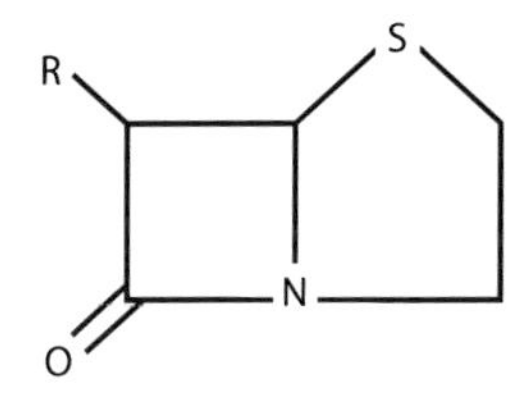

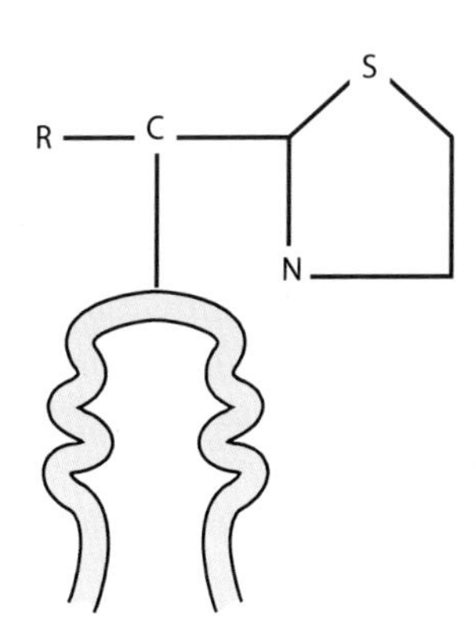

Figure 4.24 Penicillin derivatives on their own are not antigenic. However, on binding to a host protein they act as a hapten and antigenic determinant to which antibiotics are produced.

While these drug reactions can involve other organs, cutaneous manifestations predominate. These include:

- Urticarial rash (usually associated with an accelerated reaction).
- Maculopapular exanthems.
- Bullous lesions.
- Pustules.
- (Stevens Johnson syndrome).
- (Toxic epidermal necrolysis).

It should be noted that a significant proportion of maculopapular or urticarial reactions labelled as drug reactions are secondary to the underlying infection itself without any contribution by the suspected agent.

CROSS-REACTION BETWEEN THE β-LACTAMS IN THE SETTING OF THE NON-IMMEDIATE ALLERGIC REACTION

Key to the cross-reaction between β-lactams is the structure of the R1 side chain, which acts as an antigenic determinant of the penicilloyl hapten. **Figure 4.25** shows the structure of amoxicillin, along with representatives of first, second and third generation cephalosporins, and the carbapenem, meropenem. Note that amoxicillin and cefalexin have essentially the same R1 side chain, and thus in the patient with an amoxicillin allergy, there is likely to be cross-reactivity with cefalexin. For the other cephalosporins, the R1 side chain is different. Such information about these structures should be used when assessing the use of a β-lactam in the patient with a history of allergy to a penicillin.

OBTAINING THE HISTORY

In order for the patient not to be deprived of penicillin and other β-lactam antibiotics it is important to gain a detailed history of the circumstances of the allergic reaction and have this fully documented. There may appear to be an excessive number of questions below, but this is of particular importance in the patient who has a reaction in adulthood, as both recall of the event, and its documentation, can provide reliable information. The information that should be obtained is:

- The name of agent implicated.
- Date (approximate) the reaction took place.
- Time from administration to the reaction taking place.
- The symptoms experienced (what, where on the body, duration).
- Descriptions of skin lesions:
  - Urticarial (pale red, itchy papules [nettle burn]).
  - Maculopapular (small discoloured spots and bumps, frequently erythematous).
  - Bullous (blistering).
- Involvement of mucous membranes, rhinitis, bronchospasm.
- Other agents given at the same time.
- History of other drug allergies and reactions.
- Concurrent bacterial or viral infection at the time the allergic reaction took place.
- History of the reaction without known exposure to a contributing agent.

Where there is doubt, particularly in the setting of an infection emergency (sepsis of unknown source), and

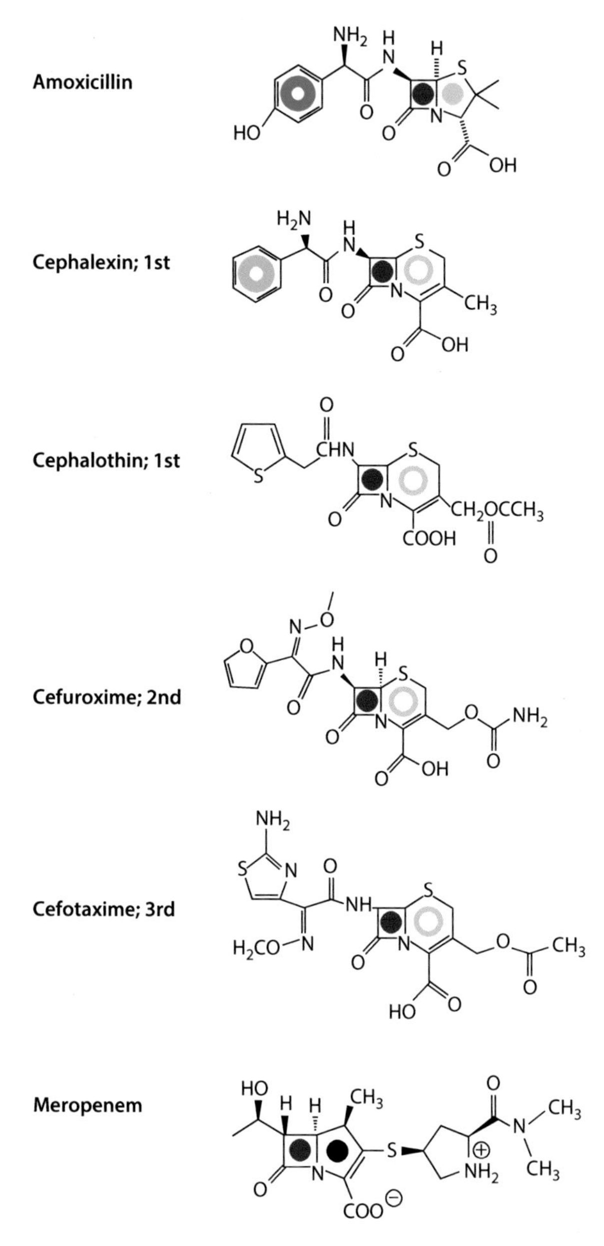

Figure 4.25 The structure of a range of β-lactams. The R1 side chains of amoxicillin and cefalexin are very similar, but this is not shared with the other agents shown (also see **Figure 4.4**).

a history of 'collapse' after a β-lactam was prescribed previously, it is reasonable to give an initial dose of a glycopeptide or daptomycin, along with gentamicin or ciprofloxacin, and metronidazole. This gives broad combination cover for sufficient time to allow the option to use a β-lactam antibiotic to be reviewed.

VACCINATION

Immunity to infection can be classed as passive or active.

ACTIVE IMMUNITY

An outline of various vaccines is shown in **Figure 4.26**. Vaccines can be grouped into live, killed, toxoid and conjugated. Live viral vaccines are obtained by multiple passage of a virus in cell culture. This results in attenuation of the virulence of the virus, but its antigenic properties are maintained. When a live vaccine is given, the virus replicates at a low level, and an efficient immune response is mounted. Immunocompromised individuals are not given vaccines with attenuated viruses due to the risk of disseminated infection; pregnancy is also a contraindication for 'live' vaccines.

With 'killed' vaccines, the organism is inactivated by chemical treatment such as formalin, but its antigenicity is maintained. As there is no reproduction of the organism, the number of cells recruited to the site of the injection is lower. The immune memory is not as marked as that with a live vaccine and antibody levels can wane over years. Toxoids are bacterial toxins that have also been inactivated, but maintain their immunogenicity.

Conjugated vaccines are those where the antigen has been linked to a carrier molecule. The 13-valent pneumococcal vaccine (PCV) is an example, containing 13 of the more common serotypes of *Streptococcus pneumoniae* conjugated to the diphtheria toxin. In this form the relatively less immunogenic capsule polysaccharide becomes a T-cell-dependent antigen, improving its immunogenicity. The 23-valent pneumococcal antigen is a polysaccharide preparation (PPV), and repeat doses are given every 5 years.

The vaccination programme in the UK in 2015 is shown in **Figure 4.27**. These schedules are regularly updated, and the latest recommendations must be obtained. This applies to individuals in the 'at risk' groups below.

PASSIVE IMMUNITY

Maternal IgG antibodies cross the placenta in the later stages of a normal pregnancy and protect the child against a wide range of infections for several months. Passive immunity can also be given by using hyper-immune globulin (IG). Although the effect is short term, susceptible individuals who have been exposed to either HBV or VZV can be protected by hepatitis B immunoglobulin (HBIG) and VZV immunoglobulin (VZIG). Non-immune individuals at risk of serious chickenpox infection, such as those who are pregnant, the new born or immunocompromised, are given VZIG.

VACCINATION OF PATIENTS IN 'AT RISK' GROUPS

Individuals in the 'at risk' groups below require specific vaccinations:

- Chronic medical conditions (respiratory, heart, renal, liver):
  - Influenza vaccine.
  - PPV.
- Immunosuppression:
  - Influenza vaccine.
  - PPV.
- Asplenic or splenic dysfunction:
  - Hib vaccine.
  - Influenza vaccine.
  - Meningococcal C vaccine, then A, C W135, Y 1 month later.
  - PPV.
- Haemodialysis:
  - HBV vaccine.
- Haemophilia:
  - HAV vaccine.
  - HBV vaccine.
- Injecting drug users:
  - HAV vaccine.
  - HBV vaccine.
- Men-who-have-sex-with-men (MSM):
  - HAV vaccine.
  - HBV vaccine.

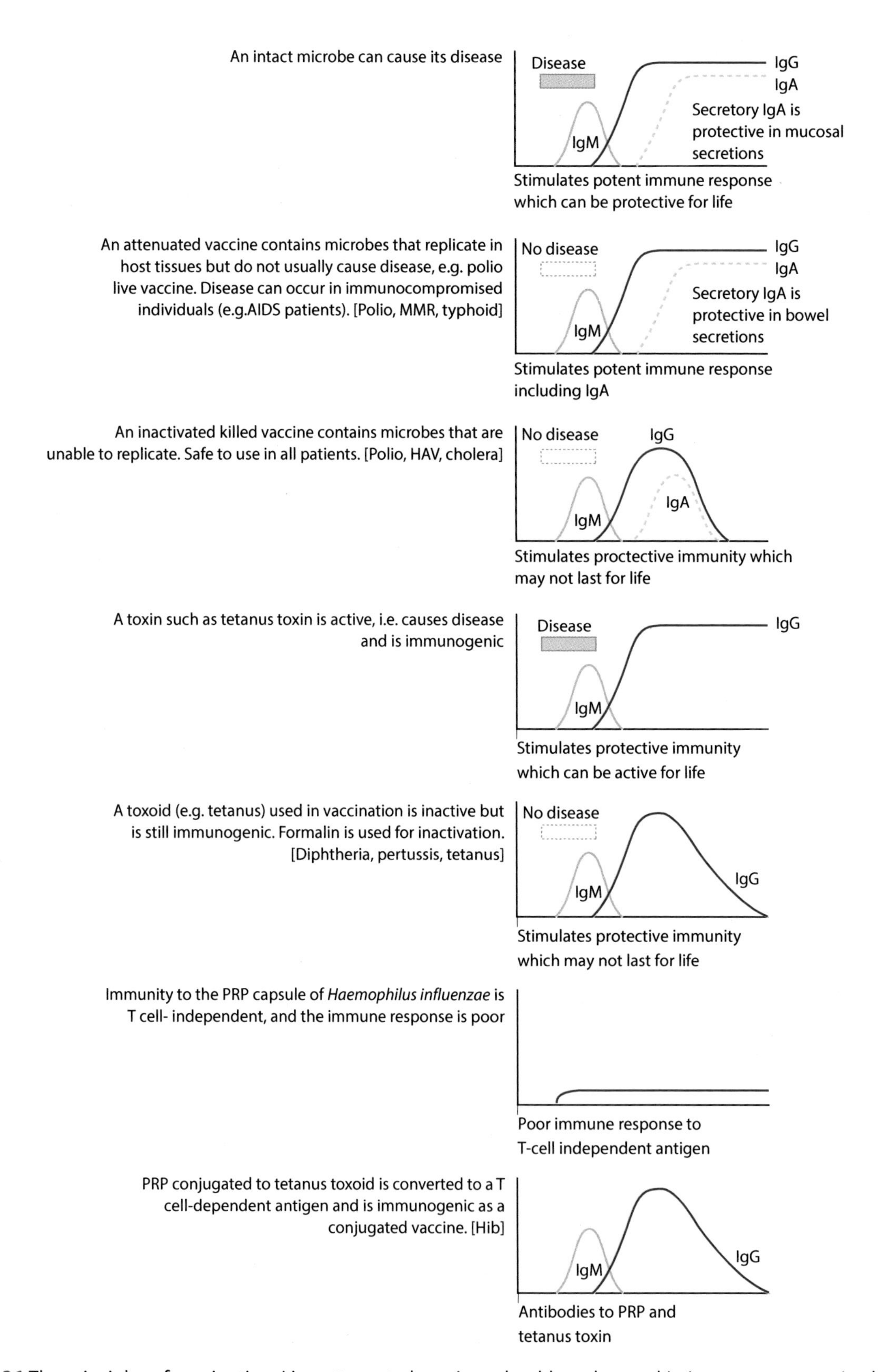

Figure 4.26 The principles of vaccination. Live attenuated vaccines should not be used in immunocompromised patients.

	2 months	3 months	4 months	12-13 months	2, 3, 4 years	School years 1, 2	3.3 years*	12-13 years	13-18 years	65 +	70 years
Diphtheria	Y	Y	Y				Y		Y		
Pertussis	Y	Y	Y				Y		Y		
Polio (IPV)	Y	Y	Y				Y		Y		
Tetanus	Y	Y	Y				Y				
Hib	Y	Y	Y	Y (booster)							
Meningococcus B	Y (1st)		Y (2nd)	Y (3rd)							
Meningococcus C		Y		Y (booster)							
Meningococcus ACWY									Y		
Pneumococcus (PCV-13)	Y (1st)		Y (2nd)	Y (3rd)							
Pneumococcus (PPV)										Y	
Rotavirus	Y (1st)		Y (2nd)								
Measles				Y (1st)			Y (2nd)				
Mumps				Y (1st)			Y (2nd)				
Rubella				Y (1st)			Y (2nd)				
Annual influenza virus A, B					Y	Y				Y	
HPV								Y			
VZV											Y (shingles)

* or soon after

Figure 4.27 The vaccination schedule for the UK, 2015. Yellow: '5 in 1'; blue: '4 in1'; green: '3 in 1' vaccines. Red: live attenuated vaccines. Note the childhood influenza vaccine, given pernasal, is live attenuated. (PCV: pneumococcal conjugated vaccine; PPV: pneumococcal polysaccharide vaccine.) The latest Department of Health guidance must be referred to.

Chapter 5

The Microbiology Laboratory

INTRODUCTION

The microbiology department provides a clinical and laboratory service for users in the hospital and community and is also an integral part of the hospital's infection control team (ICT). The work centres on timely processing of specimens, to identify organisms by a range of tests and issue a report, with interpretative comments, to be used in patient management. In the case of bacteria (and yeasts), appropriate antibiotic susceptibility testing is done. Concise clinical details must be included on the request form, as this information is a key part of the overall diagnostic process. It also alerts staff to the possibility of organisms that pose a risk to them (e.g. a travel history, possible enteric fever, *Salmonella* Typhi/Paratyphi), so that the specimen is processed in the correct containment room. Examples of the range of organisms and tests are shown in *Tables 5.1, 5.2*.

THE LABORATORY

Laboratories have to comply with national standards to be accredited to do diagnostic testing. Safety centres on directives of the Advisory Committee for Dangerous Pathogens (ACDP). This classes organisms of medical significance into three relevant groups:

Group 2: These can cause human disease and may be a hazard to employees; they are unlikely to spread to the community and there is usually effective prophylaxis or treatment available.

Group 3: These can cause severe human disease and may be a serious hazard to employees; they can spread to the community, but there is usually effective prophylaxis or treatment available.

Group 4: These cause severe human disease and are a serious hazard to employees; they are likely to spread to the community and there is usually no effective prophylaxis or treatment available.

Examples of organisms within these groups are given in *Table 5.3*. This classification can also be related to the situation in the clinical setting, and the infection control alerts they activate.

Haemorrhagic fever viruses, such as Ebola virus, highlight the clinical management and infection control procedures when a patient is admitted with suspected disease. There is one maximum containment medical facility for nursing such patients, usually with confirmed disease, in the UK (Royal Free Hospital, London). However, initial management is done at the admitting hospital, and there are set protocols for doing this safely. NHS laboratories must be able to perform the tests needed for the initial management of these patients. These include blood tests (full blood count [FBC], urine and electrolytes [U+E], clotting profile), as well as microscopy and antigen testing for malaria. Blood cultures, urine and stool specimens need to be processed in the hospital's microbiology laboratory. In addition, specimens are prepared for safe despatch to the Rare and Imported Pathogens laboratory at Public Health England Porton Down, where the molecular diagnostic tests for the haemorrhagic fever viruses are done.

The layout of a microbiology laboratory is shown in **Figure 5.1**. The laboratory is required to be secure with access restricted to the necessary staff in the microbiology and pathology departments. Specimens are delivered in safe transport boxes by the portering service, or by pneumatic tube in which they are sent from clinical areas to the laboratory in 'pods'. To minimize any risk of breakage and leakage in 'pods', all specimen collection systems are plastic. It is essential to ensure after collecting a specimen that the container lid is tightly closed, before sealing it in the plastic envelope. Specimens from a patient with a suspected group 4 organism are transported from a clinical area to the laboratory by a designated secure portering system, and never by pneumatic tube.

General areas of the laboratory are where specimens likely to contain organisms in ACDP group 2 are processed. Specimens that are likely to contain group 3 organisms are processed in a 'category 3' room. These have restricted access for designated staff only, with particular reference to their occupational health department (OHD) clearance, including tuberculosis. These rooms have an ante-room to the negative-pressure specimen processing room, which has ventilated hoods to give maximum protection

Table 5.1 Bacteriology section

Blood culture
Bacteroides fragilis
Escherichia coli
Enterobacter cloacae
Haemophilus influenzae
Klebsiella pneumoniae
Neisseria meningitidis
Pantoea agglomerans
Pseudomonas aeruginosa
Salmonella Typhi
Serratia marcescens
Enterococcus faecalis
Peptostreptococcus magnus
Staphylococcus aureus
Staphylococcus epidermidis
Staphylococcus haemolyticus
Streptococcus pyogenes
Streptococcus agalactiae
Streptococcus dysgalactiae
Streptococcus anginosus
Streptococcus pneumoniae
Streptococcus salivarius
Candida albicans
Candida tropicalis

Wound swab	**Faeces specimen**
Staphylococcus aureus	*Escherichia coli* **O157**
Streptococcus pyogenes **(GAS)**	*Campylobacter jejuni*
Streptococcus dysgalactiae **(GCS)**	*Salmonella* Enteritidis
'Coliforms'	*Salmonella* Typhimurium
Pseudomonas aeruginosa	*Shigella flexneri*
Mixed anaerobes	***Clostridium difficile toxin****
	Cryptosporidium parvum
	Giardia lamblia

Urine specimen	**Respiratory specimen**
Escherichia coli	***Streptococcus pneumoniae***
Klebsiella oxytoca	*Haemophilus influenzae*
Proteus mirabilis	*Klebsiella pneumoniae*
Pseudomonas aeruginosa	*Moraxella catarrhalis*
Enterococcus faecalis	*Pantoea agglomerans*
Enterococcus faecium	
Streptococcus agalactiae	*Mycobacterium avium*
Candida albicans	*Mycobacterium tuberculosis*

Genital specimen	**CSF**
Gardnerella vaginalis	***Streptococcus pneumoniae***
Neisseria gonorrhoeae	*Neisseria meningitidis*
Streptococcus agalactiae	***Listeria monocytogenes***
Streptococcus pyogenes	*Cryptococcus neoformans*
Candida albicans	
Trichomonas vaginalis	

Table 5.2 Virology and molecular section

Serology	Molecular
Serum	**Whole blood/plasma (EDTA)**
CMV IgM/IgG antibodies	CMV DNA viral load by PCR
EBV IgM/IgG to virus capsid antigen (VCA)	EBV DNA viral load by PCR
EBV IgG to Epstein–Barr nuclear antigen (EBNA)	
VZV IgG antibodies	HBV DNA viral load by PCR
	HCV RNA (RT-PCR)
Measles virus IgM/IgG antibodies	HCV RNA viral load (RT-PCR)
Mumps virus IgM/IgG aantibodies	
Rubella virus IgM/IgG antibodies	HIV RNA (RT-PCR)
	HIV RNA viral load (RT-PCR)
Parvovirus B19 IgM/IgG antibodies	
	Respiratory virus panel PCR
HIV1, HIV 2 antigen/antibody test	Adenovirus
	Influenza virus A, B
HAV IgM/IgG antibody	Metapneumovirus
HBV surface antigen/antibody (HBsAg; HBsAb)	Parainfluenza virus
HBV early antigen/antibody (HBeAg; HBeAb)	Respiratory syncytial virus
HBV core IgM/IgG antibody (HBcAb)	Rhinovirus
HCV antibody	
HEV IgM/IgG antibody	**CSF virus PCR panel**
	HSV 1, HSV 2, VZV PCR
Legionella urine antigen test (Sg1)	Enterovirus PCR
Pneumococcal urine antigen test	
	CSF bacterial PCR panel
Adenovirus stool antigen (children <5 years)	*Neisseria meningitidis*
Rotavirus stool antigen (children <5 years)	*Streptococcus pneumoniae*
	Streptococcus agalactiae (GBS)**
Borrelia burgdorferi (Lyme disease) IgM/IgG antibodies	*Listeria monocytogenes*
Leptospira IgM/IgG antibodies	
Treponema pallidum antibodies (VDRL, IgM/IgG, TPPA)	**Other**
	Aspergillus galactomannan antigen
Entamoeba histolytica (IgG) antibodies	*Pneumocystis jirovecii* PCR (BAL, plasma)
Schistosoma (IgG) antibodies	

Clostridium difficile EIA (antigen) test

** Usually a test requested in Neonatology

BAL: bronchoalveolar lavage; CSF: cerebrospinal fluid; EDTA: ethylenediamine tetra-acetic acid; Ig: immunoglobulin; PCR: polymerase chain reaction; RT: reverse transcription; VDRL: Venereal disease research laboratory.

Certain tests are sent to NHS/PHE or commercial reference laboratories, as departments usually do not offer the full range of (usually molecular) tests.

Table 5.3 Examples of organisms in relevant ACDP groups

Group 4	Group 3	Group 2
Congo Crimean HF virus	*Escherichia coli* O157	*Escherichia coli*
Ebola virus	*Histoplasma capsulatum*	Influenza virus A, B
Lassa fever virus	*Mycobacterium tuberculosis*	*Salmonella* Enteritidis
Marburg virus	Rift Valley fever virus	*Staphylococcus aureus*
	Salmonella Typhi	*Streptococcus pyogenes* (GAS)

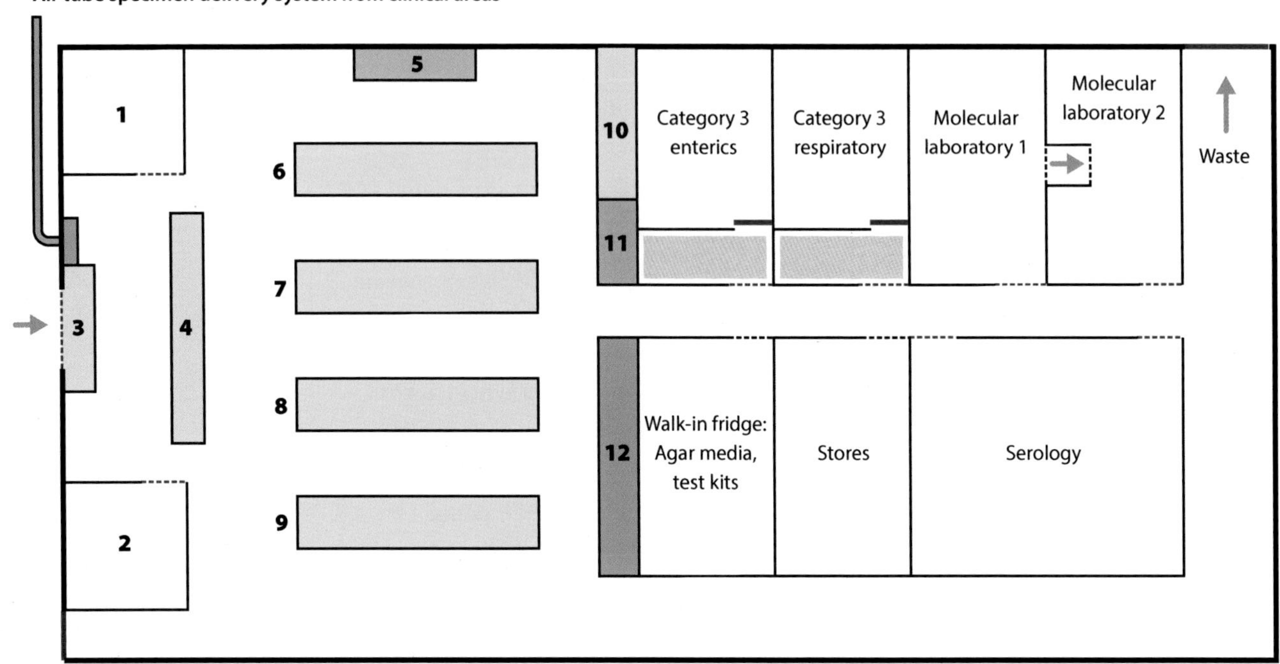

Figure 5.1 Schematic layout of a microbiology laboratory. 1, 2: administration offices; 3: specimen reception; 4: specimen booking-in station; 5: blood culture machine; 6: blood culture and sterile sites/tissues bench; 7: identification and antibiotic susceptibility bench; 8: urine, wound swab and ENT bench; 9: genital specimen bench; 10: incubators; 11: anaerobic cabinet; 12: fridges. Red lines: restricted access; dashed lines: locked when not in use (key pad on doors); category 3 rooms have an additional restricted level of access for designated staff only.

for staff. Air that leaves this negative-pressure system is filtered through high-efficiency particulate assisting (HEPA) filters. All respiratory specimens received in the laboratory are processed in the category 3 room, and laboratories can have a separate category 3 room for faecal specimens.

It is essential that the request form has all pertinent information recorded. For example, the words 'travel history' and 'fever of unknown origin' (FUO) are examples of general alerts at the laboratory's specimen reception station for the specimen to be processed in the category 3 room. In addition to clinical information, other details of importance include exposure risk (food, vectors, animals) and country where the exposure took place. When necessary, specimens containing suspected group 4 organisms must be processed in a category 3 room, with the staff using additional personal protective equipment (PPE). Each section of the laboratory has a dress code for the type of PPE required. A laboratory coat is the basic requirement.

Molecular (usually polymerase chain reaction [PCR]-based) tests are done in a suite of two rooms. Specimen preparation is done in one room, and by a one-way flow system are then moved to the second room where

PCR-based testing is done. This is to prevent cross-contamination at the PCR amplification stage in these very sensitive testing systems.

All clinical specimens and the resulting waste generated in the laboratory process have to be disposed of according to strict guidelines. Waste from the category 3 laboratory has to be sterilized in the Pathology department's secure autoclave system before disposal through the hospital's waste management process, and this is the preferred method for all laboratory waste. However, while adhering to guidelines, safely contained non-category 3 laboratory waste can also be removed off-site by designated contractors for incineration. Representative laboratory facilities are shown in **Figure 5.2**.

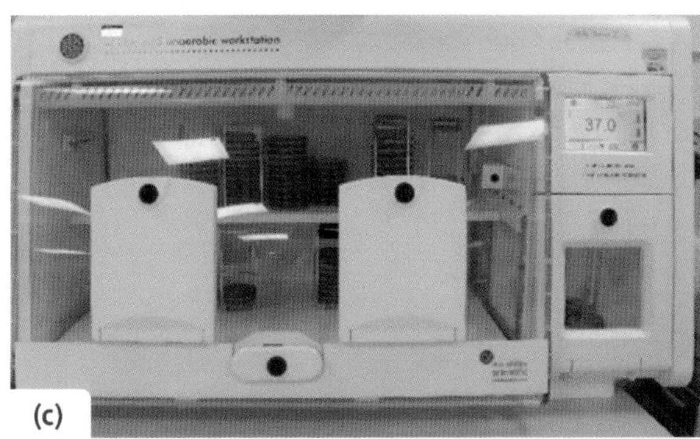

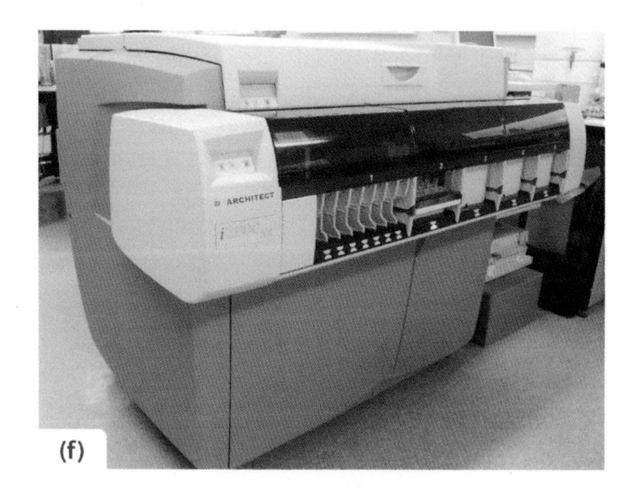

Figure 5.2 (a) A general bacteriology bench; (b) access to a category 3 room is restricted (note the [negative] pressure gauge (red); (c) a Whitley™ anaerobic workstation/cabinet where agar plates are incubated in an atmosphere of CO_2, nitrogen and hydrogen; (d) the Brucker MALDI-TOF™ machine; (e): automated antibiotic susceptibility testing machine (Vitek, Biomerieux™); (f) The Abbott Architect™ chemiluminescent microparticle immunoassay (CMIA) serology platform.

BACTERIOLOGY LABORATORY TESTS

BASIC PROCESSING OF A SPECIMEN

To illustrate how a specimen is processed in the bacteriology laboratory, the example of a specimen of pus aspirated from a liver abscess is used. The most likely bacteria present would be members of the 'normal' bowel flora comprising enterococci, streptococci, 'coliforms' and anaerobes. The first tests performed on the specimen are a Gram stain and white cell count (WCC) differential stain (**Figure 5.3**). A specimen of pus will usually show a large number of neutrophils, and may also reveal several gram-staining types of bacteria. Gram-positive cocci in chains can be enterococci, facultative streptococci or anaerobic streptococci. The gram-negative rods may be 'coliforms' or *Bacteroides*.

The specimen is plated out onto a range of solid (agar) media in order to optimize the growth of all the bacteria. The media used are shown in **Figure 5.4**. A portion of the specimen is also inoculated into an 'enrichment' liquid such as brain–heart infusion (BHI) broth, enabling small numbers of bacteria to reproduce. Broths are usually incubated for at least 5 days and examined daily for turbidity. Growth in the broth will not be numerically representative of bacteria in the original specimen of pus, but enrichment acts as a back-up, especially if antibiotics have damaged organisms. When the broth is turbid, indicating growth of bacteria, it is subcultured onto solid media. In the setting of all 'unrepeatable' specimens such as liver abscess pus and tissue biopsies, an enrichment broth must be used. Note that laboratories use a fairly narrow range of culture media to grow most bacteria and yeasts, as shown in **Figure 5.5**. Incubation is usually at 37°C.

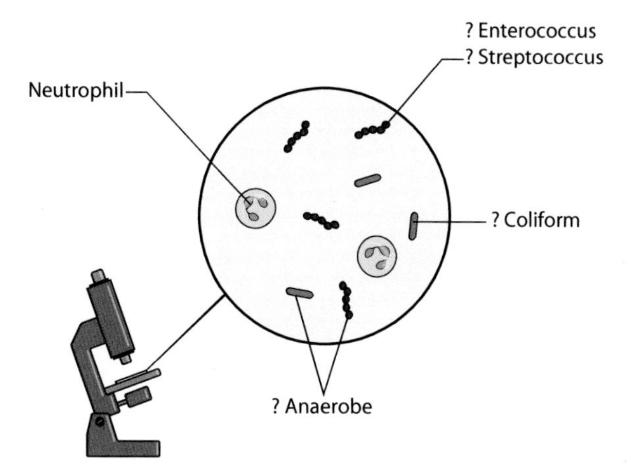

Figure 5.3 The bacteria that may be seen in a Gram stain of pus from a pyogenic liver abscess. In the laboratory, media that will grow all these bacteria are inoculated.

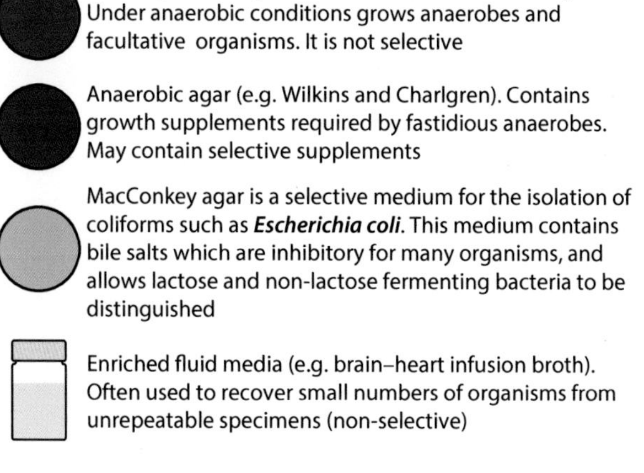

Figure 5.4 A range of solid and liquid media that are commonly used in the bacteriology laboratory.

Bacteria	Atmosphere	Culture medium
Enterococci	O_2/ANO_2	Blood agar/MacConkey agar
Beta-haemolytic streptococci	O_2/ANO_2	Blood agar
Alpha-haemolytic streptococci	O_2/ANO_2	Blood agar
Streptococcus pneumoniae	O_2/ANO_2	Blood agar
Staphylococci (all types)	O_2/ANO_2	Blood agar, MacConkey agar
Haemophilus influenzae	$O_2 + CO_2$	Chocolate agar
Neisseria meningitidis	$O_2 + CO_2$	Blood agar, chocolate agar
Neisseria gonorrhoeae	$O_2 + CO_2$	Chocolate agar or enriched blood agar
Coliforms (*Escherichia coli, Klebsiella* spp.)	O_2/ANO_2	Blood agar, MacConkey agar
Salmonella and *Shigella* spp.	O_2/ANO_2	Blood agar, MacConkey agar, DCA, and XLD agar (selective media)
Pseudomonas spp.	O_2	Blood agar, MacConkey agar
Legionella pneumophila	$O_2 + CO_2$	Legionella selective agar
Bacteroides spp.	ANO_2	Blood agar, anaerobic agar
Clostridium spp.	ANO_2	Blood agar, anaerobic agar
Peptostreptococcus spp.	ANO_2	Blood agar, anaerobic agar
Candida spp.	O_2/CO_2	Sabouraud agar

Figure 5.5 Most of the bacteria considered in clinical practice can be grown on blood or chocolate agar. MacConkey (enteric organisms) and Sabouraud (fungi) are examples of selective media. Special media are needed for an organism such as legionella. (O_2: aerobic [air]; ANO_2: anaerobic; CO_2: 5% CO_2 present.)

When solid culture media are inoculated, a plastic loop is used to produce several 'streak' zones (**Figure 5.6a**). This process dilutes out the organisms in the specimen so that bacteria are sufficiently separated; following multiplication, individual colonies are visible after 18–24 hours of incubation (**Figure 5.6b**). Each colony represents reproduction of one organism and is in essence a 'pure' clone, which can be used for identification and for antibiotic susceptibility testing. Using such parameters as colonial morphology and atmospheric growth conditions, the biomedical scientist (BMS) will be able to make an initial assessment of the type(s) of bacteria (and yeasts) present and will make a decision as to how far the organisms will be processed. From a liver abscess, up to three different types of bacteria would warrant each isolate being identified and antibiotic susceptibility testing. Four or more different bacterial types would often be reported as 'faecal flora'. It is important at this stage that the microbiologist discusses these findings with the clinical team, in order to determine what additional work would aid management of the patient.

Two different coliforms growing on CLED agar, each having a distinct colonial morphology, are shown in **Figure 5.7**. Single colonies can be 'picked off' to determine the identity of the organism and its antibiotic susceptibility profile. In the following sections, basic tests are discussed, as well as the rapid identification MALDI-TOF machine.

As highlighted in Chapter 3, when infection is a consideration, a portion of any tissue biopsy for histology is also placed in a sterile container with saline; *Mycobacterium tuberculosis* is an example of a relevant

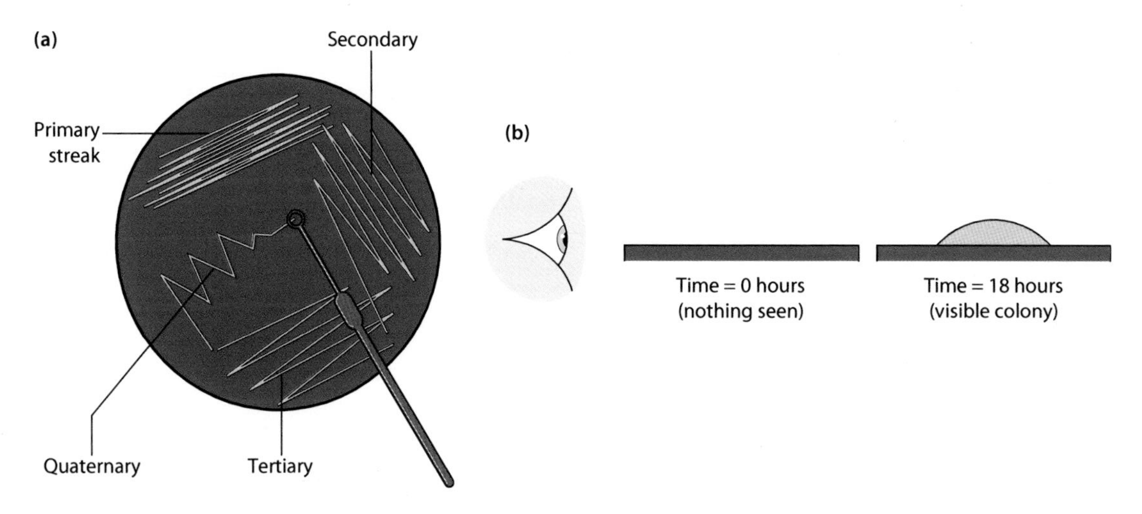

Figure 5.6 (**a**) A 'loop-full' of specimen is inoculated onto agar medium. Usually four 'streaks' are made; (**b**) the progeny of a single bacterium will be visible as a single colony after overnight incubation at 37°C.

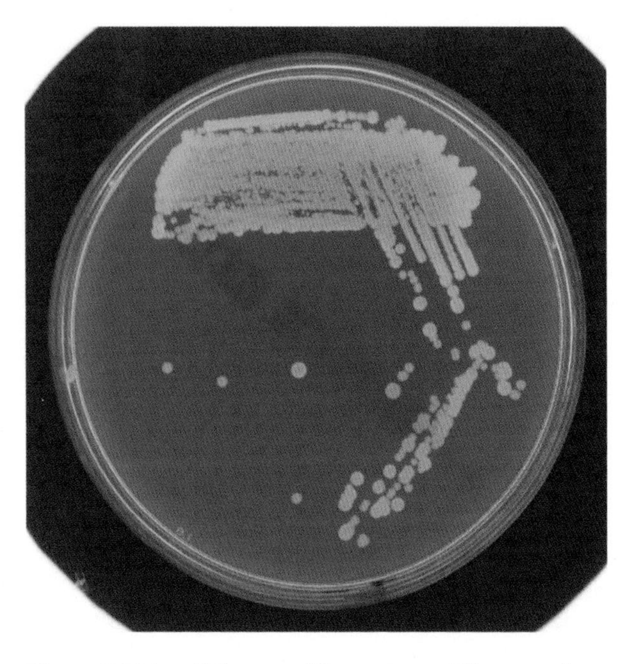

Figure 5.7 Two different coliforms have different colony morphologies as shown here with CLED agar.

organism. Laboratories also retain any residual sample for 7 days. This is important when a specimen is culture negative after prolonged incubation. It can be sent to a reference NHS/commercial laboratory for the detection and speciation of organisms by tests such as ribosomal ribonucleic acid (RNA) gene amplification and sequencing. (Specimens of pus, or other samples from normally sterile sites, are also retained for this purpose.)

THE TIME SCALE OF EVENTS IN THE LABORATORY

Most specimens are 'routine' and will be received in the laboratory throughout the working day. There should be systems in place to enable all specimens to have their initial processing on the day of receipt (day 0). All blood culture sets should be placed in the blood culture machine within 4 hours of being taken from the patient. This is required to optimize the growth of bacteria and yeasts, and obtain the 'time to positivity' in the shortest period. Critical specimens such as cerebrospinal fluid (CSF) and aspirates from sterile sites are processed immediately on receipt in the laboratory. (Full details for processing blood cultures are given in Chapter 7.)

The following examples are used to show the time scale of events in the microbiology laboratory; the blue down arrow is time of receipt and booking in to the laboratory.

BLOOD CULTURES (FIGURE 5.8)

Blood culture bottles are placed in the blood culture machine (BM) and incubated at 37°C. Bacteria (and yeasts) reproduce in the liquid medium, and when a set growth titre is exceeded, a signal alerts that the bottle is positive (BC+), usually within 12–18 hours. A Gram stain (GS) is done on fluid from positive bottles and this result is reviewed by the microbiologist, who will discuss this with the clinical team either on the ward or by phone (Organisms are usually seen in the Gram stain.) The blood culture fluid is cultured on agar plates (C), and direct antibiotic susceptibility tests are set up using the fluid (D-AS).

After incubation for 18–24 hours, plates are examined for growth, and if pure, an individual colony is picked off (PO) for identification by systems such as MALDI-TOF (ID-MT); the direct antibiotic susceptibility results are also reported (AS-R).

Blood culture sets are usually incubated for 5 days. Prolonged incubation (up to 10 days) is done on occasion to allow 'slow growers' all opportunities to multiply; endocarditis caused by nutritionally deficient streptococci is an example, as is *Brucella*. When the clinical team are considering such organisms in the differential diagnosis, they should be in discussion with the microbiologist at the earliest opportunity to ensure that the correct laboratory process is followed.

CSF (FIGURE 5.9)

Microscopy is performed as a priority. The WCC, its differential and Gram stain result are reported by phone, and reviewed in conjunction with the CSF protein and CSF/blood glucose values. The CSF is cultured on agar plates (C). After 18–24 hours' incubation, plates are examined for growth, and if present an individual colony is picked off (PO) for identification by MALDI-TOF (ID-MT). Antibiotic susceptibility tests are set up (AS), and after incubation are reported after 18–24 hours (AS-R).

If no organisms are seen on the initial Gram stain and/or the CSF is culture negative, a decision is taken, based on the clinical circumstances, to process the CSF for bacterial PCR (e.g. pneumococcus, meningococcus) and viral PCR (HSV1, HSV2, VZV, enterovirus). Both positive and negative PCR results should be reported by telephone to the clinical team. Laboratories use commercial or reference laboratories for these tests.

WOUND SWAB (FIGURE 5.10)

Gram stains are not usually done on superficial wound swabs, as a range of bacteria are often seen. Swabs are cultured (C), and after 18–24 hours' incubation, individual colonies are picked off (PO) and identified by MALDI-TOF (ID-MT). Antibiotic susceptibility tests are set up (AS), and reported after 18–24 hours' incubation (AS-R). Not infrequently, there is a mixed growth (MiG) that may harbour a likely pathogen, e.g. β-haemolytic streptococcus amongst mixed skin flora. Colonies of the streptococcus are picked off (PO) and cultured (C). After 18–24 hours' incubation, further 'POs' for identification (ID-MT) and antibiotic susceptibility tests are done (AS) and read the next day (AS-R).

MIDSTREAM URINE (FIGURE 5.11)

Most laboratories have automated urine analysers (A MIC), that use flow cytometry for 'microscopy', identifying white blood cells (W) and particles the size of bacteria (O). If preset 'cut-off' values are exceeded for these parameters, the urine specimen is cultured (C). (The presence of [vaginal] epithelial cells (E) is also reported in the 'microscopy' result.) Samples are cultured on selective media, and after 18–24 hours' incubation, plates are

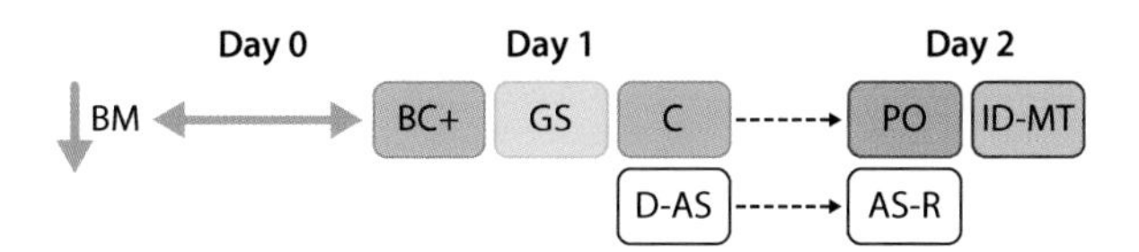

Figure 5.8 Timescale for blood cultures (see text for explanation).

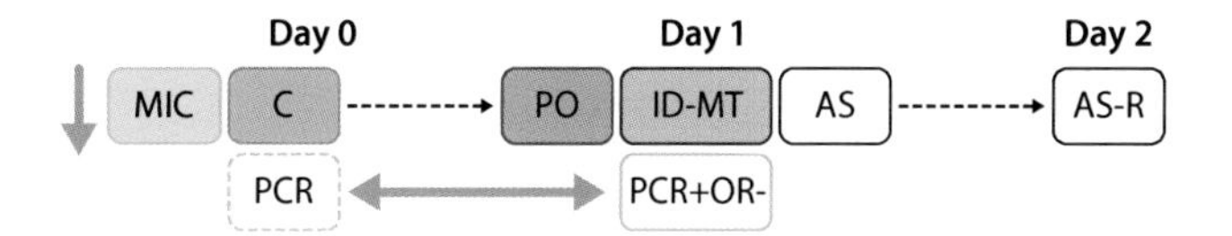

Figure 5.9 Time scale for CSF cultures (see text for explanation).

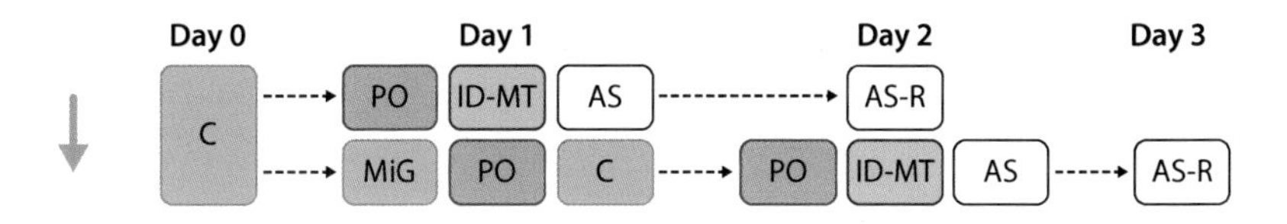

Figure 5.10 Time scale for wound swab (see text for explanation).

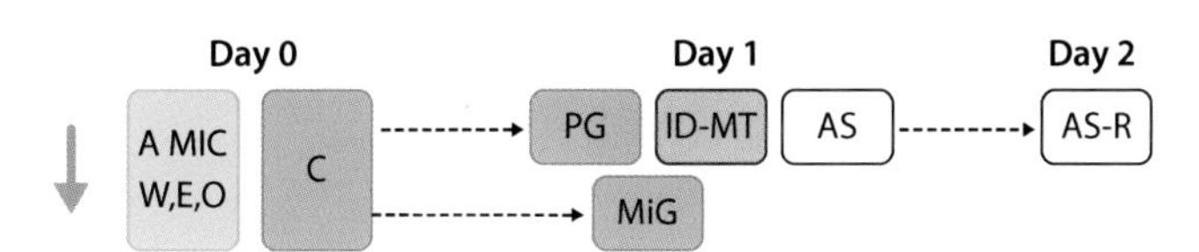

Figure 5.11 Time scale for midstream urine (see text for explanation).

examined for growth. A pure growth (PG) of an organism is the expected result. 'Selective' media enable direct identification of *Escherichia coli* and other common uropathogens, or the MALDI-TOF (ID-MT) system can be used. The number of bacterial colonies/L of urine is also recorded, with $>10^8$/L being considered significant. Antibiotic susceptibility tests are set up (AS) and reported (AS-R).

A urinary tract infection (UTI) is caused by one species, and if two or more grow, this is usually reported as mixed growth (MiG). Further work may be done on the bacteria. On occasion a report can include additional information, particularly if there is an infection control alert, e.g. mixed growth with $<10^8$/L colonies of extended-spectrum β-lactamase (ESBL)-producing *Escherichia coli*.

STOOL CULTURE (FIGURE 5.12)

Stool specimens are examined for *Escherichia coli* O157, *Campylobacter*, *Salmonella* and *Shigella*, using selective media for these enteric pathogens. The *Escherichia coli*, *Salmonella* and *Shigella* plates are incubated at 37°C for 24 hours, while the *Campylobacter* plate is incubated for 48 hours in a microaerophilic environment to optimize recovery. In addition, a portion of the stool sample is inoculated into selenite broth, a selective enrichment broth for *Salmonella*. After 18 hours this is subcultured onto agar medium to recover *Salmonella* that has grown.

Antibiotic susceptibility tests are done but are usually not reported. Antibiotics may be used in treatment, with the exception of shiga toxin-producing *Escherichia coli* (e.g. O104 and O157), where antibiotics are currently contraindicated. Stool extracts are also examined for the *Cryptosporidium* parasite by microscopy or enzyme immunoassay (EIA). These tests are also done to detect *Giardia* and other parasites.

CLOSTRIDIUM DIFFICILE TOXIN TESTING (FIGURE 5.13)

Laboratories routinely do this test once each day, providing specimens have been 'booked in' by 0900–1000. (The same booking-in protocol is used for norovirus and the respiratory virus PCR tests.) An extract of the stool specimen (Ex) is tested for the *Clostridium difficile*-specific glutamate dehydrogenase enzyme (GDH) by EIA. GDH-positive specimens are then tested for toxins A and B by an EIA (E), and toxin-positive results are communicated to the clinical team and the infection control team. Certain laboratories test GDH+/toxin− samples for the presence of the toxin genes by PCR. This identifies organisms that have the genes to code for toxin, but were not producing them at the time the specimen was collected.

HOW BACTERIA ARE IDENTIFIED

It is useful to refer to the classification of bacteria and yeasts, presented in Chapter 1 (**Figures 1.1, 1.31**). In order to speciate these organisms, single tests such as those outlined below are used. In addition the Biomerieux API system uses panels of tests in one strip. The MALDI-TOF machine, for example manufactured by Brucker, is being increasingly used by laboratories.

GRAM-POSITIVE BACTERIA

Tests for the initial identification of gram-positive bacteria are outlined below.

Differentiation of streptococci and staphylococci

The catalase test is used to differentiate streptococci (catalase-negative) from staphylococci (catalase-positive). When exposed to hydrogen peroxide, catalase-positive

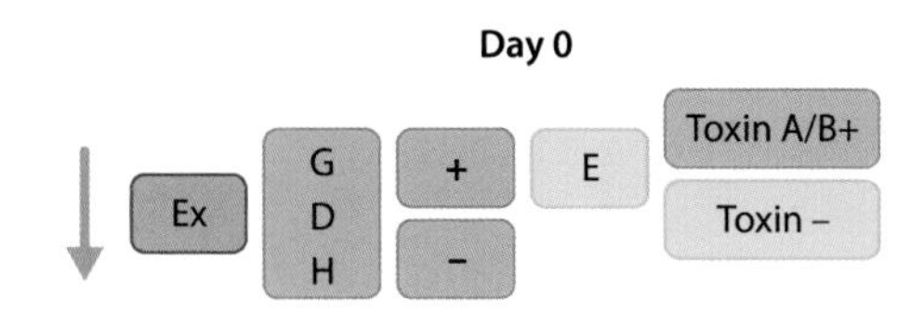

Figure 5.13 *Clostridium difficile* toxin testing (see text for explanation).

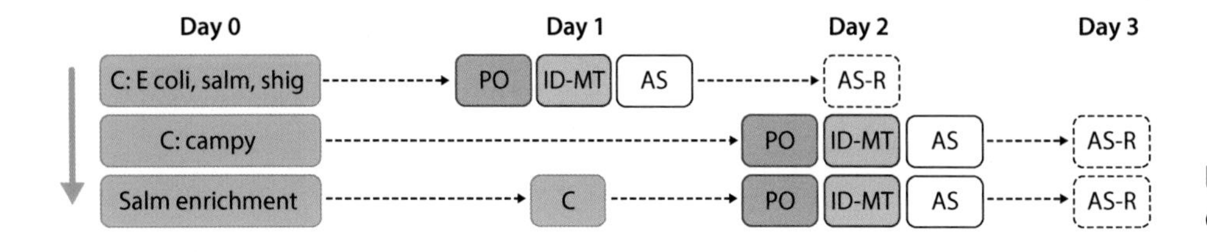

Figure 5.12 Time scale for stool culture (see text for explanation).

bacteria convert the peroxide to water and gaseous oxygen. There are various ways of determining catalase activity. In the example shown in **Figure 5.14**, a capillary tube containing hydrogen peroxide solution is carefully dipped into a single colony, and if catalase is present, oxygen gas is released and the bubbles observed.

Differentiation of staphylococci

Coagulase test

The test that is usually done on staphylococcal isolates is the coagulase test. Coagulase is an enzyme that is able to clot plasma in a fashion similar to the thrombin-catalyzed conversion of fibrinogen to fibrin. The test is important in differentiating *Staphylococcus aureus* from the coagulase-negative staphylococci such as *Staphylococcus epidermidis*, the common skin commensals. The presence of a coagulase-negative staphylococcus in blood culture would usually be considered a skin contaminant, whereas *Staphylococcus aureus* prompts a reassessment of the patient to determine the possible source and confirm the correct antibiotic treatment.

There are two types of coagulase enzyme, cell bound and free, which are detected by the 'slide' and 'tube' test respectively (**Figure 5.15**). Commercial kits are available to do tests such as the coagulase. The Oxoid Staphytect test uses blue coloured latex beads coated with protein A, fibrinogen and antibodies to the cell wall polysaccharide of *Staphylococcus aureus*. Coagulase-negative staphylococci will not react with these beads. However, *Staphylococcus aureus* cross-links the beads by the interaction of the coagulase with fibrinogen, cell wall components with the antibodies and the fact that protein A binds the Fc portion of the antibodies. Clumping of the beads indicates a positive result (**Figure 5.16**).

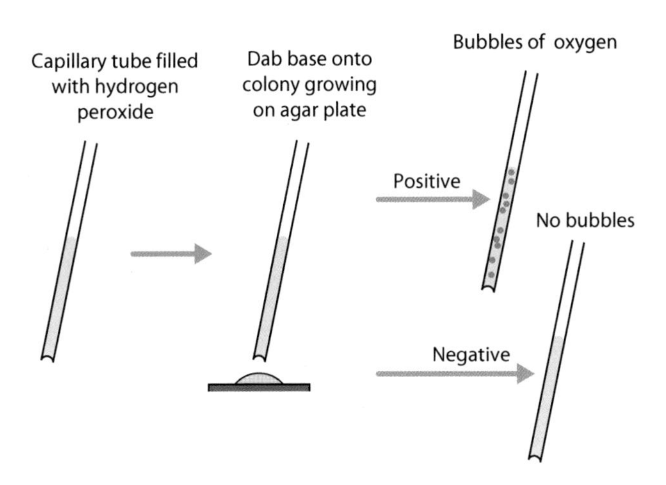

Figure 5.14 An outline of the catalase test.

DNase test

The identification of *Staphylococcus aureus* can be confirmed by the deoxyribose nuclease (DNase) test, as this organism is DNase-positive and the coagulase-negative staphylococci are DNase-negative. The test relies on the fact that unhydrolyzed native deoxyribonucleic acid (DNA) is insoluble and precipitates in strong acid. Agar containing DNA is 'stabbed' with staphylococci and incubated overnight at 37°C. If bacteria in the developing colony secrete the enzyme, DNA will be degraded into soluble nucleotides. After incubation, the plate is flooded with hydrochloric acid. A clear area around a colony, indicating hydrolysis of the DNA, identifies the organism as *Staphylococcus aureus* (**Figures 5.17, 5.18**).

Identification of streptococci

α- and β-haemolysis

The first step in the classification of streptococci is their haemolytic nature as exhibited on blood agar. Many bacteria produce haemolysins, which are extracellular proteins secreted by the cells that degrade lipid membranes. The membrane of red blood cells is also degraded and the lysis of these cells can be seen on blood agar plates. There are two types of haemolysis, α and β. α-haemolysis produces a green discolouration of the agar as a result of incomplete haemolysis of blood cells; β-haemolysis produces a clear zone of haemolyzed cells around the bacterial colony (**Figure 5.19**). α-haemolytic streptococci are further subdivided on the basis of the optochin test. *Streptococcus pneumoniae* is sensitive to the chemical optochin, while the remainder of the α-haemolytic streptococci are resistant to this compound. This test is also shown in **Figure 5.19** and is another example of a simple method to classify an organism to species level.

Lancefield grouping of the β-haemolytic streptococci

The β-haemolytic streptococci are further subdivided on the basis of their cell wall polysaccharide. This 'Lancefield grouping' relies on the extraction of the polysaccharide

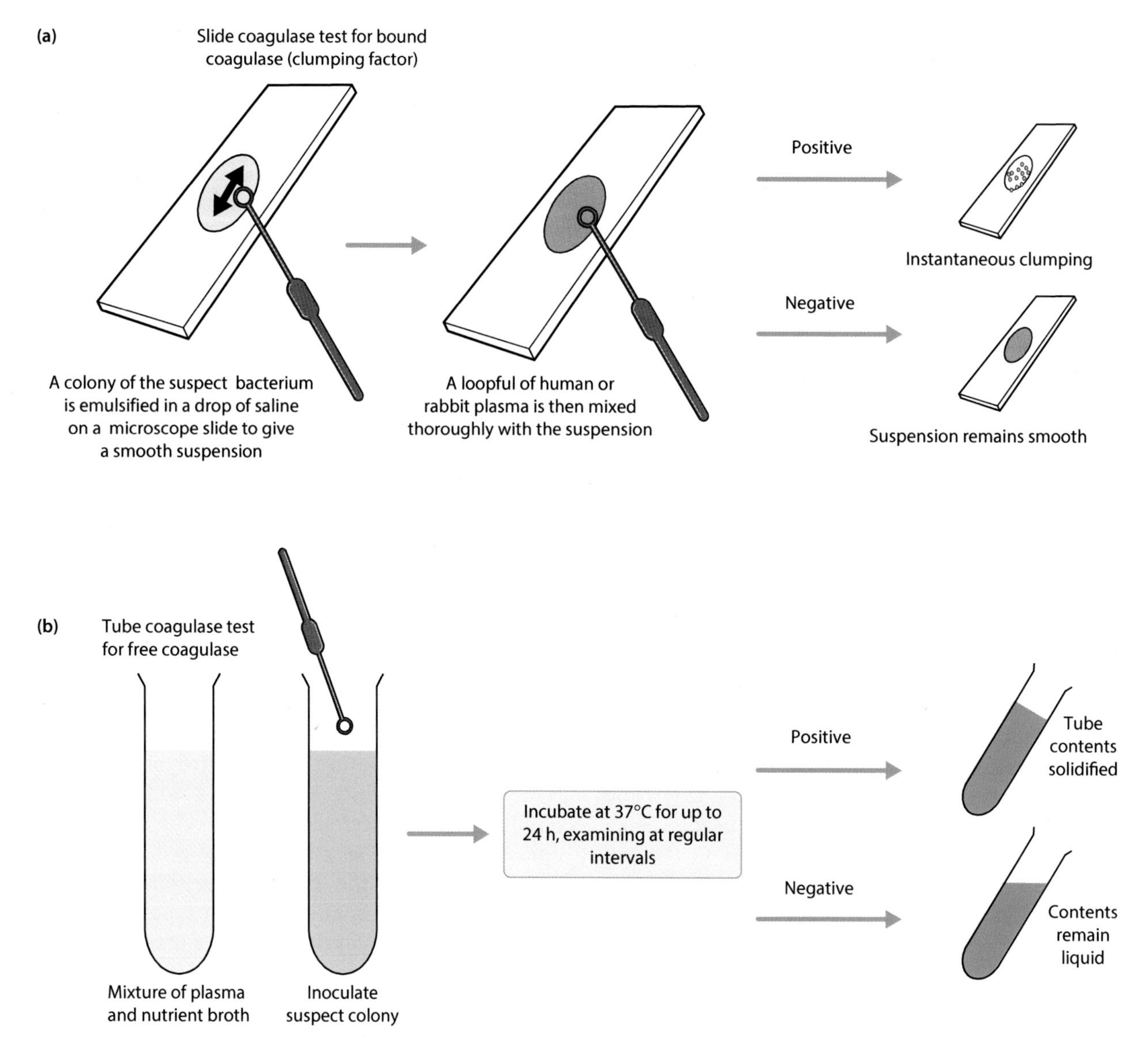

Figure 5.15 An outline of (**a**) the slide coagulase test; (**b**) the tube coagulase test.

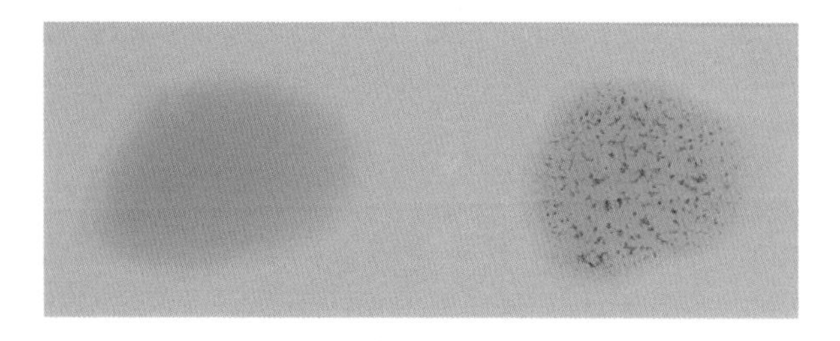

Figure 5.16 A 'loop-full' of two staphylococcal isolates is mixed with Staphytect latex beads. Clumping of the blue latex beads identifies one isolate as *Staphylococcus aureus*.

by enzyme or acid. The procedure used is outlined in **Figure 5.20**. The soluble extract is mixed with latex beads coated with group-specific antibodies; an agglutination reaction identifies the group the isolate belongs to. There are six major groups of β-haemolytic streptococci, termed A, B, C, D, F and G, and within these there are a number of important pathogens, including group A streptococcus, *Streptococcus pyogenes*, group B streptococcus, *Streptococcus agalactiae* and group C streptococcus, *Streptococcus dysgalactiae*.

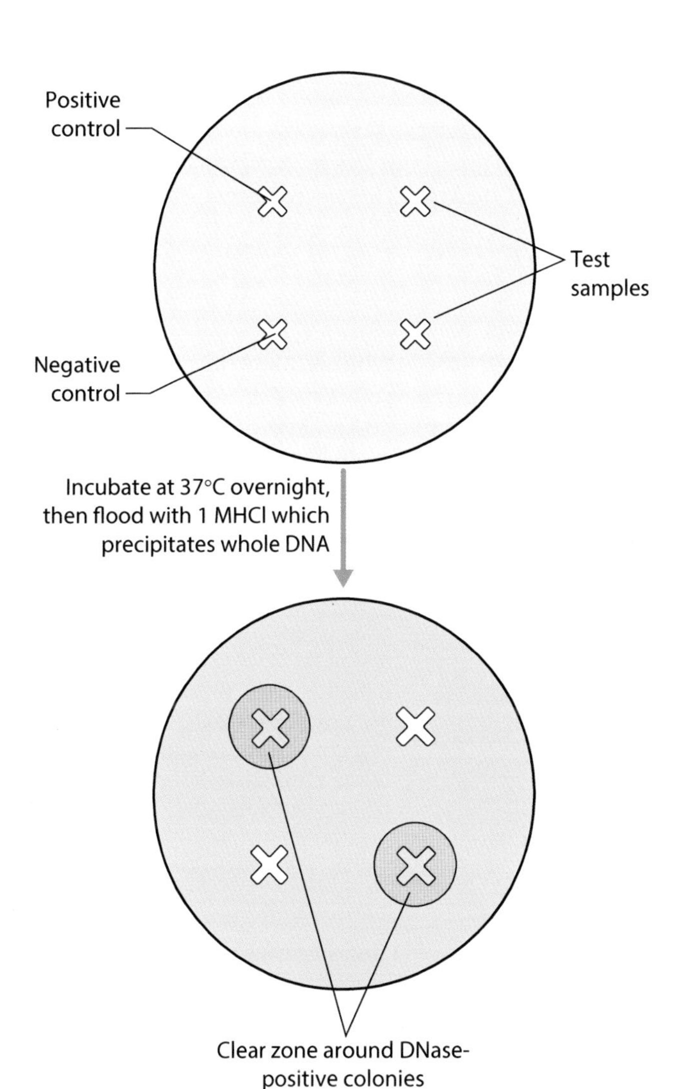

Figure 5.17 The deoxyribose nuclease (DNase) test. Staphylococci are inoculated into agar containing macromolecular DNA. Following incubation, addition of hydrochloric acid (HCl) precipitates macromolecular DNA. Around colonies producing DNase, there is no intact DNA and a clear area is seen (see **Figure 5.18**).

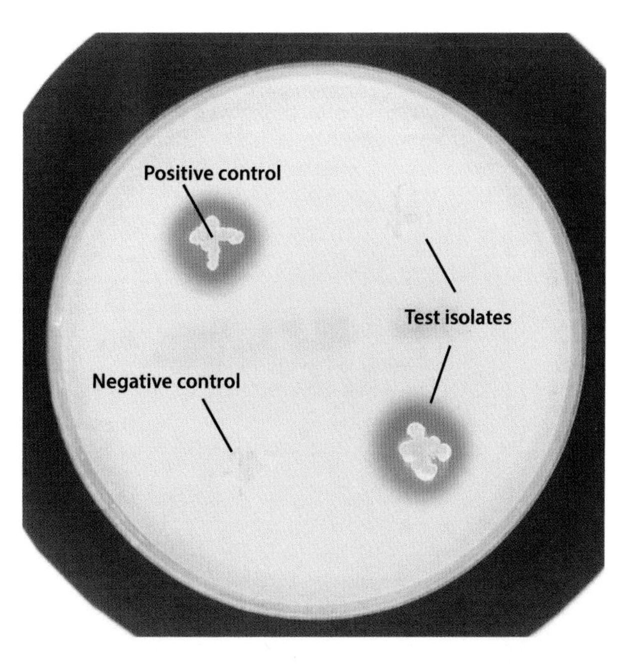

Figure 5.18 A DNase nuclease plate showing both positive and negative results.

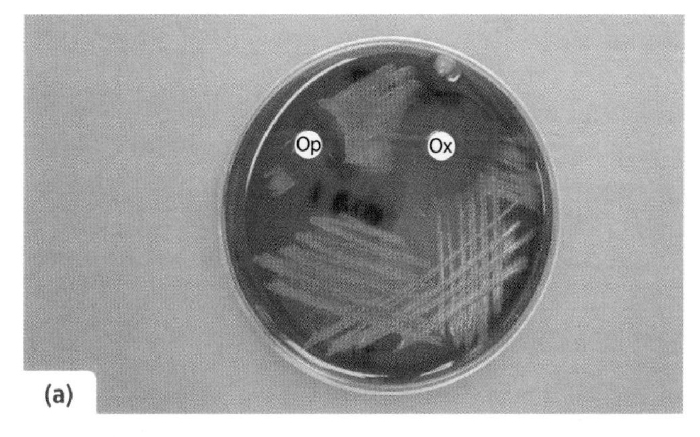

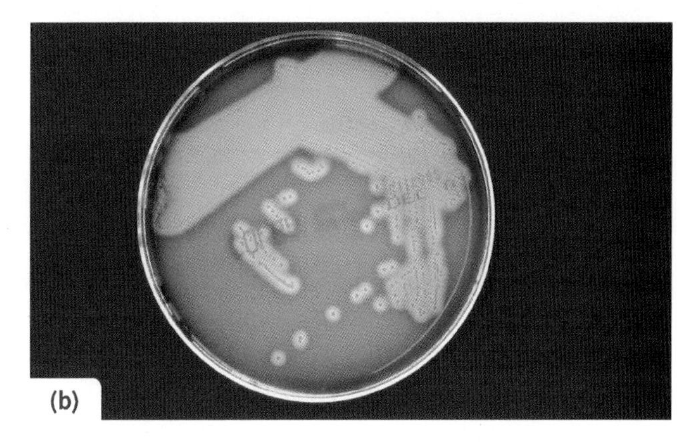

Figure 5.19 Blood agar plates showing (**a**) α- and (**b**) β-haemolysis. The α-haemolytic organism is sensitive to optochin (Op) and to oxacillin (Ox), identifying the isolate as a penicillin-sensitive pneumococcus.

GRAM-NEGATIVE BACTERIA

The Gram stain outline of these bacteria as a rod, coccus, coccobacillus or diplococcus is useful in making a presumptive identification. The finding of gram-negative diplococci in a CSF specimen collected from a previously well individual with meningitis is essentially diagnostic of meningococcal meningitis. *Vibrio* is a curved organism, while *Campylobacter* and *Helicobacter* are 'seagull' shaped.

Once the bacteria have produced colonies on solid media, the BMS will visually differentiate 'coliforms', *Pseudomonas*, *Haemophilus*, and other bacterial types

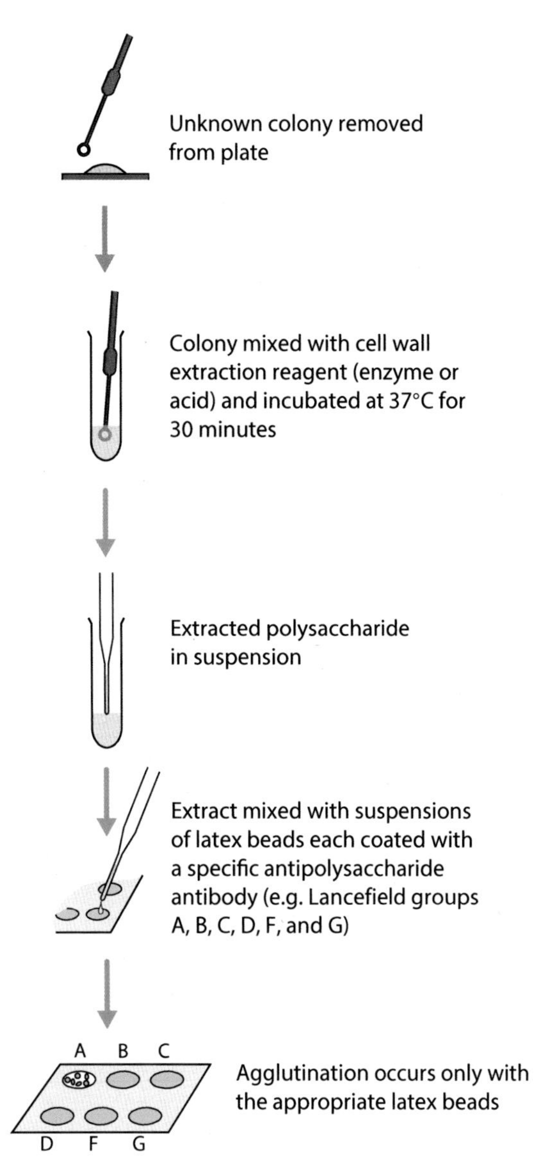

Figure 5.20 Lancefield streptococcal grouping of the β-haemolytic streptococci. A positive antibody–antigen reaction is shown by clumps and identifies the group; in this example, group A streptococcus or *Streptococcus pyogenes*.

from each other. As with gram-positive bacteria, gram-negative organisms can also be classified using simple tests.

Oxidase test

Gram-negative bacteria can be divided into oxidase-positive and oxidase-negative. Oxidase-positive bacteria have cytochrome c (part of the electron transport chain, situated in the cytoplasmic membrane), which is able to convert the colourless agent tetramethyl phenylenediamine to a blue compound. This is an important test as it differentiates oxidase-positive bacteria, such as *Pseudomonas aeruginosa*, *Neisseria meningitidis* and *Neisseria gonorrhoeae*, from the large group of oxidase-negative 'coliforms'. The outline of the procedure is shown in **Figure 5.21** and a positive result in **Figure 5.22**.

X and V test

This simple test is used to differentiate *Haemophilus influenzae* from other haemophilus species. *Haemophilus influenzae* optimally grows on 'chocolate' agar. Chocolate agar is blood agar where the medium has been heated to 80°C

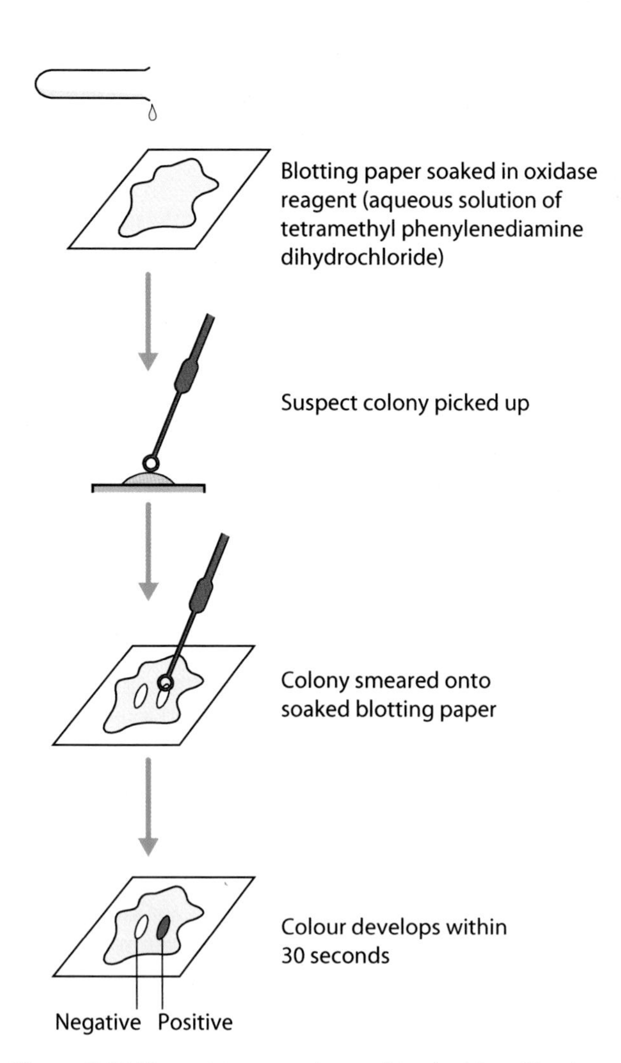

Figure 5.21 The oxidase test is used in the identification of gram-negative bacteria. Cytochrome c is able to convert the colourless oxidase reagent to a blue colour, identifying the organism as oxidase positive.

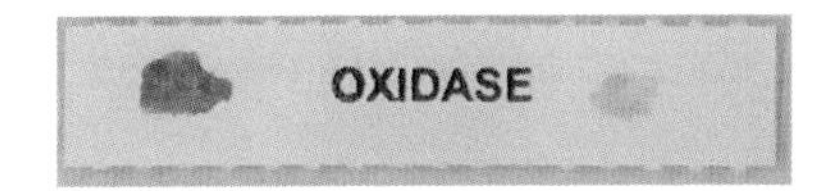

Figure 5.22 A purple discolouration on oxidase paper indicates an oxidase-positive gram-negative isolate.

before the plates are poured. This process releases haemin from the lysed red cells into the medium. Haemin (factor X) and NAD (factor V; found in the serum) are essential for the growth of *Haemophilus influenzae*. The test relies on placing paper discs containing either X or V factor and a disc containing both factors on a basic medium such as nutrient agar. *Haemophilus influenzae* will only grow around the XV disc as it has a requirement for both factors. *Haemophilus parainfluenzae* grows around both the XV and V disc, as it only requires the V factor for growth (**Figure 5.23**). Identification of *Haemophilus influenzae* by the X and V test is shown in **Figure 5.24**.

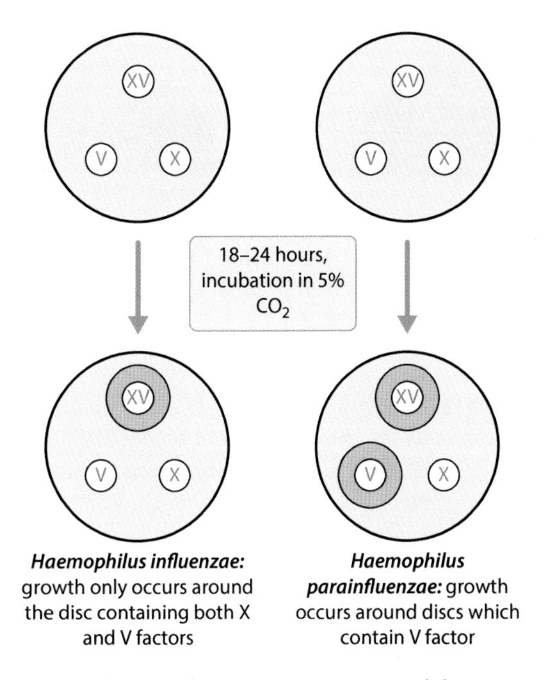

Figure 5.23 The X and V (NAD) test. *Haemophilus influenzae* is identified by growth around the disc that contains both X and V.

ANAEROBIC BACTERIA

As with other bacteria, the Gram stain result is part of the initial identification process. Members of the gram-positive clostridia produce heat-stable spores. The position of these spores within the cell is also useful in identification.

By their very nature, anaerobes do not survive in the presence of oxygen, as they are unable to 'detoxify' O_2- radicals. If several millilitres of pus is present, this should be collected. While modern transport swabs do keep a wide range of labile organisms viable, pus is the better sample. If further testing is required, for example for mycobacteria, pus is a superior specimen to process for this purpose. In most laboratories a metronidazole antibiotic disc is placed at the junction of the primary and second streak on the anaerobic blood agar plate (**Figure 5.25a**). Plates are incubated in an anaerobic cabinet, and examined at 48 hours. A zone of inhibition around the metronidazole disc identifies anaerobes (**Figure 5.25b**). Further identification is done by the testing systems described below.

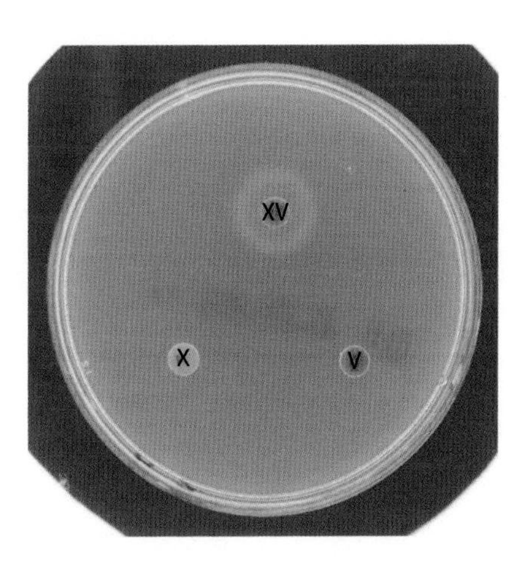

Figure 5.24 An X and V plate identifies *Haemophilus influenzae*.

SPECIATION OF BACTERIA AND FUNGI

BIOCHEMICAL TESTS

The full identification of gram-negative bacteria usually relies on a series of biochemical tests or use of MALDI-TOF technology. As bacteria are able to metabolize (ferment) certain sugars to acids, growth (turbidity) or a colour change in the medium from alkaline to acid can be used to determine which sugars a particular organism utilizes. The expression of an enzyme such as urease and the detection of end-products of metabolism such as indole, a product of tryptophan metabolism, are other tests used. A full range of these tests is available commercially, for example the API system produced by Biomerieux. The wells of the 'API strips' are inoculated with a preparation of bacteria from pure culture, incubated overnight, and the biochemical tests either read

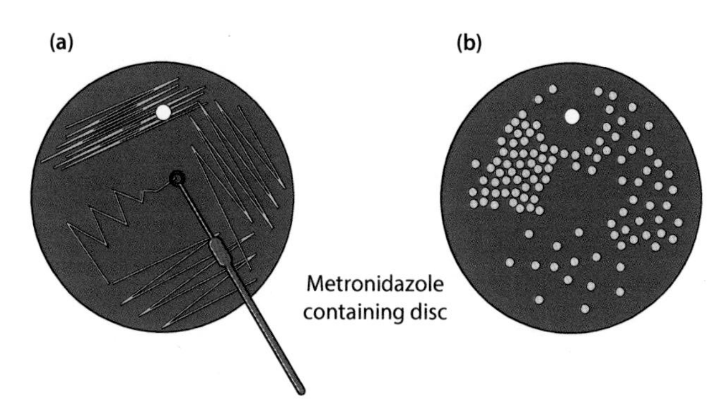

Figure 5.25 (**a**) A metronidazole disc is usually placed near the junction between the primary and secondary 'streaks' on the anaerobic blood agar plate. (**b**) A zone of inhibition after 48 hours, incubation around the metronidazole disc gives a preliminary identification of an anaerobe.

directly or completed by the addition of various agents. The number code that is produced gives the identification of the organism as outlined in **Figure 5.26**. Three API strips are shown in **Figure 5.27** after inoculation (**a**), after overnight incubation (**b**), and after addition of test reagents (**c**). The organism here is *Escherichia coli*, which ferments all but two of the sugars, and amongst other test results is indole-positive.

MASS SPECTROMETRY: THE MALDI-TOF SYSTEM

Mass spectroscopy has revolutionized the identification of organisms, including bacteria and yeasts, by matrix-assisted laser desorption ionization time of flight (MALDI-TOF; Figure **5.2d**). A sample of a single colony of

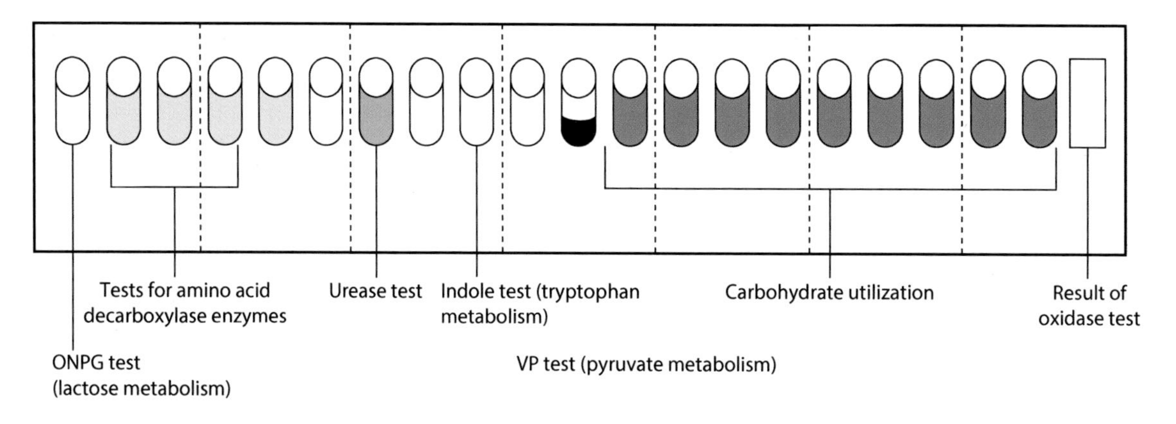

For each set of three tests, positive tests are scored (from left to right): 1;2;4. The code obtained identifies a particular bacterium

Examples of strips after incubation

1 0 4 | 1 0 0 | 0 0 4 | 0 0 4 | 1 0 4 | 1 2 4 | 0 2 0
=5 | =1 | =4 | =4 | =5 | =7 | =2

Identification profile = 5144572 = ***Escherichia coli***

1 2 0 | 1 2 0 | 0 0 0 | 1 0 4 | 1 0 4 | 1 2 4 | 1 2 0
=3 | =3 | =0 | =5 | =5 | =7 | =3

Identification profile = 3305573 = ***Enterobacter cloacae***

Figure 5.26 The API 20E™ (produced by Biomerieux) is an example of a commercially available identification system for bacteria such as coliforms. (ONPG: orthonitrophenyl-β-D-galactopyranoside; VP: Voges–Proskauer.)

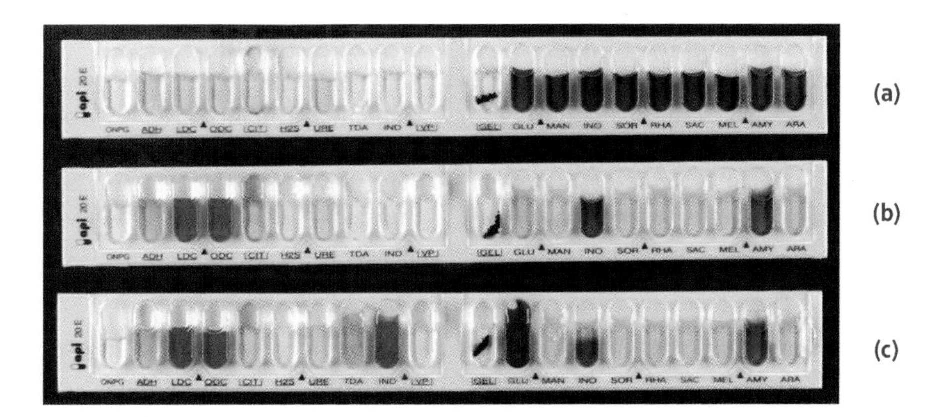

Figure 5.27 Three API 20E strips: (**a**) immediately after inoculation; (**b**) after 24 hours, incubation; (**c**) that in (**b**) after the addition of reagents to certain wells. The organism here is *Escherichia coli*. Here the first carbohydrate well (glucose) is also used for the nitrate reduction test.

the organism from a culture plate is mixed with a matrix substance such as sinapinnic acid, a phenyl propanol. When this mixture is bombarded with an ultraviolet (UV) laser, this ablates the matrix, which 'explodes' into an ionized gas. The proteins from the organism are ionized into a gaseous form too. This process takes place in a vacuum, and the ionized components are accelerated through an electric field generator and travel through the vacuum at a rate proportional to their mass. The time of flight and amount of each protein are measured by the detector (**Figure 5.28**).

The detector gives a qualitative and quantitative readout of the proteins that were in the sample. The complex aggregation of proteins that make up the ribosomes of the cell are dominant. This profile is compared with a database that has the protein profile of most of the bacteria (and yeasts) found in clinical practice, and speciation of the organism is obtained.

This methodology provides the identification of the organism within minutes, significantly improving the turn-around time. The sample plate has 96 wells that are read sequentially, so that the identification of dozens of organisms can be achieved within 30 minutes.

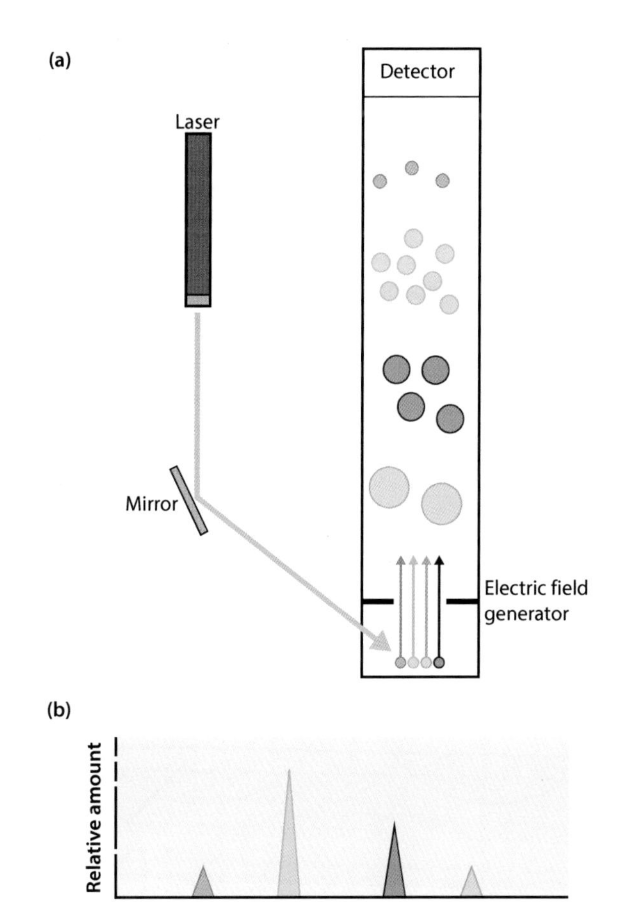

Figure 5.28 (**a**) The MALDI-TOF process involves vapourizing the proteins of an organism in a matrix, which are accelerated through the electric field; 'time of flight' is proportional to mass. (b) The mass and relative amount of each protein is recorded and this profile is compared with a database that gives the species of the organism.

ANTIBIOTIC SUSCEPTIBILITY TESTING

The tube MIC and minimum bactericidal concentration (MBC) tests as discussed in Chapter 4 are cumbersome and are not suitable for routine high-volume work. Many laboratories use the disc susceptibility testing, described below. There are also automated antibiotic susceptibility testing systems such as the Biomerieux Vitek machine (**Figure 5.2e**). A cassette containing a number of cells, each with one concentration of an antibiotic, is inoculated with a dilute suspension of the organism. In the machine, growth in each cell is monitored by electronically measuring turbidity (growth) at regular intervals. The rate of growth is inversely proportional to the susceptibility to the antibiotic. This rate is used to record if the organism is sensitive or resistant to the antibiotic. A MIC value can also be obtained.

ETEST®

The ingenious Etest employs a strip of material incorporating an exponential gradient of an antibiotic from one end to the other. The strip is placed on an agar plate inoculated with the organism (**Figure 5.29a**). As the organism multiplies, the antibiotic diffuses out of the strip. The rate of diffusion is proportional to the concentration at a particular point, and this will determine the zone of inhibition. The MIC can be read from the scale, as shown in **Figure 5.29b**. A photograph of an Etest is shown in **Figure 5.30**. While the test is more expensive and labour intensive than the disc method, it is useful for determining the MIC of selected organisms. For example, laboratories may monitor the MIC values of *Streptococcus pneumoniae* isolates in a region where the prevalence of isolates with reduced susceptibility to penicillin is increased. In the setting of infective endocarditis, this test is used to determine the MIC of streptococci and enterococci to a number of antibiotics, including penicillin and (synergistic) gentamicin.

DISC SUSCEPTIBILITY TEST

Here, cellulose discs impregnated with a standard amount of antibiotic are placed on agar plates inoculated with the organism to be tested. As the bacteria multiply during overnight incubation, the antibiotic diffuses into the agar. A zone of inhibition is produced proportional to the susceptibility of that organism to that antibiotic (**Figures 5.31, 5.32**). Using standardized criteria, the size of the zone of inhibition equates to the organism's resistance or sensitivity to the antibiotic. Results are usually reported as 'resistant' or 'sensitive'; occasionally the term 'intermediate' may be used. Note that in the examples given here, isolate B is resistant to amoxycillin. This is due to production of β-lactamase such as the TEM1 and SHV1 of *Escherichia coli* and *Klebsiella pneumoniae*.

The zone sizes used in disc testing that relate to susceptibility or resistance of one organism to an antibiotic are determined by 'breakpoints'. These are derived from MIC values that distinguish wild-type, susceptible populations of the bacterium from those that have acquired resistance mechanisms. This is done by determining the MIC value of a large number of isolates of that organism, and the MIC values are used to separate sensitive from resistant populations. These can then be equated to zones of inhibition of decreasing diameter to distinguish an organism as being susceptible or resistant to that antibiotic. The pharmacokinetics of the antibiotic is a critical parameter used in setting breakpoints.

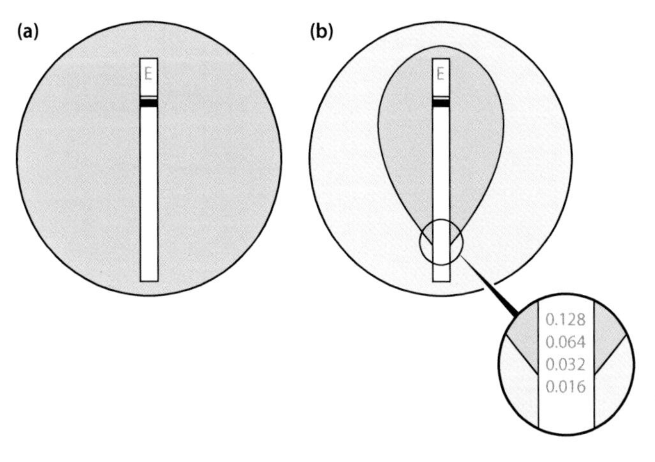

Figure 5.29 The Etest™. (**a**) A strip containing a logarithmic range of an antibiotic on an inoculated plate. (**b**) After 18–24 hours' incubation, the minimum inhibitory concentration is determined by reading the scale, here 0.032 mg/L.

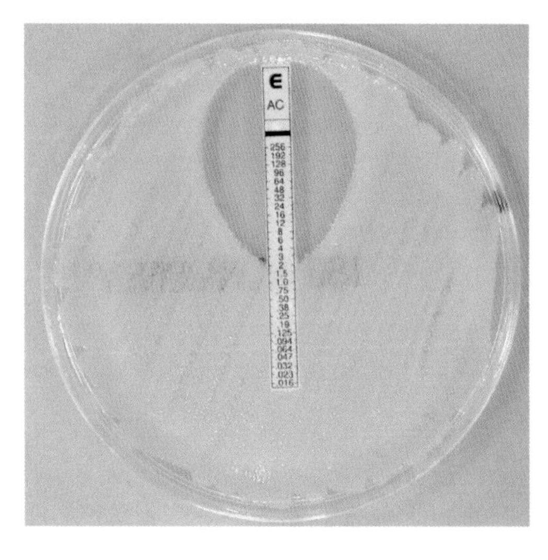

Figure 5.30 An amoxicillin Etest™ strip tested determines an MIC of 2 mg/L for this 'coliform'.

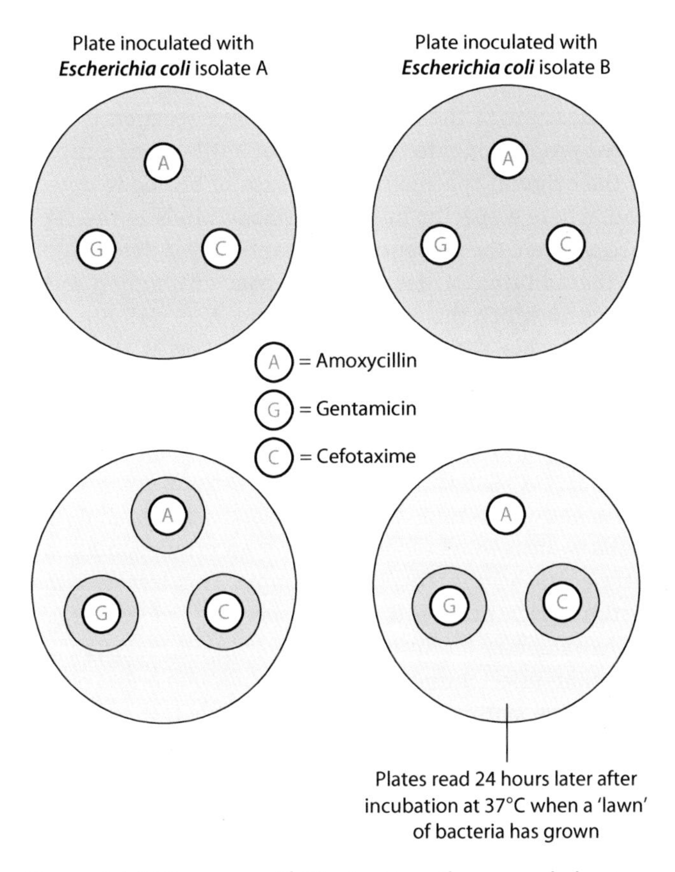

Figure 5.31 Disc susceptibility testing. The example here shows that isolate B is resistant to amoxicillin due to a (penicillinase) β-lactamase.

THE VIROLOGY LABORATORY (SEROLOGY AND MOLECULAR TESTING)

The classification of viruses (Chapter 1, **Figure 1.24**) relates to the diseases they cause, and the diagnostic tests used to identify them.

SEROLOGICAL TESTS

Serological tests are based on the interaction of an antibody that is specific for an antigen and tests are used to detect either antigen or antibody in the patient's serum. As discussed in Chapter 2, the body produces immunoglobulin (Ig) M and then IgG antibodies following exposure to an organism, either as a result of infection or vaccination. The presence of the antigen, or IgM antibodies, is a marker of acute infection. IgG antibodies are markers of past infection. Examples of the practical use of these tests are given in Chapter 12 (the hepatitis viruses) and Chapter 17 (HIV).

ENZYME IMMUNOASSAY

EIA is the fundamental serological test, and in addition to manual or semiautomated test systems, laboratories also use automated chemiluminescent microparticle immunoassay platforms (**Figure 5.2f**).

An outline of the EIA for detecting antigen or antibodies is shown in **Figure 5.33**. A specific antigen or antibody attached to wells of plastic plates is used to capture the

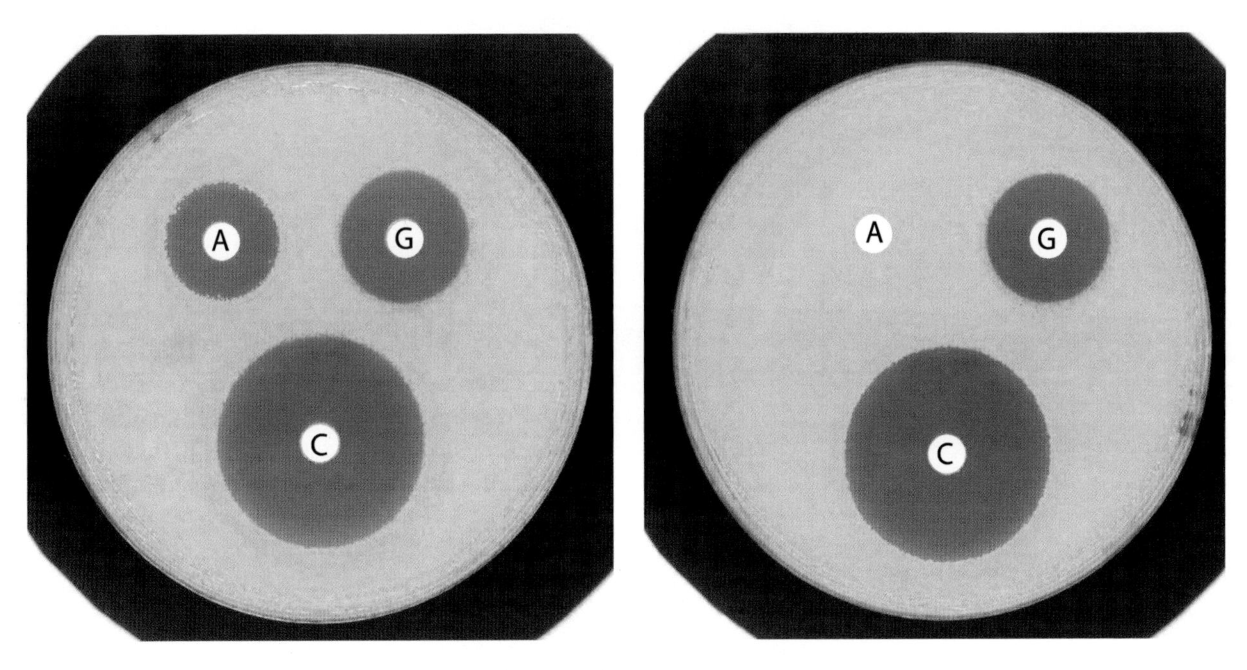

Figure 5.32 The photomicrograph shows the appearance of zones for isolates A and B.

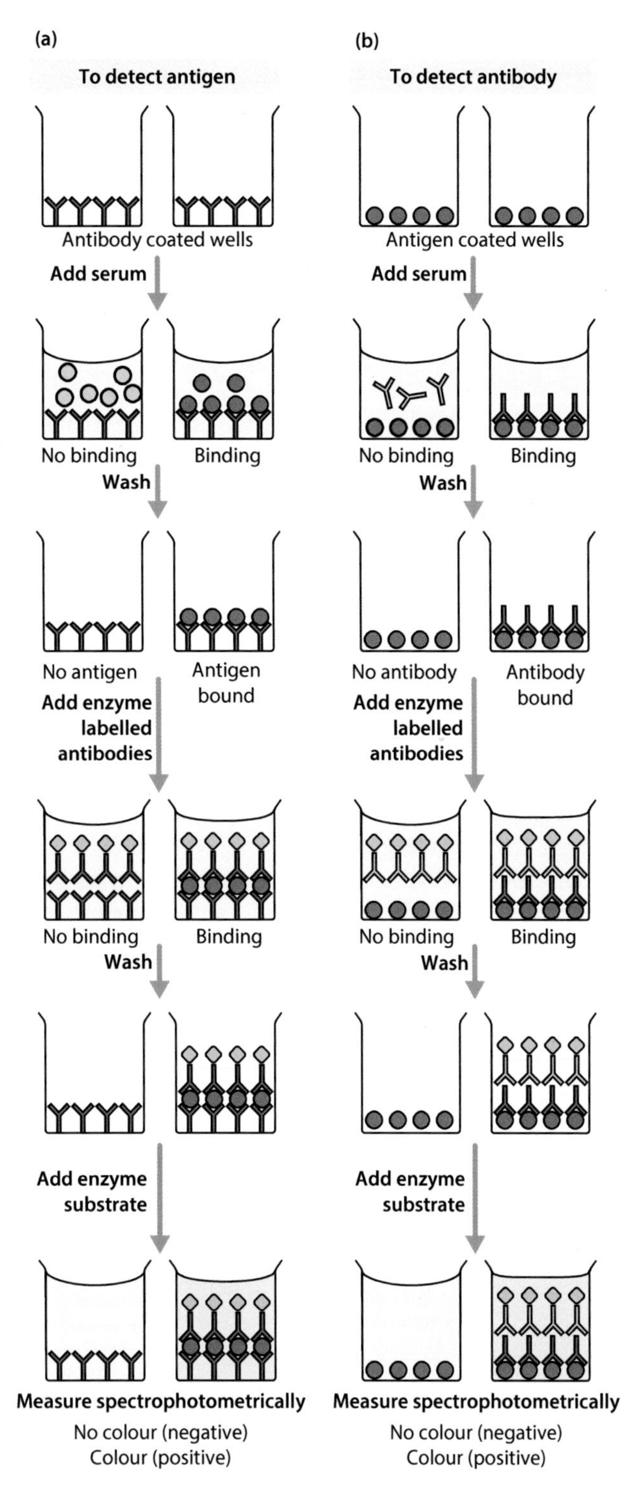

Figure 5.33 An outline of the enzyme immunoassay used to detect (**a**) antigen; (**b**) antibody.

corresponding antibody or antigen in a specimen, usually serum. After incubation and stringent washing, an indicator antibody is added to which an enzyme such as alkaline phosphatase is attached at the Fc end. Following further incubation and washing, the amount of bound indicator antibody is determined by adding the colourless substrate of the enzyme. The colour reaction is produced proportional to the amount of antibody or antigen in the original specimen. In the case of antibody detection, where a specific human antibody binds to the capture antigen, the presence of this antibody is determined by the addition of, for example, goat antihuman antibodies, to which the enzyme is bound. A 96 well EIA test for *Chlamydia* antigen in genital specimens is shown in **Figure 5.34**, with positive coloured wells.

MALARIA ANTIGEN TEST

Rapid tests to detect malaria in blood include microscopic examination of thin and thick blood smears and the malaria-antigen test, both done on an ethylenediamine tetra-acetic acid (EDTA) blood sample. This test uses the histidine-rich protein II (HRPII) specific for *Plasmodium falciparum*, and an antigen common in all malaria species that infect humans (*P. falciparum, malariae, ovale, vivax*). The outline of a malaria antigen test is shown in **Figure 5.35**. This consists of a nitrocellulose strip to which are bound bands of antibodies to human IgG (C; the test control), monoclonal IgG antibodies to *Plasmodium* species antigen (Pan P) and monoclonal IgG antibodies to the *Plasmodium falciparum*-specific HRPII protein (P fal). There is also a pad containing monoclonal antibodies to these antigens, which are conjugated with gold nanoparticles (Au con), as well as a sample and buffer pad. Gold-conjugated antibodies against IgG are included to react with the control band C.

A sample of the patient's blood is added to the sample pad, and elution buffer is added to the buffer pad. The elution buffer moves through the sample and Au con pads, and any antigen will complex with its corresponding monoclonal antibody. These complexes are washed up the nitrocellulose strip, and when they reach the relevant monoclonal antibody band, are arrested in a 'sandwich' effect at that site. As the bound complexes increase in number in a band, the proximity of gold nanoparticles to each other emits a red colour and a positive result is visualized. A positive control band of human IgG reacting with antihuman IgG validates the test. The result differentiates non-falciparum from falciparum infection.

MOLECULAR TESTS

There are a range of molecular tests in clinical microbiology, most of which are based on the PCR and nucleic acid sequencing. Other methods include ribotyping, which is used for epidemiological typing of *Clostridium difficile* isolates. DNA is extracted from a pure growth

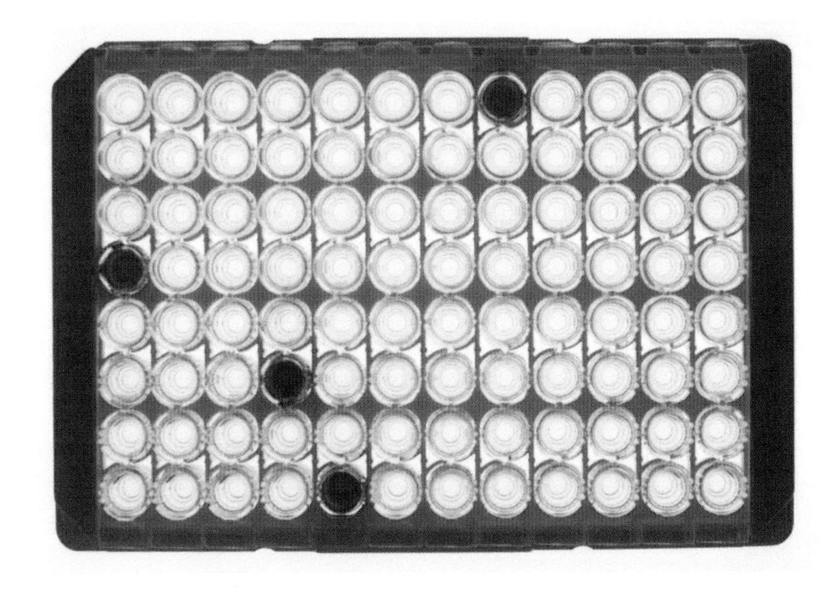

Figure 5.34 A chlamydia enzyme immunoassay, clearly showing the reaction of positive (coloured) samples.

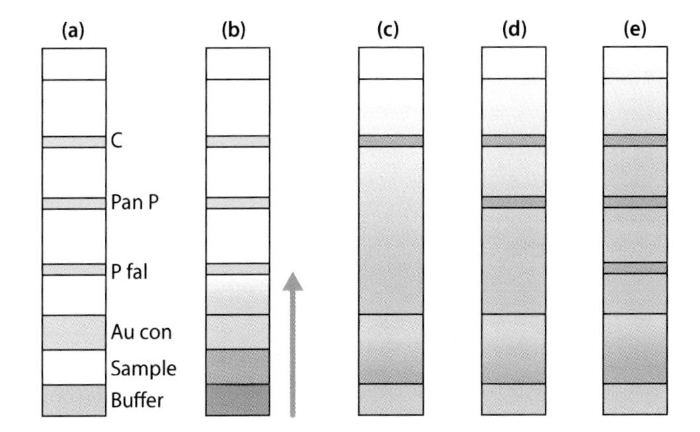

Figure 5.35 The malaria antigen test. (**a**) Immobilized bands of monoclonal antibodies to human IgG; control (C), common malaria antigen (Pan P), and *Plasmodium falciparum* HRPII antigen (Pfal). (**b**) After adding the blood sample and then the elution buffer, antigens combine with specific gold-conjugated antibodies (Au con) as they are washed up the strip, to be captured by the corresponding antibody in a strip; (**c**) C band (test control) positive only: negative result; (**d**) C and Pan P bands positive: malaria other than *Plasmodium falciparum* present; (**e**) C, Pan P and P fal bands positive: *Plasmodium falciparum* present.

of the organism, and is cut with a specific restriction enzyme. The DNA fragments are separated by gel electrophoresis, and transferred to a nitrocellulose membrane by Southern blotting. The DNA in the membrane is subjected to an annealing process with a radioactive-labelled DNA probe specific for the 16S ribosomal RNA gene of *Clostridium difficile*. The membrane is then exposed to an X-ray sheet and the bands are revealed. The ribotypes of *Clostridium difficile* have a different restriction enzyme fragment pattern. This restriction fragment length polymorphism (RFLP) is used to determine the ribotype of isolates from a ward outbreak in hospital. Identical isolates show that cross-transmission has occurred.

POLYMERASE CHAIN REACTION

PCR is the fundamental diagnostic molecular test, and relies on the fact that all organisms have unique sequences making up their genomes, be they DNA or RNA. If an organism is present in a specimen in very small numbers, the PCR will amplify a specific sequence into a detectable signal. To do this, short primer sequences, a heat-stable DNA polymerase and an excess of nucleotides are needed. The primer sequences are the key to PCR. These are short (10–20 nucleotides) lengths of 1-DNA whose sequence enables them to bind to the complementary nucleotide sequence in the target DNA. There are two sets of primers that bracket the target. Using repeated cycles of primer annealing, DNA synthesis and denaturation, an exponential reaction is produced. The end-product, produced within several hours, can be identified, for example, by agar gel electrophoresis (**Figure 5.36**). DNA molecules have the same charge to mass ratio, and migrate through an agar gel under the influences of an electric current on the basis of size. By using control DNA molecules and molecular weight markers, a specific amplified DNA sequence can be identified when the gel is stained.

Gel-based systems are not routinely used in diagnostics nowadays, and all the necessary reactions take place in liquid in a cassette. PCR is used for RNA viruses such

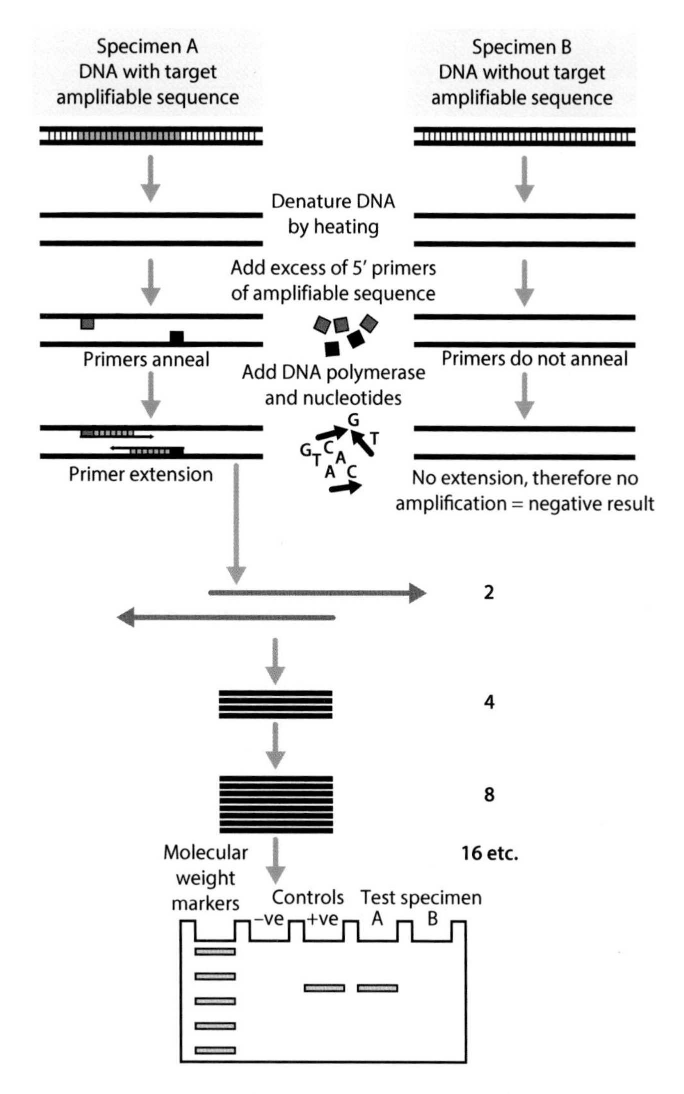

Figure 5.36 The polymerase chain reaction. Repeated cycles of denaturation, primer annealing and deoxyribonucleic acid (DNA) synthesis lead to exponential production of the target DNA that can be identified by gel electrophoresis. Specimen A is positive; specimen B is negative.

as HCV, HIV and the RNA viruses in the 'respiratory virus PCR panel'. The first step is to convert the extracted RNA to DNA by use of a reverse transcriptase enzyme.

Quantitative PCR is used in the management of virus infections such as CMV, HBV, HCV and HIV, to determine the viral load in blood of the patient and the effectiveness of antiviral therapy. Real-time PCR monitors the amplification of a target DNA molecule during the reaction. The more copies of the target nucleic acid in the original sample extract, the quicker the (exponential) reaction rate will be. One method of detecting the PCR product is to include a fluorescent dye that intercalates into 2-DNA. The amount of fluorescence detected is directly proportional to the amount of 2-DNA product in the reaction.

MULTIPLEX PCR TESTS

Multiplex PCR systems have revolutionized diagnostic microbiology. Even if the nucleic acid of two or more organisms is present in a sample, they will be identified. The PCR reaction has primers specific for each organism, so if the target nucleic acid is present, it will be amplified. In order to detect each organism, a further organism-specific nucleic acid probe is included in the reaction mixture. Each probe has a fluorescent reporter molecule that emits light of different wavelength so that the presence of each amplified target is identified.

Multiplex systems include:

- The multiplex respiratory virus PCR panel for adenovirus, influenza A/B, respiratory syncytial virus (RSV), paramyxovirus, metapneumovirus, rhinovirus (and bocavirus). The availability of the test on a daily basis is essential for the management of a hospital in the winter 'influenza season'. Prompt identification of patients with one (or more) of these viruses enables judicious use of single room facilities, which are often in short supply. When influenza virus is identified in this test, the use of antiviral treatment and prophylaxis and identification of vaccination status of contacts is done with clarity.
- The multiplex faecal pathogen panel. This includes *Campylobacter*, *Salmonella*, *Shigella*, *Clostridium difficile*, norovirus, *Cryptosporidium*, *Entamoeba* and *Giardia*. An extract of the stool specimen is screened for these organisms, and if one or more pathogens is identified, standard culture, microscopy and EIA methods are then used. This circumvents the labour-intensive examination, usually by culture, for pathogens on every stool specimen.
- The multiplex CSF panel. This includes *Streptococcus pneumoniae*, *Neisseria meningitidis*, *Haemophilus influenzae*, *Listeria monocytogenes* and *Streptococcus agalactiae*. Combined with CSF virus PCR (HSV1, HSV2, VZV and enterovirus), this testing system is an essential adjunct to standard bacterial culture methods.

PCR FOR DETERMINING ANTIBIOTIC SUSCEPTIBILITY

The carbapenemase-producing Enterobacteriaceae (CPE) pose a critical challenge in clinical diagnosis, treatment and infection control. There are also two important laboratory diagnostic issues:

- Conventional culture and antibiotic susceptibility testing takes several days for the final result.

Conventional susceptibility testing determines the phenotypic susceptibility profile of the organism. Isolates of *Klebsiella* can appear to be producing a carbapenemase, but on further examination are producing an ESBL and also have reduced susceptibility to carbapenems due to porin restriction, which gives the false carbapenemase-producing phenotype.

Multiplex PCR systems are available for the specific carbapenemase genes. The laboratory test takes several hours, and not only detects the presence of the carbapenemase gene, but using specific primers is able to identify specific types of enzyme, termed KPC, OXA-48, VIM and NDM. The use of such a test system has clear implications for infection control, and the management of patients with suspected colonization or infection with a CPE.

16S RIBOSOMAL RNA DNA AMPLIFICATION AND SEQUENCING

The bacterial 16S ribosomal RNA gene is 1542 base pairs in length, and contains nine hypervariable regions, V1–V9, each containing signature sequences that are unique to species level. DNA is extracted from a pure growth of the organism, and the 16S ribosomal gene DNA is amplified by PCR using primers specific for amplification of the variable regions. The DNA produced is sequenced, and matched against a known sequence database to identify the organism.

This methodology is also useful when specimens, such as joint fluid aspirates, fail to grow an organism, despite clear evidence of infection. This may be due to previous antibiotic use, or the organism has fastidious growth requirements. A centrifuged concentrate of the fluid is subjected to the same DNA extraction, amplification and sequencing steps and can then reveal the identity of the likely culprit causing the infection.

NUCLEIC ACID SEQUENCING

In recent years there has been a revolution in nucleic acid sequencing technology. New methodologies using nanopore technology underline this. This magnificent system uses bacterial proteins such as a streptococcal α-haemolysin, which act as the nanopore. This is inserted into and traverses a lipid membrane or graphene sheet. An ion current can be created that flows through the nanopore, and any change to this current can be measured.

In order to sequence DNA, a DNA polymerase molecule is attached to the haemolysin. The polymerase feeds through a single strand of DNA through the pore under the influence of the electrical gradient across the pore. Each of the four nucleotides has a different effect on the electrical gradient, and by measuring this, each nucleotide is recorded as it moves through the pore, and the nucleotide sequence in the nucleic acid is determined. Remarkably, once one strand of the DNA molecule has been sent through for sequencing, the polymerase can then feed through the complementary strand for comparative sequencing.

In order to obtain a whole genome sequence (WGS) of an organism, DNA is purified, fragmented and dispensed in a machine with multiple nanopores. Individual DNA molecules are fed through the nanopores, the sequences obtained are aligned, matched and compared to known sequence databases to provide the WGS. All of this can be done within hours.

Because a whole genome sequence provides the highest degree of resolution it is likely to become the best way to determine if organisms in an outbreak are identical or not, as resolution down to single nucleotide polymorphisms (SNP) can be obtained.

WGS is likely to become a central methodology in routine microbiology laboratories. The national tuberculosis reference laboratories are doing this. In addition to the WGS of the organism, the genotypic markers of resistance to rifampicin and the other antibiotics are determined, and used in treatment, infection control and public health management.

This technology can also be used to sequence RNA directly.

Chapter 6

Infection Control

INTRODUCTION

Infection control is an integral part of the patient's care in hospital, and as discussed in Chapter 3, it is an essential process that needs to be initiated from admission.

The patient can be admitted with an infection control alert, for example known methicillin-resistant *Staphylococcus aureus* (MRSA) carriage, or residence in a nursing home where there is an outbreak of norovirus. If the patient has been in another hospital, especially outside the UK, this must prompt the alert for carbapenemase-producing Enterobacteriaceae (CPE).

After the initial risk assessment, the first action is to determine the need for nursing the patient in a single room, in order to protect other patients and staff. Appropriate signs identifying the infection control risk need to be placed outside the single room, only essential equipment retained in the room and single-use items used where possible. All health care professionals (HCPs) must be directed to use the appropriate personal protective equipment (PPE) and reminded of the correct disposal of waste.

Each infection control situation sets in train a management process for that patient, which is modified depending on confirmation of the infection control risk/diagnosis, the treatment required and whether or not the infection control risk can be revoked during the hospital stay. The patient who is admitted with infectious diarrhoea can usually be moved out of the single room onto the open ward when they are asymptomatic for 72 hours. The single room is then 'deep cleaned', including use of a hydrogen peroxide 'fogging' machine, which ensures destruction of *Clostridium difficile* spores. The room is then safe for the next patient to use.

Certain infection control situations, e.g. carriage of a CPE, require that the patient is nursed in a single room for the duration of their stay. When this patient is due for discharge to a nursing home, this is done in full agreement with the staff there, so that the correct process is followed.

While clinical staff on the ward have direct responsibility for patients with an infection control alert, the overall process is managed by the infection control team (ICT). This nursing team is led by a senior nurse, with the Director of Infection Prevention and Control (DIPC) responsible for the service to senior management level. The DIPC is often a consultant medical microbiologist.

The ICT is responsible for overseeing the delivery of infection control practices throughout the hospital, and ward 'link nurses', with additional training in this discipline, coordinate practice on their wards. Policies and procedures for specific organisms, e.g. MRSA and CPE, are regularly updated, and are the subject of mandatory training. New threats, such as Ebola virus, direct additional practices and procedures, and while these are determined at national level, each hospital must produce policies specific for its own circumstances in order to manage patients safely.

All HCP have a professional responsibility to ensure that organisms are not transferred from themselves, or from one patient to another, by poor practice. 'Bare-below-the-elbow' attire, correct hand-hygiene procedures, centred on the World Health Organization's (WHO) '5 moments', as well as any additional infection control precautions required for a specific organism, make the process safe.

OCCUPATIONAL HEALTH COMPLIANCE

All individuals working with patients need to comply with occupational health department (OHD) directives and have proof that they have immunity to certain infectious agents. HCP who are to perform exposure-prone procedures (EPPs) require additional screening tests. EPPs are those invasive procedures where a risk of injury to the worker may result in exposure of the patient's open tissues to the blood of the worker. HCP who are found to be hepatitis B virus (HBV) carriers (HBsAg+), can do EPP provided they are HBeAg negative, their HBV viral load is less than 10^3 genome equivalents/mL (or less than other recommended values) while on continuous therapy and are regularly monitored by the OHD physician and liver expert.

HCP who are hepatitis C virus (HCV) antibody positive can do EPP if they are shown to have HCV ribonucleic acid (RNA) below detectable levels (e.g. <50 IU/mL), having cleared the acute infection. When the HCP is identified as HCV RNA positive, identifying chronic carriage and infectiousness, the person is restricted from doing EPP. This may be revoked if there is the appropriate response to treatment, with a sustained virological response (SVR) 24 weeks after treatment has stopped. The HCP with human immunodeficiency virus (HIV) infection can do EPP if antiretroviral therapy (ART) is effective, with the viral load in plasma at <200 copies/mL (or less than other recommended values). In addition to OHD input, these HCP are reviewed regularly by the relevant specialist. (More details on HBV, HCV and HIV infection and diagnosis are given in Chapters 12 and 17).

An occupational health screening record for a HCP is shown in **Figure 6.1**; this person has satisfied the necessary OHD clearance regulations. The OHD should use anonymizing codes for staff confidentiality.

When a new employee has no (or inconclusive) evidence of bacillus Calmette–Guérin (BCG) vaccination, a Mantoux tuberculin skin (or interferon gamma) test should be performed. If the test is negative and the employee has not been previously vaccinated, BCG vaccination is advised if work involves contact with patients and/or clinical specimens. A risk assessment for HIV is done before BCG vaccine is administered.

Influenza vaccination is offered every year before the 'flu season' starts, and is often seen as a voluntary process. This is an unusual contradiction of infection control practice, as all staff should be vaccinated, using an 'opt-out' process in order to maximize protection for themselves and patients.

DRESS CODE AND PERSONAL PROTECTIVE EQUIPMENT

Unless wearing a hospital staff uniform, the practical dress code for medical staff in clinical areas is 'bare-below-the-elbows' (**Figure 6.2a**). Jewellery should be limited to a wedding ring and the identification badge is attached to the belt or waist band (not on a lanyard).

Before entering a single room, a disposable apron and gloves are applied, as the basic PPE for direct patient contact (**Figure 6.2b**). After completing the necessary interventions, these are removed, placed in the designated clinical waste bin and the hands washed and dried before exiting the single room. Application of alcohol-based hand wash then completes the process.

Face masks are worn when there is likely to be direct contact with respiratory secretions. Respiratory secretions can be broadly classified into droplets and aerosols. Droplets are propelled from the nasopharynx and tracheal/bronchial tree and range in size from 100 to 10 µm in diameter. These are mostly spread within a radius of 1 m from the patient. In general, a waterproof surgical face mask is worn, with eye protection as well. A combined mask/eye shield is shown in **Figure 6.2c**. An important reason for doing this is to reduce the chance of secretions

Occupational Health Staff identity code: ABCDEF		
Date of report: 01/03/20XX		
Hepatitis B virus surface antibody (HBsAb)	DETECTED (>100 mIU/mL)	Satisfactory level
Measles IgG antibody	DETECTED	Satisfactory level
Rubella virus IgG antibody	DETECTED	Satisfactory level
Varicella zoster virus IgG antibody	DETECTED	Satisfactory level
TB/BCG check	BCG scar visible	Consider protected
Exposure-prone procedure screen:		
Hepatitis B virus surface antibody (HBsAb)	DETECTED (>100 mIU/mL)	Satisfactory level
Hepatitis B virus surface antigen (HBsAg)	NOT DETECTED	No evidence of chronic infection
Hepatitis C virus antibody	DETECTED	Evidence of past infection
Hepatitis C virus RNA	NOT DETECTED	No evidence of chronic carriage
HIV 1 and 2 antigen/antibody	NOT DETECTED	No evidence of HIV infection

Figure 6.1 Example of laboratory results of a health care professional who satisfies the occupational health department (OHD) criteria for safe working, including exposure-prone procedures. The negative HBsAg, HCV RNA and HIV tests are confirmed on two separate blood samples to exclude any chance of error from collection to complete laboratory processing of the specimen. The OHD should use codes to maintain staff confidentiality.

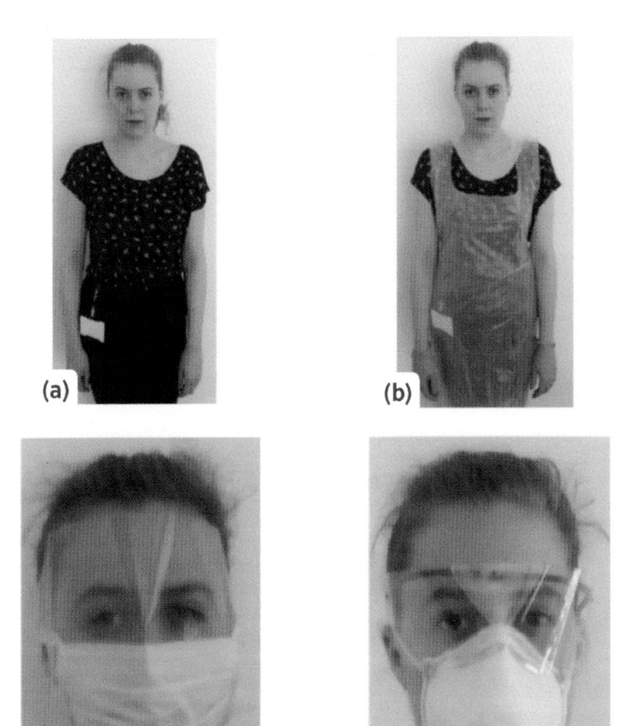

Figure 6.2 (**a**) 'Bare-below-the-elbow' is the basic standard for dress code in clinical practice. (**b**) Disposable apron and gloves are applied before entering a single room and direct patient contact. (**c**) Use of a combined face mask and eye shield is appropriate when managing the patient with a respiratory tract infection. (**d**) FFP3 masks are used in aerosol-generating procedures, and must always be used when managing patients with multi-drug-resistant tuberculosis and extensively-drug-resistant tuberculosis. Eye goggles are an additional protection.

landing on the face, and later being inadvertently brushed into the mouth.

Aerosols are produced from the alveoli and have a diameter of <10 µm. During aerosol generating procedures, such as bronchoscopy, the HCP wears a filtering face piece mask number 3 (FFP3). These masks filter out particles down to 0.6 µm in diameter. When properly fitted, these provide at least 99% protection. Every HCP who uses FFP3 masks has to undergo 'fit testing' to ensure that they know how to apply the mask so that there is no leakage of air around the outside (**Figure 6.2d**). An FFP3 mask is used at all times when managing patients with multi-drug-resistant tuberculosis (MDR-TB) or extensively-drug-resistant tuberculosis (XDR-TB), who are always nursed in a negative-pressure single room.

HAND WASHING

The correct process for hand washing is shown in **Figure 6.3**, and ensures clean hands. This should include the wrists, which cannot be done correctly if the HCP is not bare-below-the-elbow. Hand basins with soap dispensers should be used where possible. In most circumstances at the bed-side, alcohol-based hand rub is used. Cuffs of jackets and cardigans clearly interfere with hand washing.

THE 5 MOMENTS OF HAND HYGIENE

The WHO '5 moments' direct the stages when hand hygiene needs to be performed (**Figure 6.4**). This is usually done with an alcohol-based hand rub, which is available at each patient's bed side.

THE WARD LAYOUT AND ENVIRONMENT

It is important for the HCP to note the layout of each ward where they work, including the number of bays and beds and single rooms (including those single rooms under negative-pressure control). **Figure 6.5** shows the key areas of a hypothetical orthogeriatric ward for female patients. There are two 4-bedded bays, A and B. Staff rest areas and the kitchen for preparing patient food are situated away from the ablutions, single room and sluice room. The latter are in close proximity so that clinical waste from the single room is transported over the shortest distance to the sluice room. The patient in the one single side room (with patient ablutions) in this example, has an extended-spectrum β-lactamase (ESBL)-producing *Escherichia coli* identified in the catheter urine.

The ward has card-key access outside visiting times, and staff-only areas have key pad access doors. It is important to be aware of the hand-washing stations, as well as the siting of clinical and domestic waste bins. All HCP, including medical staff, need to be aware of the state of cleanliness on the ward. Waste bins that are overflowing, as well as clutter, are signs that there are problems with ward management and medical staff, especially consultants, need to address these.

The sluice room is central to the safe functioning of the ward, as it is here that relevant fluid waste from patients is disposed of, and where bed pans and commodes are cleaned and decontaminated. Not infrequently there can be lapses in cleaning that put patients at risk. A commode appeared spotless when viewed from above, but when turned upside down, large areas of dried faecal material were revealed (**Figure 6.6**). The spores of *Clostridium*

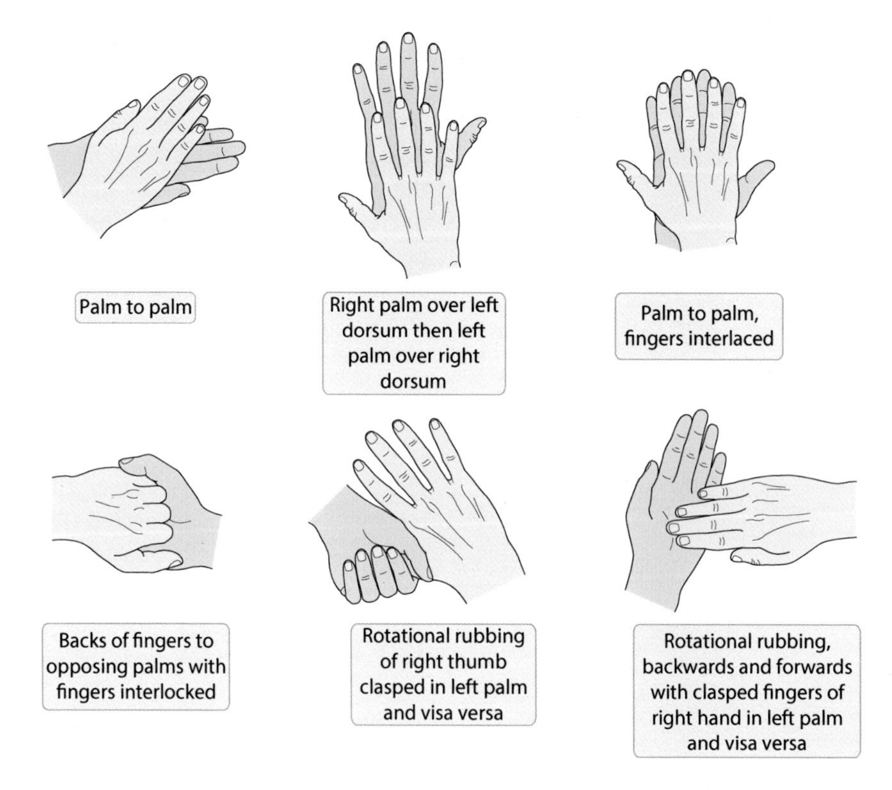

Figure 6.3 The correct method of washing the hands. Sleeves to the wrist inhibit the process being done correctly. From: Ayliffe's Control of Healthcare-Associated Infection Fifth Edition, Adam P. Fraise and Christine Bradley (Eds), Copyright 2009 Edward Arnold (Publishers) Ltd, reproduced by permission of Taylor & Francis Books UK.

difficile will be present here, a situation that is unsafe for patients.

NEGATIVE-PRESSURE VENTILATION SINGLE ROOMS

Patients with an infection control alert should be nursed in an en-suite single room that is preferably under negative-pressure ventilation (**Figure 6.7**). This is essential for patients with confirmed or likely MDR-TB or XDR-TB. A negative-pressure room has an anteroom, where staff can apply the necessary PPE before entering the bedroom. Patients with MDR-TB or XDR-TB will often be nursed here for a prolonged period, and require the necessary facilities for their wellbeing.

MDR-TB and XDR-TB patients must be nursed in negative-pressure isolation and FFP3 respirator masks must be worn.

THE SOURCE OF ORGANISMS AND HOW THEY ARE SPREAD

A list of organisms to consider is shown in **Figure 6.8**. Several are listed under a number of headings, reflecting various sources, such as airborne and indirect contact, e.g. influenza virus. Door handles contaminated with hands that are unwashed after sneezing are a route of transmission. Norovirus is found in large concentrations in the projectile vomit of the infected individual. This settles on surfaces, and is a source for others when contaminated hands are inadvertently put into their mouths.

AIR-BORNE

The organisms in this group come from human or environmental sources.

DROPLET

Droplets range from 100 to 10 μm in diameter, and are produced from the upper respiratory tract (URT) following sneezing or coughing. When inhaled by another individual, droplets will usually be trapped in the secretions of the URT, and organism will be retained here. This is however, an important route of transmission of respiratory viruses, where they gain access to the respiratory epithelium of the nasopharynx. Droplets of vomit breathed in through the mouth will enable norovirus to be swallowed.

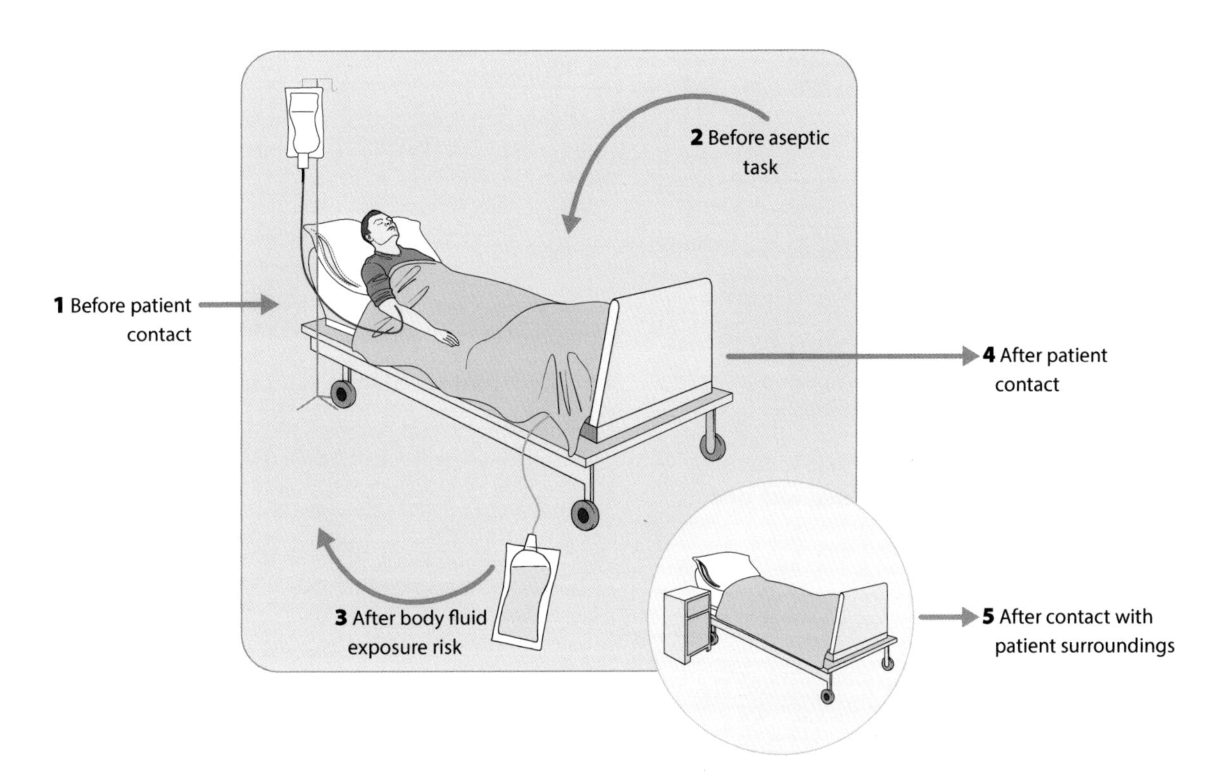

Figure 6.4 The World Health Organization's (WHO) '5 moments' of hand hygiene that must be used in the 'bed space' of the patient. Appropriate personal protective equipment must be placed first, and then removed in the correct sequential steps after the interaction with that patient. (Reproduced with permission from the WHO Guidelines on Hand Hygiene in Health Care: a summary [2009].)

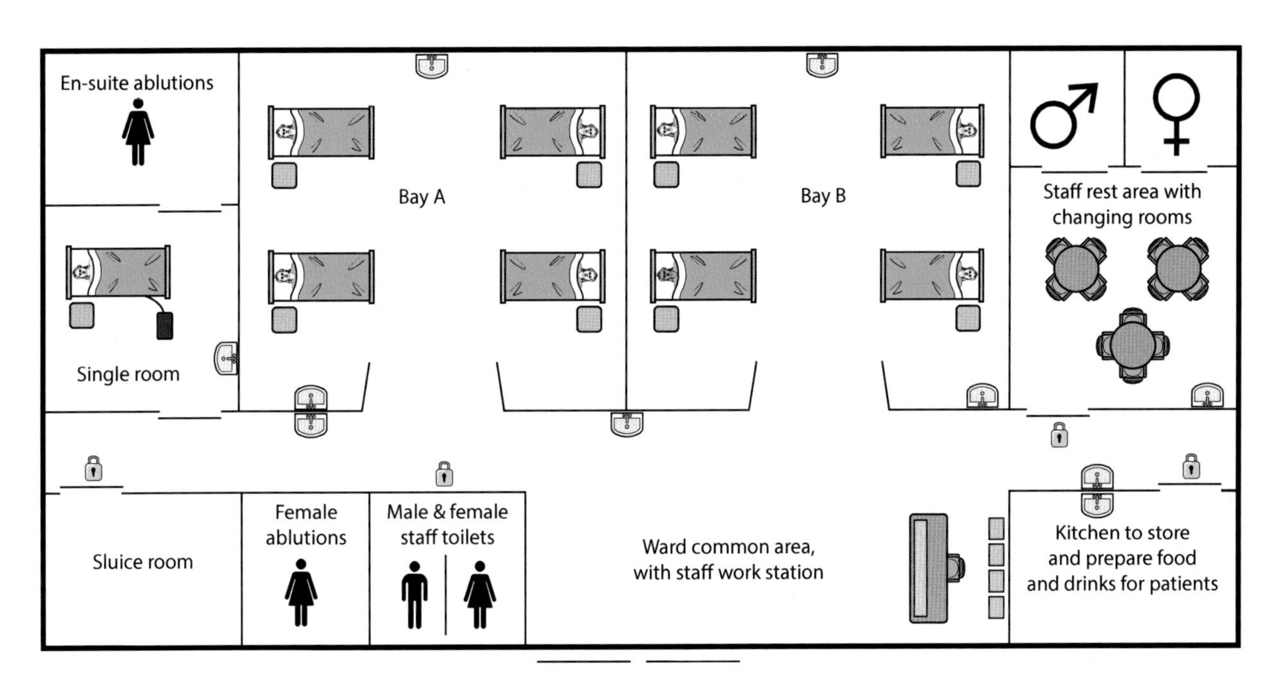

Figure 6.5 The organization of a hypothetical orthogeriatric ward for female patients. The patient in the single room has an extended-spectrum β-lactamase (ESBL)-producing *Escherichia coli* in her catheter specimen of urine. Note the position of the sluice room, single room, as well as patient and staff facilities. Staff areas and the ward kitchen (for patient catering only) are locked.

Figure 6.6 The underside of a commode that appeared spotless from above is heavily contaminated with dried faecal material when turned over.

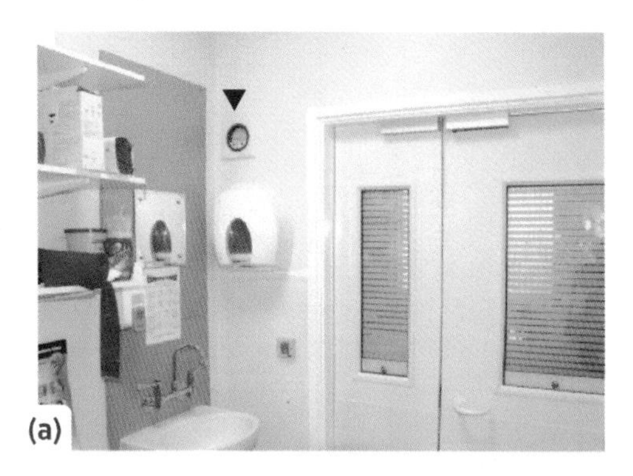

Figure 6.7 (a) The anteroom of a negative-pressure single room (note the pressure gauge [arrow]; the appropriate setting must be monitored). (b) The patient area showing the door to the en-suite ablutions.

AEROSOL

These droplets are less than 10 µm in diameter and can reach the terminal bronchioles and alveoli. This is how *Mycobacterium tuberculosis* and *Legionella pneumophila* reach the lung to establish infection. (The spores of *Aspergillus* will gain access directly from the air.)

- **Airborne**

- *Mycobacterium tuberculosis*, influenza virus, parainfluenza virus, rhinovirus, RSV, VZV (chickenpox) [person to person], *Aspergillus fumigatus* (airborne spores), *Legionella pneumophila* (aerosolized water from showers, taps).

- **Faecal–oral**

- Norovirus, *Clostridium difficile*, *Salmonella* Enteritidis, vancomycin-resistant enterococci (VRE)

- **Direct contact (skin to skin) [person to person]**

- MRSA, *Streptococcus pyogenes*, VRE, ESBL-producing "coliforms", CPE, multi-resistant *Acinetobacter baumannii* (MRAB).

- **Indirect contact (the contaminated environment; fomites)**

- Norovirus, influenza virus, MRSA, *Streptococcus pyogenes*, *Clostridium difficile*, VRE, ESBL-producing 'coliforms', CPE, multi-resistant *Acinetobacter baumannii*, *Pseudomonas aeruginosa*, HBV.

- **Blood-borne**

- HIV, HBV, HCV as 'sharps injuries', and blood/fluid splashes onto mucous membranes such as mouth and eye. Note risk in haemodialysis units.

- **Water-borne**

- *Pseudomonas aeruginosa* (in tap water, sinks), *Legionella pneumophila* (aerosolized water from showers, taps and baths)

- **Reusable surgical equipment sterilized by standard procedures**

- Creutzfeld–Jacob disease (CJD), variant CJD (vCJD) prion disease.

Figure 6.8 The important sources of organisms and the way they are spread. Note that a number are spread by several routes including *Clostridium difficile*, *Streptococcus pyogenes* and norovirus.

FAECAL–ORAL

Here organisms in the faeces of one person reach the mouth of another person, usually via contaminated fomites or food.

Particular attention needs to be taken in the management of the confused patient with diarrhoea. It may not be possible to monitor these patients at all times, especially at night, when staff numbers are reduced. These patients can contaminate ward areas and communal patient ablutions with faecal material. They unwittingly enter unlocked staff toilets, rest rooms or the ward kitchen, inadvertently contaminating the environment there. A ward outbreak of *Salmonella* Enterica was transmitted in these circumstances.

DIRECT CONTACT

Preventing this method of spread of organisms is central to routine infection control practice. From one individual, either a patient or a HCP, material such as skin scales is transferred to another patient by direct contact; MRSA is an example of an organism spread by this route. Bacteria such as *Klebsiella pneumoniae* can produce massive amounts of extracellular capsule material. This sticky material can attach to the cuff of a HCP who is not bare-below-the-elbow.

INDIRECT CONTACT

A number of the organisms listed here can survive for varying lengths of time on skin scales or in drying secretions on a surface. They are thus a potential source of organisms that can be transferred to another patient by the unwashed hand. The respiratory viruses, including influenza virus, as well as norovirus, and the sporulating *Clostridium difficile* can all be transmitted by this route. It is for this reason that during outbreaks, 'touch points' such as door handles are regularly decontaminated, to remove material that may have been deposited. Computer key boards are another item that can be contaminated.

Streptococcus pyogenes is not an infrequent infection control problem on maternity units, and transmission events do occur between mothers who are recently post-partum. This is usually a lower genital tract infection, and there is the potential for contamination of toilet seats to occur, with transmission from this source. Toilet and wash room facilities must be kept scrupulously clean and patients encouraged to report breaches in their own hygiene standards, so that prompt environmental decontamination can take place.

BLOOD-BORNE

HIV, HBV and HCV are the main viruses considered in blood-borne infections, and are of particular relevance in 'sharps injuries'. General prevention relies on good infection control practices when performing venepuncture or the insertion of a peripheral venous cannula (PVC) or long-term venous access system. Provided usual safe practice is employed, and 'sharps' are correctly disposed of, there is no risk.

While the vast majority of HCP will be immune to HBV, they will be susceptible to HIV and HCV. The risk of transmission in a significant 'sharps' injury for each virus is:

HBV: 1 in 3.
HCV: 1 in 30.
HIV: 1 in 300.

Initial precautions centre on washing the site of the injury with copious amounts of warm running water, and actively expressing blood from the site over several minutes, but not scrubbing. When the mucous membranes of the mouth are exposed, copious amounts of water should be used to rinse, ensuring that water is spat out and not swallowed. Eyewash stations are situated in all clinical areas. With HIV, urine, saliva, stool and vomit are considered low-risk exposures unless blood stained.

A sharps injury is a medical emergency, and the HCP who sustained the injury then becomes a patient. The key part of the management process is the assessment of risk, in terms of the nature of the injury sustained, and the patient who was the 'source'. During working hours, management of sharps injuries will be done through the OHD, while 'out of hours' incidents are usually coordinated through the emergency department. Every hospital has guidelines for this process, with senior clinical staff on duty having designated responsibility. This information should be part of induction or mandatory training.

The immediate actions relate primarily to HIV. Testing a blood sample of the source patient provides the key information to direct the process. Most patients agree to this being done, and with prompt testing in the laboratory, the correct management of the incident can be ensured. If the source patient was already known to be newly diagnosed with HIV infection, prophylactic antiretroviral therapy (ART) would have been started immediately. The HCP who is not immune to HBV, but who sustains an injury where the source patient is a chronic carrier of the virus is given hepatitis B virus immunoglobulin (HBIG) as passive immunity, and a rapid course of the vaccine is initiated.

WATER-BORNE

Legionella is an ubiquitous resident of natural water supplies. It is not surprising that it is found in the water of domestic, commercial, educational and health care facilities. While there is no specific guidance to limit risk in the domestic situation, water quality standards in commercial and health care facilities are governed by national legislation.

Legionella reproduces between temperatures of 20°C and 45°C. The organism adheres to the inner wall of pipework, where it can form biofilms. From here planktonic bacteria enter the water, and when a tap or shower is opened, the organism can enter the environment within aerosolized water (and droplets).

Institutions such as hospitals have an obligation to control any potential for legionella to grow in hot and cold water supplies. This relies on regular weekly flushing of water outlets in routine use on wards, and monitoring of water temperatures at exit positions. Cold water must be below 20°C and hot water above 50°C 2 minutes after an outlet has been opened. Correct temperature control and regular flushing minimize organism growth and its ability to form biofilms, which, aided by flushing, keeps the concentration of *Legionella* below the accepted lower limit.

When a patient develops a hospital-acquired pneumonia, and there are no clear predisposing factors, *Legionella* must be considered, especially if routine bacterial culture, and respiratory virus polymerase chain reaction (PCR) results are negative. The *Legionella* urine antigen test can be used; however, this test detects *Legionella pneumophila* serogroup 1, which accounts for approximately 85% of all *Legionella pneumophila* isolates. Lower respiratory tract secretions should be sent for culture and PCR.

Pseudomonas aeruginosa is a rather hardy organism; it can be found in tap outlets and water and sink plugholes. From these sites it can be inadvertently transferred to the patient. Examples have included washing babies on a neonatal unit with tap water, or using tap water when shaving male patients on the intensive care unit (ICU); this should never be done. Shaving may not seem an obvious risk, but here, with the face in close proximity to central venous catheter (CVC) line sites or an endotracheal tube, *Pseudomonas* can be introduced with relative ease. The patient is now at risk of a pseudomonal ventilator-associated pneumonia or bacteraemia. The finding of this organism in specimens from the ICU or high-dependency unit (HDU) patient is an alert as to its source. *Pseudomonas* can grow in dilute detergent to high numbers, and when a contaminated spray was used for cleaning on an ICU, the organism was directly sprayed onto the patients' bed space!

REUSABLE SURGICAL EQUIPMENT

Reusable surgical equipment is sterilized at 121°C for 20 minutes to kill all potential pathogens. The exception is prions, the cause of transmissible spongiform encephalopathies (TSE) and the agent of Creutzfeldt–Jakob disease (CJD) is an example. Because of their unique structure, these proteinaceous infectious agents are not inactivated by standard sterilization procedures.

The prion of classical CJD is restricted to the central nervous system (CNS) (brain, spinal cord and ganglia). Patients who have suspected or confirmed CJD need to be identified before a surgical procedure is performed on the CNS, and (potentially) reusable equipment employed during operation is then destroyed by incineration.

Between 1995 and 2005 there was a prolonged outbreak of a new variant form of CJD (vCJD) in the UK that was directly attributed to the consumption of beef obtained from cattle that had been fed offal from sheep infected with the ovine TSE, scrapie.

vCJD was a challenging public health issue, and in addition to the 178 plus cases of clinical vCJD, all of whom died, certain persons were identified as being at increased risk of vCJD, including those who had received a large volume of blood products over the relevant time. The reason for this concern was that the tissue distribution of vCJD is also outside the CNS, with infectivity associated with organized lymphoid tissue. All patients undergoing a surgical procedure have to be asked if they have been informed if they are at risk of a TSE for public health purposes. If this is the case, where equipment has been in contact, or likely contact, with organized lymphoid tissue, it should be quarantined, and is usually destroyed by incineration.

Endoscopy is one important area, as procedures will often breach a mucosal surface (e.g. biopsy, diathermy), and likely enter submucosal lymphoid tissue. In these circumstances the endoscope is considered contaminated, and is either quarantined for use in that patient only, or destroyed by incineration.

VENOUS ACCESS LINES

Venous access systems include the PVC, peripherally inserted central catheters (PICC), midline catheters and CVC; access to the jugular or subclavian vein is made directly or via a tunnel (**Figure 6.9**). While midline catheters usually terminate before the subclavian vein, PICC and CVC terminate at the junction of the superior vena cava and right atrium. These are used to deliver hypertonic solutions and total parenteral nutrition reliably and safely, and can be in place for weeks or months.

Access by PVC is usually done using the veins of the hand or forearm, and is suitable for short-term access, e.g. 5–10 days, often for the delivery of intravenous (IV) antibiotics. PVC should be replaced and re-sited by 72 hours to minimize the risk of phlebitis and infection. There were considerable problems with poorly managed PVC to 2010 at least, with these being the source of numerous MRSA and methicillin-sensitive *Staphylococcus aureus* (MSSA) bacteraemias. Clear guidance and training are in place for the insertion, monitoring and removal of these cannulas to prevent infection.

Nursing staff use the visual infusion phlebitis score (VIP) to assess PVC insertion sites. The scoring system is shown in **Figure 6.10**. VIP scoring should be done every time the PVC is used. Note that a VIP score of 1 is a record of possible phlebitis, while a VIP score of 2 is early phlebitis and directs re-siting the cannula promptly. A VIP

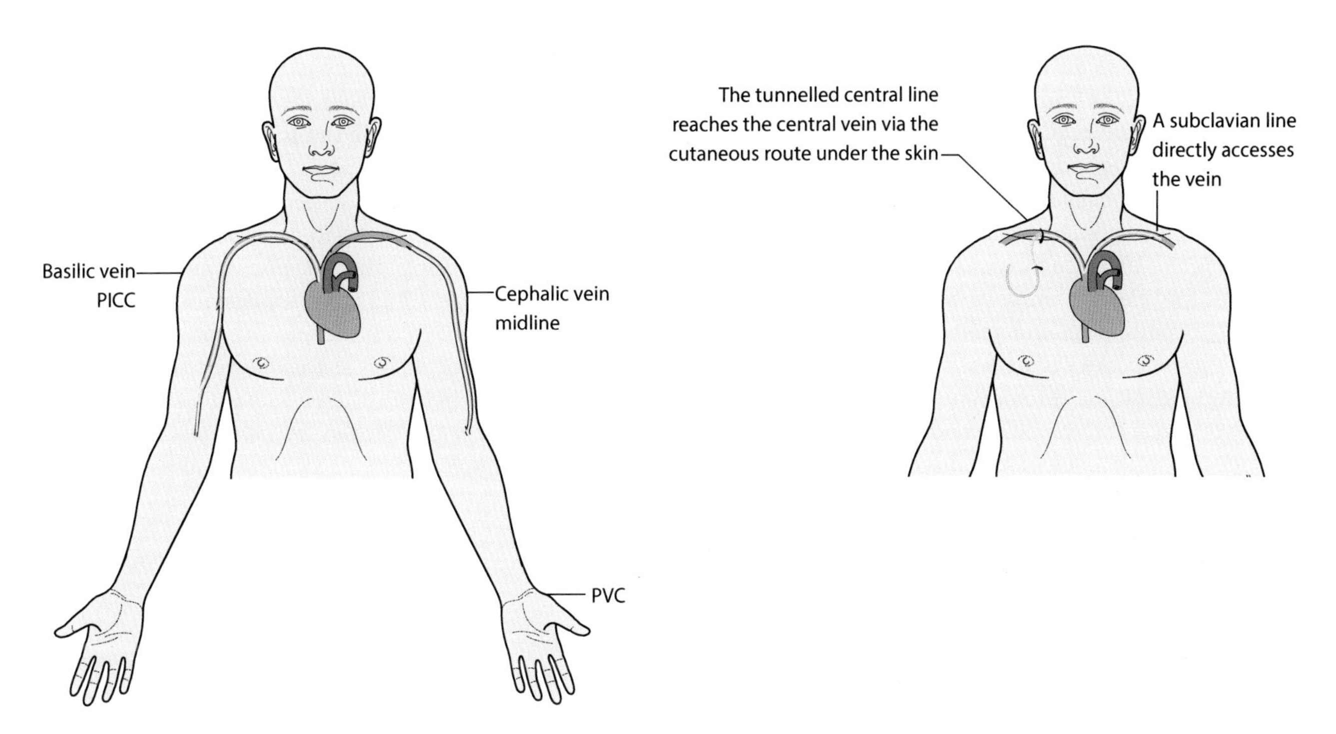

Figure 6.9 The positioning of lines that access the venous system. A peripheral venous cannula (PVC) is usually placed in a vein of the hand. The peripherally inserted central catheter line and central venous catheters terminate in the superior vena cava.

Visual Infusion Phlebitis Score	Score	Action
IV site appears healthy	0	No signs of phlebitis OBSERVE CANNULA
One of the following is evident: • Slight pain at IV site • Redness near IV site	1	Possible first sign of phlebitis OBSERVE CANNULA
Two of the following are evident: • Pain • Erythema • Swelling	2	Early stage of phlebitis RESITE THE CANNULA
All of the following signs are evident: • Pain along the path of the cannula • Erythema • Induration	3	Medium stage of phlebitis RESITE THE CANNULA CONSIDER TREATMENT
All of the following signs evident and extensive: • Pain along the path of the cannula • Erythema • Induration • Palpable venous cord	4	Advanced stage of phlebitis or start of thrombophlebitis RESITE THE CANNULA CONSIDER TREATMENT
All of the following signs are evident and extensive: • Pain along the path of the cannula • Erythema • Induration • Palpable venous cord • Pyrexia	5	Advanced stage of thrombophlebitis INITIATE TREATMENT RESITE THE CANNULA

Figure 6.10 The visual infusion phlebitis score (VIP) used to monitor a PVC. (Reproduced with the permission of Andrew Jackson, IV Nurse Consultant, The Rotherham NHS Foundation Trust UK.)

score of above 2 should not occur, and direct the need for antibiotics according to the hospital's guidelines, as well as the collection of a blood culture set. If the patient is bacteraemic, a positive culture directs the appropriate treatment. It is essential that the previous insertion site is also monitored.

Incidents do occur where the previous 'venflon' in the antecubital fossa, hidden by a pyjama sleeve, has not been removed. It is later identified as the source of a MRSA bacteraemia. These situations are 'never events'.

Infection control is an essential part in the management of venous access systems, with scrupulous attention paid to preparation, insertion and monitoring. Medical staff must be fully engaged in all aspects of their management. The site of insertion needs to be reviewed regularly, involving the patient too. The transparent dressing over the insertion site enables clear observation. Pain or erythema must be immediate triggers to assess and remove the PVC, and if access is still needed, siting it on the other arm.

SCREENING FOR INFECTION CONTROL 'ALERT' ORGANISMS

There are a number of organisms that are routinely screened for in order to identify agents that will be a risk to that patient or another individual. These include blood-borne viruses, antibiotic resistant bacteria and, via exposure history, the TSEs. Examples include:

- All patients undergoing haemodialysis are screened every 3 months to determine their HBV, HCV and HIV status. Additional screening is done when the patient has had haemodialysis at another institution and outside the UK in particular.
- Screening elective surgery patients for MRSA carriage. Usually nose and groin swabs are is collected. Patients who are screen-positive for MRSA undergo a 'decolonization' protocol using mupirocin nasal cream and a chlorhexidine body wash to reduce carriage. The MRSA status also directs appropriate antibiotics used in prophylaxis and treatment.
- All patients who have been in another medical facility in the previous year should be screened for CPE. This applies in particular to countries and hospitals where there is an increased incidence of these organisms. Countries include the Indian subcontinent and those in southern Europe.

For infection control incidents or outbreaks on a ward, screening is also done to determine whether other patients are carrying the specific organism such as *Streptococcus pyogenes* and multi-resistant *Acinetobacter baumannii* (MRAB). In addition to the patients, staff are also screened on occasion, as they can be the source. For example, HCP have been shown to be the source of incidents and outbreaks caused by *Staphylococcus aureus* (MSSA and MRSA) and *Streptococcus pyogenes*.

Chapter 7

Infections of the Blood

INTRODUCTION

This chapter is primarily concerned with the presence of bacteria in the blood. However, many other organisms enter the blood to cause disease. These include those transmitted by various arthropod vectors, such as mosquitoes of the genera *Anopheles* (malaria), *Aedes* (yellow fever virus, dengue fever virus), *Culex* (West Nile virus) and the tsetse fly, *Glossina morsitans* (African trypanosomiasis/sleeping sickness). Ticks of the genus *Ixodes* transmit Lyme disease (*Borrelia burgdorferi*), while rickettsial diseases are transmitted by *Dermacentor* (Rocky Mountain spotted fever) and *Amblyomma* (African tick-bite fever) ticks. *Wucheria bancrofti*, a tissue nematode, is distributed in parts of the tropics and subtropics. Adult worms reside in the lymphatics, and their microfilaria enter the blood to be taken up by blood-feeding female mosquitoes of the genera *Aedes*, *Anophyles* and *Culex*. Passing through the insect's midgut, they reach the thoracic muscles, and mature into larvae that infect the human when the mosquito next feeds.

There are various routes that organisms take to reach the blood. Pneumococcus colonizing the upper airways can be aspirated into the lungs during sleep and go on to cause a lobar pneumonia; from here it can enter the blood. The alimentary canal is the route for a vast array of organisms. The human is the definitive host of the pork tapeworm *Taenia solium*. If eggs in the faeces of an infected individual contaminate food, the person consuming that food is at risk of cysticercosis. Upon hatching from the egg, larvae cross the wall of the bowel into the blood, and the cysticerci can be transported to various organs. The first manifestation of neurocysticercosis can be a seizure. Examples of some sources of exogenous organisms that enter the blood are shown in **Figure 7.1**.

The presence of bacteria in the blood requires identification of the likely source. There is the obvious association of *Escherichia coli* in blood and an ascending urinary tract infection (UTI). When native valve endocarditis is identified it can be straightforward to determine the likely source of the organism. The patient with endocarditis caused by a streptococcus of the mouth flora, such as *Streptococcus sanguinis*, can have poor dentition, and this needs to be addressed as part of the patient's management, usually involving the maxilla-facial surgical team. More unusual situations occur, and one is the identification of *Streptococcus gallolyticus* in blood culture. This organism is a minor member of the normal flora of the colon. However, it is recognized that there is an association that can develop between it and a large bowel malignancy, likely due to a specific interaction between the organism and these malignant cells. The streptococcus gains a selective growth advantage, from where it accesses the blood. Once in the blood it has the potential to initiate infective endocarditis. The finding of *Streptococcus gallolyticus* in blood culture, often in the setting of endocarditis, is an alert to investigate this malignancy; if found this is removed before any valve surgery.

ORGANISMS

As the range of organisms that can enter the blood to cause disease is so wide, a separate list is not included here. It is worthwhile to refer to the range of organisms that are identified in blood culture and by serological and molecular tests, as listed in the introduction to Chapter 5.

PATHOGENESIS

Bacteraemia defines the presence of bacteria as detected by the culture of blood. Septicaemia also defines the presence of bacteria in blood, but it signals a sense of urgency in the management of the patient. The terms sepsis and septic shock are also used and, with clinical parameters such as fever, hypotension, tachycardia, multi-organ failure and leucocytosis, alert the clinician to the severity of the situation, and the need for immediate action in the management of the patient. This was introduced in Chapter 3.

Bacteraemia can be defined as transient (a single episode lasting less than 30 minutes or so), intermittent or continuous (**Figure 7.2**). These definitions are important concepts in terms of the site from which they may arise. A transient bacteraemia can arise from a pneumococcal pneumonia, or pyelonephritis caused by *Escherichia*

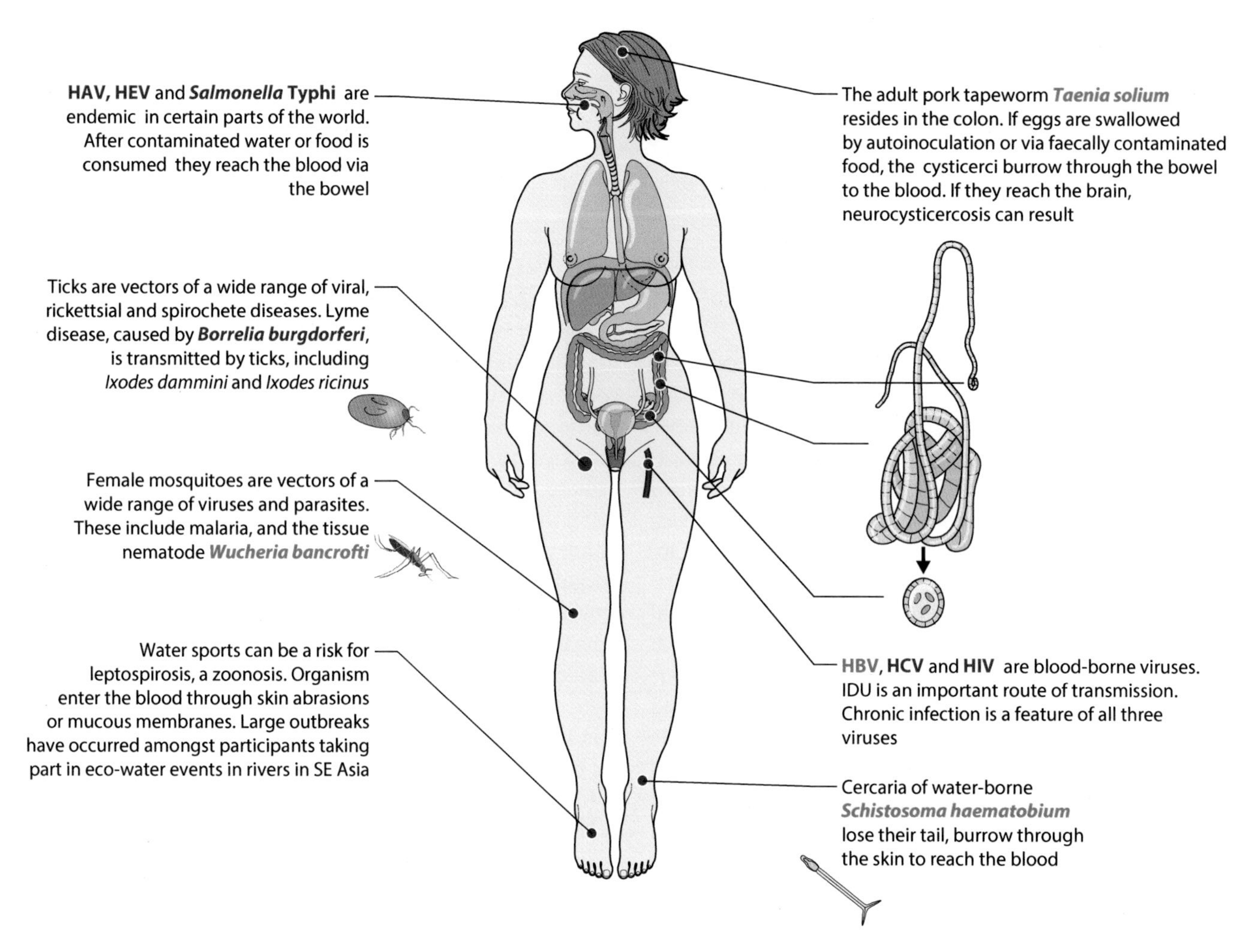

Figure 7.1 Some examples of how exogenous organisms that reach the blood from various sources.

coli. An intermittent bacteraemia implies manipulation of an extravascular site, such as a *Staphylococcus aureus* abscess, where bacteria enter the lymphatics at irregular intervals, and from there, the blood. A continuous bacteraemia implies an intravascular source, and endocarditis is the most important example.

The presence of a few bacteria in the blood for short periods is likely to be a common event. For example, it is recognized that manipulation of teeth, including brushing, enables small numbers of oral bacteria to enter the blood. The presence of these few organisms is usually of no consequence as the natural defences rapidly clear them. Even when recognized pathogens enter the blood, their effect on the host may not be serious. On occasion, the microbiologist may telephone the results of a positive blood culture to the clinician to find that the patient, admitted the day before, has been discharged home without antibiotics. The blood culture taken on admission contains for example, gram-positive diplococci, subsequently identified as pneumococcus. Here, rapid clearance of the organism from the blood by the body's defences resolved the situation. Clearly the patient needs to be carefully reviewed in light of this result, and antibiotics given appropriately. It is reasonable to assume that individuals in the community who are unwell for a short period of time may have pathogenic organisms in their blood, but the natural defences clear them. If these defences fail, the individual will seek medical advice.

Once bacteria enter the blood, they have the potential to settle in other sites of the body, and set up another focus of infection. A *Staphylococcus aureus* bacteraemia arising from an infected peripheral venous cannula (PVC) site can result in bacteria attaching to a heart valve to initiate endocarditis, or settling in the spine and causing an abscess there. The bacteria can cross the synovial membrane of a joint to initiate septic arthritis. These examples

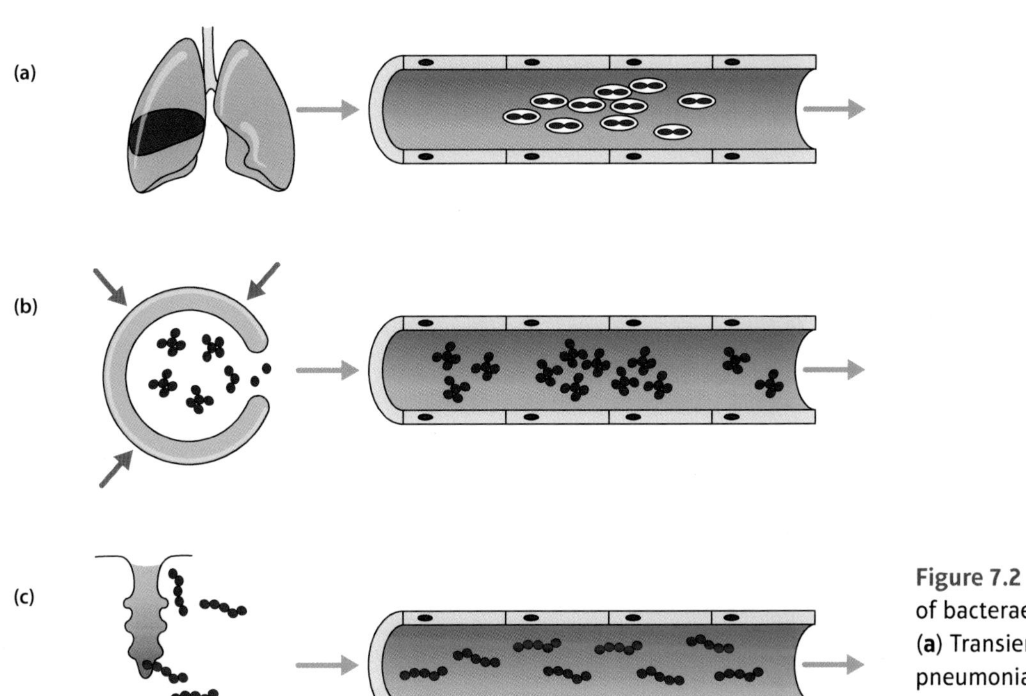

Figure 7.2 Diagrammatic examples of bacteraemias (and sources). (**a**) Transient (pneumococcal pneumonia); (**b**) intermittent (*Staphylococcus aureus* soft-tissue abscess); (**c**) continuous bacteraemia (streptococcal native valve endocarditis).

underline the critical importance of full clinical assessment of the septic or bacteraemic patient.

An important defence system of the blood is the macrophage population of the spleen and liver. Individuals who have had a splenectomy are at risk of overwhelming infection with encapsulated bacteria such as pneumococcus, meningococcus and *Haemophilus influenzae* b (Hib). With loss of the spleen's ability to remove abnormal red blood cells (RBCs) from the circulation, these individuals are also at additional risk when exposed to malaria. The related protozoal parasite *Babesia*, which is transmitted by *Ixodes* ticks in areas such as Scotland and the eastern and upper mid-western states of the USA, can result in severe infection.

INFECTIVE ENDOCARDITIS

NATIVE VALVE ENDOCARDITIS

With native valve endocarditis a predisposing anatomical abnormality of the affected valve may be present. These defects may be congenital, or they may arise as a result of acquired conditions such as rheumatic fever or, more commonly nowadays, degenerative valve disease of ageing. As the flow of blood through the valve orifice is abnormal, the resulting turbulence damages the endothelium. Fibrin and platelets are deposited on the damaged surface, and such deposits can be colonized by bacteria passing through the valve orifice in the blood. The process is outlined in **Figure 7.3**. With the injecting drug user (IDU), it is possible that particulate contaminants in the injected material directly damage the valve endothelium. Fibrin and platelets are then deposited, creating a potential site for bacteria to adhere to.

Once the bacteria have established a niche, their reproduction leads to localized valve damage and the formation of a vegetation. The damage can progress to compromise of valve function with resulting heart failure. Abscess formation in the valve root and spread of the infection into the myocardium can occur. The continuous bacteraemia arising from the infected valve stimulates the immune system. A chronic cytokine response results in weight loss and fever, and immune complexes, deposited in the kidney, can cause glomerulonephritis. Infected emboli shooting off from a vegetation can lodge in the arterial system of any organ. For example, septic emboli in the brain will produce the complication of a brain abscess, as outlined in **Figure 7.4**.

Endocarditis can present as an acute life-threatening illness requiring urgent surgical intervention to remove

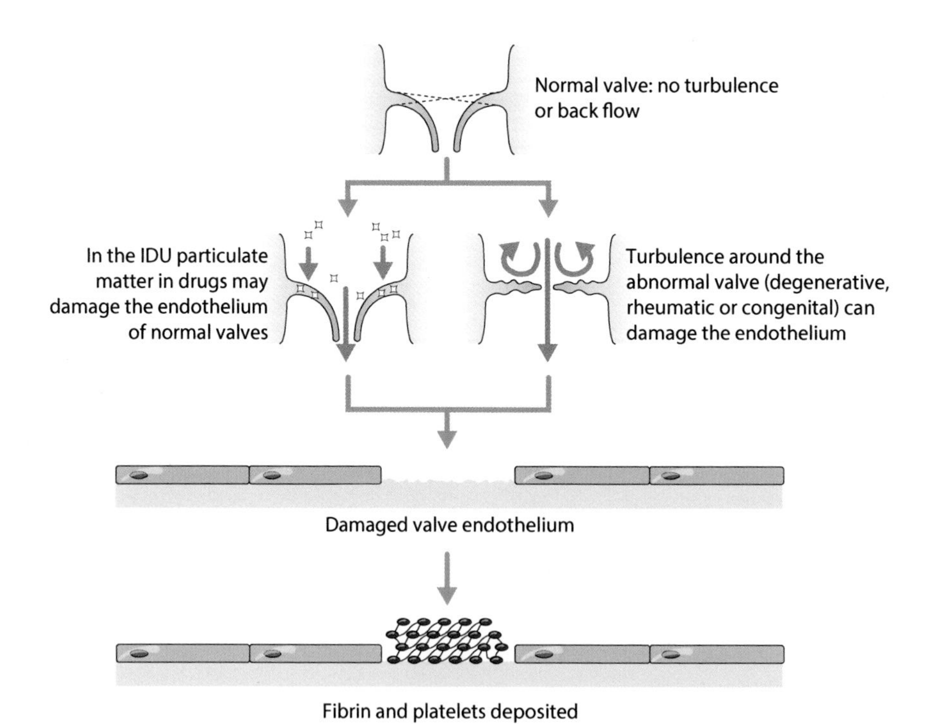

Figures 7.3 An outline of the pathogenic changes that occur in the heart valves that predispose to infective endocarditis. (IDU: injecting drug user.)

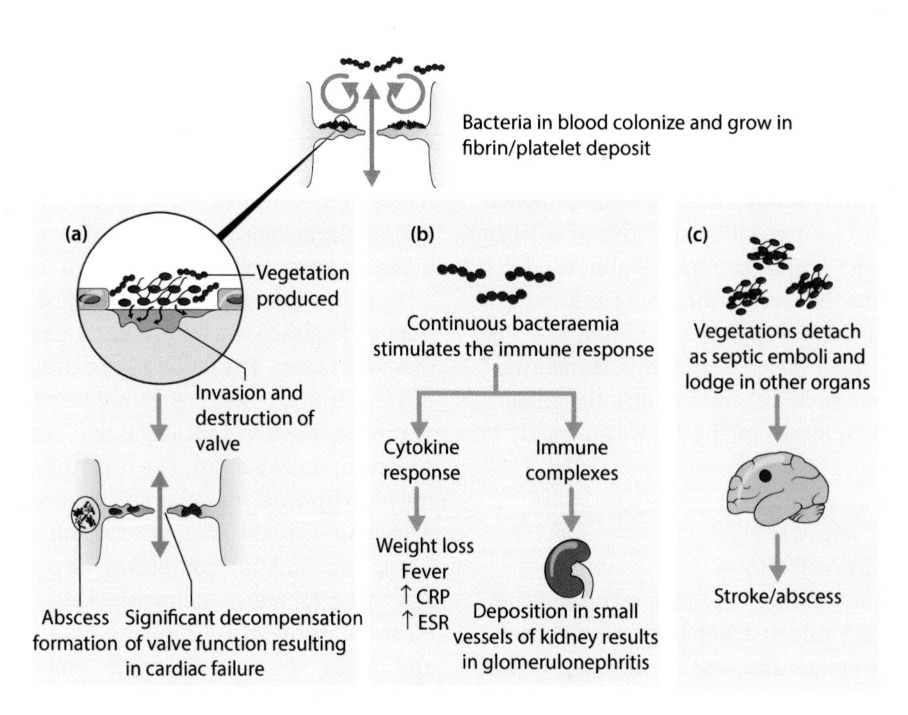

Figure 7.4 Bacteria from an occult bacteraemia can initiate endocarditis. (**a**) Destruction of the valve can occur; (**b**) the continuous bacteraemia stimulates the immune system; (**c**) vegetations can detach from the infected valve, and the resultant emboli can lodge in distant organs. (CRP: C-reactive protein; ESR: erythrocyte sedimentation rate.)

the infected valve. A patient can present weeks into a vague illness with lethargy and low-grade fever being the only symptoms. An embolic event such as a stroke can be the first evidence of endocarditis caused by an oral streptococcus, such as *Streptococcus salivarius*. A heart murmur in such a setting must alert the clinician to the diagnosis of bacterial endocarditis, and direct the collection of three sets of blood cultures before any antibiotics are given.

PROSTHETIC VALVE ENDOCARDITIS

A prosthetic heart valve will create local abnormalities in blood flow, and turbulence may damage the endothelium adjacent to the valve. Deposition of fibrin and platelets will give rise to a structure that can be colonized by bacteria, and cause prosthetic valve endocarditis (PVE) (**Figure 7.5**). PVE is divided into early and late endocarditis. Early PVE occurs within 1 year of valve replacement and is often caused by coagulase-negative staphylococci introduced at the time of operation. Late PVE endocarditis is defined as that occurring more than 1 year after valve replacement and can be caused by a wide range of bacteria, and fungi, usually *Candida*. PVE must be considered promptly in any unwell patient with a prosthetic valve, and three sets of blood cultures must be collected before antibiotics are given. This is of critical importance, as only one set of blood cultures growing a coagulase-negative staphylococcus creates a diagnostic conundrum, as this organism could be a skin contaminant. If three sets of blood cultures, separated by time, grow the identical coagulase-negative staphylococcus, this directs the correct antibiotic therapy with certainty. Conservative treatment with antibiotics can be successful; however, replacement of the prosthetic valve may be required.

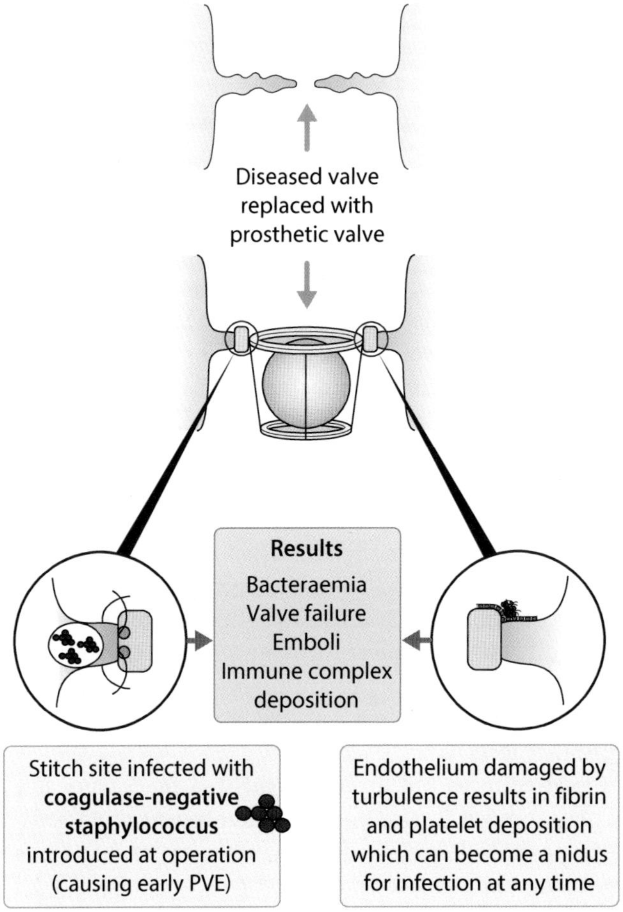

Diseased valve replaced with prosthetic valve

Results
Bacteraemia
Valve failure
Emboli
Immune complex deposition

Stitch site infected with **coagulase-negative staphylococcus** introduced at operation (causing early PVE)

Endothelium damaged by turbulence results in fibrin and platelet deposition which can become a nidus for infection at any time

Figure 7.5 Replacement of a native valve with a prosthetic valve can predispose to prosthetic valve endocarditis (PVE). Bacteria such as the coagulase-negative staphylococci are important.

INFECTIONS OF IMPLANTABLE CARDIAC ELECTRONIC DEVICES

Implantable cardiac electronic devices (ICEDs) are used to manage cardiac conduction disorders that require defibrillation and resynchronization. They consist of the generator device or box, inserted into the subcutaneous tissue of the upper chest. From the box leads pass into the subclavian vein, and from there to the right heart, terminating at the medial side of the atrium and base of the right ventricle (**Figure 7.6**).

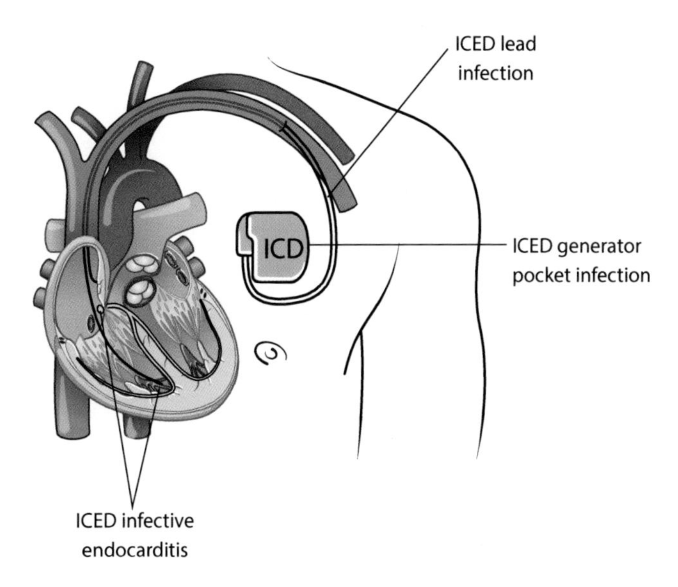

Figure 7.6 An implantable cardiac electronic device in the subcutaneous tissue. Leads pass into the subclavian vein, and terminate at the medial wall of the right atrium and ventricle. The sites where infection can occur are indicated.

Infection is a serious complication following ICED insertion. Early infection occurs within 6 months of insertion. Risk factors include previous temporary pacing, inadequate antibiotic prophylaxis at the time of insertion, haematoma formation and wound dehiscence. Renal failure, chronic obstructive pulmonary disease, anticoagulation and chronic steroid therapy are important risk factors too, emphasizing the 'at risk' surgical patient discussed in Chapter 4.

Gram-positive bacteria account for the majority of organisms isolated from ICED infections, with the coagulase-negative staphylococci being predominant, followed by *Staphylococcus aureus*. Other skin commensals such as *Propionibacterium* can also be involved. Gram-negative bacteria such as 'coliforms' and *Pseudomonas aeruginosa* can account for up to 15% of infections.

The clear role of the common skin commensals such as the coagulase-negative staphylococci and *Propionibacterium* underlines the need for scrupulous skin decontamination and surgical procedure at the time of the operation when the device is inserted. The appropriate antibiotic and dose must be given for perioperative prophylaxis. Recommendations in the UK are the administration of a single dose of teicoplanin (e.g. 800 mg), which can be given with a single 160 mg dose of gentamicin.

DIAGNOSIS AND MANAGEMENT

This section centres on the collection of blood cultures, their processing in the laboratory and interpretation of results. It is useful to review Chapter 3, **Figure 3.15**, and the other tests used to diagnose infections. The important parasitic infections to exclude in the patient who has visited tropical and subtropical regions are the malaria parasites of the genus *Plasmodium*. This emphasizes the need to obtain a vector exposure risk in the patient with a travel history and fever.

BLOOD CULTURES

WHEN TO TAKE BLOOD CULTURES

Blood cultures should be collected from the hospitalized patient who has a fever (usually ≥38.2°C) and there is a focus of infection. However, fever is not always a reliable sign in the older patient, and confusion, or increased confusion, is important. The clinical assessment of the patient and use of the modified early warning score (MEWS) were discussed in Chapter 3.

A minimum of one set of well-taken blood cultures should be collected. Not infrequently, and particularly in the older patient, obtaining a specimen from the likely focus of infection can be difficult. For example, the elderly patient may be incontinent of urine or demented, making it difficult to collect a clean-catch midstream urine (MSU) sample. When endocarditis or infection of an ICED is a consideration, three sets of blood cultures, separated by time, are required (see below).

Central venous access systems are integral to the management of several groups of patients, including those requiring total parenteral nutrition, the haematology/oncology patient and those with renal failure who require ongoing haemodialysis. These lines can become a nidus for infection with a range of bacteria, including environmental mycobacteria, and *Candida*. Even if there is no obvious sign of infection at a central venous catheter (CVC) insertion site, in the setting of sepsis, there must be prompt consideration of collecting a set of blood cultures from each lumen of the CVC, as well as a peripheral set. The site of each set must be clearly labelled.

WHAT BLOOD CULTURES DO

When blood is collected from the bacteraemic patient, there are relatively few bacteria present, less than 100/mL. The aim of blood cultures is to amplify these bacteria to numbers that can be used for laboratory purposes and **Figure 7.7** outlines the process. The commercial medium and blood of the patient provide a rich nutrient source, and organisms multiply at 37°C; when they reach a critical concentration, circa 10^5 organisms/mL, a metabolic signal generates an alert that the bottle is 'positive'.

In standard blood culture systems, it is dissolved metabolic CO_2 of the reproducing organisms that is the chemical basis for detection. This combines with water to form carbonic acid that dissociates into hydrogen ions and carbonate ions. When a set concentration of hydrogen ions is reached, the sensor at the base of the bottle triggers the electronic signal. When the Gram stain is performed on the fluid from the bottle, organisms will usually be seen. The Gram stain result is reviewed with the details on the request form, and discussed with the clinical team. This result also influences the range of agar plates that are inoculated, and the 'direct' antibiotic susceptibility tests done.

Blood cultures are usually incubated for 5 days, but this is extended in settings such as infective endocarditis or possible *Brucella* infection, when more fastidious organisms are being considered.

HOW TO TAKE A BLOOD CULTURE

Every hospital will have a training programme for blood culture collection, and it is important that this training

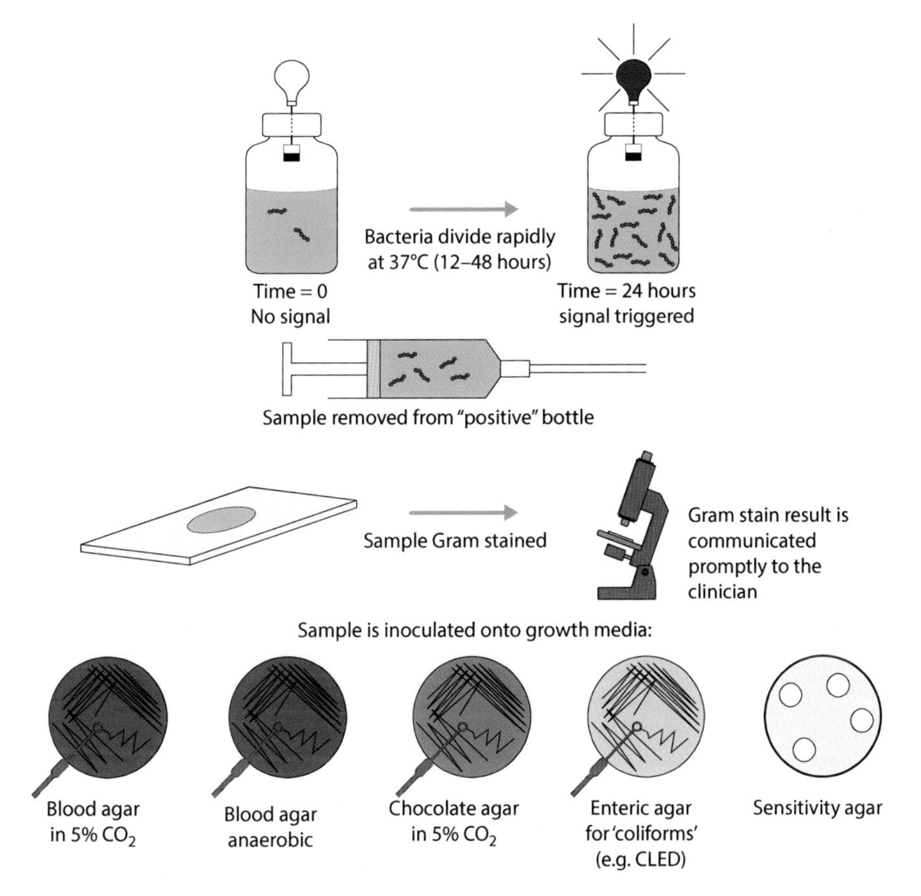

Figure 7.7 Reproducing organisms in the blood culture bottle produce a metabolic signal that alerts it as positive. The example here shows the emission of gaseous CO_2 produced by respiration or fermentation of the organisms in the blood culture 'broth'.

is provided, and adhered to. An example of the equipment required for collecting a blood culture is shown in **Figure 7.8**. The key to collection of blood cultures is the adherence to guidance to ensure that the blood is not contaminated with organisms from the skin of the patient or the taker. This is emphasized with collection of blood from a CVC, where it is essential not to contaminate the lines during the collection process.

All the necessary materials should be assembled in a tray, with a container to dispose of sharps. The bottles should be checked to see if they are in date and that the colour at the bottle base is green/blue. After hands have been washed and dried, the plastic caps are removed and decontaminated with the alcohol–chlorhexidine wipe. After locating by palpation the patient's vein from where the blood will be collected, the collector puts on disposable surgical gloves and the area of venepuncture is carefully cleaned for 30 seconds with the disinfection applicator and the area allowed to dry for a further 30 seconds. Without palpating the site again, venepuncture is achieved and the plastic sleeve/needle is clamped over the aerobic bottle. The flow of blood removes air from the collecting system into this bottle, so on transfer to the anaerobic bottle, no oxygen will enter that bottle. Approximately 10 mL of blood is collected into each bottle. Further blood samples can then be taken using the plastic adapter insert. **Figure 7.9** shows several points relating to blood culture collection.

Blood culture bottles must be correctly labelled, ensuring that the barcodes are not obscured, and the request form fully completed and a barcode copy placed on the form too. Legible details are essential, including the clinical information, and whether the set is peripheral or from the named lumen of a CVC. A clear record must be made in the patient's notes of the date, time, access site (peripheral or CVC), and indication for taking blood cultures.

It is important that blood cultures reach the laboratory promptly to be placed in the blood culture machine; the accepted time is within 4 hours of collection. The reason

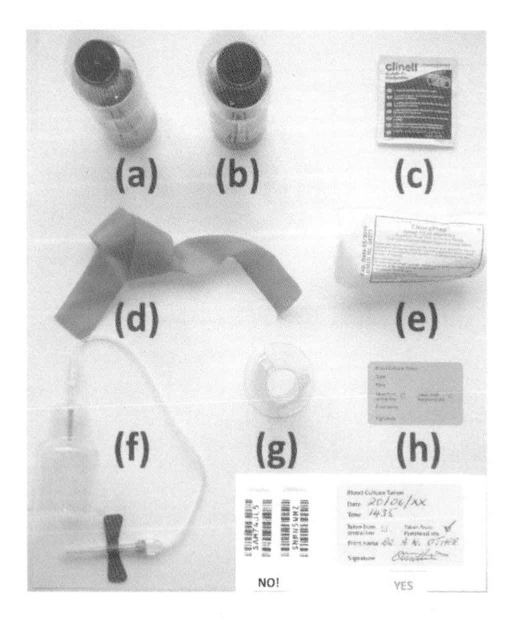

Figure 7.8 Equipment needed to collect a blood culture set: (**a**) aerobic bottle, (**b**) anaerobic bottle, (**c**) alcohol wipe to decontaminate bottle septum after cap removal, (**d**) disposable tourniquet, (**e**) alcohol/chlorhexidine skin disinfector, (**f**) butterfly needle and connector, (**g**) insert to collect other blood samples, (**h**) sticker to record collection time and date in the notes. The barcode copies go on the request form, and not in the notes.

for this is that the sensor system indirectly monitors for an exponential growth of organisms. Organisms in bottles left at room temperature for prolonged periods will multiply, and when placed in the machine, the signal will not be triggered.

WHAT HAPPENS IN THE LABORATORY

In the laboratory, the biomedical scientist will match the bottles with the request form using the barcode reader. The bottles are then placed in trays in the machine, incubated at 37°C and rocked to maintain a suspension of both organism and blood cells, thus providing a uniform growth medium. When a bottle signals as positive, usually after 12–18 hours, it is removed from the machine, and using a safety needle a sample is placed on a glass slide, Gram stained and examined by microscopy. Fluid from the blood culture bottle is also inoculated onto agar plates. The Gram stain result determines the plates inoculated for culture and identification, as well as directing which antibiotic susceptibility tests are done, using the fluid (direct testing). This process is shown in **Figure 7.10**.

The Gram stain; aiding the diagnosis and treatment

Examples of Gram stains of blood cultures in several clinical settings are shown in **Figure 7.11**. The patient

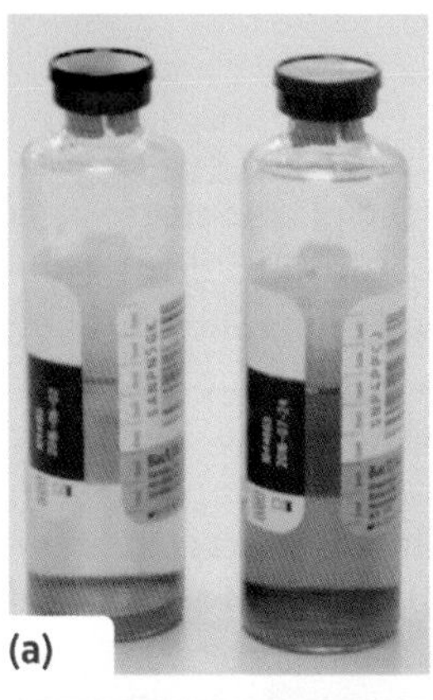

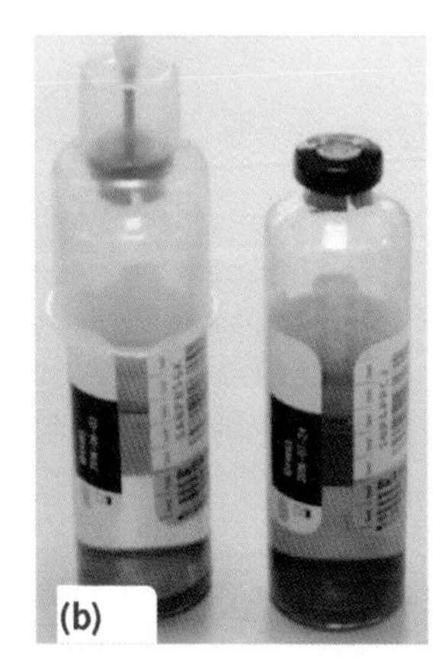

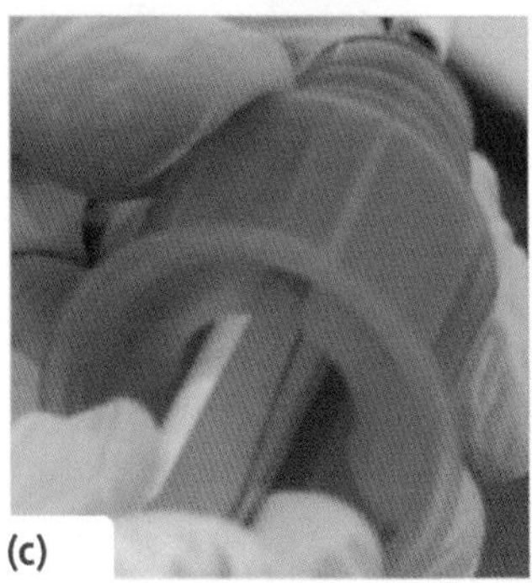

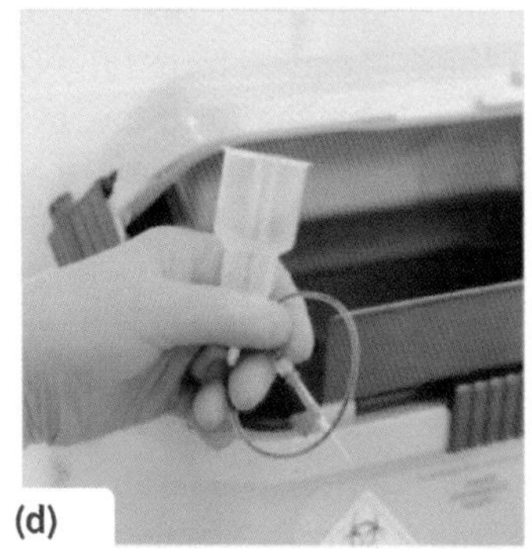

Figure 7.9 The blood culture set. (**a**) The aerobic (left) and anaerobic bottle (right) before caps are removed; (**b**) collection starts with the aerobic bottle; (**c**) after the blood culture bottles are filled, tubes for other blood tests can be collected; (**d**) dispose of the 'sharp' safely.

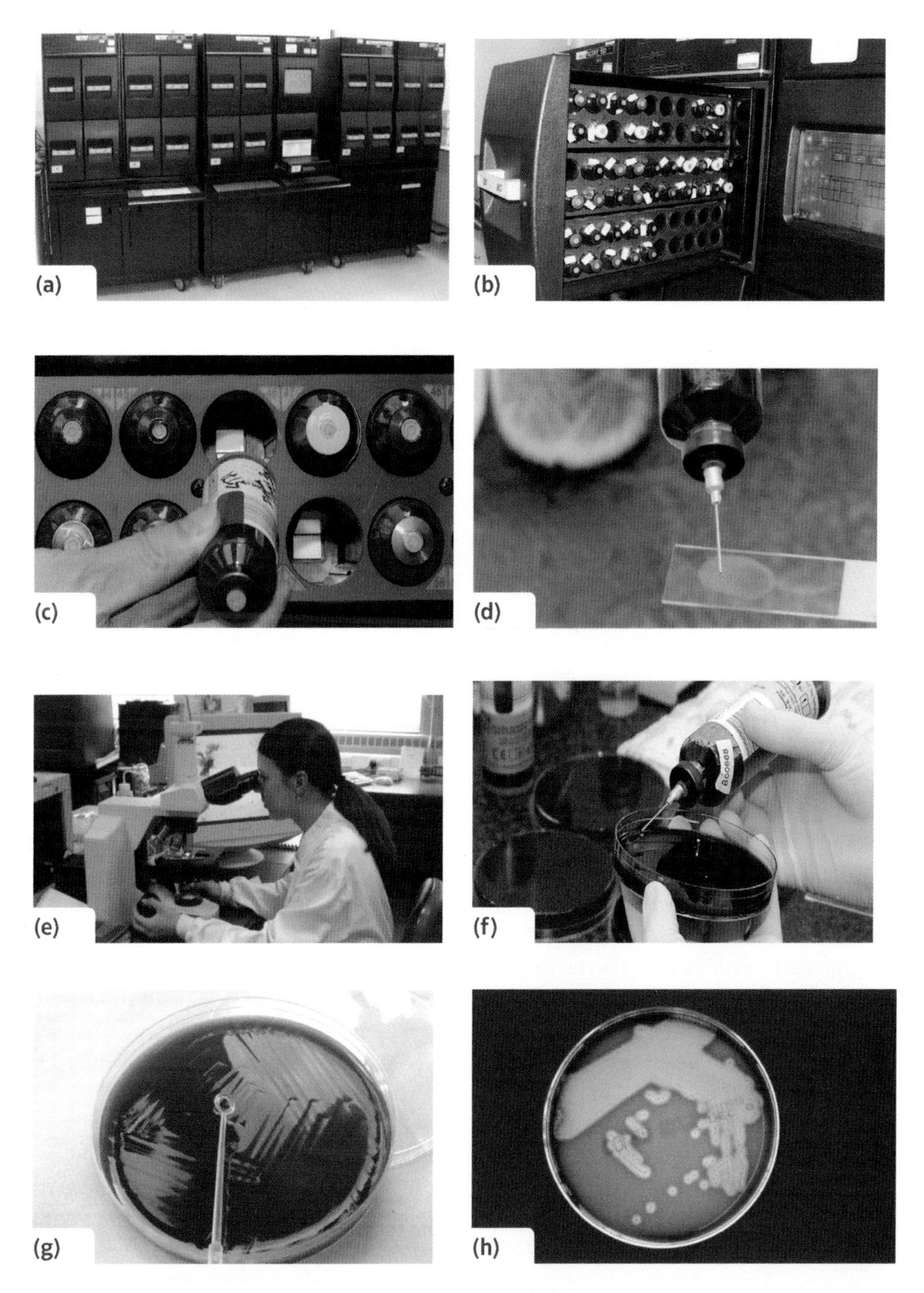

Figure 7.10 (**a**) The BacT/Alert™ Biomerieux blood culture machine; (**b**) bottles in racks (yellow is a paediatric bottle); (**c**) removing the positive bottle; (**d**) a Gram stain is prepared; and (**e**) examined under the microscope; (**f**) a sample is inoculated onto appropriate culture media; (**g**) spreading the inoculum; (**h**) after overnight incubation, plates are examined for growth and single colonies selected and 'picked off' for identification and, as required, additional antibiotic susceptibility testing (*Streptococcus pyogenes* here).

with painful induration at a recent PVC site on the hand, lymphangitis and fever has blood cultures growing gram-positive cocci in clusters, staphylococcal-like (**Figure 7.11a**). This must give the interim identification of *Staphylococcus aureus*, either methicillin sensitive (MSSA) or methicillin resistant (MRSA). The patient with spreading cellulitis following an abrasion to the right shin has gram-positive cocci in chains growing in blood

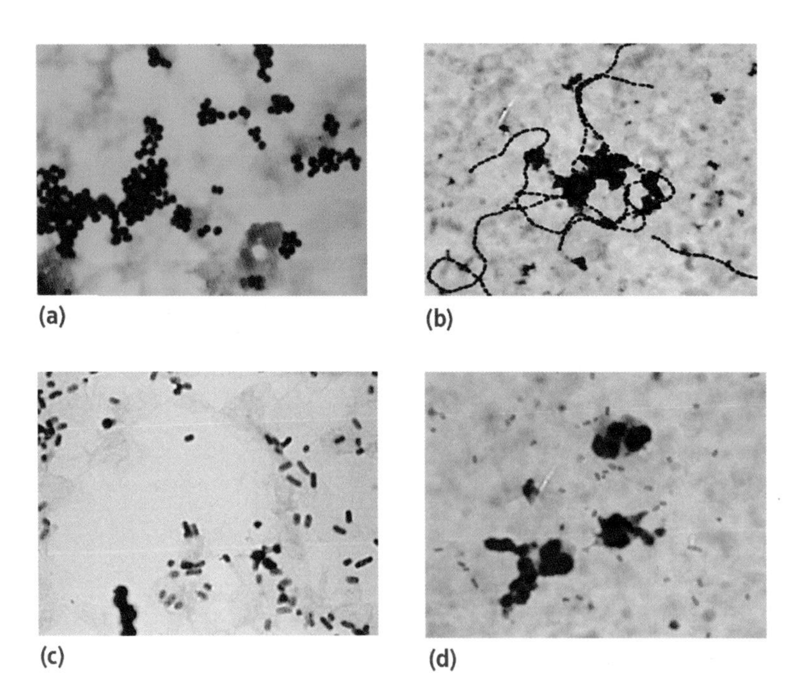

Figure 7.11 Examples of clinical situations and the Gram stain result of blood culture taken. (**a**) Infected antecubital fossa (peripheral venous cannula) site and gram-positive cocci in clusters, staphylococcal like; (**b**) cellulitis of the right shin and gram-positive cocci in chains, streptococcal like; (**c**) from a young woman with cystitis, pyelonephritis, sepsis and gram-negative rods; (**d**) from a young man who returned from India 10 days ago with fever, malaise and gram-negative rods.

culture (**Figure 7.11b**). This is most likely to be group A streptococcus, *Streptococcus pyogenes*. This organism can cause rapidly progressing necrotizing fasciitis.

Figure 7.11c shows gram-negative rods in the blood culture of a young woman who has had cystitis in the past, and presented with pyelonephritis, fever and hypotension; this is most likely to be *Escherichia coli*. In **Figure 7.11d** the young man with recent travel to the Indian subcontinent has fever and malaise of 1 week's duration. Scattered around RBC debris are gram-negative rods; this organism is most likely to be either *Salmonella* Typhi or *Salmonella* Paratyphi; the patient has enteric fever. It would have been essential to have included details of this travel history on the request form accompanying the blood culture set. This not only provides information for clinical management, but critically alerts the laboratory staff to do all processing of this positive blood culture in the category 3 room for their safety.

Figure 7.12 shows nine examples of the Gram stain of a blood culture, and the initial management process. **Figure 7.13** has additional management information once the identity and antibiotic susceptibility of the organism(s) are known. It is useful to go through each of these examples as they highlight key important points in the management of the bacteraemic/candidaemic patient.

When blood cultures confirm a CVC infection, with the line and peripheral sets growing the same organism, consideration must be given to the prompt removal of the line. Many organisms have a potential to settle elsewhere in the body. The candida yeasts can cause endophthalmitis, and all patients with a candidaemia must have an ophthalmology review. Bacteria that may be considered as relatively benign, such as the environmental *Stenotrophomonas maltophilia*, can colonize the CVC of a neutropenic haematology patient. The 'benign' nature of the organism should not allow a 'wait and see' decision in the setting of positive blood cultures. The subsequent appearance of embolic skin lesions that grow the organism is a serious complication, particularly if the organism is co-trimoxazole resistant, the only effective antibiotic treatment option.

INFECTIVE ENDOCARDITIS

In industrialized countries about one-half of the cases of infective endocarditis develop in patients with no known history of valve disease. Of the remainder, patients with prosthetic valves, ICED, unrepaired cyanotic congenital heart disease or a history of infective endocarditis make up the majority. Conditions such as diabetes, human immunodeficiency virus (HIV) infection, haemodialysis and IDU are also risks.

Fever is a common presentation occurring in at least 80% of cases. In addition, weight loss, haematuria, splinter haemorrhages, unexplained heart failure and renal failure may also be presenting manifestations. The C-reactive protein (CRP) is raised in about two-thirds of cases and raised white cell count and anaemia are seen in about 50% of cases.

Gram stain	Gram stain description	Specimen and patient details	Other specimen	Comment
	Gram-positive cocci in clusters; staphylococcal like.	Peripheral blood culture collected from 25-year-old female patient with first episode cystitis/pyelonephritis.	Purulent MSU specimen collected on admission growing *Escherichia coli* >10^8 CFU/L.	Improving on co-amoxiclav 1.2 g q8h IV; await antibiotic sensitivity result of *E. coli*. Blood culture likely to be a contaminant (coagulase-negative staphylococcus).
	Gram-positive cocci in clusters; staphylococcal like (see **Figure 7.11a**).	Peripheral blood culture collected from 75-year-old male patient with infected PVC site. Line was in for 5 days.	None, patient was MRSA negative on admission screening 7 days ago.	No improvement on flucloxacillin. Change to vancomycin 1 g q12h IV. Review infection control.
	Gram-positive cocci in chains; streptococcal like (see **Figure 7.11b**).	Peripheral blood culture collected from 35-year-old male patient with right shin cellulitis after trauma.	Swab of shin abrasion taken on admission growing mixed 'coliforms' and a streptococcus.	Good response to benzyl penicillin 1.2 g q6h IV. Review infection control, nurse in single room.
	Gram-positive diplococci; ? pneumococcus.	Peripheral blood culture collected from previously well 55-year-old male patient (smoker) with right middle-lobe pneumonia.	Urine antigen test taken on admission positive for pneumococcus. Advise HIV test.	Good response to benzyl penicillin 1.2 g q6h IV. Consider change to oral amoxicillin once culture and sensitivity result is available.
	Gram-negative rods (see **Figure 7.11c**).	Peripheral blood culture collected from 25-year-old female patient with first episode cystitis/pyelonephritis.	Purulent MSU specimen collected on admission growing *Escherichia coli* > 10^8 CFU/L. Sensitivities awaited.	No significant improvement on co-amoxiclav 1.2 g q8h IV. Change to ertapenem 1 g q24h IV. Review infection control.
	Gram-negative rods (see **Figure 7.11d**).	Peripheral blood culture collected from 23-year-old male patient with ongoing fever. Returned from India 10 days ago.	None. No signs or symptoms of a UTI, thus no MSU sent.	With travel history, likely enteric fever. Start ceftriaxone 2 g q24h IV. Review infection control. Lab staff alerted as to likely category 3 organism. Inform Public Health now; patient works in local restaurant kitchen.
	Gram-negative diplococci; ? meningococcus.	Peripheral blood culture collected from 25-year-old female patient with pyrexia, hypotension and non-blanching rash.	EDTA blood collected for meningococcal PCR.	No evidence of meningitis or 'seeding' elsewhere. Good response to benzyl penicillin 1.2 g q6h IV. Consider change to oral ciprofloxacin once culture and sensitivity result is available. Review infection control, inform Public Health.
	Gram-negative rods, and yeasts, candida-like.	Peripheral blood culture collected from 45-year-old male patient with AML, and currently neutropenic.	Hickman line blood cultures (red and white lumens) also gram-negative rods and yeasts. *Candida glabrata* grown from previous HL site swab.	Still pyrexial with rigors after 48 hours of piperacillin/ tazobactam 4.5 g q8h IV. Change to meropenem 1 g q8h IV. Stop prophylactic fluconazole and give caspofungin. Discuss prompt line removal.
	Gram-positive cocci in pairs and short chains, and yeasts, candida-like.	Peripheral blood culture collected from 65-year-old male patient following biliary tract surgery, on ITU and HDU before return to ward. Broad spectrum antibiotics given for 10 days.	*Candida albicans* grown from ET aspirate and CSU on ITU. No central line present now.	Spiking temperature, rising WCC; ? abdominal source. On vancomycin and meropenem since blood culture collected. Change vancomycin to linezolid 600 mg q12h IV and add fluconazole 400 mg q24h PO.

Figure 7.12 Examples of Gram stains of blood cultures showing bacteria and yeasts. Each is a scenario that outlines several management steps relevant to the organism(s) identified in the Gram stain.

Gram stain	Identification	Susceptibility profile	Comment
	Staphylococcus hominis, coagulase-negative staphylococcus.	Not reported, as staphylococcus is a contaminant.	This is a skin contaminant. *Escherichia coli* in MSU specimen: R: Amoxicillin S: Co-amoxiclav, gentamicin, ciprofloxacin, trimethoprim. Much better, change to oral co-amoxiclav, consider discharge home.
	Staphylococcus aureus, MRSA.	R: Flucloxacillin, erythromycin, clindamycin, vancomycin (MIC 4 mg/L); S: Linezolid, daptomycin (MIC 0.5 mg/L)	This is a hospital-acquired MRSA bacteraemia. Infection control alert and actions; nurse patient in side room, and screen patients on same bay. Vancomycin resistant, stop, and give daptomycin 8 mg/kg q24h IV. Must take surveillance blood culture after 48 hours of daptomycin.
	β-haemolytic group A streptococcus, *Streptococcus pyogenes*.	S: Penicillin, erythromycin, clindamycin, linezolid.	Shin swab grew *Streptococcus pyogenes*. Much better, change to oral amoxicillin 1 g q8h PO, and consider discharge home. Inform Public Health, as invasive GAS infection in the community.
	Streptococcus pneumoniae.	S: Penicillin, erythromycin, clindamycin, linezolid.	As there is a good response to therapy, change to oral amoxicillin 1 g q8h PO and consider discharge home. HIV test negative.
	Escherichia coli (ESBL-producer).	R: Amoxicillin, co-amoxiclav, piperacillin/tazobactam, ciprofloxacin, gentamicin. S: Ertapenem, meropenem.	The patient responds well to the antibiotic change, and is apyrexial within 24 hours. No suitable oral option, arrange for ertapenem via OPAT, once patient is ready for discharge. Review Infection Control.
	Salmonella Paratyphi.	R: Amoxicillin, ciprofloxacin S: Ceftriaxone, azithromycin (MIC 2 mg/L)	The patient is apyrexial within 48 hours of treatment with ceftriaxone, and treatment is changed to azithromycin 500 mg q24h PO for a further 7 days, and the patient is discharged. Laboratory confirm result with Public Health now.
	Organism failed to grow.	None available.	PCR positive for meningococcus serotype W135. This case underlines the importance of using all appropriate diagnostic tools. These organisms are fragile, and do not always grow from blood culture. Review Infection Control and inform Public Health.
	Stenotrophomonas maltophilia. *Candida glabrata*.	*Stenotrophomonas maltophilia*. R: Co-trimoxazole *Candida glabrata* R: Fluconazole; S: Caspofungin, voriconazole	There are two problems here, the *Stenotrophomonas* is resistant to the only appropriate agent used for treatment, and both organisms can seed elsewhere. Fungal endophthalmitis must be excluded. The Hickman line must be removed.
	Enterococcus faecium (VRE). *Candida albicans*.	*Enterococcus* R: Amoxicillin, vancomycin (VRE); S: Linezolid, daptomycin *Candida albicans* S: Fluconazole	It is reasonable to continue the meropenem for broad cover, with full surgical review and re-imaging of the abdomen. The vancomycin was already changed to linezolid, and fluconazole started. Review Infection Control regarding VRE. What is the status of the radiological and surgical review?

Figure 7.13 The same scenarios shown in **Figure 7.12** but with further management information that is available once the organism's identity and antibiotic susceptibility profile have been determined.

A list of the common organisms that are encountered in infective endocarditis is shown in **Figure 7.14**. This shows that streptococci and staphylococci account for the majority. The coagulase-negative staphylococci are important in prosthetic valve endocarditis, but it is increasingly recognized that native valve endocarditis can be caused by *Staphylococcus epidermidis* and *Staphylococcus lugdunensis* on occasion.

The key to the microbiological diagnosis of infective endocarditis and ICED infection is the collection of three sets of blood cultures, separated by time, and before antibiotics are administered. In most circumstances the three sets can be collected over a period of several hours. In the acute setting, where the need to administer antibiotics is considered urgent, they can be collected from different sites, over minutes.

Once an organism has been identified in serial blood cultures, further microbiological testing is done to ensure that a range of antibiotic options is available for treatment. Minimum inhibitory concentration (MIC) tests are determined for appropriate antibiotics. For streptococci and enterococci these include penicillin, vancomycin and daptomycin, as well as gentamicin, which can be used in combination (synergy) with one of these antibiotics. In the case of staphylococci, vancomycin and daptomycin MICs are determined. Other agents such as linezolid and rifampicin may also be used. Treatment varies from 2 weeks for a sensitive streptococcus in native valve endocarditis to 6 weeks for PVE.

The involvement of the cardiology team is essential in assessing all patients with suspected endocarditis. Transthoracic echocardiography (TTE), and transoesophageal echocardiography (TOE) must be done promptly in order to confirm the diagnosis, and assess the degree of valve failure and local complications such as aortic root abscess formation. The cardiology team would work closely with the cardiothoracic surgeons as well, as prompt surgery may be deemed necessary.

Streptococci 40%

Streptococcus sanguis, Streptococcus mutans (oral flora)
Streptococcus gallolyticus, Streptococcus anginosus (bowel flora)
Streptococcus constellatus, Streptococcus intermedius (bowel flora)

Staphylococcus aureus (MSSA, MRSA) 30%

Coagulase-negative staphylococci (CNS) 10%

Staphylococcus epidermidis, Staphylococcus lugduniensis
Staphylococcus haemolyticus, Staphylococcus hominis

Enterococci 10%

Enterococcus faecalis, Enterococcus faecium (VRE)

Other organisms 5%

Gram-positive bacilli
Gram-negative 'coliforms'
Anaerobes
Yeasts

Culture-negative endocarditis 5%

Figure 7.14 A list of common organisms that cause infective endocarditis. The coagulase-negative staphylococci are usually associated with prosthetic valve endocarditis, but can cause natural valve endocarditis on occasion, e.g. *Staphylococcus lugduniensis*.

CULTURE-NEGATIVE ENDOCARDITIS

About 5% of cases of infective endocarditis are culture negative. Other organisms are then considered, including 'slow growers', with fastidious growth requirements that are not cultured by routine methods (**Figure 7.15**). The appropriate diagnostic tests for these should be discussed with the microbiologist and infectious diseases physician. These include serological tests for *Coxiella* and *Bartonella*, while those for *Legionella* (including the urine antigen test for *Legionella pneumophila* serogroup 1) and *Chlamydia psittaci* may be done.

If the patient has valve surgery, it is important that valve tissue obtained at operation is submitted for histopathology, as well as culture and that molecular testing such as 16S ribosomal ribonucleic acid (RNA) polymerase chain reaction (PCR) is used as necessary.

These cases require an agreed empirical antibiotic regime that covers usual organisms (gram-positive) and certain ones listed in **Figure 7.15**. For example, culture-negative PVE could be treated with vancomycin and ceftriaxone.

INFECTIONS OF IMPLANTABLE CARDIAC ELECTRONIC DEVICES

The patient with an ICED infection can present with obvious signs of pocket disease, with swelling and cellulitis. Diagnosing ICED lead infection or infective endocarditis is not that straightforward. The patient may present many months after insertion of the device, with low-grade fever, malaise, night sweats and anorexia. Spinal osteomyelitis or discitis due to septic embolization can

Organism	Features of note
Nutritionally deficient streptococci. *Abiotrophia* spp.	May require pyridoxal hydrochloride or L-cysteine for growth in the laboratory
H: *Haemophilus*	May be slow grower, ensure chocolate agar inoculated, and incubated in 5% CO_2
A: *Aggregatibacter*	May be slow grower, ensure chocolate agar inoculated, and incubated in 5% CO_2
C: *Cardiobacterium*	May be slow grower, ensure chocolate agar inoculated, and incubated in 5% CO_2
E: *Eikinella*	May be slow grower, ensure chocolate agar inoculated, and incubated in 5% CO_2
K: *Kingella*	May be slow grower, ensure chocolate agar inoculated, and incubated in 5% CO_2
Brucella	FUO associated with travel from an endemic country (Middle East, Northern Africa)
Legionella pneumophila	Requires special culture medium; BCYE agar
Coxiella burnetii	Q fever; animal or animal milk contact
Chlamydia psittaci	Exposure to psittaccine birds

Figure 7.15 The organisms to be considered in the setting of culture-negative endocarditis. The HACEK organisms are susceptible to ceftriaxone.

be the presentation. The first important step at initial assessment is finding out that the patient has an ICED and when it was inserted.

The critical step is the collection of three sets of blood cultures, separated by time. Consultation must then take place with the cardiologist as to the appropriate antibiotic(s) to give, and the infectious diseases physician or microbiologist should also be contacted. The cardiologist will promptly assess the situation, and consider the collection of pus or fluid from an infected generator pocket, and the process for ICED removal.

In the setting of ICED infection occurring within a few days, where *Staphylococcus aureus* is most likely, flucloxacillin is appropriate, unless the patient is known to be MRSA positive or has a history of allergy, where vancomycin is an alternative. If the patient is admitted with sepsis, meropenem and vancomycin are used as initial treatment, after collection of blood cultures.

OTHER WAYS OF MAKING THE DIAGNOSIS

A number of the organisms presented in **Figure 7.1** and tests available to diagnose them have been discussed in Chapter 3, **Figure 3.15**. Micrographs of thin films of the malaria parasite *Plasmodium falciparum* are shown in **Figure 7.16**. This simple test, done on thick films too, is still a central diagnostic test to identify the species of malaria. If the first films are negative, blood should be examined at least twice more in the following 24 hours, and further sampling continued if the diagnosis is strongly suspected. The malaria antigen test is done in conjunction with blood films.

THE IDENTIFICATION OF *STAPHYLOCOCCUS AUREUS* IN BLOOD CULTURE

The identification of a *Staphylococcus aureus* bacteraemia is of major significance for the patient, in terms of both morbidity and mortality. It directs the prescription of at least 2 weeks of intravenous antibiotics, and the need to exclude a focus, of which endocarditis is a critical example. *Staphylococcus aureus* bacteraemia, endocarditis, ICED infection or spinal abscess formation secondary to a poorly managed PVC do still occur, and are essentially evidence of negligence.

In addition, diagnosis of these complications is difficult, time consuming and expensive, and when identified they add considerable time to the length of antibiotic treatment required. Problems with renal and liver function arise, and as the peripheral venous route usually becomes difficult, access by a peripherally inserted central catheter (PICC) or midline is needed, adding to the stresses that the patient has to endure.

While the length and type of treatment following a *Staphylococcus aureus* bacteraemia will be determined in discussion with the infectious diseases or microbiology teams, several key points to remember from the outset are:

- Identification of the source and its control is a priority, with removal wherever possible. An infected PVC or CVC is one example.
- Appropriate doses of antibiotic need to be given; for the adult patient of average weight, flucloxacillin is recommended as 2 g q6h.

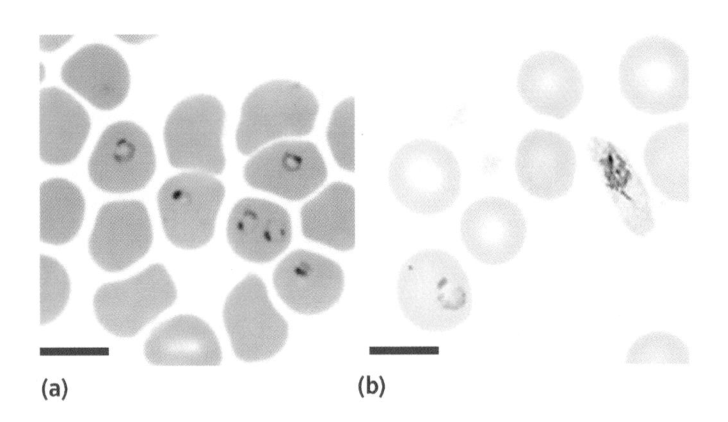

Figure 7.16 Photomicrographs of Giemsa-stained blood smears showing: (**a**) ring forms and trophozoites and (**b**) trophozoites and gametocytes of *Plasmodium falciparum*. (Bar: 10 µm.) (Photomicrographs sourced from the Public Health Image Library, CDC, Atlanta, USA.)

- The vancomycin MIC for a susceptible strain of *Staphylococcus aureus* is ≤2 mg/L. Serum predose vancomycin levels between 15 and 20 mg/L need to be maintained throughout the course of treatment to ensure that the organism is eliminated from a source such as an infected heart valve.
- If daptomycin is used, predose levels between 15 and 20 mg/L should be obtained, usually requiring a dose of 6–8 mg/kg body mass. Weekly creatinine kinase levels have to be monitored.
- After 2–3 days of appropriate antibiotics, even if the patient has improved, a set of surveillance blood cultures must be taken. If these are negative, this is reassuring. If positive, a more detailed search for the source(s) needs to be conducted, along with a multi-disciplinary discussion of the antibiotic treatment options.

Chapter 8

Infections of the Respiratory Tract

INTRODUCTION

With its direct access to the outside, and the various sources of organisms, it is not surprising that the list of organisms that can cause disease in the respiratory tract is vast. Included are pneumococcus, *Streptococcus pyogenes*, *Mycoplasma pneumoniae*, *Chlamydophila pneumoniae* and *Mycobacterium tuberculosis*. Important viruses to consider are influenza, metapneumovirus, parainfluenza, respiratory syncytial virus (RSV) and the 'common cold' rhinoviruses. In countries where vaccination programmes are in place, infection caused by *Corynebacterium diphtheriae* (diphtheria), *Bordetella pertussis* (whooping cough) and *Haemophilus influenzae* serogroup b (childhood epiglottitis) are uncommon.

It is reasonable to consider that most of the organisms named above have a human source, and that they circulate within communities by spread from the infected individual to the susceptible. While the respiratory viruses generally have a peak of activity during winter, outside this they will be maintained, albeit at low levels in the community.

Water systems in homes or in buildings such as hospitals can provide the conditions for *Legionella pneumophila* to grow. This bacterium lives in water over a wide range of temperatures, and grows optimally in water where the temperature is between 20°C and 45°C. A water supply in this range may contain the organism in significant numbers. When a shower or tap is used, aerosols are produced, which can be inhaled into the lung. *Aspergillus*, which is ubiquitous in the environment, causes the immunoglobulin (Ig) E-mediated hypersensitivity disease allergic aspergillosis, and invasive lung disease in the immunocompromised. *Pneumocystis jirovecii* is considered a colonizer of the lung, and can cause serious disease in the immunocompromised patient. Some examples of the sources of infections are shown in **Figure 8.1a–d**.

Cystic fibrosis is a condition that has a unique microbiological flora. This is a multi-organ disease with an autosomal recessive pattern of inheritance. Mutations in the gene encoding the cystic fibrosis transmembrane conductance regulator (CFTR) protein give rise to decreased transport of chloride and bicarbonate out of cells. The outcome of this defect is most pronounced in the lungs. Mucus produced by the goblet cells becomes inspissated and the ciliated epithelium is unable to clear it adequately. Colonization of the secretions with *Staphylococcus aureus*, *Pseudomonas aeruginosa* and *Burkholderia cepacia* define the complexity of this condition, with chronic infection leading to progressive deterioration in lung function.

Other respiratory pathogens can be acquired from animals (zoonoses). These include *Coxiella burnetii* (Q fever) from sheep and *Chlamydia psittaci* (psittacosis) from parrots. Coccidioidomycosis and histoplasmosis are caused by two dimorphic fungi with diseases ranging from acute pneumonia to disseminated infection. This is most pronounced in the immunocompromised patient. *Histoplasma capsulatum* is found in soil in tropical and temperate regions, and there is a strong association with bat guano; recreational 'caving' can be a risk factor. Coccidioides is associated with soil in arid regions of the south western USA and Mexico.

Pathogens also reach the lung via the blood to initiate infection there. *Staphylococcus aureus* infective endocarditis, and *Streptococcus anginosus* thrombophlebitis in the groin of the intravenous drug user are examples of such sources.

Anatomically, the respiratory system can be divided into the upper and lower tracts. The upper tract consists of the middle ear, mastoid cavity, nasal sinuses and nasopharynx, while the lower tract extends from the larynx to the lungs. All these structures can be involved in infection. Certain organisms such as *Legionella* are restricted to one anatomical site, the lungs, while others such as pneumococcus can cause middle ear infection, sinusitis and pneumonia. Some important defences of the respiratory tract, based on normal anatomy, physiology and immune function, are outlined in **Figure 8.2a**. Compromise of these defences predisposes to infection (**Figure 8.2b**).

Pneumonia is a disease of the lung parenchyma, which can be divided into community-acquired pneumonia (CAP) and hospital-acquired pneumonia (HAP), when clinical disease arises 48 hours after admission of the patient to hospital.

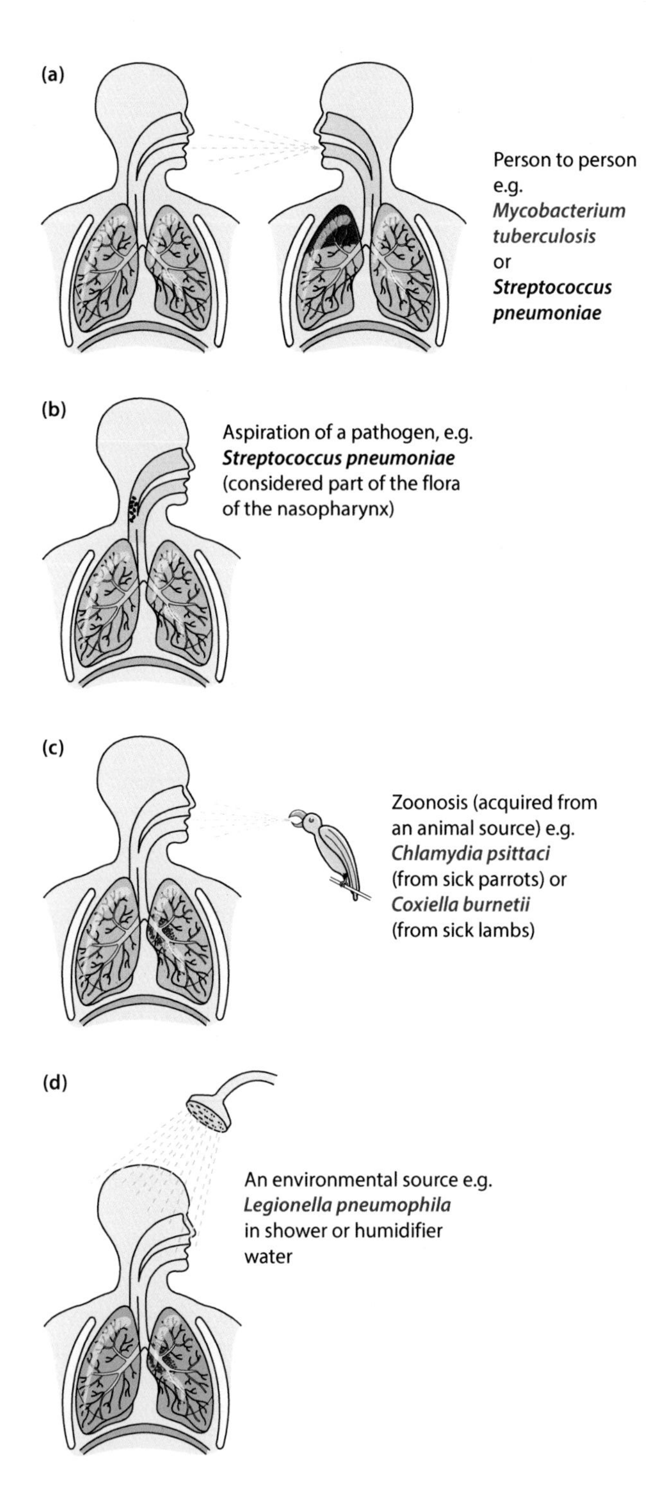

Figure 8.1 The routes whereby respiratory infections arise: (**a**) person to person; (**b**) aspiration; (**c**) acquired from an animal source, e.g. psittacosis; (**d**) acquired from an environmental source.

While some of the agents that cause CAP may be involved, HAP is usually associated with 'coliforms' including *Klebsiella pneumoniae*, *Enterobacter cloacae*, *Pantoea agglomerans*, *Serratia marcesens*, as well as *Staphylococcus aureus* (both methicillin sensitive [MSSA] and methicillin resistant [MRSA]). The lower respiratory tract is a relatively easy system for an organism to access in the hospitalized patient. This is particularly so in the patient on the intensive care/high-dependency unit (ICU/HDU). Attention needs to be given to the bacteria that are identified in respiratory secretions, irrespective of their antibiotic susceptibility profile. *Pseudomonas aeruginosa* is a cause of HAP, and when it is identified in the hospitalized patient, this should be an alert to a source from the water supply.

Klebsiella pneumoniae can be heavily capsulated, and produce vast amounts of extracellular mucoid material. This enables it to survive and colonize endotracheal tube (ETT) and central line sites, with the potential to cause ventilator-associated pneumonia (VAP), central venous catheter line infections and septicaemia. It can survive on computer keyboards and other fomites, and is then transferred with ease to the environment of the patient.

In the community, typical pneumococcal lobar pneumonia is characterized by fever, chest pain and production of purulent sputum. This infection results in a significant influx of neutrophils into the lobe, giving rise to the purulent sputum. So-called 'atypical pneumonia' is characterized by dyspnoea and cough, with minimal sputum production. Organisms associated with atypical pneumonia include *Legionella pneumophila*, *Mycoplasma pneumoniae*, *Chlamydophila pneumoniae*, *Chlamydia psittaci* and *Coxiella burnetii*.

Lung abscess can be a complication of pneumonias caused by *Staphylococcus aureus* or *Klebsiella pneumoniae*, following aspiration of oral bacteria, from septic emboli of right-sided endocarditis, or from thrombophlebitis of the great veins of the neck or the pelvis. Abscesses may also arise distal to an obstruction in the bronchial tree, such as a carcinoma.

ORGANISMS

A list of organisms that cause infections of the ear, throat, and the lower respiratory tract are shown in **Figures 8.3, 8.4**. Diagnostic tests are listed here to show the range used.

Some features of organisms associated with atypical pneumonia are shown in **Figure 8.5**. *Legionella* is an intracellular pathogen. It is phagocytosed by alveolar macrophages, but is able to escape the killing mechanism of the phagolysosome. Following multiplication, the macrophage ruptures and the released bacteria infect other cells.

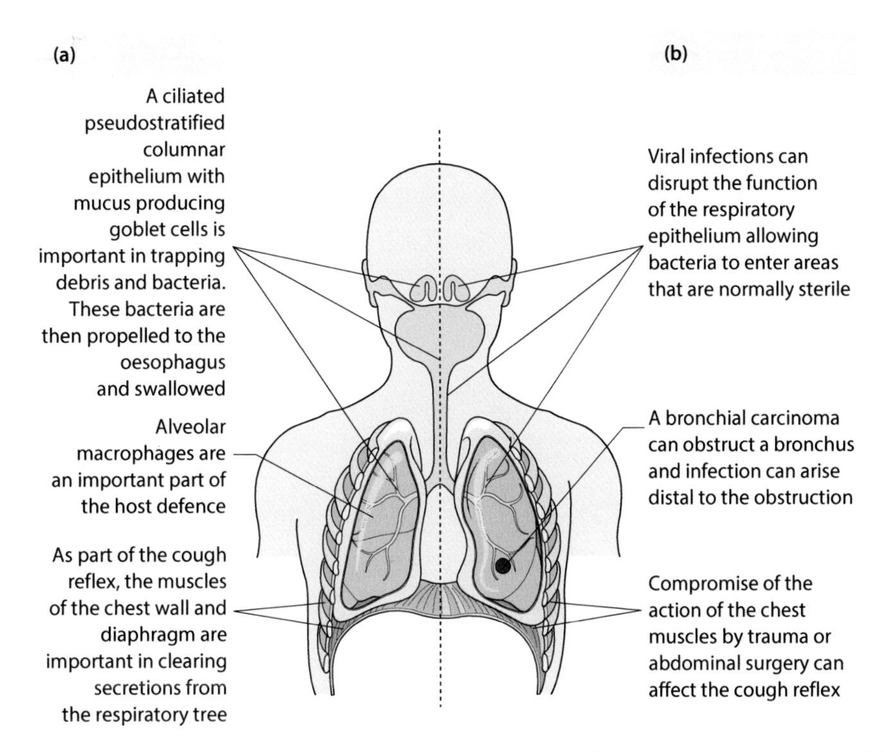

Figure 8.2 (**a**) Some of the defences of the respiratory tract that prevent infection. (**b**) Compromise of these defences increases the likelihood of infection.

	Otitis Externa	Otitis Media	Pharyngitis	Identification by
***Staphylococcus aureus* (MSSA/MRSA)**	√√	√		Culture
Streptococcus pneumoniae		√√√		Culture
***Streptococcus pyogenes* (GAS)**	√√	√√	√√√	Culture
Haemophilus influenzae		√√		Culture
Neisseria gonorrhoeae			(√)*	Culture (STD clinic)
Pseudomonas aeruginosa	√√√**			Culture
Aspergillus niger	√√			Culture
Mycoplasma pneumoniae		√	√	(PCR)
Chlamydophila pneumoniae		√	√√	(PCR)
HIV			√	Serology
CMV			√	Serology
EBV			√√√	Serology
HSV			√√√	PCR
Coronavirus			√√	PCR
Influenza virus A and B		√	√√	PCR
Metapneumovirus		√	√√	PCR
Parainfluenza viruses			√√	PCR

* Gonococcal throat carriage is usually asymptomatic

**In addition to diffuse otitis externa, *Pseudomonas aeruginosa* is the causative agent of invasive malignant otitis externa, usually affecting the older diabetic or immunocompromised patient. It is likely to have come from a (domestic) water source.

Figure 8.3 Organisms that cause infections of the ear and throat. The tests listed highlight the different methods used.

	Pneumonia	Tracheobronchitis	AECB*	COPD	Identification by:
Staphylococcus aureus******	√√				Culture
Streptococcus pneumoniae	√√√		√√√	√√√	Culture, serology***
Haemophilus influenzae	√√		√√√	√√	Culture
Klebsiella pneumoniae	√				Culture
Pseudomonas aeruginosa	√√√			√√	Culture
Mycobacterium tuberculosis	√√√				Culture, PCR
Bordetella pertussis		√√			Culture, serology, PCR
Chlamydophila pneumoniae	√√√	√√√		√√	PCR
Mycoplasma pneumoniae	√√	√√√		√√	PCR
Legionella pneumophila	√√√				Serology***, culture, PCR
Coronavirus	√√	√√√	√√	√√	PCR
Influenza virus A and B	√√	√√√	√√	√√	PCR
Metapneumovirus	√√	√√√	√√	√√	PCR
Parainfluenza viruses	√√	√√√	√√	√√	PCR
RSV	√√	√√√	√√	√√	PCR
Aspergillus fumigatus	√√				HR-CT (serum galactomannan antigen)
Pneumocystis jirovecii	√√				PCR (BAL, sputum, EDTA blood)

*AECB: acute exacerbation of chronic bronchitis.

** MSSA, MRSA (both can be Panton Valentine leucocidin [PVL] toxin producers, with haemorrhagic pneumonia)

*** Urine antigen test (for *Legionella pneumophila* this test detects serogroup 1 only).

Figure 8.4 Organisms that cause infections of the lower respiratory tract. A number are associated with specific patient groups, e.g. *Pseudomonas aeruginosa* (hospital-acquired pneumonia), *Aspergillus* and *Pneumocystis* in the immunocompromised.

Mycoplasmas are free-living bacteria that lack a cell wall. *Mycoplasma pneumoniae* attaches to the respiratory epithelium by a terminal P1 protein structure. This organism can cause pharyngitis, tracheitis, bronchitis and pneumonia.

Chlamydiae are obligate intracellular parasites. *Coxiella burnetii*, the causative agent of Q fever, is also an obligate intracellular parasite. *Coxiella burnetii* is able to survive desiccation and can be wind-borne. Outbreaks of Q fever have been recorded downwind of farms where there were infected farm animals. Most infections occur in farm workers, veterinary staff and those transporting or handling animals at abattoirs.

PATHOGENESIS

OTITIS MEDIA, MASTOIDITIS AND SINUSITIS

The middle ear, mastoid cavity and sinuses are connected either directly or indirectly to the nasopharynx. The ciliated respiratory epithelium, which lines the sinuses and Eustachian tube, pushes mucus out of these structures and trapped organisms are removed. In middle ear and sinus infection it is likely that viruses such as RSV invade this epithelium, destroy the cells and compromise the mucociliary function, allowing bacteria to enter sterile areas (**Figure 8.6a**). Although disease of the mastoid is uncommon, it is important to recognize this condition. Bacteria can spread from the middle ear to the mastoid cavity via the aditus. Because of the proximity of the mastoid cavity to the middle cranial fossa, lateral venous sinus and jugular bulb, mastoiditis can have serious complications (**Figure 8.6b, c**).

PHARYNGITIS

A range of organisms cause pharyngitis (**Figure 8.3**). In addition to *Streptococcus pyogenes*, *Streptococcus dysgalactiae* (group C streptococcus) can also cause a streptococcal throat infection. An important virulence factor of *Streptococcus pyogenes* is the surface M protein, which has antiphagocytic properties. In addition, secreted

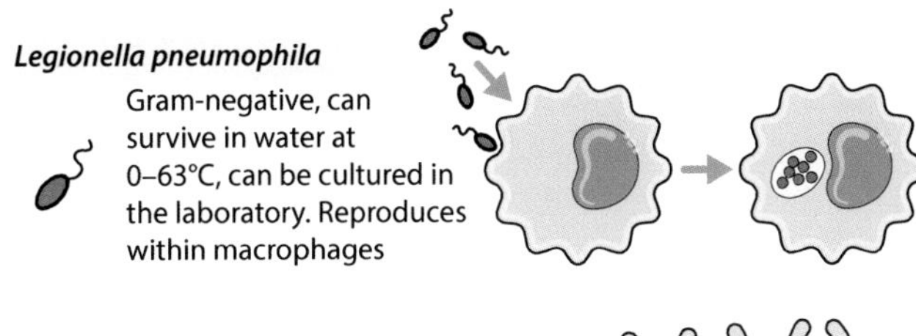

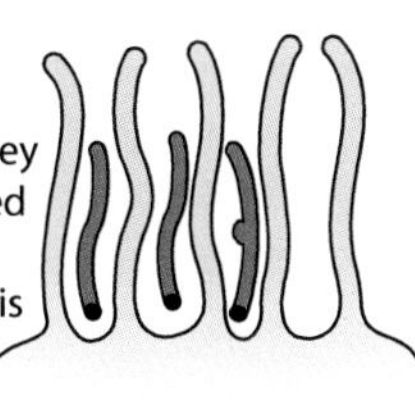

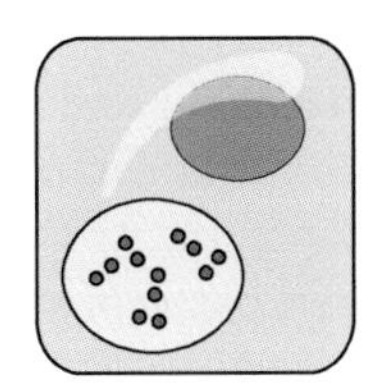

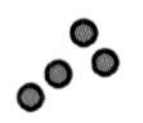
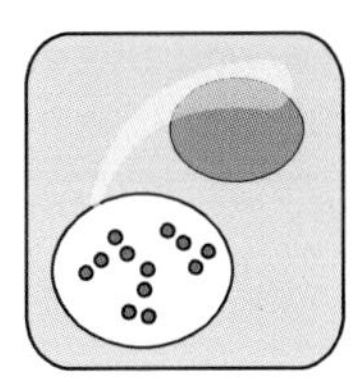

Figure 8.5 Some features of certain organisms associated with causing 'atypical' pneumonia.

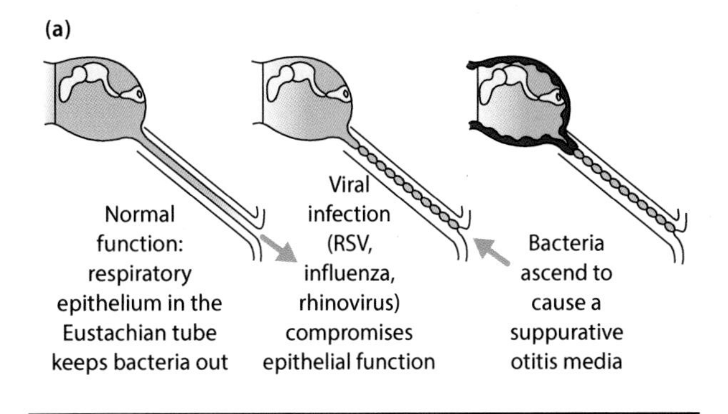

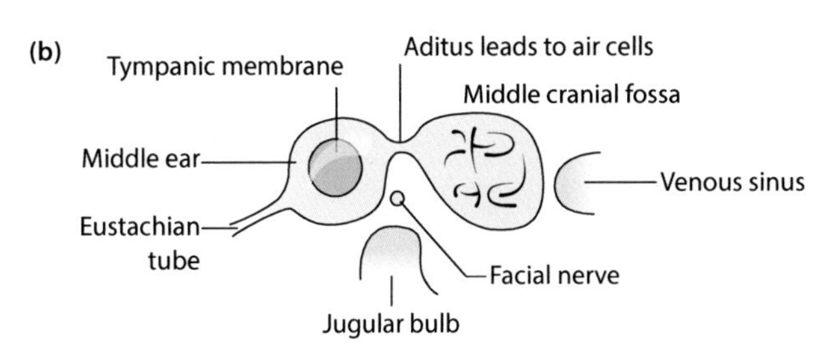

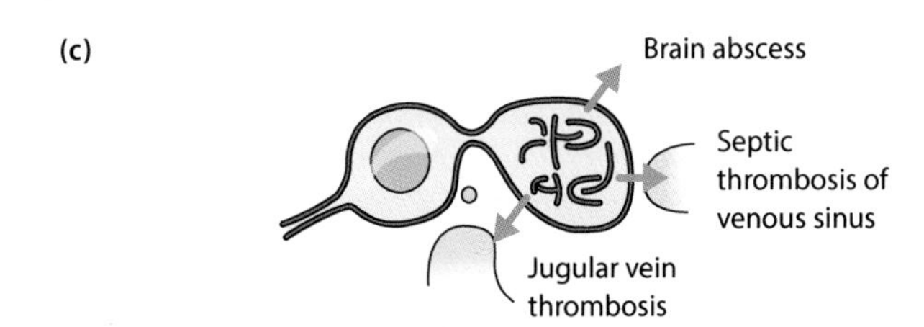

Figure 8.6 (**a**) Compromise of the respiratory epithelium in the Eustachian tube allows bacteria to enter the middle ear. (**b**) The relationship of the mastoid air cells to other structures. (**c**) Infection in the mastoid cavity can spread to neighbouring structures.

extracellular proteins such as the haemolysins O and S, DNase, NADase, streptokinase and the pyrogenic toxin account for the range of diseases that this organism can cause in addition to pharyngitis (**Figure 8.7**). These include scarlet fever, toxic shock syndrome (TSS) and cellulitis.

Streptococcus pyogenes infection can progress to peritonsillar abscess or quinsy. Such infections can involve anaerobes such as *Fusobacterium necrophilum*. A complication of the condition is septic thrombophlebitis of the jugular vein.

PNEUMONIA

Pneumonia usually arises as a result of aspiration of a pathogen such as pneumococcus into the lung where they overwhelm the local defences. The establishment of pneumococcal infection depends on several factors. These include the number of organisms aspirated and the ability of the ciliated respiratory epithelium to remove them; the smoker, and the individual with chronic obstructive pulmonary disease (COPD) are at additional risk due to compromised function of the ciliated epithelium.

Bacteria reproducing in the alveoli stimulate the local macrophages and an immune response is initiated. Classic pneumococcal lobar pneumonia is divided into four stages. The acute congestion stage, characterized by engorgement of the capillaries and recruitment of neutrophils into the lung parenchyma, is followed by red hepatization where there is indiscriminate flow of red blood cells (RBCs) from the capillaries into the alveolar space. The next stage is grey hepatization, with large numbers of dead and dying neutrophils along with degenerating RBCs. The last stage, resolution, coincides with the arrival of specific antibodies.

In the case of HAP, it is reasonable to assume that the patient is stressed, and one result of this is increased proteolytic activity in the saliva, which contributes to rapid turnover of the fibronectin layer that covers the epithelium of the pharynx. This layer is considered to have the resident colonizing flora attached and its loss removes their place of residence. The exposed epithelium can become colonized with large numbers of gram-negative bacteria, that can be aspirated into the lungs (**Figure 8.8a**). Some of the other contributory factors in HAP are shown in **Figure 8.8b**.

Group specific
carbohydrate (e.g. A) is a dimer of rhamnose and N-acetylglucosamine

M protein is a major virulence factor, with antiphagocytic properties

Extracellular products
Streptococcal pyrogenic toxin:
- scarlet fever rash
- toxic shock syndrome
Streptolysins (haemolysins):
- streptolysin O
- streptolysin S
DNase
NADase
Streptokinase

Some strains produce a hyaluronic acid capsule which makes them appear mucoid on culture

Figure 8.7 Some of the pathogenic properties of *Streptococcus pyogenes*, many of which are involved in the development of 'strep throat'.

LUNG ABSCESS

A lung abscess can arise as a complication of pneumonia, aspiration or septic emboli (**Figure 8.9a,b**). With pneumonia caused by *Staphylococcus aureus* and *Klebsiella pneumoniae*, the inflammatory response can progress to local tissue necrosis, then abscess formation. An abscess may also arise following the aspiration of fluid containing stomach contents and the bacteria of the oral cavity. This may occur in an obtunded or unconscious person, and alcohol and epilepsy can be relevant here. Any condition that compromises the swallowing function of the oesophagus or the anatomy of the trachea and bronchi allows bacteria to access the lung in significant numbers. A lung abscess may arise as a result of compression of a bronchus and can be the first indication of a bronchial carcinoma (**Figure 8.9c**). Septic emboli from right-sided endocarditis, or thrombophlebitis of the veins of the neck or pelvis, can lodge in the lung, progressing to abscess formation.

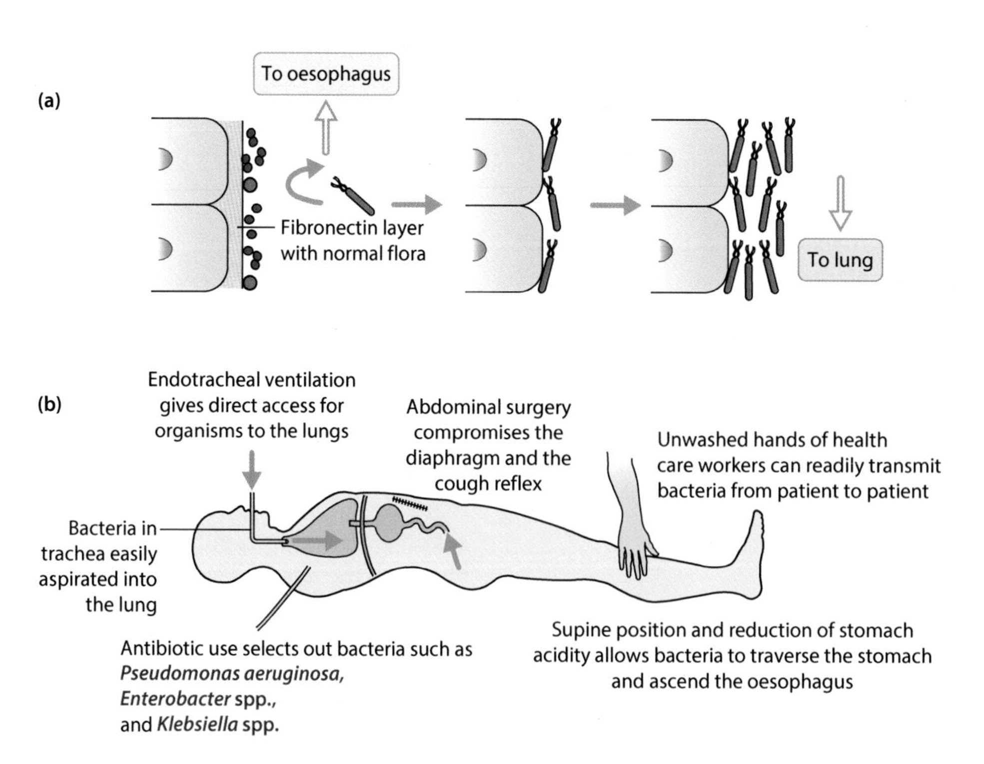

Figure 8.8 (**a**) Mucosal surfaces such as the oropharynx are coated with fibronectin and the normal flora. Removal of this protective layer allows gram-negative bacteria to colonize the oropharynx in significant numbers. (**b**) Some of the contributing factors to hospital-acquired pneumonia.

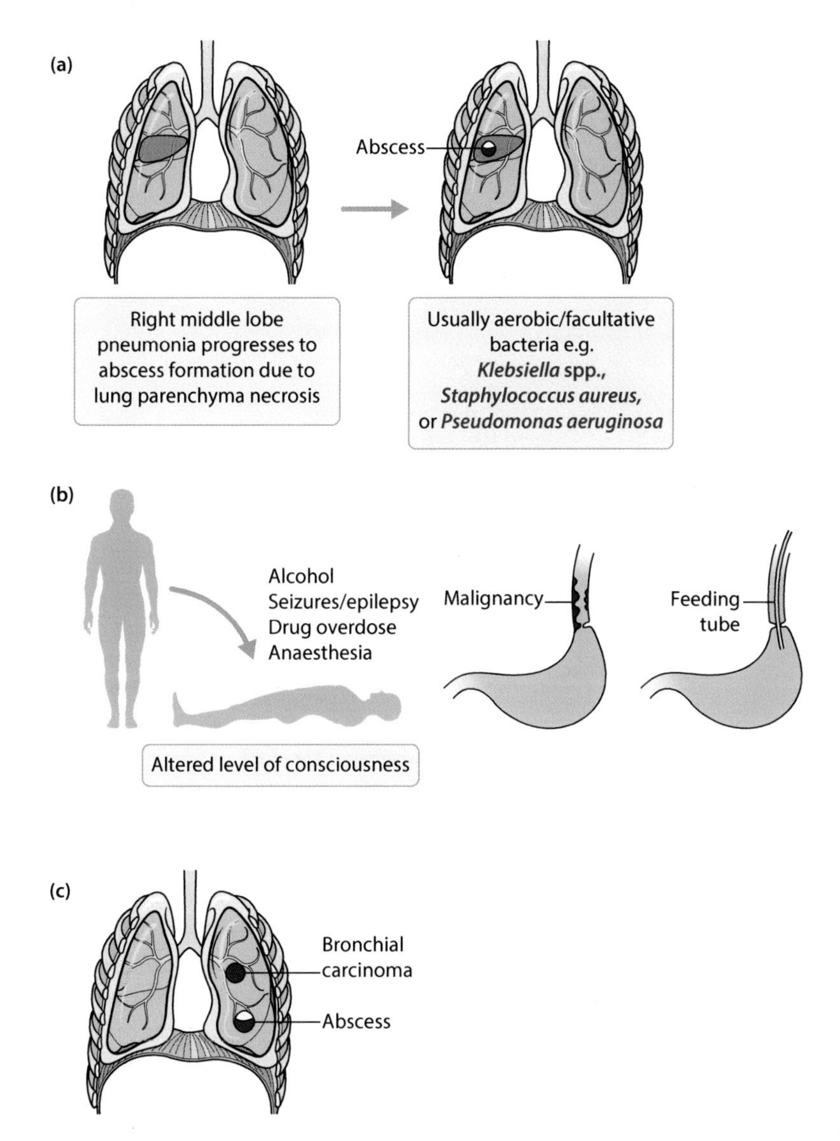

Figure 8.9 A lung abscess may arise as a result of: (**a**) bacterial pneumonia; (**b**) aspiration of stomach fluid and oral bacteria; (**c**) compression of a bronchus by a carcinoma.

DIAGNOSIS AND MANAGEMENT

Relevant diagnostic tests are referred to in **Figures 8.3, 8.4**. In the patient of 65 years-plus with CAP, the CURB-65 scoring system should be used (**Figure 8.10**). This determines how the patient is managed and which antibiotic regime is given (see Chapter 4).

The patient with a CAP must have a set of blood cultures, and sputum, if being produced, is collected too. If tuberculosis (TB) is a consideration, the first of three specimens collected on consecutive days is obtained. The pneumococcal and legionella urine antigen test is done for patients with moderate or severe CAP. In the winter period or 'influenza virus' season, testing nasal/throat swabs for the respiratory virus polymerase chain reaction (PCR) panel is done for all patients admitted with a CAP. In addition to clinical management of these patients, the PCR results determine if a patient needs ongoing care in a single room, and they are key to the hospital's bed management system. The need for annual influenza virus vaccination of 'at risk' individuals and all health care professionals underlines the importance of this public health intervention.

Purulent sputum should be collected for microbiological investigation. The best quality specimen is that coughed-up after waking; salivary specimens are of little use and should not be sent to the laboratory

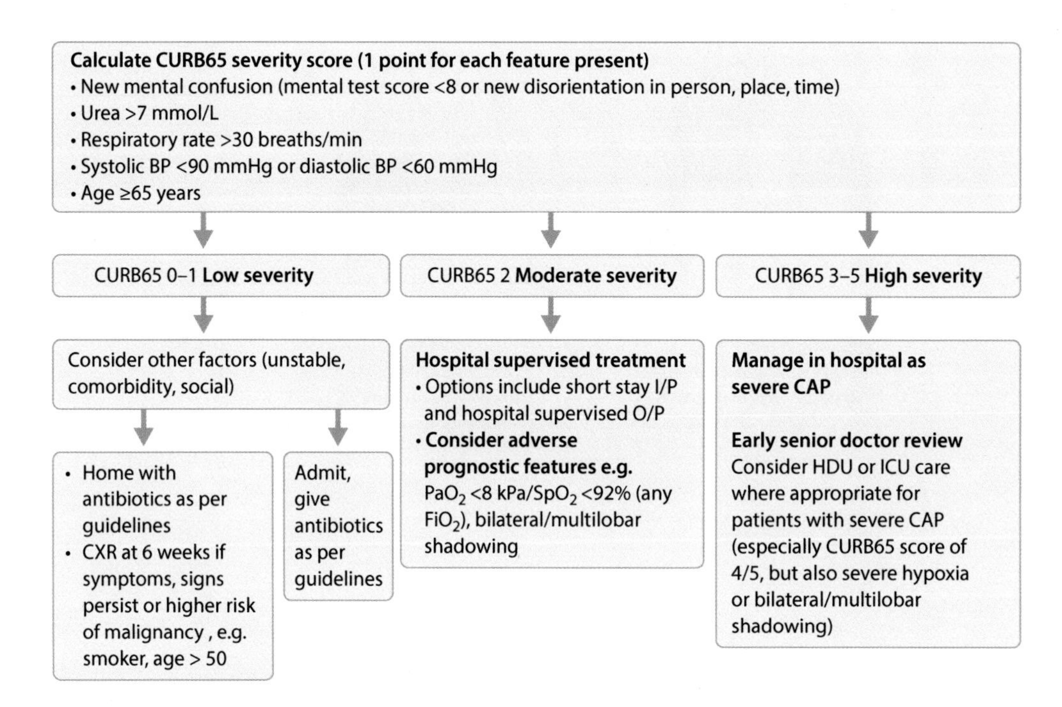

Figure 8.10 The assessment of the patient with a community-acquired pneumonia includes use of the CURB-65 criteria. The scoring system also relates to the antibiotics that are recommended for mild, moderate and severe infection (Chapter 4, **Figure 4.20**). (Adapted with kind permission of Dr Das Pillay, PHE laboratory, Heart of England NHS Trust, Birmingham, UK.)

(Chapter 3, **Figure 3.17**). In the ventilated patient on the ICU, specimens can be taken as ETT, bronchial washes or protected brush samples (**Figure 8.11**). Particular attention should be taken to optimizing the tests on a specimen such as a bronchoalveolar lavage (BAL), to ensure that all the necessary ones are done at the first opportunity. Examples include processing the specimen for *Pneumocystis jirovecii* or *Mycobacterium tuberculosis*. The same applies to pleural effusions, which should be aspirated, drained as needed and a pleural biopsy considered.

It should be noted that the legionella urine antigen test is most sensitive for detecting *Legionella pneumophila* serogroup 1. As this serogroup accounts for approximately 85% of legionella cases, a negative result does not exclude other serogroups or other species of *Legionella*. In the ICU setting in particular, sputum or ETT secretions must be cultured for this organism.

For the treatment of otitis media and sinusitis, supportive treatment with decongestants and analgesia is usually appropriate. However, antibiotics such as amoxycillin can be given on occasion when clinically indicated. Group A streptococcal pharyngitis can be treated with oral penicillin V or erythromycin.

The diagnosis of pulmonary TB can be straightforward when the clinical and radiological evidence is confirmed by a positive Ziehl–Neelsen (ZN) stain of the sputum. The diagnosis is usually not so obvious and it is always important to ask the question 'does this patient have TB?' It is essential to consider and then discuss with the patient who has a CAP the need for the HIV test.

Treatment of CAP and HAP has been outlined in Chapter 4, **Figure 4.20**.

LUNG ABSCESS

Where there is a lung abscess caused by a single organism, the antibiotic regime is specifically directed. In the case of a lung abscess arising as a result of aspiration of oral bacteria, agents effective against streptococci and anaerobes are used. Metronidazole with penicillin or co-amoxiclav are two options. Antibiotics are given for at least 6 weeks.

INFECTION FROM HOSPITAL WATER SYSTEMS

The management of water systems in the hospital is an important part of safe patient care, and two bacteria that can pose a significant risk are *Legionella pneumophila* and *Pseudomonas aeruginosa*.

LEGIONELLA PNEUMOPHILA

Cold water in the hospital is provided by gravity feed from a tank at the top of the building. This has to be at

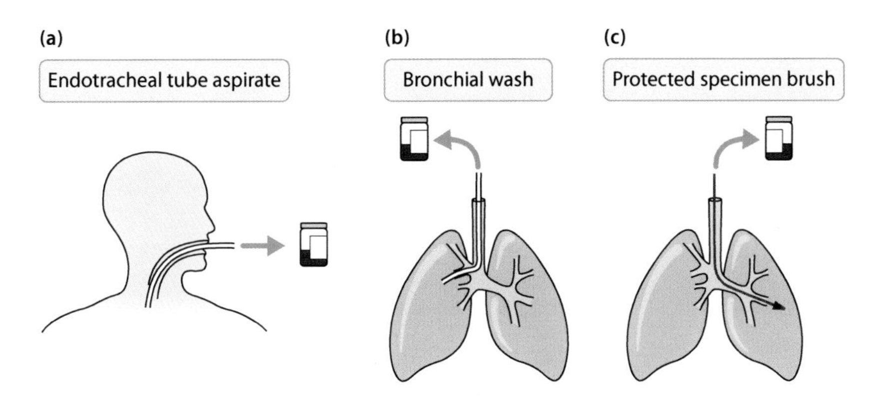

Figure 8.11 With the intensive care unit patient, various invasive diagnostic procedures can be used to obtain a specimen: (**a**) endotracheal tube aspirate; (**b**) bronchial wash/lavage; (**c**) protected specimen brush.

a temperature of 20°C or below after the fully opened 'sentinel' tap has been running for 2 minutes. Hot water is heated in a basement calorifier, and leaves this at a temperature of 60°C. This has to be at a temperature of 50°C or above after a fully opened 'sentinel' tap has been running for 2 minutes. (Users are protected from scalding by a thermostatic mixer valve, which delivers water at an appropriate temperature for hand washing.)

The reason for having these checks is to ensure that the temperature in the water system is unsuitable for the growth of legionella. Water temperatures between 20°C and 45°C and nutrients from sludge, scale, rust, algae, or other organic matter, allow growth in biofilm. When a water system is used, planktonic bacteria from the biofilm will be released in aerosols produced from shower and tap outlets.

In order to prevent colonization of water systems, institutions such as hospitals have to regularly flush outlets. Temperature recordings are monitored and water is sampled to monitor for growth of *Legionella*. On occasion, lapses in protocol do occur. From a clinical perspective, when the hospitalized patient develops an unexplained HAP, *Legionella* has to be considered, and appropriately investigated.

PSEUDOMONAS AERUGINOSA

Clinical staff should have an awareness of the appearance of *Pseudomonas aeruginosa* in the specimens collected from a patient, especially on ICU/HDU. *Pseudomonas* is an environmental organism that colonizes water systems in homes and institutions. It can form biofilms in taps, piping and is found in sink outlets and drains. When *Pseudomonas aeruginosa* is identified, the clinical team need to question where it has come from. This would involve the infection control team.

Chapter 9

Tuberculosis

INTRODUCTION

It is estimated that over 2.0 billion people, or one-third of the world's population, are infected with *Mycobacterium tuberculosis* and there are over 9 million new cases each year. Tuberculosis (TB) is responsible for more deaths than any other infectious agent. In countries such as the UK, the overall annual rate of the disease is about 10 cases/100,000, while in Africa this rate is in excess of 200 cases/100,000. A breakdown of the incidence in different ethnic groups in the UK reflects the rate in the countries of origin. For ethnic minorities who are classified as Black African, Indian subcontinent and Caribbean, the rates per 100,000 are >150, >50 and 20–30 respectively.

Treatment of TB is prolonged, lasting at least 6 months. If the services to supply and monitor treatment in each patient are under-resourced, compliance can become a major issue, creating the circumstances for selection of antibiotic resistant bacteria. Isoniazid (INH), rifampicin, pyrazinamide and ethambutol are used for 2 months, and then INH and rifampicin are continued together, for 4–10 months depending on the site of infection.

Multi-drug-resistant TB (MDR-TB) is resistance to at least rifampicin and INH (not infrequently there is also resistance to ethambutol and/or pyrazinamide). Extensively drug-resistant TB (XDR-TB) is resistance to at least INH, rifampicin and two second-line agents, a fluoroquinolone (e.g. moxifloxacin) and an injectable agent (either amikacin, kanamycin or capreomycin). Patients with MDR-TB or XDR-TB are of critical concern, in terms of treatment and infection control in the hospital and community.

TB has complex medical and social aspects, not just for one individual, but also for family, work, hospital and social contacts. An important risk factor for progression to active tuberculosis is human immunodeficiency virus (HIV) infection. Consequently, the greatest case rates are in countries where the prevalence of this virus is high. Prompt identification of the person with open pulmonary disease is essential to prevent spread of the organism to other individuals. In the patient with a chronic cough, weight loss and fever, and whose chest X-ray is characteristic of TB, the clinical diagnosis can be obvious. A positive microscopy result can confirm the diagnosis. Unfortunately, patients are still admitted to hospital for other reasons, and the diagnosis of pulmonary TB is made days later when a chronic cough is noted or a chest X-ray is belatedly reviewed. TB must be considered in any individual with a chronic cough. Whilst there may be risk factors, including ethnic origin, living in an area of high incidence, HIV disease or alcoholism, there are patients with no clear risk. Not infrequently a health care professional with a chronic cough is identified with open pulmonary TB, and patients and staff can have been exposed for weeks before the infection is considered and diagnosed.

ORGANISMS

A number of important bacteria belong to the genus *Mycobacterium*, and a classification of these organisms with some of the diseases they cause is shown in **Figure 9.1**. All mycobacteria are obligate aerobic, rod shaped, non-spore-forming organisms that are neither gram-positive nor gram-negative. Mycobacteria are 'acid-fast' and stain a deep magenta colour with the Ziehl–Neelsen (ZN) stain; for this reason they are referred to as acid-fast bacilli (AFB). The procedure for ZN staining is outlined in **Figure 9.2**. The acid-fast nature is due to the structure of the cell wall, whose constituents include complex lipids and mycolic acids. The heating process forces the carbol fuchsin stain into the cell, it complexes with the mycolic acid so that the organism is not decolourized by acid alcohol. A photomicrograph of *Mycobacterium tuberculosis* in a sputum specimen stained by the ZN method is shown in **Figure 9.3**.

An outline of the cell wall structure of *Mycobacterium tuberculosis* is shown in **Figure 9.4**. Constituents of the cell wall are responsible for the pathogenic features of disease caused by *Mycobacterium tuberculosis*.

PATHOGENESIS

When a pathogen such as pneumococcus invades the lung, the infection can be terminated when cells such as activated macrophages phagocytose the bacteria and

Organism	Diseases usually associated with
Tuberculous – these are a public health risk *Mycobacterium tuberculosis*	Tuberculosis (pulmonary, miliary, etc.).
Mycobacterium bovis	Bovine-derived tuberculosis.
Non-tuberculous – are not a public health risk *Mycobacterium avium/intracellulare*	**All these organisms are ubiquitous in the environment** Pulmonary disease in patients with pre-existing lung disease (COPD, bronchiectasis), plus disseminated disease in HIV patients; AIDS-defining illness.
Mycobacterium chelonae	Can cause disease in the lung, skin, soft tissue, or bone in individuals who are immunocompromised or following trauma. A 'rapid grower'.
Mycobacterium kansasii	Chronic pulmonary lung disease simulating tuberculosis, can be disseminated in HIV patients.
Mycobacterium malmoense	Pulmonary disease of patients with pre-existing lung disease.
Mycobacterium marinum	Skin and joint infections from salt or fresh water, especially of the hands in those who keep tropical fish. Considered to be more common in life guards as well. Can disseminate to other joints in immunosuppressed individuals.

Figure 9.1 Mycobacteria are broadly grouped into tuberculous and non-tuberculous groups. Some features of important members are outlined.

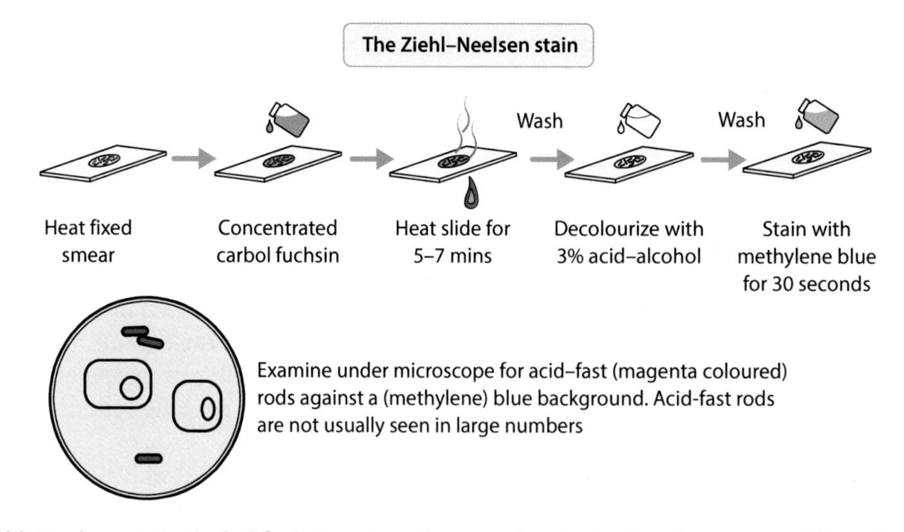

Figure 9.2 The Ziehl–Neelsen stain. Carbol fuchsin enters the organism in the heating process. The subsequent decolourization step is unable to remove the stain.

destroy them. The ability of the macrophages to take up the bacteria is enhanced by certain acute phase proteins, complement and antibodies neutralizing the pathogenic potential of the encapsulated pneumococcus. In TB, a strong antibody response is mounted; it is unclear what role this has in the disease process.

TB is a problem for the immune system to deal with. This is due to unusual components of the mycobacterial

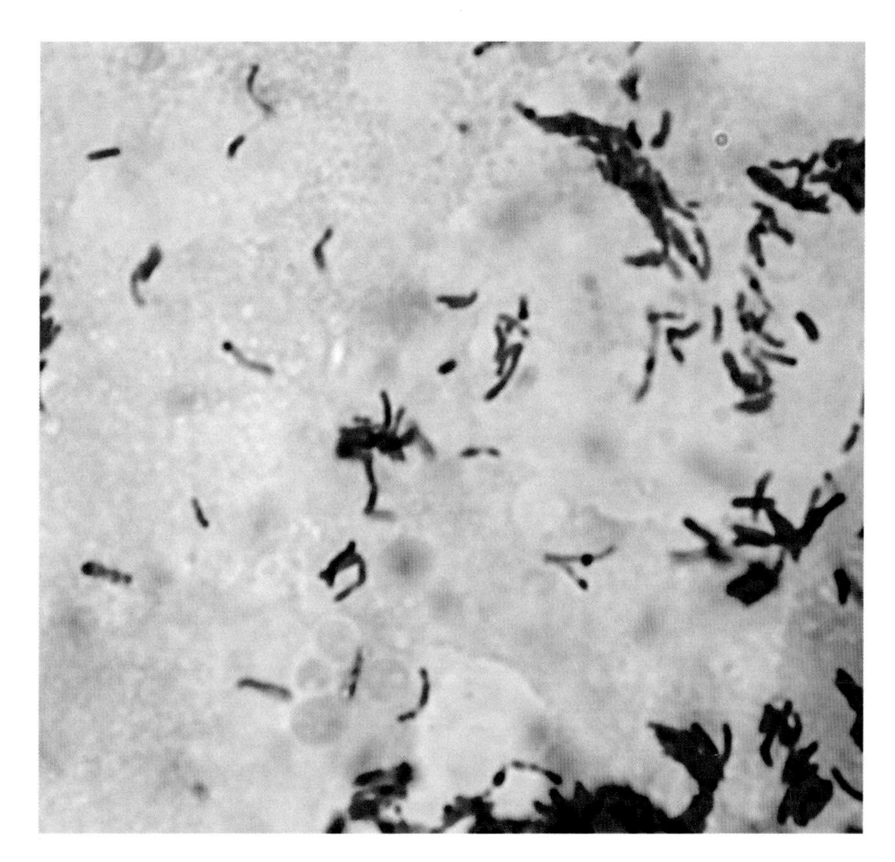

Figure 9.3 A photomicrograph of a sputum specimen stained by the Ziehl–Neelsen method. Numerous acid-fast bacilli are seen against the blue background.

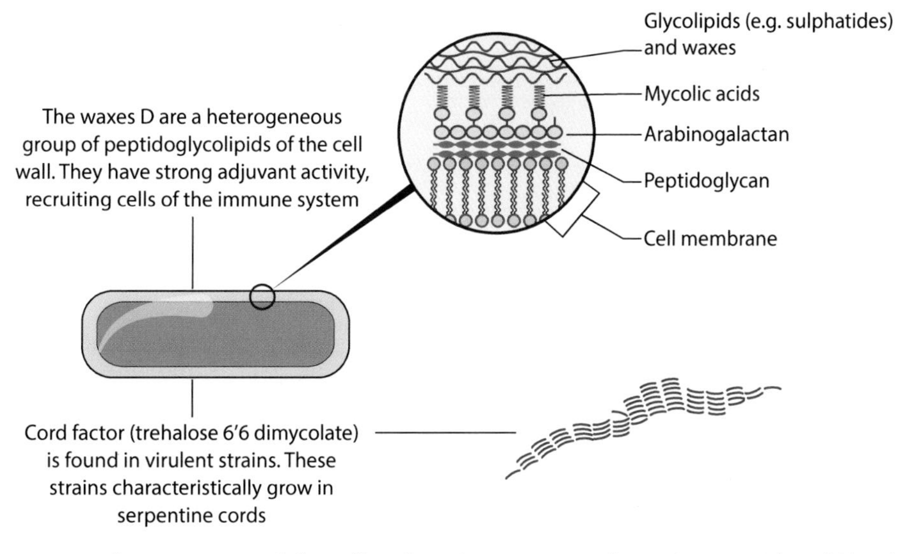

Figure 9.4 The structural components of the cell wall are important pathogenic properties of *Mycobacterium tuberculosis*.

cell wall, which interfere with cytokine activation of T cells and macrophages. Once inside the macrophage, mycobacteria inhibit the fusion of the phagosome and lysosome, and they escape into the cytoplasm, evading the intracellular killing machinery.

As stated in the introduction about one-third of the world's population is infected with *Mycobacterium tuberculosis*, but clearly only a minority have active disease. An outline of the process that occurs in the exposed individual is shown in **Figure 9.5a**. Aerosol-borne organisms are

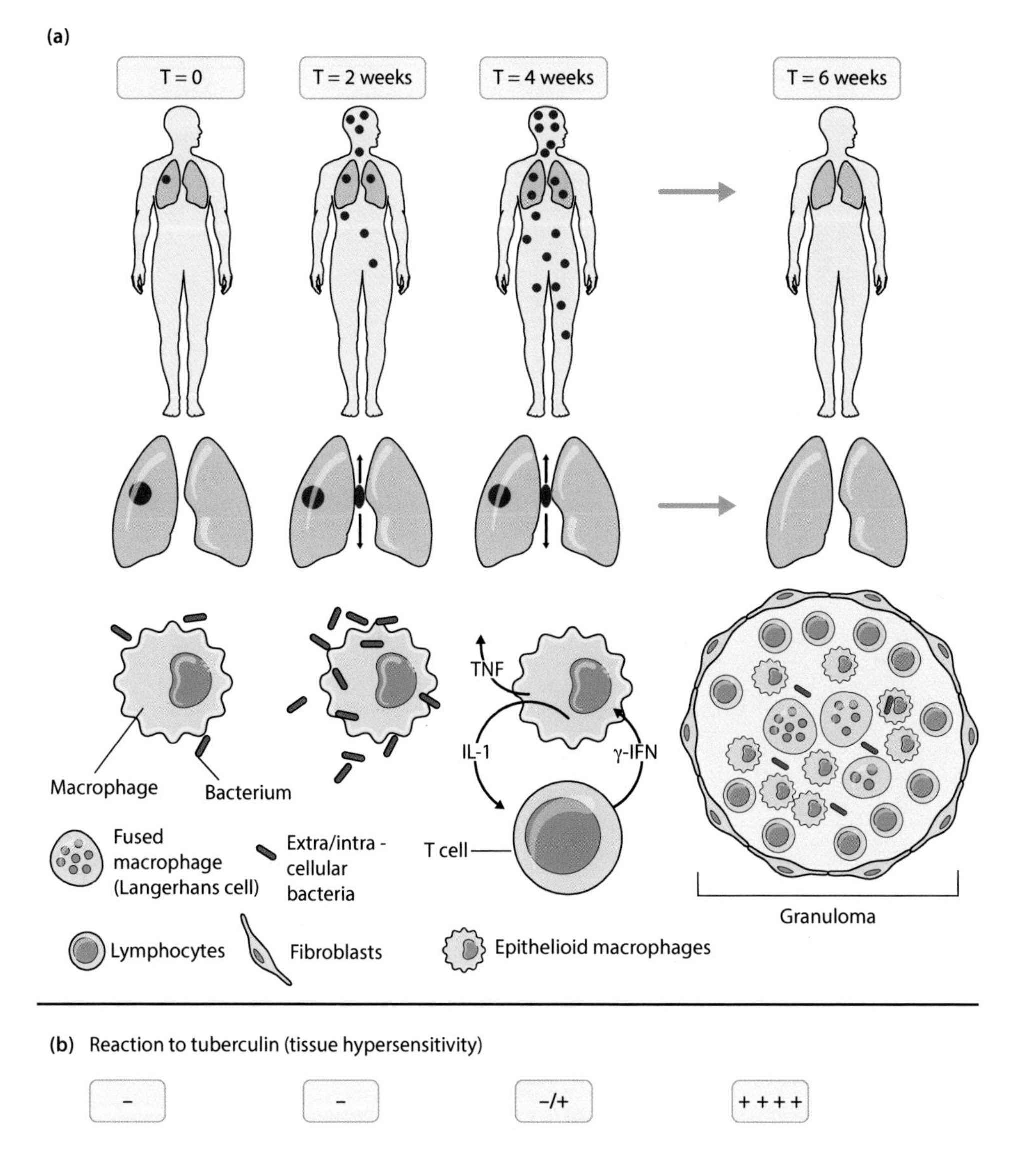

Figure 9.5 The first exposure to *Mycobacterium tuberculosis* usually results in a cell-mediated immune response (delayed type hypersensitivity) that kills the bacteria.

inhaled and settle in the alveoli of the lung. The bacteria are phagocytosed by alveolar macrophages, which at this stage are unable to kill the bacteria. Mycobacteria reproduce both in and outside macrophages, and as their numbers increase, they spread, via infected macrophages, to the perihilar lymph nodes. Entering the blood organisms distribute throughout the body. This process takes several weeks and reflects the slow division rate of mycobacteria, about 18 hours.

After about 4 weeks, a cell-mediated immune (CMI) response develops. Macrophages, activated by T cells, now contain high enough concentrations of enzymes and other metabolites that enable them to destroy the bacteria inside them. These activated macrophages are called epithelioid cells, which organize into a granuloma. In the centre they fuse as multinucleate giant cells or Langerhans cells. T cells and fibroblasts surround this effective killing machine. The whole process of cooperation between the T cells and macrophages, and the degree of macrophage killing activity, is tightly controlled by cell-to-cell contact and cytokines. The end result is the termination of the infection at all sites in the body that the organism has reached. This is a successful outcome. However, it is recognized that a few dormant bacteria survive in macrophages for decades. If immunological control is compromised, these bacteria have the potential to reproduce and cause reactivated disease.

The development of the cell-mediated response in TB is also referred to as the delayed type hypersensitivity (DTH) response, which can be identified several weeks after infection by the Mantoux and Heaf tests (**Figure 9.5b**). Injection of purified protein derivative (PPD), an extract of boiled bacteria, into the dermis of the skin incites a local inflammatory response in those individuals who have a CMI response to the organism. The Heaf and Mantoux tests can be used to identify those individuals who have been infected.

Interferon gamma tests (IGT) measure the presence of *Mycobacterium tuberculosis*-specific macrophage stimulating effector memory T cells in the patient's blood, and identify past exposure to the organism. The two antigens used, ESAT-6 and CFP-10, are proteins secreted by *Mycobacterium tuberculosis* that interfere with the primary response of T and B cells, and are critical for survival of the organism. The result obtained by the IGT is better correlated with latent infection and the presence of dormant organisms, and importantly, the result is not affected by prior bacillus Calmette–Guérin (BCG) vaccination, unlike the Mantoux or Heaf test.

A poor outcome of infection at primary exposure, in reactivated disease or re-infection, depends on many factors but central to this is some compromise of the cell-mediated immune system. It is thus not surprising that immunosuppression, such as that arising from HIV infection, pushes the infection towards the poorer outcome. Malnutrition and a very young age are other predisposing features. This emphasizes the fact that the ability to kill this organism relies on a fully functioning and integrated cell-mediated immune system.

The 'good' outcome of infection, outlined above, is shown in comparison with a 'poor' outcome in **Figure 9.6**. In the 'poor' situation, the degree of hypersensitivity is

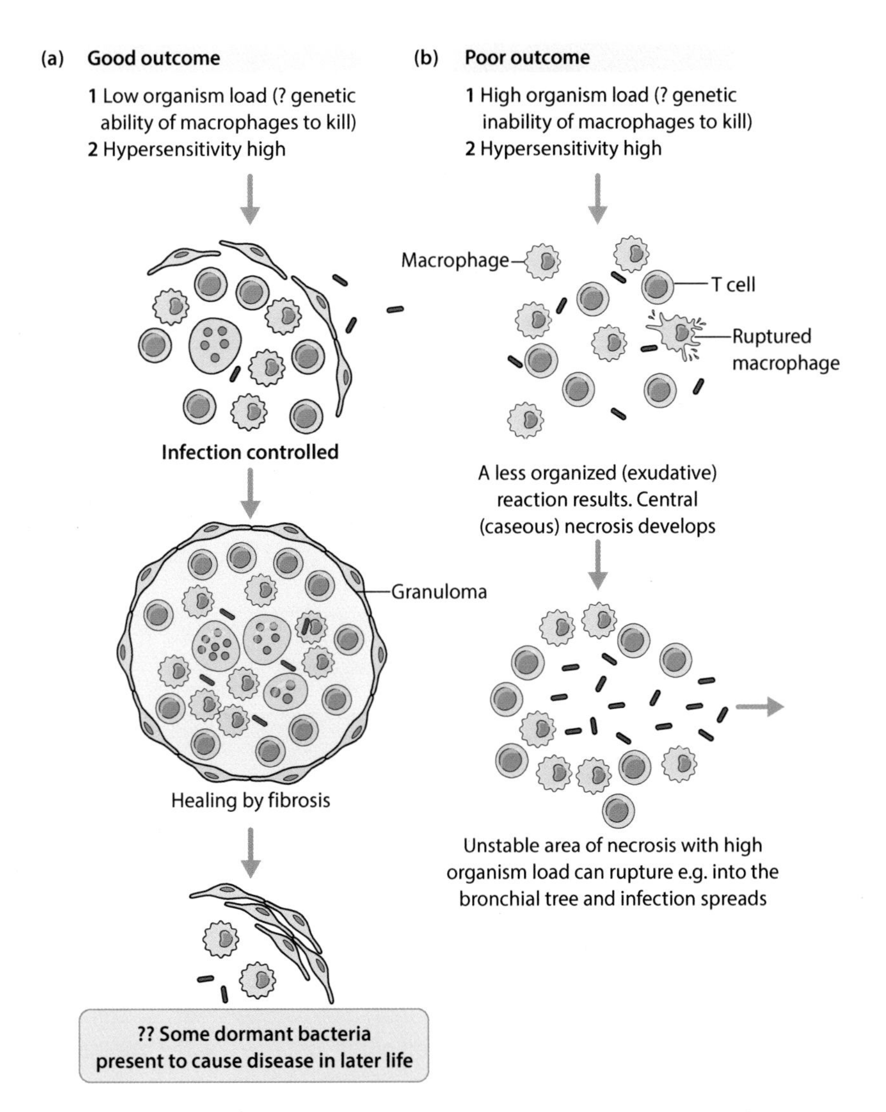

Figure 9.6 (**a**) The end result of a good outcome is effective organization in a granuloma. (**b**) A poor outcome arises as a result of some defect in the cell-mediated response.

increased. When the cell-mediated immune system is compromised, effective killing is not achieved and organism numbers are high. The problem is exacerbated by the fact that cell wall components of mycobacteria act as adjuvants, recruiting T cells and macrophages to the site of the infection. Uncontrolled lysis of macrophages releases large amounts of enzymes and metabolites that destroy local tissue. In the lung, unstable areas of caseous necrosis liquefy into adjacent tissue. Cavities form, fuse and break down, and organism-laden necrotic material spreads to other parts of the lung. The individual who is coughing up this material in aerosols is highly infectious.

Two broad types of clinical disease can be defined. Primary TB occurs following the initial exposure to the organism and secondary or reactivated disease occurs years later. An outline of infection and disease over time is shown in **Figure 9.7**. If the outcome of the primary infection is favourable, the resulting cell-mediated response controls the infection wherever the organism is in the body. If the outcome is unfavourable, growth of the organism is not controlled and primary disease may manifest. In the hilar lymph nodes, inflammation can be significant, especially in young children, resulting in compression of the bronchi with symptoms of cough and stridor. Organisms seeding the meninges from the blood give rise to meningitis. If a granuloma in the lung ruptures into the pleural space, the resulting hypersensitivity reaction results in a pleural effusion. Miliary TB is a manifestation of disseminated primary disease and derives its name from the numerous 'millet seed'-sized

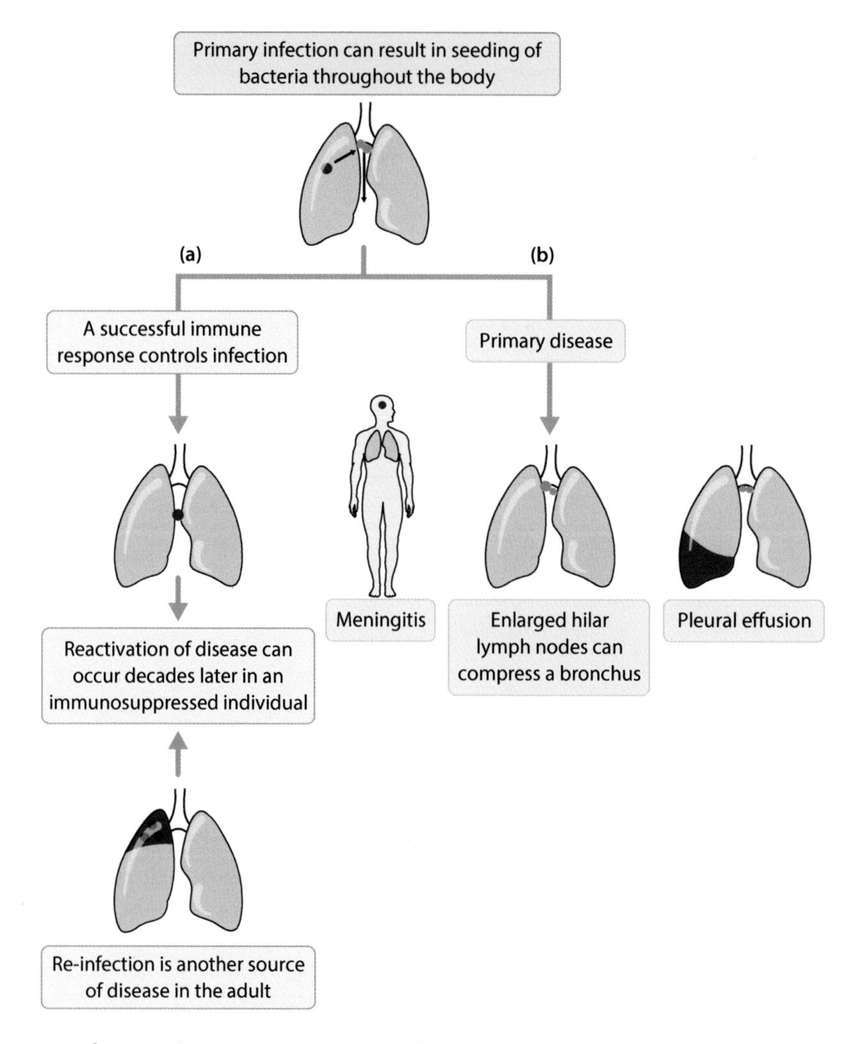

Figure 9.7 During primary infection the organism can spread to all organ systems. (**a**) The optimum outcome of the hypersensitivity response is control of the organism at all sites. (**b**) During the primary infection stage, clinical disease such as meningitis and pleural effusion can occur.

lesions seen on chest X-ray. The organism has reached the lungs via the systemic circulation from a focus of infection.

When the primary infection is controlled, organisms can reside in tissue for many years where they have the potential to reactivate. Reactivation or secondary disease occurs as a consequence of an individual's CMI waning and dormant bacteria reproduce to overwhelm the compromised defences. Reactivated disease often occurs in the upper, well-oxygenated lobes of the lung. Inability to control the infection leads to the breakdown of lesions, which fuse and develop into cavities of a size that can be seen on the chest X-ray. It is important to appreciate that reactivation is the likely mechanism of disease in older patients in areas where the incidence of TB is low. In areas where the incidence is high, re-infection is the likely mechanism. Here again it will be the immunocompromised patient who is likely to develop active disease.

DIAGNOSIS AND MANAGEMENT

Laboratory diagnosis of pulmonary TB requires an early-morning sputum specimen to be collected on 3 consecutive days (**Figure 9.8**). In the microbiology laboratory both the ZN stain and the auramine stain are used. Auramine is a fluorescent dye that binds to cell wall components of mycobacteria and as it is more sensitive than the ZN stain it is used for screening large numbers of smears. Auramine-positive smears are then confirmed with the ZN stain (**Figures 9.2, 9.3**).

Even if the first specimen is microscopy positive, two further specimens are still needed, as all three are processed to ensure that the organism is grown. Bacteria from the mouth flora can occasionally survive the decontamination process (below) and overgrow the mycobacteria present. Three negative smears do not exclude pulmonary TB, as microscopy is less sensitive than culture.

Other specimens that should be collected from the patient with suspected (miliary) TB include a 'first-of-the-day' midstream urine specimen (MSU) collected into a white-topped container, and blood in two citrated 'blue-topped' (citrated) bottles, which are used to inoculate specific blood culture bottles at the Reference laboratory. The frequency of involvement of various organs and sites in TB infection, as well as specimens to collect, is shown in **Figure 9.9**. *Mycobacterium marinum* causes 'fish tank-associated' infection of the joints of the hand and grows best at 30°C in the laboratory; a recreational exposure history is an important laboratory alert to include on the request form.

Patients with confirmed or suspected TB are referred to the appropriate TB specialist and Public Health are notified. Patients must have an HIV test done.

It is essential that a confirmed or suspected case of 'open' pulmonary TB is nursed in a single room. These rooms should have a negative-pressure ventilation system in order to reduce the organism load in the environment of the patient, and prevent spread to other patients and staff. Nursing the patient with MDR-TB or XDR-TB in a negative-pressure single room is essential.

An outline of the laboratory process is shown in **Figure 9.10**. Sputum specimens are decontaminated with sodium hydroxide to inactivate other bacterial contaminants. As mycobacteria are relatively hardy organisms, they resist this decontamination process. After neutralization of the alkaline step, sputum samples and all other specimens are centrifuged in order to concentrate any mycobacteria

Figure 9.8 It is important that the sputum specimens collected are 'first of the day', in order to obtain secretions that have accumulated in the lungs during sleep. Samples must be collected on 3 consecutive days to optimize isolation of the organism.

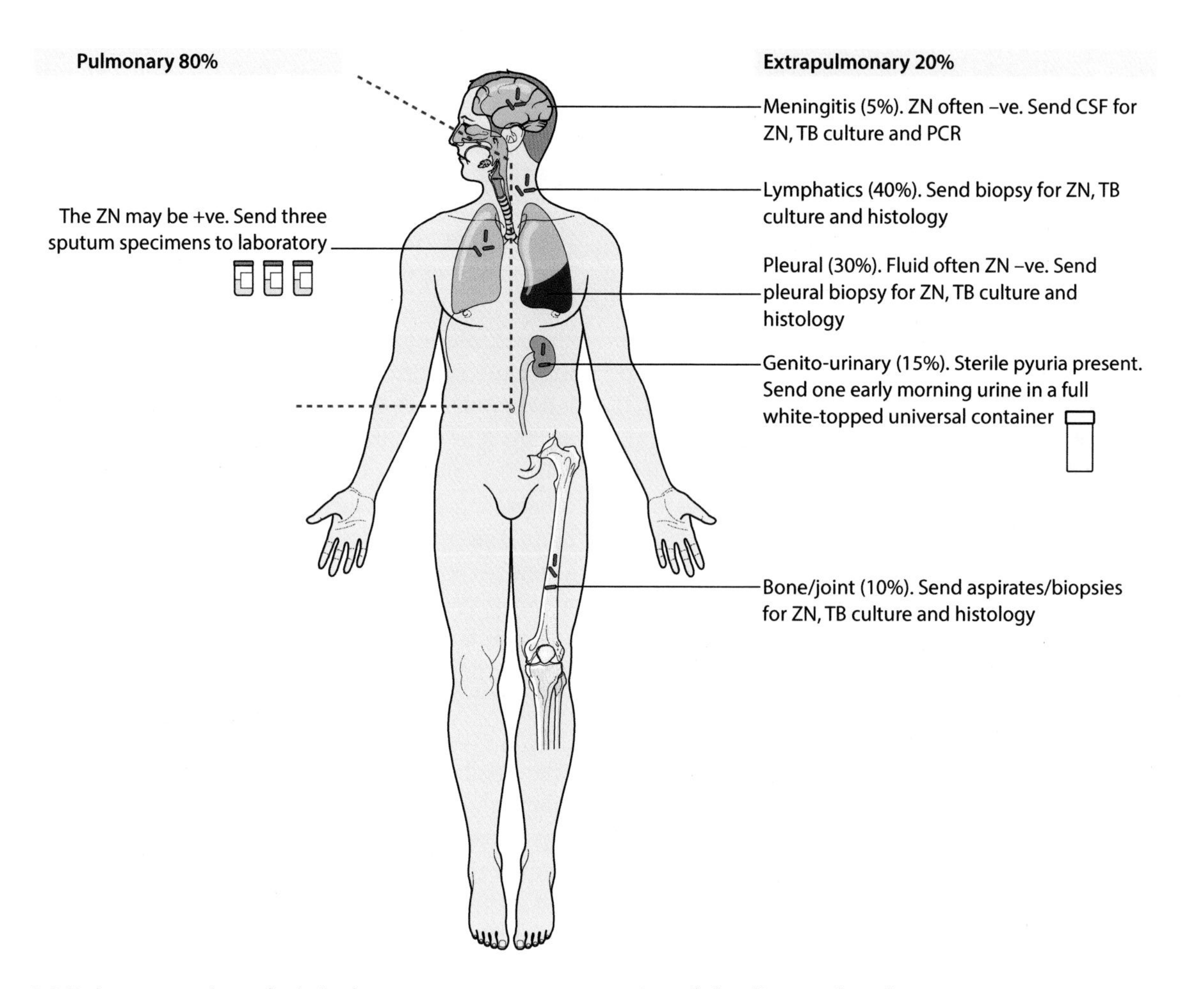

Figure 9.9 Pulmonary tuberculosis is the most common presentation of the disease, but disease may arise at any site. Some examples and specimens to be collected are shown here.

present. The centrifuged deposit is used to inoculate agar media (Lowenstein–Jensen [LJ]). Mycobacteria divide every 15–20 hours or so, and the time taken to produce visible growth is slow in comparison to other bacteria. Inoculated agar 'slopes' are thus examined weekly for up to 10 weeks.

'Rapid' liquid culture systems are used in conjunction with standard culture. The BD Bactec™ system includes a fluorescent compound in the base of the bottle. This compound is quenched by oxygen in the liquid. As the aerobic mycobacteria reproduce, oxygen levels fall, the quenching effect of oxygen is progressively reduced and a florescent signal is generated. On average it takes about 3–4 weeks before growth is seen on solid LJ medium, compared to 1–2 weeks in liquid culture. Isolates of mycobacteria are sent to the Reference laboratory, where full identification and susceptibility tests are done. The first-line antibiotics tested are INH, rifampicin, pyrazinamide and ethambutol. Whole genome sequencing (WGS) is likely to become the routine method for identification and antibiotic susceptibility testing.

MOLECULAR TESTING

It is now recommended that at least one sputum sample submitted for TB investigation is examined for the presence of *Mycobacterium tuberculosis* and rifampicin resistance by a molecular method, irrespective of the ZN/auramine stain result. The Cepheid Xpert® MTB/rif polymerase chain reaction (PCR) assay amplifies a portion of the ribonucleic acid (RNA) polymerase (rpoB) gene, including the nucleotide where a point mutation confers resistance to rifampicin. About 60% of microscopy-negative, culture-positive sputum specimens are positive by the molecular test.

THE HISTOPATHOLOGY SPECIMEN

When a biopsy is taken at surgery, and TB is a consideration, it is essential that a portion of the sample is sent to the microbiology laboratory. This should be placed in a white-topped sterile container with a few millilitres of saline. If the entire specimen is fixed in

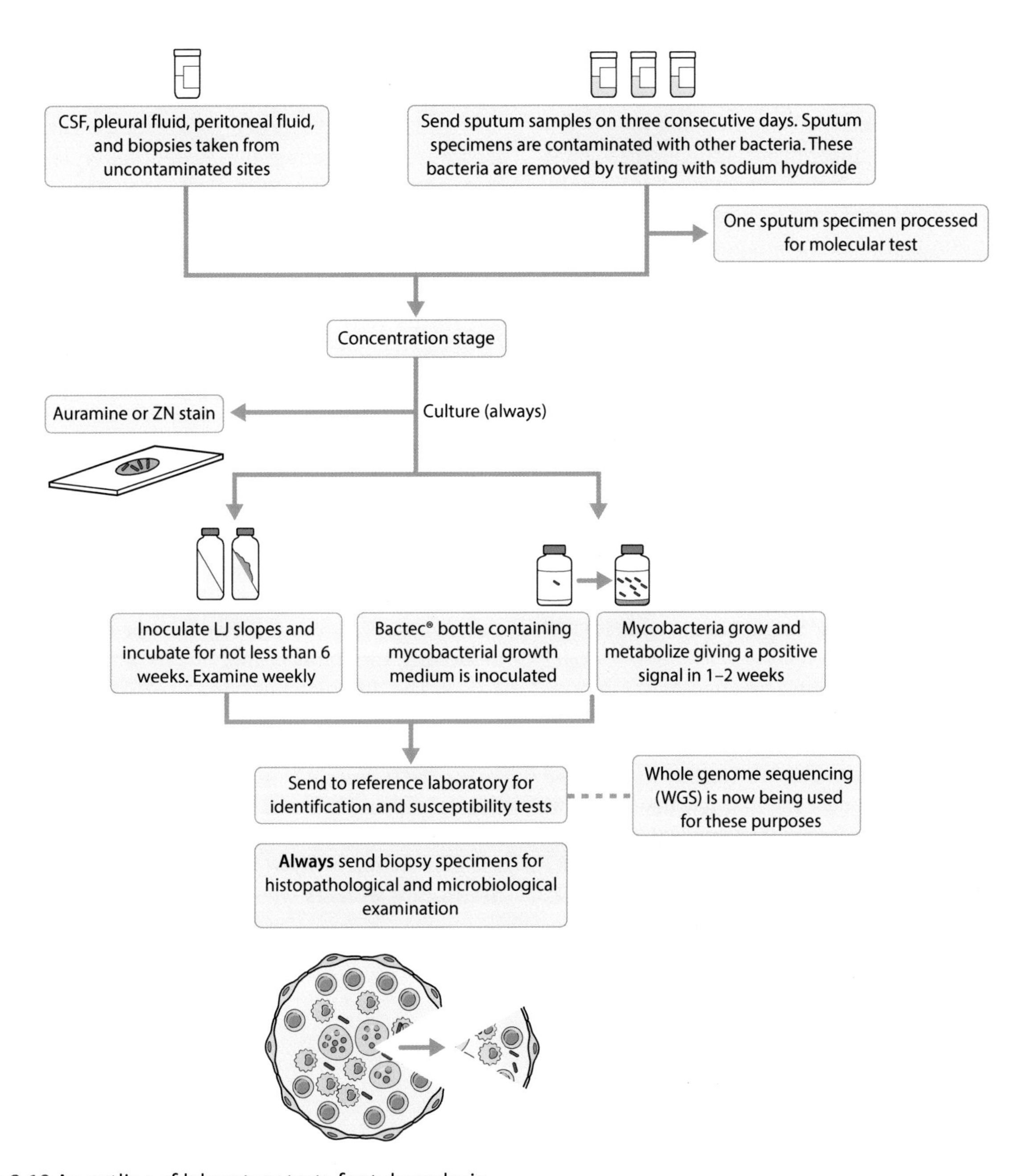

Figure 9.10 An outline of laboratory tests for tuberculosis.

formalin, the essential requirement to grow, identify and determine the susceptibility profile of an organism is lost (see Chapter 3, **Figure 3.19**).

TREATMENT

The treatment of pulmonary TB in the patient with a sensitive organism is quadruple therapy of INH, pyrazinamide, rifampicin and ethambutol for 2 months, followed by a further 4 months of INH and rifampicin. In the case of TB meningitis, INH and rifampicin are continued for 10 months, completing a 12-month course. To counter INH-induced vitamin B6 deficiency, pyridoxine is given daily.

It is essential that the patient on anti-TB treatment is closely monitored. Compliance over a long period is a problem, and directly observed therapy (DOT) can be used. Here the patient reports daily to a clinic and is observed taking the tablets. INH, pyrazinamide and rifampicin are potentially hepatotoxic, and baseline liver function tests (LFTs) are performed before treatment is started. It is important to have an index of suspicion and the patient with nausea and right upper quadrant tenderness must have liver functions checked. A bilirubin of >60 μmol/L and raised levels of aspartate transaminase (AST) and alanine transaminase (ALT) must prompt assessment of the patient and review of treatment. An alternative combination in the setting of deranged LFTs is ethambutol and streptomycin, bearing in mind that both these agents can be nephrotoxic. Optic neuritis is a side-effect associated with ethambutol. All patients on TB treatment are screened for hepatitis B and C infection.

INFECTION CONTROL

Patients with suspected or confirmed pulmonary or laryngeal (open) TB must be nursed in a single room, which should preferably be under negative pressure. MDR-TB or XDR-TB is an absolute requirement for nursing in a negative-pressure single room, where FFP3 masks are worn by staff.

In the absence of underlying immunocompromise or concerns about resistance, the compliant patient should be non-infectious within 2 weeks and can be nursed on the open ward if they need to remain in hospital. Resolution of symptoms over the first few days of treatment is reassuring. Other factors that enable this decision to be made are the absence of cavitation on chest X-ray or computed tomography (CT) scan and scanty numbers of organisms seen on the initial stains. The patient who remains unwell must be reassessed in terms of compliance and the possibility of drug resistance. The Reference laboratory performing the susceptibility tests should relay the results of any drug resistant isolate promptly. Red coloured urine shows that rifampicin is being taken.

For both clinical and infection control reasons, it is important to have an appreciation of individuals who are at increased risk of reactivated or acquired tuberculosis (**Figure 9.11**).

PUBLIC HEALTH ISSUES

There are various recommendations for the use of the tuberculin skin test (TST) and IGT, which include initial testing with a TST, and then using the IGT test in those who are TST positive.

Skin tests using PPD are used for screening programmes to determine the annual infection rate in a community. They can aid diagnosis or they can be used to identify the status of individuals who are contacts of a case. The different reactions to PPD are then used to decide whether a contact needs BCG vaccination, chemoprophylaxis or further clinical investigation.

The BCG vaccine is a live attenuated strain of *Mycobacterium bovis* and its use and efficacy throughout the world is variable. In the UK, recommendations for its use in individuals who do not have a positive TST include health care workers who work in high-risk clinical and

High risk individuals*		Medium risk individuals*
HIV positive	Jejuno-ileal bypass	Chronic renal failure or receiving dialysis
Aged <5 years	Haematological malignancy	Gastrectomy
IDU	High-dose steroids**	Diabetes mellitus
Solid organ transplant recipient	Cancer/transplant chemotherapy	Head and neck cancer
Receiving anti-TNF treatment	Silicosis	Significantly underweight/malnourished
Alcoholism		Radiographic findings of previous TB
		Chronic malabsorption

* In infection control incidents, individuals in both categories are risk assessed.
** >15 mg of prednisolone, or equivalent/day, for >2 weeks.

Figure 9.11 Individuals in these categories are at increased risk of *Mycobacterium tuberculosis* infection, and these risk factors can be used in the clinical assessment of patients, and staff, as relevant.

laboratory settings, prison staff, contacts of an individual with active pulmonary disease and immigrants from areas of the world where the incidence of TB is high. An HIV-negative status must be confirmed before BCG is administered.

With the Mantoux test, PPD is injected into the dermis of the anterior forearm (**Figure 9.12a**). The degree of hypersensitivity is determined by examining the injection site 48–72 hours later (**Figure 9.12b**). As previous BCG vaccination can give a positive result, the response to PPD is most useful in patients who have not been vaccinated. A reaction of >15 mm identifies individuals who are certainly infected and who need further investigation, and this would include a chest X-ray. Those individuals who have no reaction to PPD have not come into contact with the organism, or they could be in the early stages of primary exposure, before DTH develops. In the setting of exposure to a case of TB, it is these latter individuals who could be offered prophylaxis with INH for 6 months.

An outline of public health actions that should be taken when the index case is diagnosed as having active pulmonary TB is shown in **Figure 9.13**. For the purposes of this example, none of the five contacts of the case has had BCG vaccine. The results of the TST and further investigations such as chest X-ray determine the course of action for each contact. In the cases shown in **Figure 9.13** there are two children under 16 years of age. The 1-year-old child had no response to TST, and in view of the possibility of serious primary disease at this age INH prophylaxis is given. Six weeks later the TST is repeated and is still negative. It is reasonable to assume that this child has not been infected. The INH can stop and BCG vaccination is given. The 12-year-old child has a suspicious reaction. A normal chest X-ray rules out the likelihood of pulmonary disease and INH prophylaxis is given. The logic here is that the single agent will help the CMI response to control infection at this stage.

The 30-year-old adult's TST changes from 3 mm to 12 mm over the 6-week period, indicating recent infection and developing CMI. INH prophylaxis is appropriate here as well. The 31-year-old adult has a florid hypersensitivity reaction, and further questioning elicits the symptoms of TB, which is confirmed by chest X-ray and the examination of sputum. The 40-year-old adult has a positive Mantoux and normal chest X-ray. She has clearly been infected in the past and is given advice about symptoms and the need for a repeat chest X-ray 3 and 12 months later.

An essential aspect of the control of TB is the notification of any case to the public health authorities, including patients started empirically on a 'trial of treatment' before laboratory results are available. Contacts are appropriately advised to be HIV tested.

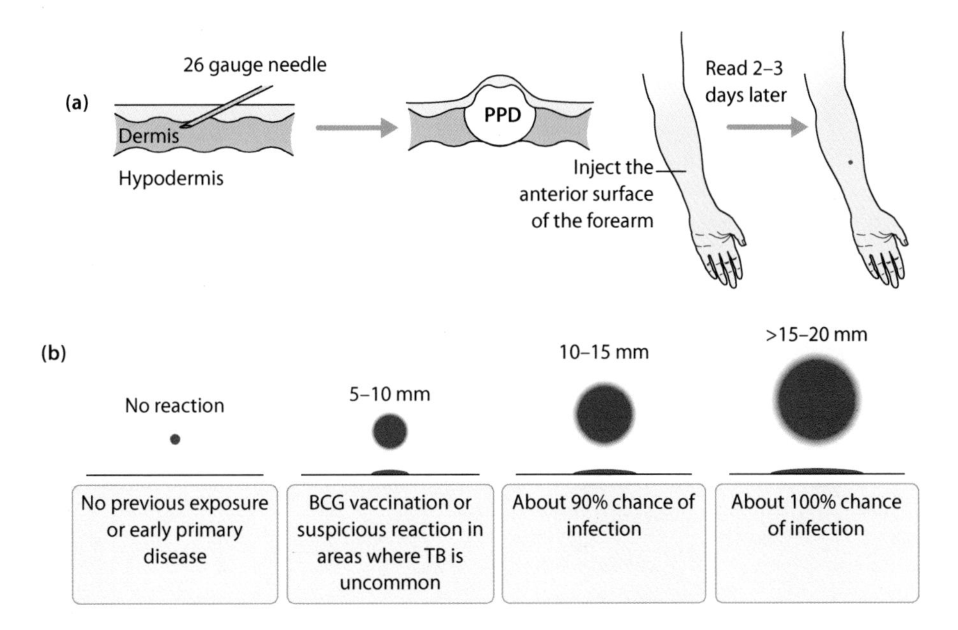

Figure 9.12 (**a**) The Mantoux test involves injecting purified protein derivative (PPD) into the anterior forearm. The extent of induration is then recorded. (**b**) The size of the reaction to PPD is used in management of contacts.

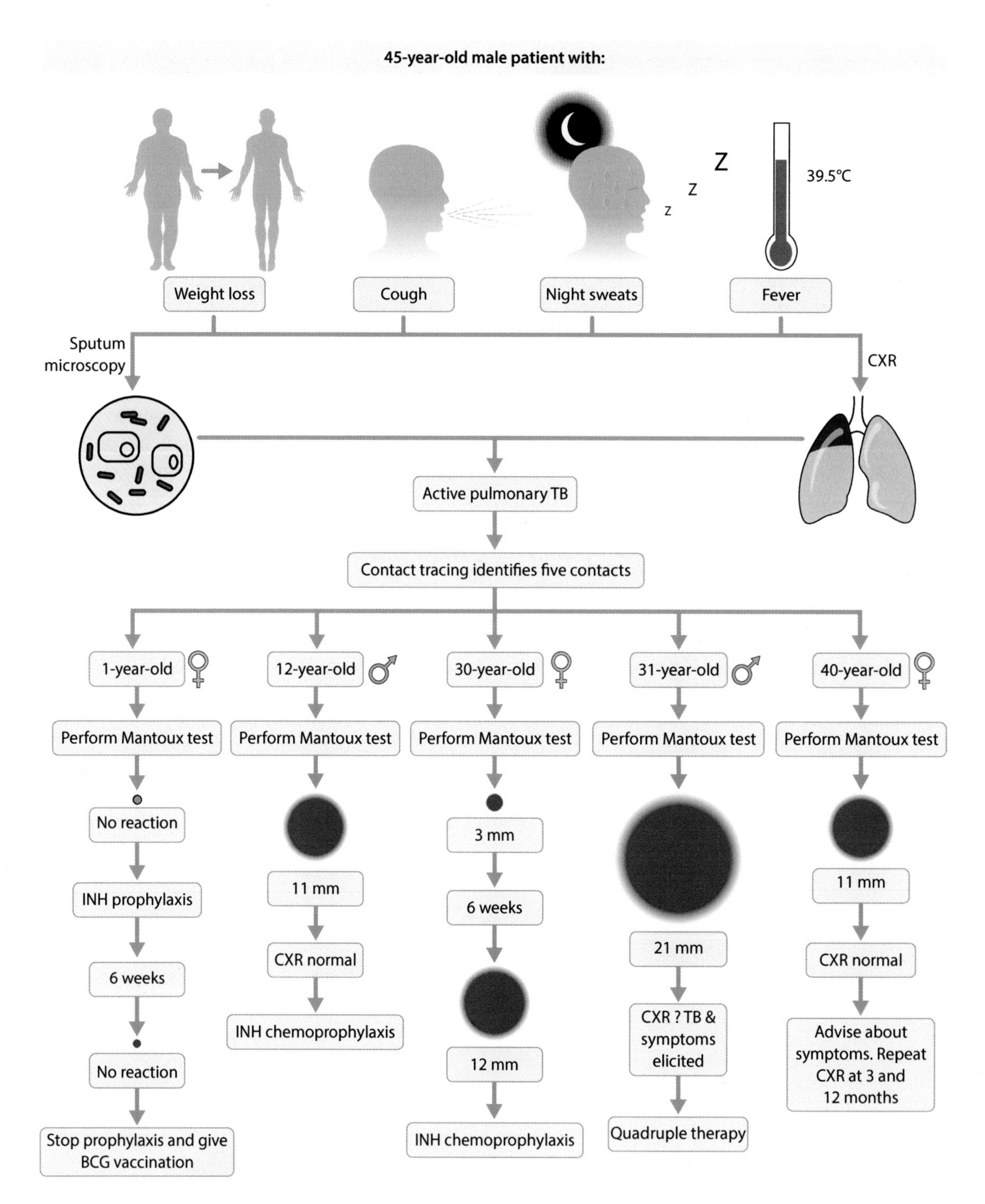

Figure 9.13 A 45-year-old patient is identified as having active pulmonary tuberculosis, with profuse numbers of mycobacteria seen on the Ziehl–Neelsen stain. None of the contacts has had the bacillus Calmette–Guérin vaccine; examples of the management of such contacts are shown here.

Chapter 10

Infections of the Urinary Tract

INTRODUCTION

The anatomical and physiological defences of the urinary tract, as well as routes that organisms use to access and travel through the system, are shown in **Figure 10.1**. The majority of organisms enter via the urethra, while *Mycobacterium tuberculosis*, *Salmonella* Typhi and cytomegalovirus (CMV) gain access from blood through the renal artery. Eggs produced in the venules of the urogenital system by the adult female trematode, *Schistosoma haematobium*, are pushed through the bladder wall into the lumen.

Bacteria can survive and reproduce in urine, despite the fluctuations in pH, osmolality and high urea levels. Defence of the bladder relies on a number of factors, including the length of the urethra as well as regular and complete emptying at micturition. In addition, the uroepithelium is covered with mucopolysaccharide, which inhibits bacterial binding, and 'surveillance' neutrophils secrete defensin peptides, which have antibacterial properties.

In the premenopausal female, a few bacteria are likely to enter the bladder on occasion, and if they do not bind the bladder surface, complete emptying removes them. However urinary tract infection (UTI) in this patient group is strongly associated with sexual activity; it is likely that physical manipulation around the periurethral area of the introitus forces more bacteria up the urethra into the bladder, where they are more likely to bind to the epithelium and establish themselves. The male is protected by the length of the urethra, but this anatomical defence can be compromised if a partner's vagina is colonized with a uropathogenic strain of *Escherichia coli*, or insertive anal intercourse is practised.

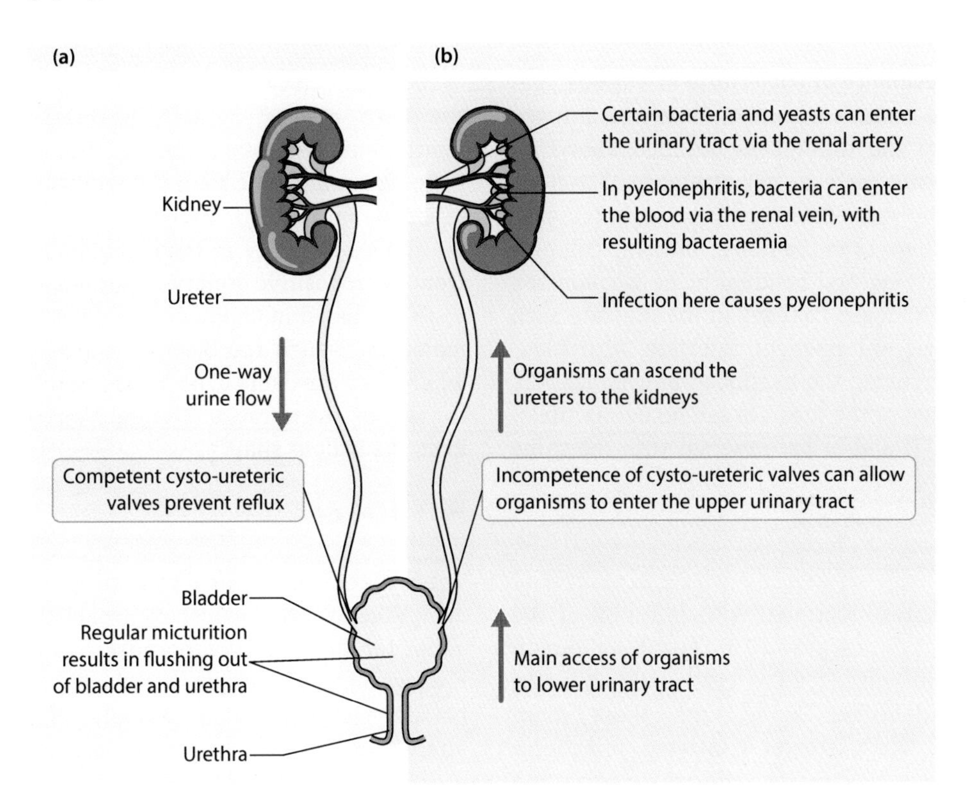

Figure 10.1 The urinary tract. (**a**) Normal structure and physiology maintain a sterile tract. (**b**) Bacteria can enter the urinary tract via the urethra or blood.

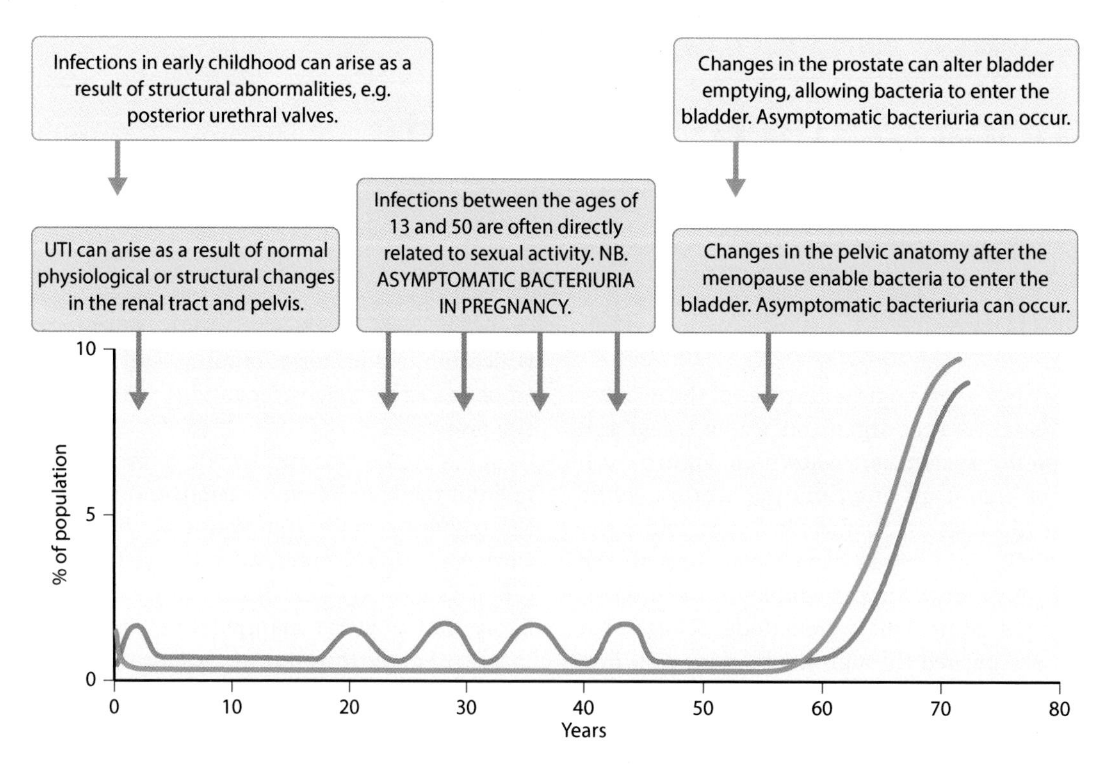

Figure 10.2 The prevalence of bacteriuria, in percent, in males (blue text/line) and females (pink text/line) in the stages of life. Note that asymptomatic bacteriuria is a common finding in the patient over 60 years, and there is no benefit in diagnosing and treating this.

The general prevalence of bacteriuria at various ages in the female and male is shown in **Figure 10.2**, highlighting certain of the differences outlined above. Of note is the significant increase in prevalence of bacteriuria at the age of 60 years-plus in both sexes. In females, postmenopausal changes in the pelvic anatomy compromise bladder emptying and residual urine is colonized with bacteria. In the male, changes in the prostate such as benign prostatic enlargement interfere with bladder emptying, increasing the likelihood of colonization. Chronic colonization of the prostate can occur, manifesting as relapsing UTI and/or bacteraemia with the same organism.

ORGANISMS

A range of organisms associated with infection of the urinary tract is shown in **Figure 10.3**. The more common bacteria associated with cystitis and pyelonephritis, 'coliforms', are identified, along with those more frequently found with catheter-associated infections. *Mycobacterium tuberculosis* reaches the urinary tract via the blood from the primary infection, usually the lung, and infects the renal parenchyma, spreading to the ureters and bladder; strictures of the ureter can occur.

BK virus (BKV), a polyoma virus, and adenovirus are both associated with haemorrhagic cystitis in neutropenic patients, including those who have undergone haematopoietic stem cell transplantation. BKV also causes nephropathy in kidney transplant patients. CMV is an important entity in renal transplantation, especially from a seropositive donor to a seronegative recipient.

As discussed in Chapter 1, the eggs of *Schistosoma haematobium* transit the bladder wall to reach the lumen, where they are likely to be passed into fresh water, hatching into motile miracidia that seek out their intermediate host, the *Bulinus* snail.

PATHOGENESIS

The bladder is innervated by both sympathetic and parasympathetic nerves, which control bladder emptying and sphincter function. Compromise of nerve function following injury to the spinal cord or in diabetes can affect bladder emptying. Some examples of physiological and anatomical abnormalities that predispose to infection are shown in **Figure 10.4**. In a patient with recurrent UTI, anatomical abnormalities need to be excluded. There is a correlation between the presence of stones in the renal pelvis and UTI caused by *Proteus*. Recurrent infection

Bacteria	Fungi	Viruses	Parasites
Escherichia coli	*Candida albicans*#*	Adenovirus	*Schistosoma haematobium*
Klebsiella oxytoca	*Candida tropicalis*#*	BKV	
Proteus mirabilis		CMV	
*Pseudomonas aeruginosa**			
Enterococcus faecalis			
Enterococcus faecium			
Staphylococcus aureus (MSSA/MRSA)*#			
Streptococcus agalactiae; GBS			
(*Streptococcus pyogenes*; GAS**)			
Mycobacterium tuberculosis (via blood)			

* Usually associated with long-term catheterization

** Can be found in a CSU, and in the male, balanitis

Can reach the urinary tract from the blood

Figure 10.3 Examples of the range of organisms that can infect the urinary tract.

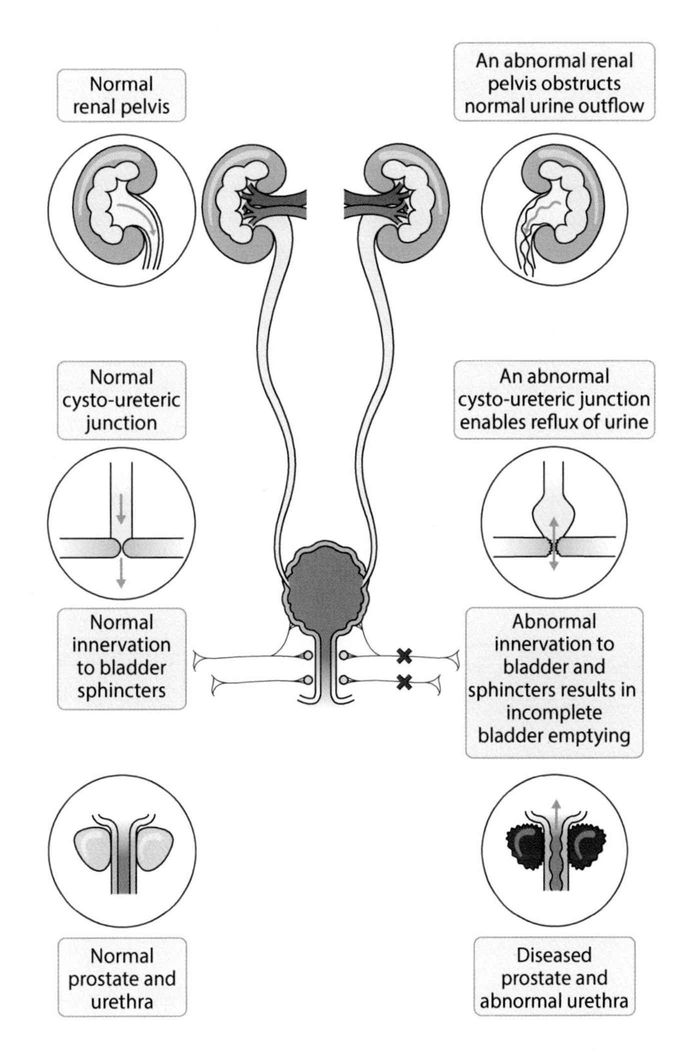

Figure 10.4 Some details of the anatomy and innervation of the urinary tract in preventing and predisposing to urinary tract infection.

is likely to cause irreversible damage to the developing kidneys of the young child. Urethral catheters compromise the structural and physiological barriers of the urethra and bladder. Bacteria and yeasts can ascend the outside or the lumen of the catheter to reach the bladder (**Figure 10.5**).

An outer layer of bladder's transitional epithelium consists of umbrella cells, whose surface is covered with the transmembrane uroplakin protein. This layer of cells provides a highly efficient permeability barrier to the wide variations in pH, osmolality and pressure. However, uropathogenic strains of *Escherichia coli* expressing type P pili adhere to uroplakin, and invade the cells. The induction of apoptosis and an intense neutrophil inflammatory response give rise to the local symptoms of dysuria, frequency and urgency.

UNCOMPLICATED AND COMPLICATED UTI

UTIs in otherwise healthy premenopausal women usually involve cystitis, with the typical symptoms of dysuria, frequency and urgency, reflecting the pathological process outlined above. Occasionally the organism ascends to the kidney, with pyelonephritis and sepsis. In these patients, infections related to sexual activity should be identified. The patient should be advised to completely empty the bladder after coitus, and postcoital prophylactic antibiotics can be considered. Re-infection is associated with use of a diaphragm with a spermicide, and an alternative method of contraception should be discussed.

Complicated UTI include infection with resistant bacteria, failed antibiotic treatment or persistent symptoms, pyelonephritis (except in otherwise healthy females) and infections associated with functional, metabolic or anatomical abnormalities of the urinary tract. Here as well, the clinical spectrum ranges from mild cystitis to life-threatening urosepsis. By definition, a relapse refers to infection with the same organism, usually within 2 weeks of completing treatment. Here an occult source of infection, which can be associated with a urological abnormality, should be excluded. Re-infection refers to recurrent infection, often with a different strain or species. This is usually more than 2 weeks after completion of therapy.

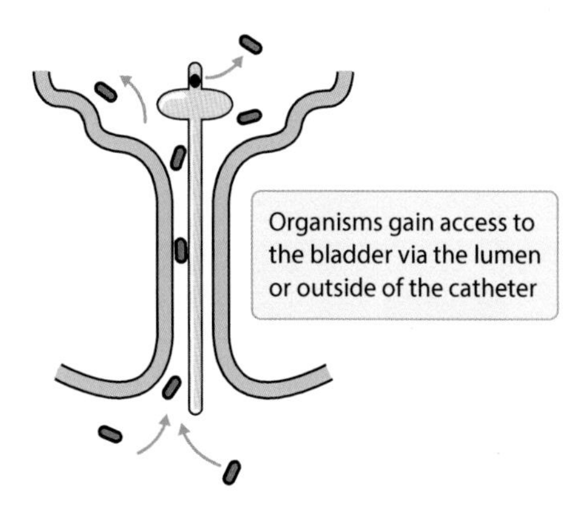

Figure 10.5 The insertion of a urinary catheter compromises the structural integrity of the urethra.

Asymptomatic bacteriuria is a common finding in the population group of 60 years-plus, and is a physiological consequence of the ageing process; because of this, its diagnosis and treatment is not recommended. Unnecessary antibiotics select resistant bacteria and put the person at increased risk of *Clostridium difficile* infection, and re-colonization of the bladder occurs following treatment. However, individuals in this age group are at risk of developing symptomatic UTI, pyelonephritis and sepsis, which require prompt recognition. As shown in Chapter 3, **Figure 3.5**, UTI were the second and fourth most common reasons for hospital admission in females and males in this age group.

DIAGNOSIS AND MANAGEMENT

The clinical diagnosis in the younger female patient presenting with sepsis, fever, rigors, urinary frequency, dysuria and renal angle tenderness is usually straightforward; an organism in an ascending UTI has invaded the blood. This is confirmed when blood culture and a midstream urine sample (MSU) subsequently grow a 'coliform' such as *Escherichia coli*. Making a clear diagnosis of a UTI in the older patient is often not as straightforward. New incontinence and dysuria may be noted, but confusion or increased confusion, with minimal temperature response, may be the presenting symptom. Not infrequently, the diagnosis made in A&E is a community-acquired pneumonia; however, when blood cultures grow a 'coliform', subsequently identified as *Escherichia coli* or *Proteus mirabilis*, it is clear that the source of the infection is the urinary tract.

Microbiological diagnosis centres on collecting a MSU, and in the setting of pyelonephritis and sepsis, blood is cultured. The usefulness of laboratory examination of urine depends on the standard of the specimen collected; a 'clean-catch' urine specimen is important (**Figure 10.6a**). When such a specimen is obtained from a patient with a suspected UTI, neutrophils (white blood cells, WBC) and a single species of bacteria will be identified. In incorrectly collected specimens, vaginal epithelial cells and the bacterial flora attached to these cells will also be present. Vaginal epithelial cells indicate a contaminated specimen, subsequently confirmed when culture identifies a mixed growth of bacteria (**Figure 10.6b**).

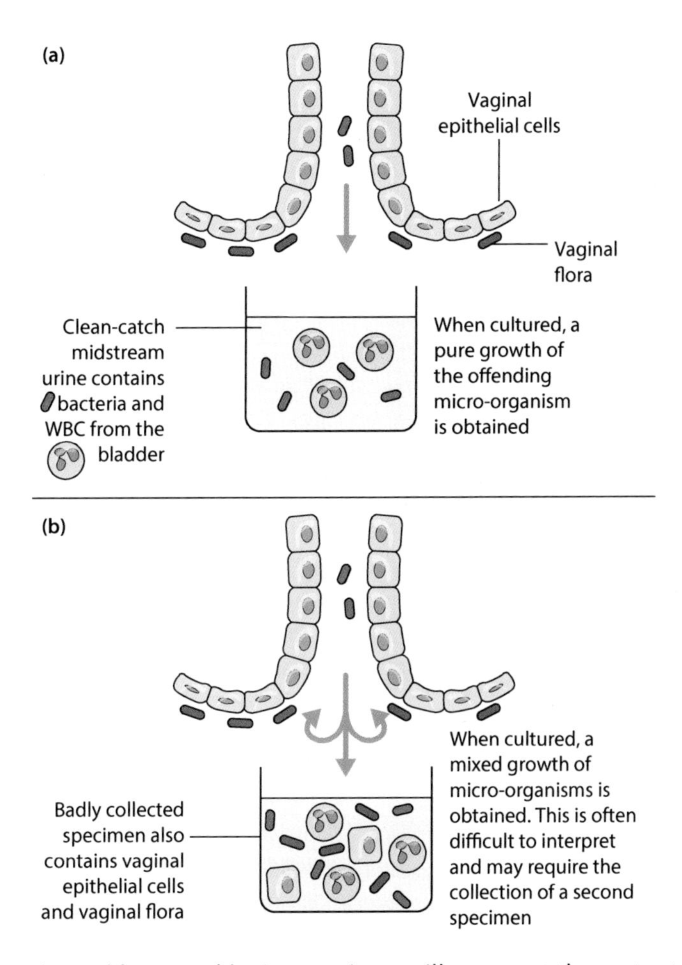

Figure 10.6 (**a**) In the female patient a 'clean-catch' urine specimen will represent the contents of the bladder. (**b**) A contaminated specimen will contain vaginal epithelial cells and attached bacteria. (WBC: white blood cell.)

In order to obtain a clean-catch MSU the patient should be given the appropriate advice and collection system. The 'collection area' of a boric acid urine specimen container is clearly impractical for the female patient to use, and suitable disposable 'pots' should be provided (**Figure 10.7**). After hand washing, the female patient should be advised to open the labia with the fingers of one hand, establish a full urine stream for several seconds, and then collect the 'midstream' sample.

Urine contains growth substances for bacteria (and yeasts), and they will multiply in urine stored at room temperature, giving a falsely elevated organism titre. Most laboratories use boric acid sample containers; when the container is filled to the correct level, the dissolved boric acid inhibits bacterial growth. This inhibitory effect is diluted out when the specimen is 'plated out' in the laboratory. As higher concentrations of boric acid can kill bacteria, smaller volumes of urine should be collected in a white-topped sterile container (**Figure 10.8**). These samples need to reach the laboratory promptly, and if this is not possible, stored in a designated fridge at 4°C.

Figure 10.7 The patient should collect the midstream urine specimen into an appropriate 'sterile' receptacle, which is then transferred to the specimen container.

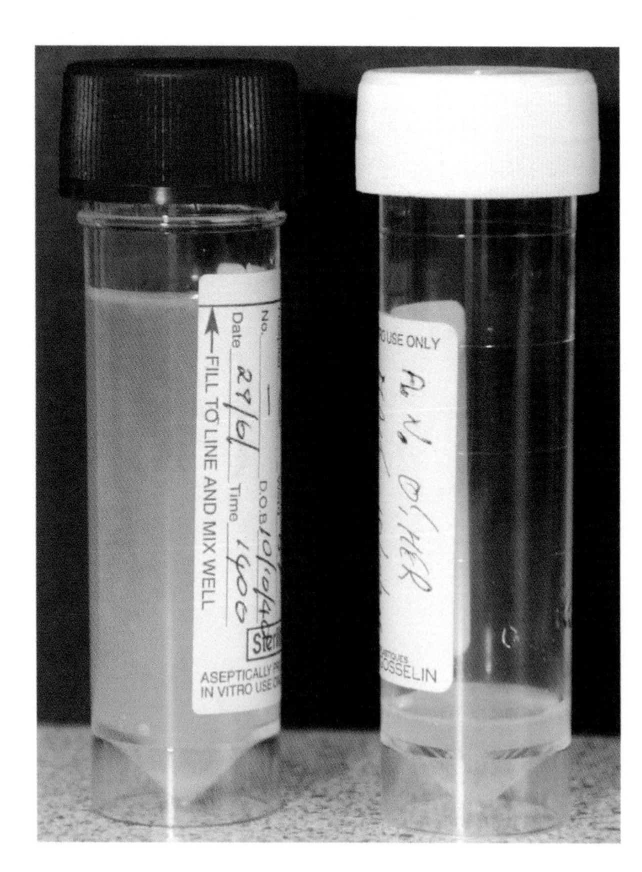

Figure 10.8 Boric acid containers must be filled to the arrow mark with the specimen to obtain the correct, inhibitory concentration of boric acid. If there is a smaller volume of urine, a white-topped sterile container is used, which must be sent to the laboratory promptly, or stored in a designated fridge.

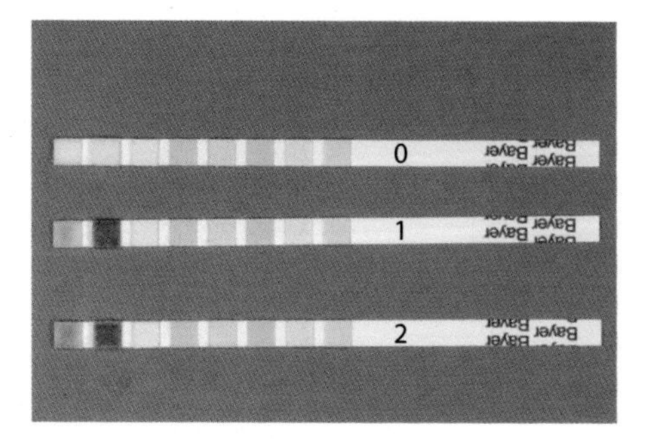

Figure 10.9 After doing a 'dipstick', the panels must be developed for the necessary time; leucocyte esterase is read after 2 minutes.

EXAMINATION OF THE SPECIMEN; THE URINE DIPSTICK

When a urine specimen has been collected, it should be examined by eye; a crystal clear urine indicates that a UTI is unlikely. If there is 'cloudiness', the specimen can be 'dipsticked' for leucocyte esterase and bacterial nitrites, closely following the manufacturer's instructions (**Figure 10.9**). The detection of leucocyte esterase and nitrites shows the presence of inflammatory cells and bacteria. It should be noted that nitrites are not produced by gram-positive bacteria such as streptococci and enterococci. For the majority of sexually active women with their first episode of cystitis, a positive dipstick essentially confirms the clinical diagnosis, and empirical antibiotic treatment for 'coliforms' (*Escherichia coli*) prescribed.

In the patient of 60 years-plus, the absence of leucocytes (and nitrites) in an MSU can be useful to exclude a UTI. However, a positive dipstick test only confirms a condition (asymptomatic bacteriuria) that is not uncommon. A dipstick result on its own must not be used to direct antibiotic treatment.

At least 20–30% of patients with a long-term urinary catheter will have the system colonized by bacteria and it is essentially pointless doing a dipstick on a catheter specimen of urine (CSU). If the urinary tract is the likely source, an appropriately collected CSU specimen (and blood culture as indicated) should be obtained.

THE LABORATORY REPORT

As the urine report will be the most common result that needs interpretation, it is important to review all the information recorded.

Laboratories perform microscopy on urine specimens, quantifying white blood cells (white cell count, WCC), red blood cells (red cell count, RCC) and (vaginal) epithelial cells (**Figure 10.10**). Because of the large numbers of urine specimens processed, laboratories use automated flow cytometry/particle counting machines that detect the presence of particles (bacteria) and WBC at and above a set limit. 'Cut-off' values are set, and urine specimens that meet one or both of these criteria will be 'selected' by the automated microscopy machine for culture. If neither of these criteria is met, the report is issued with the comment 'No microscopical evidence of a UTI'. For certain groups including children, intensive care/high-dependency unit, renal, transplant and haematology/oncology patients, laboratories will process all specimens for culture irrespective of an automated microscopy result.

WHITE CELL COUNT

The presence of WBCs in urine (pyuria) is used as a marker of inflammation, usually (but not exclusively) indicating infection. WCC are frequently in excess of 100×10^6/L, and

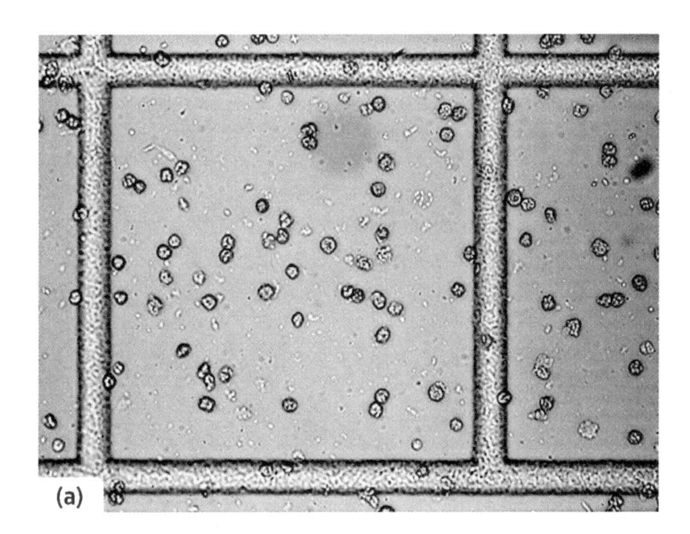

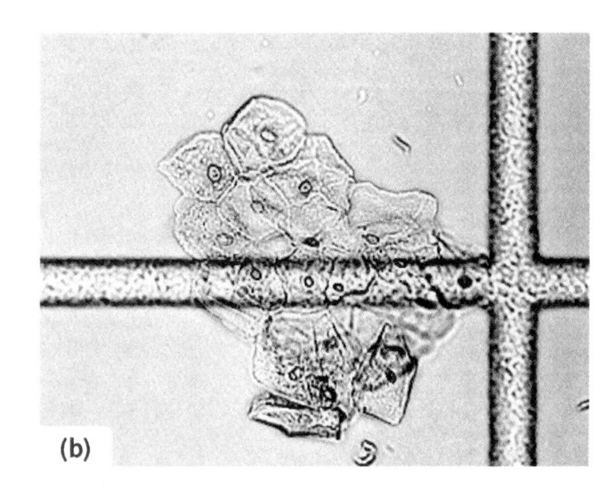

Figure 10.10 Urine microscopy. (**a**) Numerous white blood cells and bacteria (smaller background particles). (**b**) A clump of vaginal epithelial cells.

the local laboratory may set a minimum cut-off with its users, so, for example, a count of 25 × 10^6/L or more can be considered as 'significant'. Note that WBC in a urine specimen may originate from anywhere in the urinary tract (urethra, bladder, kidney). They may also contaminate the specimen by coming from either the vagina or a balanitis.

WBC are usually recorded on reports within a range of:

WCC/mL: 0–100 or >100 × 10^3/mL; OR

WCC/L: 0–100 or >100 × 10^6/L.

RED BLOOD CELL COUNT

RBCs are not a useful marker of a UTI. Their presence can be an indication of other pathologies. RBC are usually recorded on reports in a range of:

Number of RBC/mL: 0–100 or >100 × 10^3/mL; OR

Number of RBC/L: 0–100 or >100 × 10^6/L.

For clinical purposes, when automated microscopy machines are used, the level below which the RCC is not considered significant is agreed with the local urology team. Reports should state for example that "The presence of 25 × 10^6/L RBC or less, is not regarded as significant in renal pathologies". This value may be higher than is accepted for standard urine microscopy, but it should be noted that automated machines also detect empty 'ghost' RBC.

EPITHELIAL CELLS

The presence of epithelial cells indicates the presence of vaginal squamous epithelial cells, and shows contamination from this source; the specimen is not a clean-catch MSU. Adherent to these cells will be resident members of the vaginal flora that will then be represented in the final culture result, often being reported as 'mixed growth'. The relative number of epithelial cells must be used to interpret the culture report. When epithelial cells are not present this shows that the specimen can be regarded as 'clean catch', representing the bladder contents.

Epithelial cells: Not present.

Epithelial cells: Scanty OR Moderate OR Profuse.

THE BACTERIAL COUNT

Most younger women with acute onset cystitis will have more than 10^8 bacteria/L of urine, and this is regarded as a 'significant bacteriuria'. However, many have a lower bacterial count, and the term 'acute urethral syndrome' is used to define this entity, where counts of between 10^5 and 10^8/L are considered to represent the early stage of cystitis.

Sexually transmitted infections such as with *Chlamydia* and gonococcus should be considered in the sexually active woman with urethritis, and a lower titre of, or absent, uropathogen.

INTERPRETING THE URINE REPORT

A summary of the general criteria used to interpret a urine report is shown in **Figure 10.11**.

When detected by microscopy, organisms such as *Trichomonas vaginalis* and the threadworm *Enterobius vermicularis* are sometimes reported. These are usually introduced by contamination of the urine specimen with vaginal secretions at the time of collection, and direct appropriate management and treatment. In males, *Streptococcus pyogenes* is occasionally identified in a MSU.

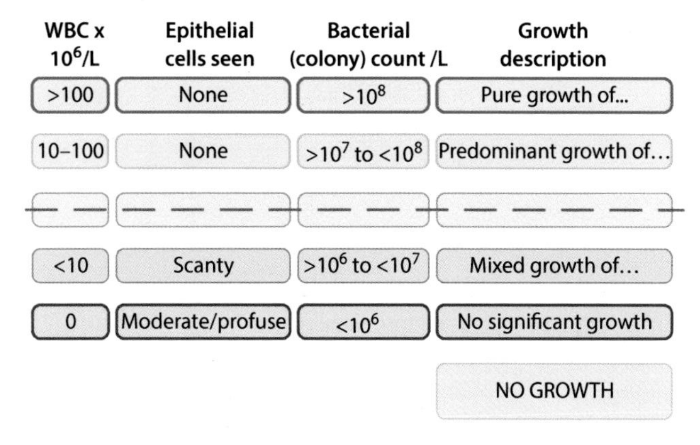

Figure 10.11 Four parameters are used in the interpretation of the urine microscopy and culture report in the symptomatic patient. If there is one or more of the four parameters below the grey dotted line, the specimen is less likely to represent a urinary tract infection. The combination of parameters must be interpreted with the clinical presentation. The acute urethral syndrome in the premenopausal female patient can be associated with a lower titre of 10^5–10^8/L.

The organism is likely to be associated with a balanitis or foreskin infection and it is appropriate to consider and review the possible source.

THE CATHETERIZED PATIENT

Insertion of a urinary catheter breaches the normal defences of the bladder and urethra, and the majority of patients with a long-term catheter will develop asymptomatic colonization (bacteriuria). There is a risk of a catheter-associated UTI, with bacteraemia and sepsis. The decision to insert a long-term catheter has to be based on clear clinical need, including:

- Acute and chronic urinary retention.
- Bladder outlet obstruction.
- Selected perioperative patients.
- To accurately monitor urine output in the critically ill patient.
- To assist healing of pressure sores in incontinent patients.
- For the management of long-term incontinence in selected patients as a last resort after underlying causes have been assessed and addressed.

The catheter should be removed as soon as clinically practical in order to minimize the risk of a UTI. When inserting a urinary catheter, key points to remember are:

- Scrupulous attention to aseptic non-touch technique (ANTT) is required.
- Always use the most appropriate size and type of catheter.
- After insertion, connect the catheter to a sterile closed system.
- Never break the connection unnecessarily.
- Always keep the catheter, and the bag, below bladder level, and never have these on the bed. Particular attention should be paid when the bed-bound patient is being moved to another ward for example.
- Keep the catheter bag on the stand and never on the floor.
- Using ANTT, the drainage system should be changed every 5–7 days.

The management of the urinary catheter must not be seen as the domain of the nursing staff. The need for, and risks associated with, a catheter must be part of the daily review by medical staff, who should direct safe removal at the earliest opportunity. When breaches in practice are seen, such as the catheter bag on the floor, medical staff must raise their concerns.

STERILE PYURIA

The reasons for a significant WCC but "No growth" in a urine report should be reviewed with the possible causes of sterile pyuria (**Figure 10.12**). Recent antibiotic use can account for this, and in the male patient a urethritis is likely to be caused by a sexually-transmitted infection (STI). When considering a STI, the appropriate swabs need to be collected (see Chapter 16). The patient should preferably be referred to the STD clinic/team, where an HIV test would be advised.

With ongoing symptoms in the setting of sterile pyuria and no other identified cause, fastidious organisms such as *Ureaplasma* can be considered. These require special culture conditions.

RELAPSING UTI

Attention needs to be given to the older patient with relapsing UTI, particularly when bacteraemia occurs with the same organism. This may be due to an inadequate length of antibiotic treatment for the previous infection, and a longer course of antibiotic is then considered appropriate. However, it is important for both female and male patients to exclude structural abnormalities, including stones, by use of the appropriate imaging.

In the older male patient, relapsing UTI can be due to chronic prostatitis, and this source needs to be investigated. Correct investigation of the prostate requires the following samples, including gentle massaging to obtain expressed prostatic secretions (EPS):

- The patient should have the need to pass urine, so that there is sufficient urine to complete the test.

Treatment/non-infective causes

- Recent antibiotic therapy
- Recent surgery
- Analgesic nephropathy
- Foreign bodies in the genito-urinary tract, including catheters
- Tumours of the genito-urinary tract

STD

- Male patients in particular, obtain detailed history and do full examination of genitalia
- For males send a 'first catch' specimen for gonococcal and chlamydia PCR

Vaginal discharge

- Pus cells from vaginal contamination; review previous examination. Has an HVS been done?

Tuberculosis

- An early morning urine (EMU) should be sent to exclude this infection in the renal tract

Other possible causes

- Fever in children
- Uncommon infection with fastidious organism such as haemophilus or rarely anaerobes
- Uncommon infection with fastidious organism such as mycoplasma; consider empirical clarithromycin if other options excluded
- Patient on recent trimethoprim or co-trimoxazole. Consider nutritionally deficient, thymidine-dependent 'coliforms'

Figure 10.12 A range of causes to consider when a sterile pyuria is detected. The opportunity to do an HIV test must always be highlighted, in the adult patient in particular. (HVS: high vaginal swab.)

- As necessary, retract the foreskin, then carefully clean the prepuce and foreskin area.
- Collect the first 5–10 mL of the voided urine (VU1).
- Collect the MSU sample (VU2).
- Following gentle digital palpation of the prostate *per rectum*, EPS are collected by the patient applying repeated downward massage of the penis to the urethral meatus, where the fluid is collected. (The prostate should not be palpated in acute prostatitis as there is a risk of introducing bacteria into the blood).
- Collect 5–10 mL of the urine stream (VU3).

These investigations would usually be done by the urologist, in close cooperation with the microbiology laboratory. Samples are collected into white-topped sterile containers, and transported to the laboratory promptly for processing. Before culturing, the EPS need to be examined microscopically; the presence of neutrophils and large lipid-laden macrophages, or oval fat bodies, is a feature of inflammation.

Bacterial numbers are quantified, and where the count in VU3 exceeds that in VU1 by at least 10-fold, this is strongly supportive of the diagnosis of prostatitis.

Making a diagnosis of prostatitis is important as the usual course of antibiotic is at least 4 weeks. The prostate can be a difficult organ to effect clearance of bacteria and yeasts, as penetration of antibiotics is variable, and prostatic calculi can act as foci.

The prostate should be considered a source of infection in the immunocompromised male patient. Organisms usually considered to be associated with other organs can be found here, and the yeast *Cryptococcus neoformans* is an example.

Chapter 11

Infections of the Alimentary Canal

INTRODUCTION

A wide range of bacteria, viruses and parasites gain access to the alimentary canal via the mouth. Outbreaks of 'food poisoning' at social gatherings are regularly reported. Infections transmitted from animals (zoonoses) are important; *Escherichia coli* O157 has been transmitted from the intestinal flora of bovines to humans at 'petting' farms. In higher-income parts of the world, *Campylobacter* is the commonest cause of bacterial gastroenteritis, usually acquired from contaminated, undercooked, fresh chicken carcasses. In tropical and lower-income areas of the world, organisms such as *Vibrio cholerae* assume importance.

In any outbreak, for example food poisoning, it is important to identify the source of the organism involving a family, school, institution or the wider community, so that this can be removed at the earliest opportunity. The key to this detective work is obtaining a detailed history from the affected individuals, which not only includes where and when food was consumed but listing every food item. An outbreak of gastroenteritis in Germany in 2011 caused by a shiga toxin-producing strain of *Escherichia coli* O104 affected nearly 4000 individuals, 800 of whom had haemolytic uraemic syndrome, and 51 died. This outbreak was traced to contaminated fresh bean sprouts, which were a minor ingredient of salad consumed by only certain individuals in the first groups affected.

Gastrointestinal infections can be acquired within hospitals. Antibiotic-associated diarrhoea is usually linked to *Clostridium difficile*, and is a particular problem in the older hospitalized patient, in whom it can produce a life-threatening diarrhoeal illness. Norovirus is brought into hospital via infected patients, visitors and staff, and every effort must be made to limit its spread. Although this is usually a self-limiting infection, the virus has a significant impact on the functioning of hospitals. Not infrequently wards are closed for extended periods and elective surgery cancelled for days.

Perhaps the most unusual organism causing disease in the alimentary canal is *Helicobacter pylori*. This bacterium can inhabit the inhospitable environment of the stomach and is a cause of gastric and duodenal ulcer disease and malignancy.

The endogenous flora of the alimentary canal are common culprits. The oral streptococci and anaerobes of the mouth are the agents of dental caries and periodontal infection. The bacterial flora of the intestine is involved in appendix and diverticular abscesses and biliary tract sepsis. 'Coliforms', anaerobes, streptococci and enterococci would always be considered in these situations. The streptococci of the mouth and bowel can cause infective endocarditis, and the specific association of *Streptococcus gallolyticus* and a large bowel malignancy has been discussed (Chapter 7).

The bacterial contents of the lumen, the structural integrity of the mucosal and submucosal layers of the bowel and the bowel-associated lymphoid aggregates are three broad defences. Some examples of how changes to the natural defences of the alimentary canal can predispose to infection are shown in **Figure 11.1**.

ORGANISMS

A wide range of exogenous organisms cause disease in the alimentary canal and key examples are listed in **Figure 11.2**. Food, water and animals are a common source of exogenous bacteria, but *Clostridium difficile* can be acquired from the nursing home or hospital environment, where it survives as a heat-stable spore. *Helicobacter pylori* is likely to be spread in childhood from person to person via the oral–oral and faecal–oral routes. In the case of enterotoxigenic *Staphylococcus aureus* and *Bacillus cereus*, these organisms multiply in contaminated food that is stored incorrectly, for example at room temperature rather than in a refrigerator. Heat-stable exotoxins are secreted, and when the food is consumed, nausea and vomiting occur within 1–4 hours.

PATHOGENESIS

DENTAL INFECTIONS

Bacteria making up the normal flora of the mouth colonize particular parts of the tooth and gingival surface (**Figure 11.3**). The acid metabolic by-products of these

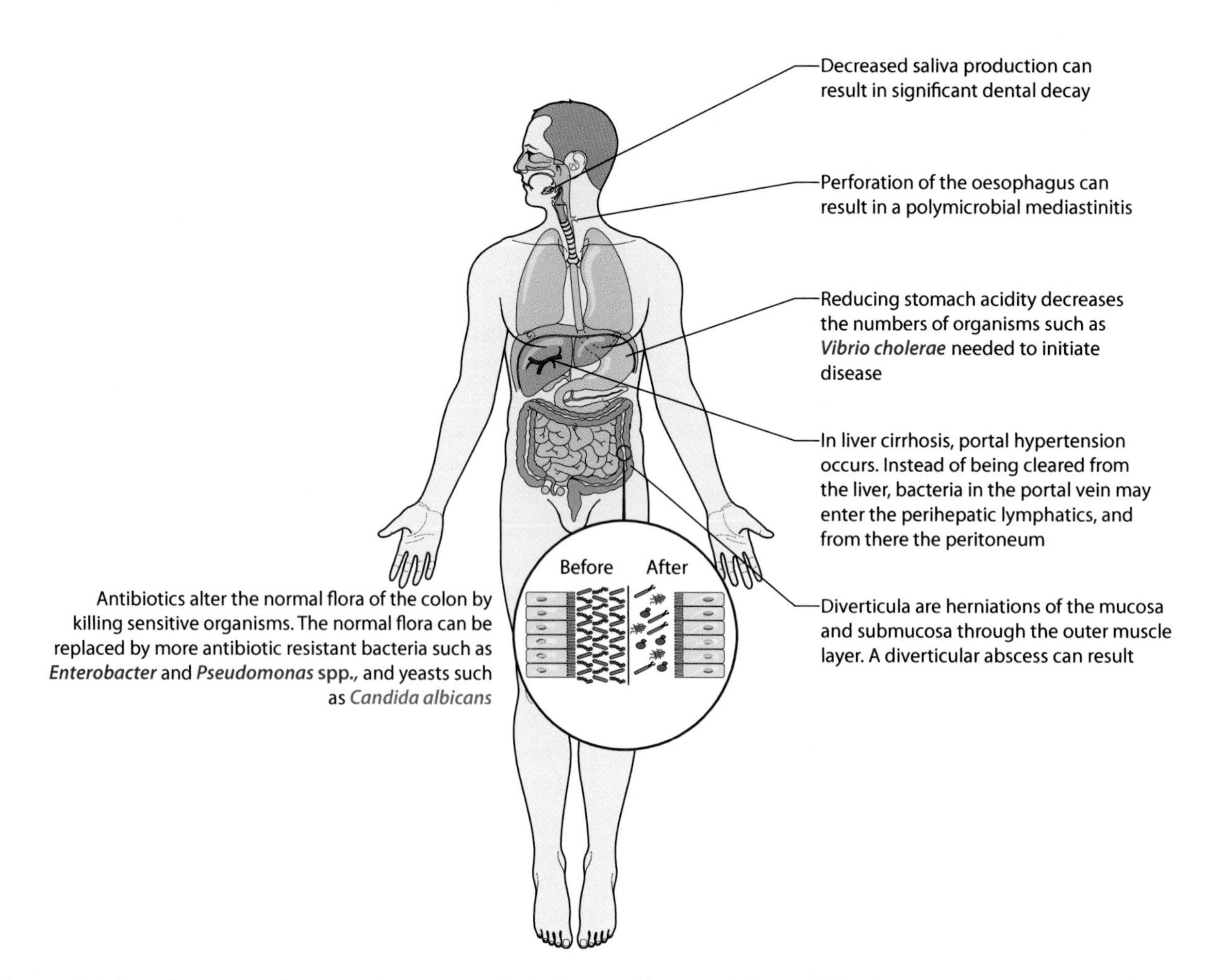

Figure 11.1 Any changes to the normal anatomy, physiology and bacterial flora of the alimentary canal may predispose to infection.

Bacillus cereus	Preformed toxin in food (usually rice) causes vomiting (with diarrhoea) within 1–4 hours
Staphylococcus aureus	Enterotoxin-producing strains grow in food, causes vomiting (with diarrhoea) within 1–4 hours
Clostridium difficile	Antibiotic-associated diarrhoea, colitis caused by toxin (A and B) producing strains
Helicobacter pylori	Gastritis, gastric and duodenal ulceration, malignancy
Escherichia coli **O104/ O157**	Diarrhoea, haemolytic uraemic syndrome (HUS), acute renal failure (ARF)
Campylobacter jejuni	Diarrhoea, sometimes bloody; zoonosis often associated with undercooked chicken
Salmonella **Enteritidis**	Diarrhoea and vomiting
Shigella sonnei	Diarrhoea and vomiting
Vibrio cholerae	Watery diarrhoea, associated with travel to tropical and subtropical regions
Vibrio parahaemolyticus	Diarrhoea associated with consumption of shellfish
Cryptosporidium parvum	Diarrhoea can be persistent, especially in the immunocompromised host
Entamoeba histolytica	Asymptomatic cyst carriage, diarrhoea, colitis, invasion of mucosa, liver, lung, brain abscess
Giardia lamblia	Asymptomatic cyst carriage, acute and chronic diarrhoea

Figure 11.2 A list highlighting the range of organisms that can cause infection in the alimentary canal. (*Continued*)

Ascaris lumbricoides	Asymptomatic carriage, also (mainly children) malnutrition, bowel obstruction
Enterobius vermicularis	Asymptomatic carriage, also (mainly children) pruritis, vulvovaginitis
Schistosoma mansoni	Asymptomatic carriage, Katayama fever, fatigue, abdominal pain, hepatomegaly
Taenia saginata	Asymptomatic carriage, abdominal discomfort, weight loss, 'moving' proglottids in faeces
Taenia solium	As for ***Taenia saginata,*** but note cysticercosis, e.g. neurocysticercosis and seizures
Trichuris trichiura	Asymptomatic carriage, abdominal pain, diarrhoea
Adenovirus	Diarrhoea
Norovirus	Vomiting and diarrhoea
Rotavirus	Vomiting and diarrhoea, prolonged diarrhoea in the immunocompromised
CMV	Diarrhoea, colitis in the immunocompromised.

Figure 11.2 A list highlighting the range of organisms that can cause infection in the alimentary canal.

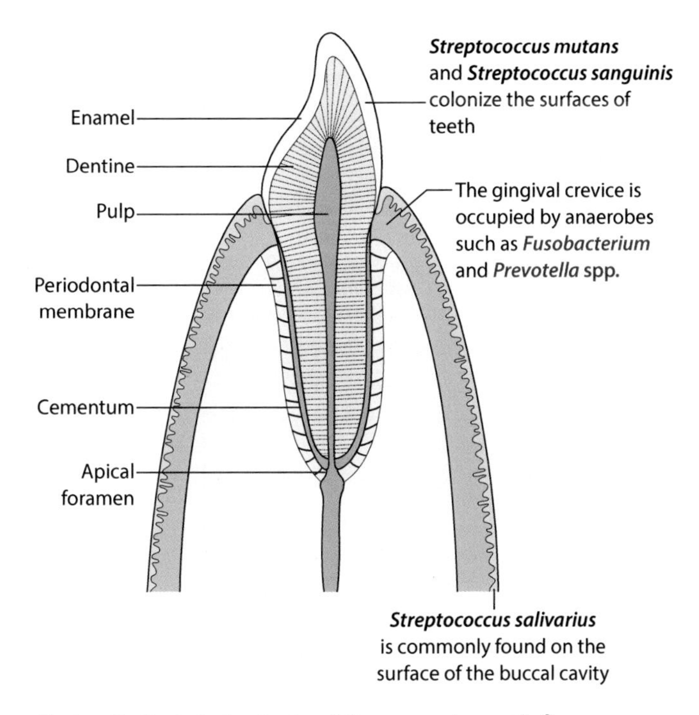

Figure 11.3 The cross section of a tooth. Certain bacteria of the normal mouth flora occupy certain ecological niches around the tooth and have the potential to cause dental infection.

bacteria, arising from the fermentation of dietary sugar, cause tooth decay. Teeth are protected by the cleaning action of the tongue, the buffering effect of saliva and the acquired pellicle, derived from saliva, which coats the surface of teeth. This coating, which becomes colonized with bacteria, is removed by regular cleaning and is replaced by new pellicle, preventing decay. In the setting of poor oral hygiene, where for example crevices are not cleaned properly, a nidus of infection may arise, with the development of dental caries. The infection can progress to pulpitis and apical abscess formation. Infection of the gums begins as subgingival plaque, progressing to periodontitis. The anaerobes of the mouth are important here (**Figure 11.4**).

Infections of the teeth can spread into adjacent bone to initiate osteomyelitis, cause local soft tissue abscesses and reach the facial sinuses. In the setting of widespread tooth decay, aspiration of mouth flora can result in development of a lung abscess. Profound halitosis may be present, indicating that the oral anaerobes are growing in the

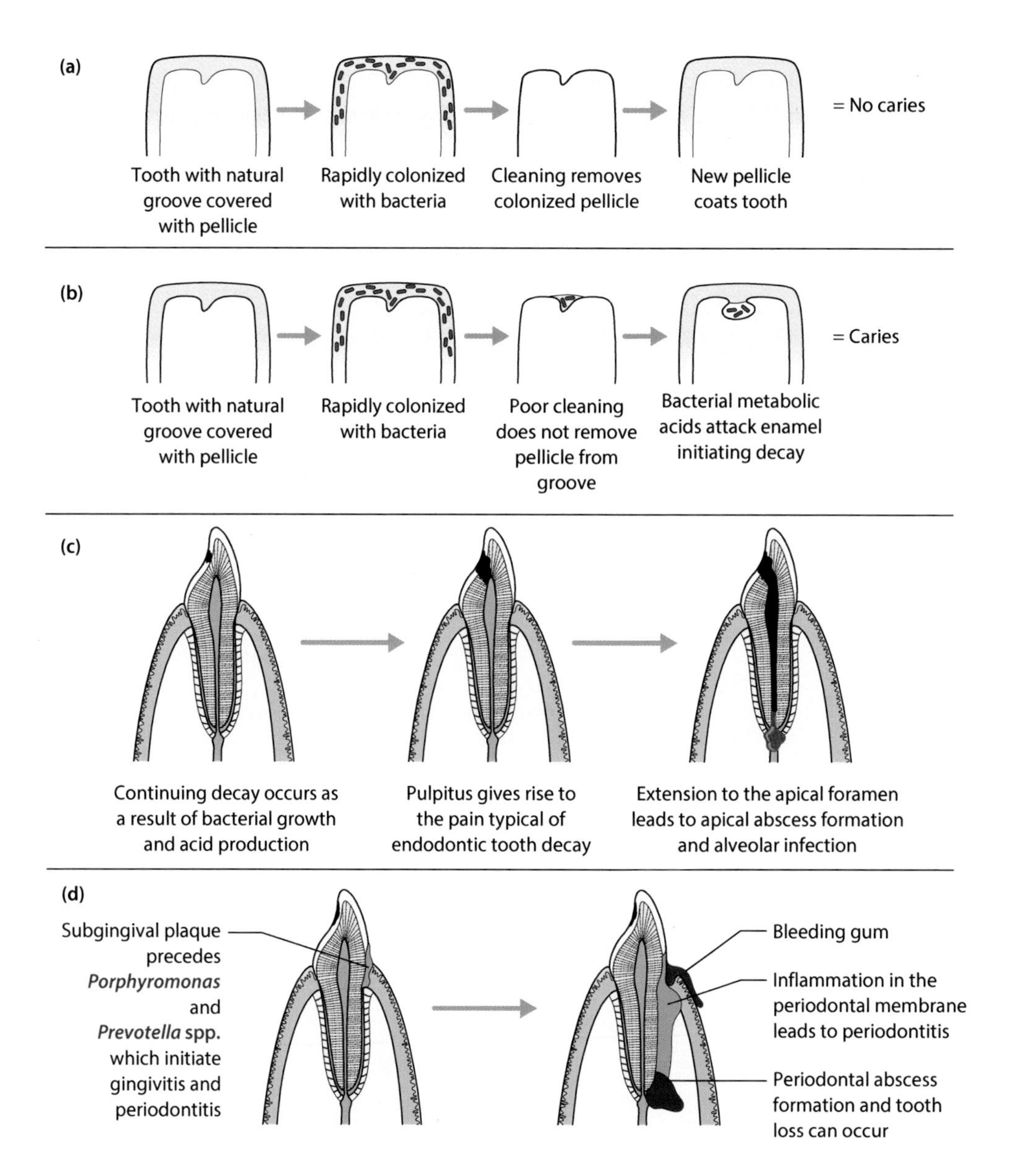

Figure 11.4 (**a**) Tooth cleaning removes the pellicle and associated bacteria. (**b**) If bacteria are not removed they initiate decay. (**c**) The decay can progress to pulpitis and alveolar abscess. (**d**) Periodontal disease leads to periodontitis and abscess formation.

lung. Some local and distant sites where the consequences of tooth decay can manifest are shown in **Figure 11.5**.

INFECTIONS OF THE UPPER ALIMENTARY CANAL

ENDOGENOUS ORGANISMS

Perforation of the oesophagus results in polymicrobial contamination of the mediastinum, which can lead to mediastinitis. Dilators used to alleviate dysphagia in the patient with oesophageal carcinoma or endoscopes can do this. Vomiting, associated with sudden pressure changes, can result in tearing of the lower end of the oesophagus (Boerhaave's syndrome). Perforation of the duodenum as a result of ulcer disease can result in spillage of bacteria into the peritoneum. In all these circumstances, there should be a low threshold for also considering *Candida*.

EXOGENOUS ORGANISMS

Helicobacter pylori resides in the inhospitable acid environment of the stomach for months or years before clinical disease is noted, colonizing gastric-type epithelial cells in the stomach and duodenum. Pathogenic properties of the organism and some features of infection are shown in **Figure 11.6**. The motility of the bacterium enables it to reside in the mucin layer and ammonia, produced by a potent urease, neutralizes stomach acid. Adhesion, a cytotoxin, the toxic effect of ammonia and the strong inflammatory response lead to gastritis and then peptic and duodenal ulceration. The long-term consequences of the infection can be gastric malignancies such as adenocarcinoma and lymphoma.

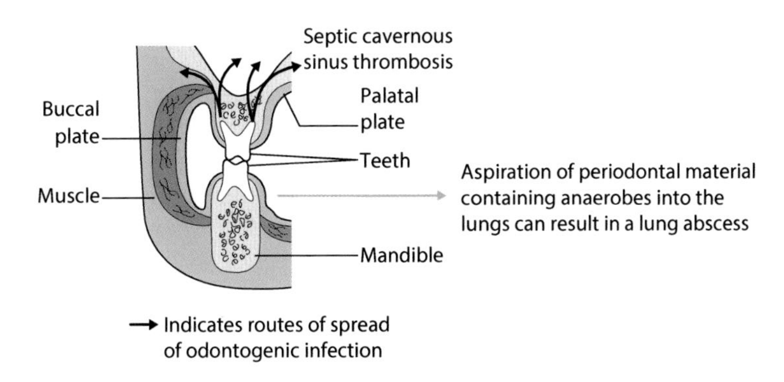

Figure 11.5 Dental disease can spread to adjacent tissues and organs.

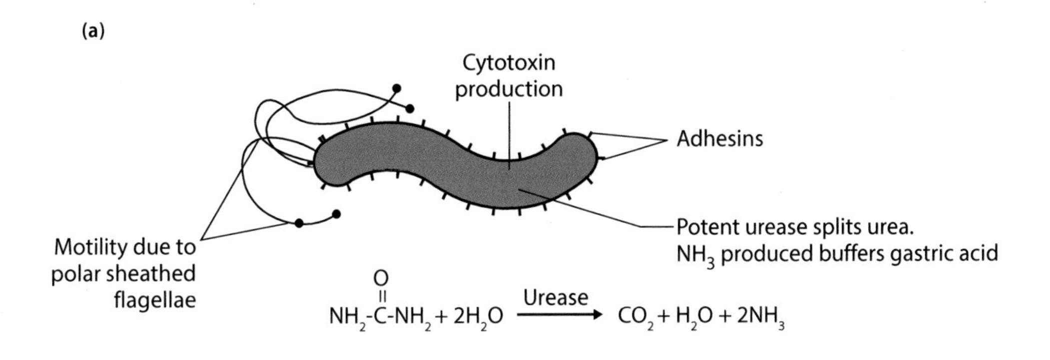

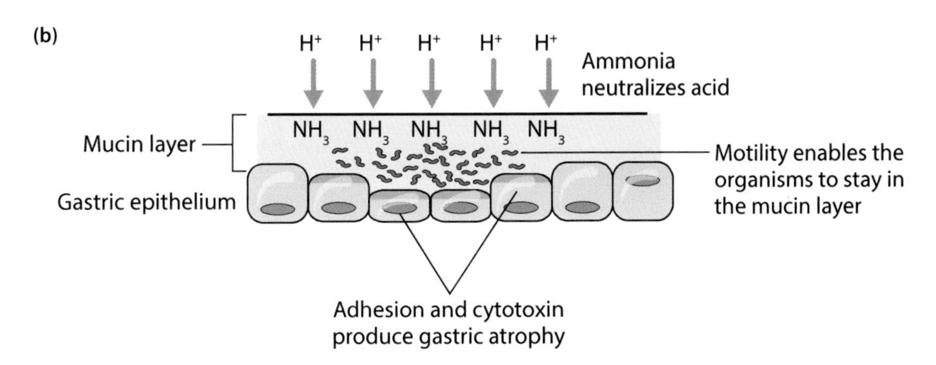

Figure 11.6 (**a**) Some of the pathogenic properties of *Helicobacter pylori*. (**b**) Some features of the mechanism of *Helicobacter* infection.

INFECTIONS OF THE LOWER ALIMENTARY CANAL

EXOGENOUS ORGANISMS

In order to cause disease here, pathogens such as *Shigella*, *Salmonella* and *Vibrio cholerae* have to traverse the inhospitable lumen of the stomach. There are significant differences in the infecting dose of these bacteria, being about 10^2, 10^4 and 10^7 organisms/dose respectively. They then have to establish themselves in the biofilm of endogenous bacteria, overcoming 'colonization resistance', usually precipitating symptoms 24–48 hours after ingestion of contaminated food or water.

There are three broad mechanisms whereby bacteria cause disease. These are interference with the secretory and absorptive properties of the intestine by adherence and/or enterotoxin, invasion of the mucosal enterocytes, or penetration of organisms through the mucosa, where they multiply in cells of the bowel-associated lymphoid tissue. Exotoxin production characterizes the diseases caused by *Clostridium difficile* and *Vibrio cholerae*.

Loss of enterocyte microvilli, effacement, caused by adherence of the bacteria, alters the secretory and absorptive properties of the enterocyte, disturbing normal water and electrolyte balance of the small intestine in particular (**Figure 11.7a**). Enteroadhesive *Escherichia coli* is an example, but it is likely that most of the bacteria discussed here will exact some degree of effacement in the course of their interaction with the surface of the enterocyte. *Vibrio cholerae* binds to the microvilli causing effacement, and then the A component of the exotoxin enters the enterocyte, being the major cause of the secretory diarrhoea (**Figure 11.7b**; see **Figure 2.25**).

Invasion of the mucosal layer of the colon is a characteristic of *Shigella* spp. (**Figure 11.8**). Here effacement of enterocytes occurs, cells are killed and the bacteria spread to neighbouring cells. With the inflammatory response, neutrophils accumulate, microabscesses form and local bleeding occurs. A bloody stool containing pus can be a feature of shigella infection. Penetration through the mucosa and submucosa into lymphoid tissue is characteristic of *Salmonella* Typhi (**Figure 11.9**). Diarrhoea is not commonly a feature of enteric fever, which is

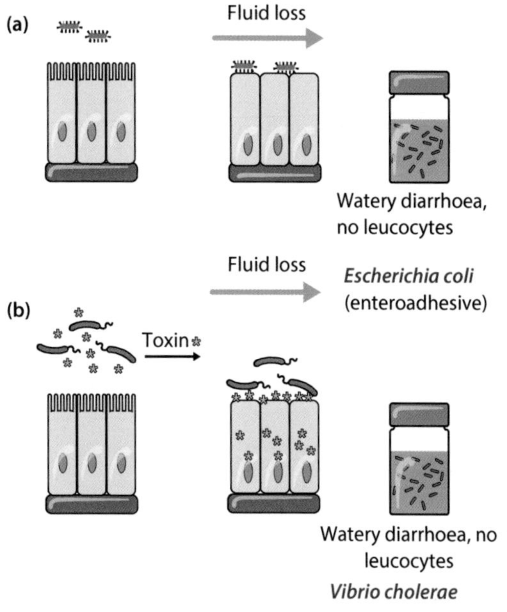

(a) Fluid loss

Watery diarrhoea, no leucocytes

Escherichia coli (enteroadhesive)

(b) Toxin Fluid loss

Watery diarrhoea, no leucocytes

Vibrio cholerae

Figure 11.7 (**a**) Bacteria such as enteroadhesive *Escherichia coli* adhere to the surface of the enterocyte, causing effacement. (**b**) While also causing effacement, *Vibrio cholerae* produces an exotoxin that results in the secretory diarrhoea.

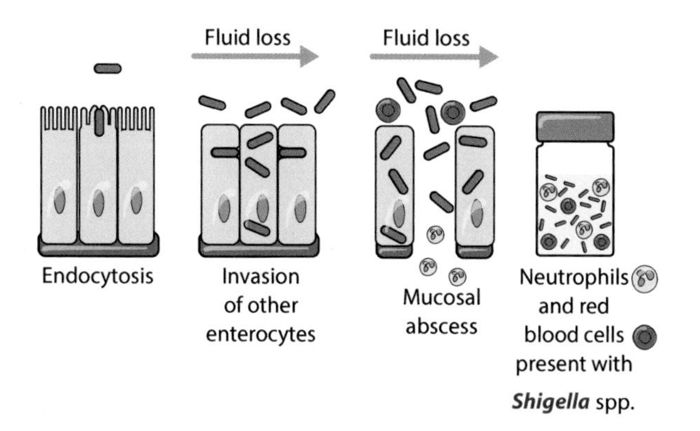

Figure 11.8 Organisms such as *Shigella* invade the mucosa and destroy enterocytes; microabscess formation occurs. The stool contains blood and pus cells.

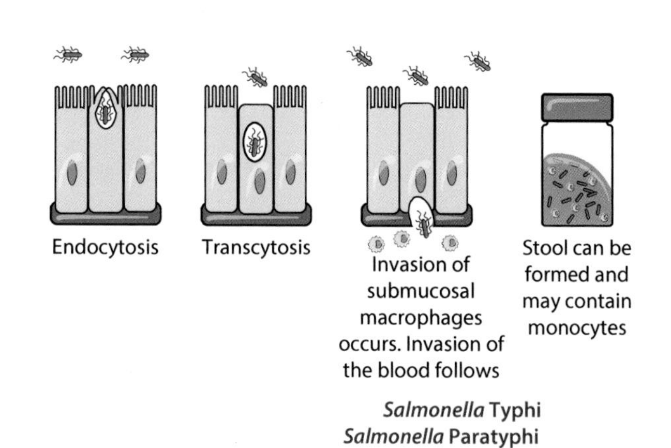

Figure 11.9 *Salmonella* Typhi/Paratyphi penetrates the mucosa and submucosa.

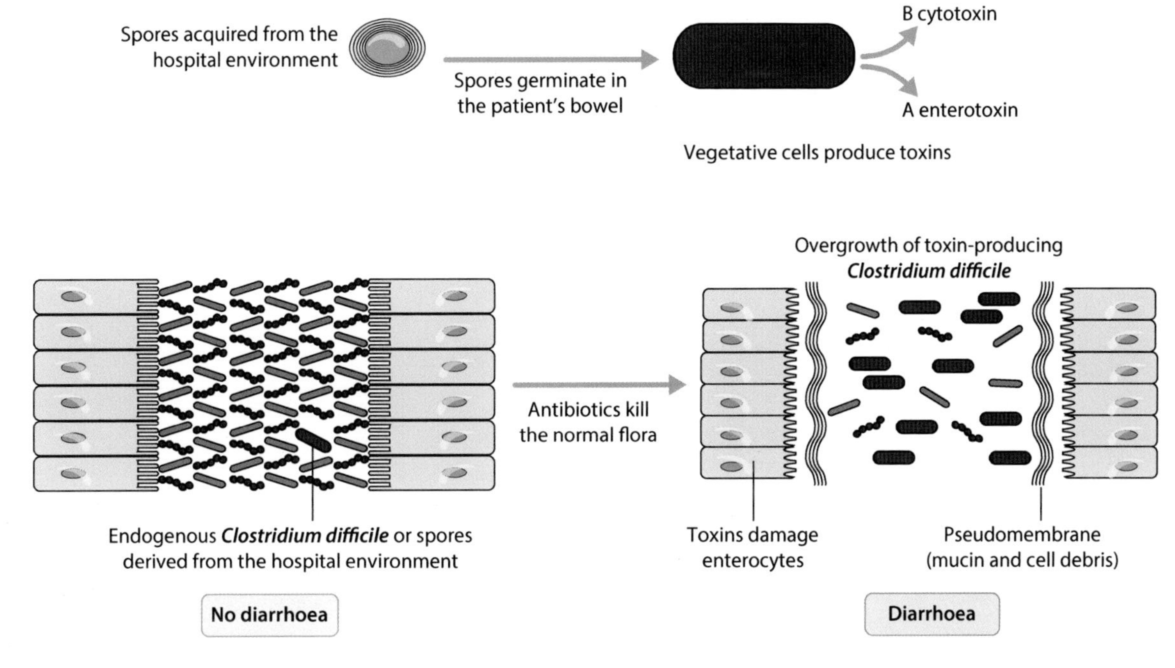

Figure 11.10 Spores of *Clostridium difficile* germinate to produce vegetative cells, which synthesize toxins A and B. Overgrowth of the organism can result in antibiotic-associated diarrhoea.

characterized by fever and malaise. The organisms enter the blood, and thus in suspected typhoid or enteric fever, the collection of blood cultures is essential.

Antibiotic-associated diarrhoea is an important entity in many hospitals and is a significant cause of morbidity in the older patient. *Clostridium difficile* is the main agent of the condition; the patient may have low numbers of organisms amongst their bowel flora on admission to hospital, or spores may be acquired from the hospital environment. Multiplying bacteria produce two proteins, A and B, which (unlike the toxin of *Vibrio cholerae*) both act as toxins; the resulting diarrhoea may progress to pseudomembranous colitis (**Figure 11.10**).

Escherichia coli O157 (and O104) shows the complexity of the pathogenic properties of the intestinal pathogens. It produces an adhesin necessary for attachment to the enterocyte, a haemolysin and verotoxin or shiga-like toxin. Once absorbed into the blood the toxin interacts with a wide range of cells, including the capillary endothelium and cells of the kidney. Haemorrhagic colitis and haemolytic uraemic syndrome (HUS) can result. These complications are more likely to occur at the extremes of age.

As discussed in Chapter 1, norovirus replication in the enterocytes results in cell death by apoptosis, leading to blunting of the villi, with loss of secretory and absorptive capacity of the intestine, resulting in diarrhoea. Cytomegalovirus should be considered in any immunocompromised individual with ongoing diarrhoea, especially when the more usual organisms have been excluded.

The life cycle of *Entamoeba histolytica* is shown in **Figure 11.11**. This infection is usually associated with residence in a tropical area. Excystation occurs in the small bowel, and each trophozoite divides to give rise to eight progeny trophozoites. Reproduction occurs in the large intestine, giving rise to diarrhoea. Trophozoites can invade the epithelium, giving rise to bloody diarrhoea, and from there they enter the systemic circulation via the portal vein, with the potential to lodge in the liver, lung and brain to cause an abscess.

Ascaris lumbricoides infects over 500 million individuals in tropical and subtropical areas of the world. It has a complex life cycle typical of the roundworms. As shown in **Figure 11.12a**, first- and second-stage larvae mature within the egg passed in faeces into the environment, and are swallowed (1, 2, 3), hatching out in the duodenum and small intestine. These larvae then burrow through the epithelium, entering the portal vein (4, 5). They travel to the liver and lung (6, 7). Larvae mature through the third and fourth stages in the alveoli, and are coughed up and swallowed (**Figure 11.12b**) (8, 9). Larvae then develop into adult worms that pair and mate in the duodenum and upper small intestine (10). The female produces vast

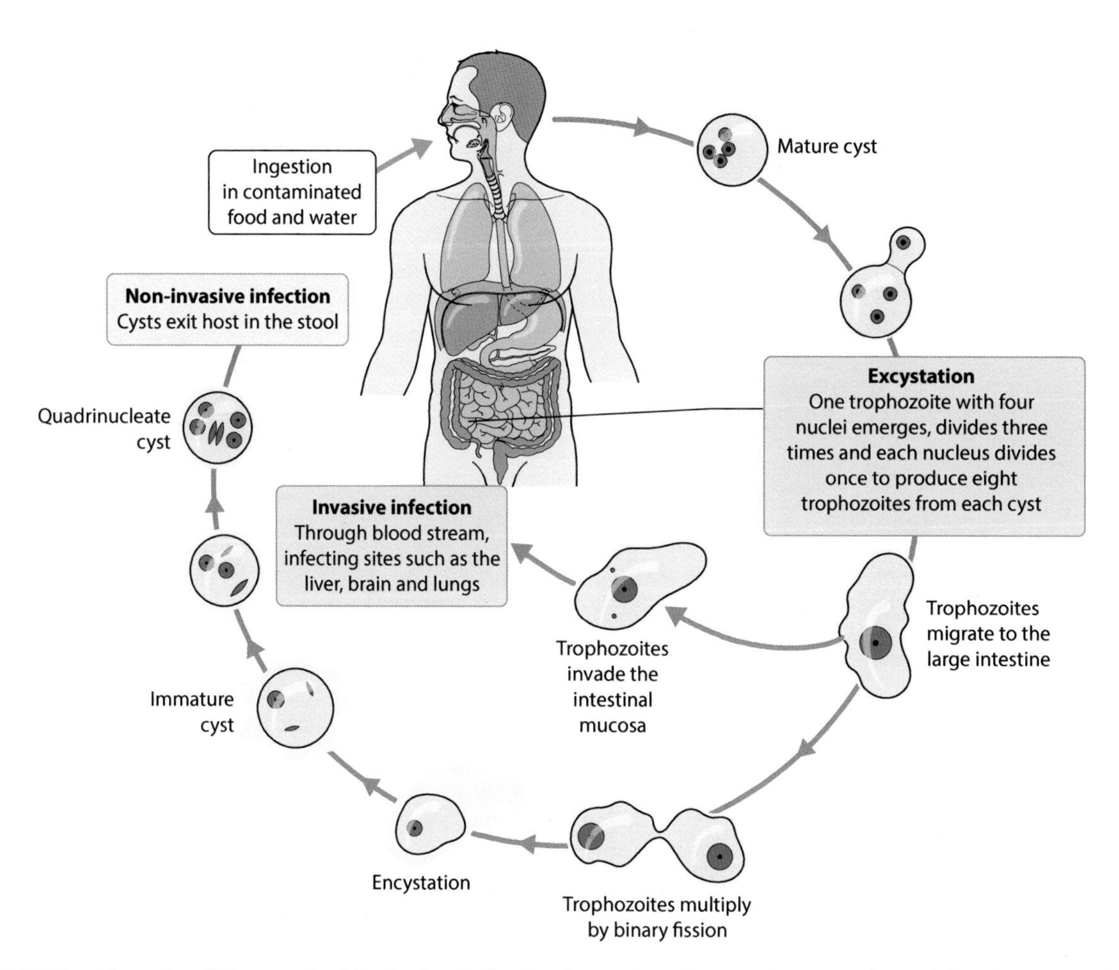

Figure 11.11 The life cycle of *Entamoeba histolytica*. Following ingestion of cysts via contaminated food or water, the usual reproductive cycle takes place in the alimentary canal. Trophozoites can also invade the intestinal mucosa, and via the portal vein reach the liver, lung or brain, where abscess formation can occur.

numbers of eggs that are then passed into the environment, contaminating soil and crops (11, 12, 1). Most infections are asymptomatic, but the infection can manifest with bowel obstruction, when hundreds of adult worms block the (small) intestine, usually in children, or the adult enters the common bile duct, obstructing that conduit, precipitating acute pancreatitis or cholangitis.

ENDOGENOUS ORGANISMS

When a faecolith obstructs the lumen of the appendix, bacteria multiply behind the obstruction to produce appendicitis. This can progress to an abscess, rupture and the development of peritonitis. Organisms can also travel via the portal vein to the liver and initiate a liver abscess (**Figure 11.13**).

In older patients diverticular disease of the descending colon is due to herniation of the mucosal and submucosal layers through the muscle. If the opening of the diverticulum is obstructed, trapped bacteria multiply, with the resulting inflammation progressing to abscess formation. The same consequences of an appendix abscess can result. Infections of the biliary tract are discussed in the next chapter.

PERITONITIS

The sagittal section of the abdomen shows the anatomical relations of organs, peritoneum and the peritoneal cavities (**Figure 11.14**). These are important anatomical markers in infection. Collections and abscess formation in the subphrenic region above the liver, in the lesser sac

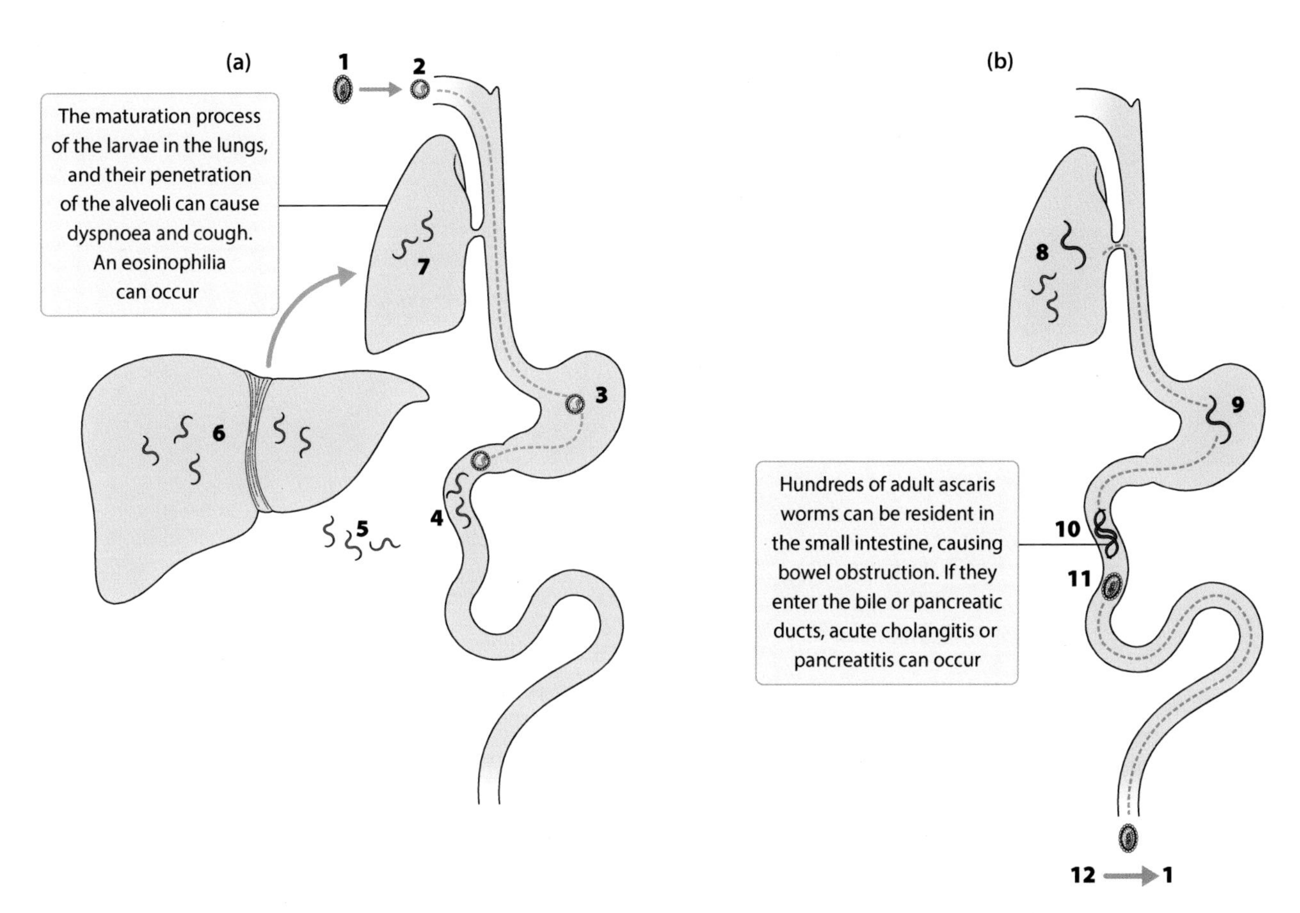

Figure 11.12 The life cycle of *Ascaris lumbricoides*. (**a**) The stages in the cycle from maturation of eggs in the environment, through to larvae reaching the lung; (**b**) the larvae develop in the lung, are swallowed and mature into adults. Following pairing in the duodenum, the adult female produces large numbers of (immature) eggs, which contaminate soil and crops. (See text for explanation of numbers.)

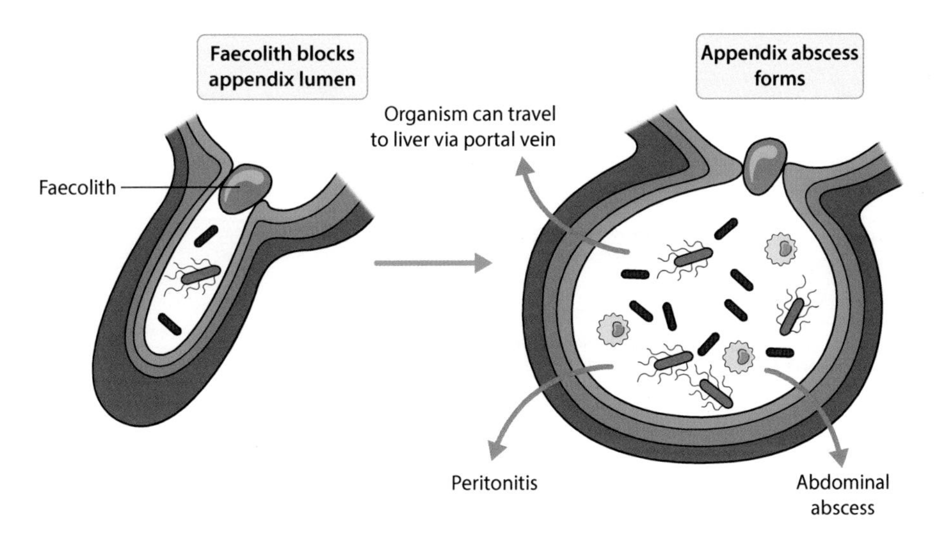

Figure 11.13 If the lumen of the appendix is obstructed by a faecolith, the trapped normal flora reproduce behind the obstruction.

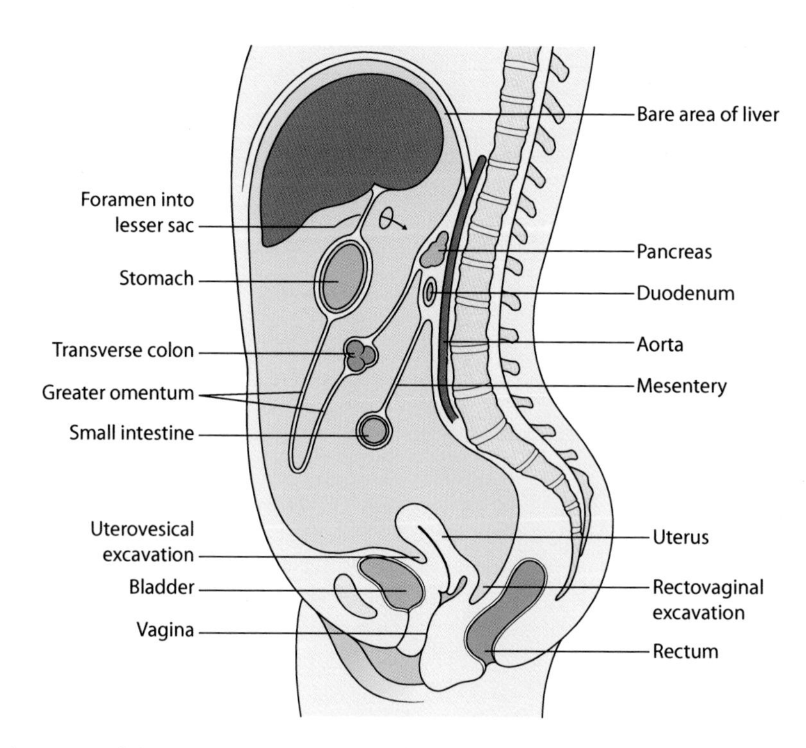

Figure 11.14 A sagittal section of the abdomen and pelvis of an adult female, showing the relations of the various organs and the peritoneal cavity. The proximity of the duodenum, pancreas and lesser sac is notable in relation to infection, including postsurgical, in the biliary system and pancreas.

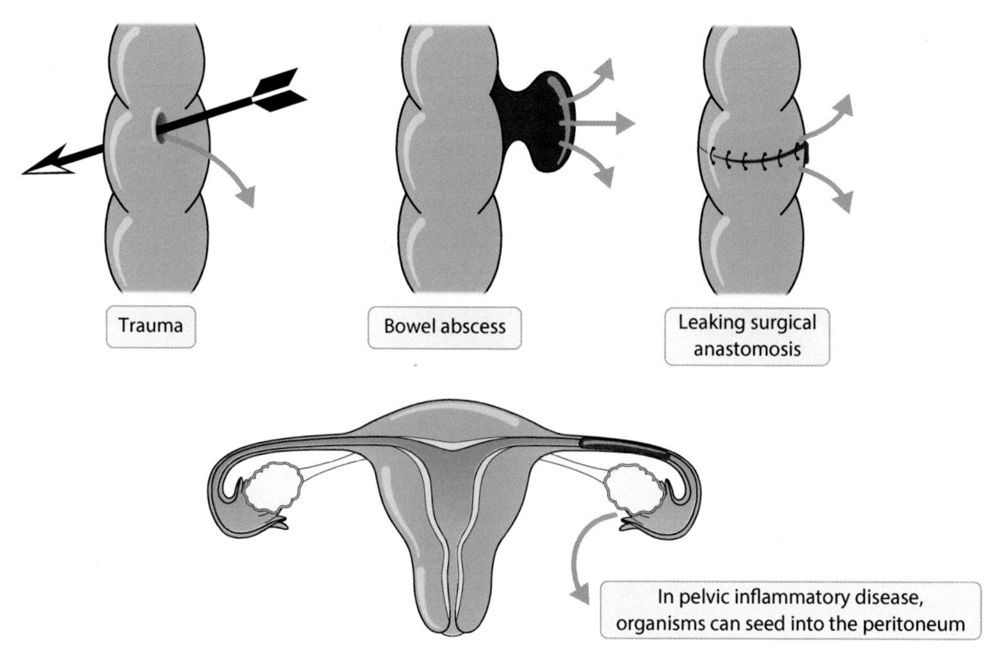

Figure 11.15 Compromise of the integrity of the bowel mucosa by a number of mechanisms can result in peritonitis. In the female patient, organisms from pelvic inflammatory disease must also be considered.

and pelvic cavity are examples. The pancreas lies in close proximity to the duodenum. Inflammation from acute pancreatitis can result in a pseudocyst in the lesser sac. Bacteria and yeasts in the duodenum can cross into the necrotic pancreatic tissue and pseudocyst.

Bacterial peritonitis can be divided into primary and secondary. In primary or spontaneous bacterial peritonitis (SBP), there is no clear source. It is likely, in the patient with liver cirrhosis (and ascites), that bacteria in the portal vein are not cleared by the compromised (Kupffer cells) macrophages of the liver. With portal hypertension, organisms are shunted via the perihepatic lymphatics, and enter the peritoneal space to initiate infection. Acute viral hepatitis, chronic active hepatitis and congestive cardiac failure are also risks for SBP.

Secondary peritonitis arises from a defined source where there is compromise of the integrity of the bowel (**Figure 11.15**). Pelvic inflammatory disease (PID) in the female can also lead to secondary peritonitis as the Fallopian tubes open into the peritoneum.

DIAGNOSIS AND MANAGEMENT

Pain typical of disease of the teeth and gums should bring the patient to the dentist. Dyspepsia and pain associated with duodenal ulcers should alert the clinician to *Helicobacter pylori* infection when other causes such as aspirin or non-steroidal anti-inflammatory drug (NSAID) use have been excluded. Disease caused by this organism can be identified or screened for by a number of tests. Endoscopy, biopsy and histopathological examination are the gold standards; culture may also be done here. The urease test relies on the fact that the potent urease produced by *Helicobacter pylori* converts ^{13}C urea to products that include $^{13}CO_2$. After swallowing labelled urea, exhaled and labelled CO_2 is measured. Antibody tests can be used to measure helicobacter immunoglobulin (Ig) G antibody levels; however, antibody can reflect past infection, and helicobacter antigen in stool is the preferred non-invasive diagnostic test.

Acute vomiting 1–4 hours after consuming food should alert the clinician to acute food poisoning by organisms producing emetic toxins, such as *Staphylococcus aureus* and *Bacillus cereus*. The patient with gastroenteritis must be asked about recent food and travel history. It is important to record this information on the request form accompanying a specimen. Travel to lower-income parts of the world would widen the tests done, and would include, for example, testing the stool for *Vibrio cholerae*. In this setting the ova and cysts of parasites should also be looked for. As listed in Chapter 3, **Figure 3.14**, the clinical diagnosis of food poisoning and infectious bloody diarrhoea is notifiable to Public Health by the attending physician.

A bloody stool can be associated with *Campylobacter* or *Shigella* infection. It is also an alert for infection by *Escherichia coli* O157 (or O104); lack of fever in the setting of frank blood in the stool can direct the clinical team to consider a non-infectious cause. However, *Escherichia coli* O157 infection must be considered, as it can present in this way.

In the patient with chronic diarrhoea, and there is no clear source, or an organism such as *Cryptosporidium* is repeatedly identified, immunodeficiency is considered, prompting an human immunodeficiency virus (HIV) test. The elderly patient on immunosuppressive therapy could have cytomagalovirus (CMV) colitis; and an ethylenediamine tetra-acetic acid (EDTA) blood for CMV polymerase chain reaction (PCR) is appropriate.

An outline of the methods used to process a stool specimen is shown in **Figure 11.16**. Laboratories test stool specimens received from adults in the community or hospital for *Clostridium difficile* toxin. In the UK, stool specimens are routinely examined for the water-borne protozoal parasite *Cryptosporidium*. It is likely that tests using multiplex PCR systems for organisms including *Campylobacter*, *Clostridium difficile*, *Escherichia coli* O157, *Salmonella*, *Shigella*, *Cryptosporidium*, *Giardia* and norovirus will become increasingly used to screen all stool specimens, and PCR-positive specimens will then be processed by the appropriate standard laboratory method.

The majority of infections caused by *Campylobacter*, non-enteric *Salmonella* and *Shigella* settle spontaneously. For ongoing infection, clarithromycin is used for *Campylobacter*, and depending on susceptibility profiles, azithromycin or ciprofloxacin are options for *Salmonella* and *Shigella*. Treatment is currently contraindicated for *Escherichia coli* O157 infection, as antibiotics are considered to increase toxin release and exacerbate HUS and renal failure.

The diagnosis of infections of the intestines, peritoneum, and hepatobiliary system will centre on clinical diagnosis. For a bacteriological diagnosis, aspirated peritoneal fluid should be sent for microscopy, culture and sensitivity. A white cell count/neutrophilia of >250 cells/μL indicates bacterial peritonitis. The specimen that must be collected is blood for culture.

Empirical treatment of 'abdominal sepsis' usually employs antibiotics active against streptococci, 'coliforms'

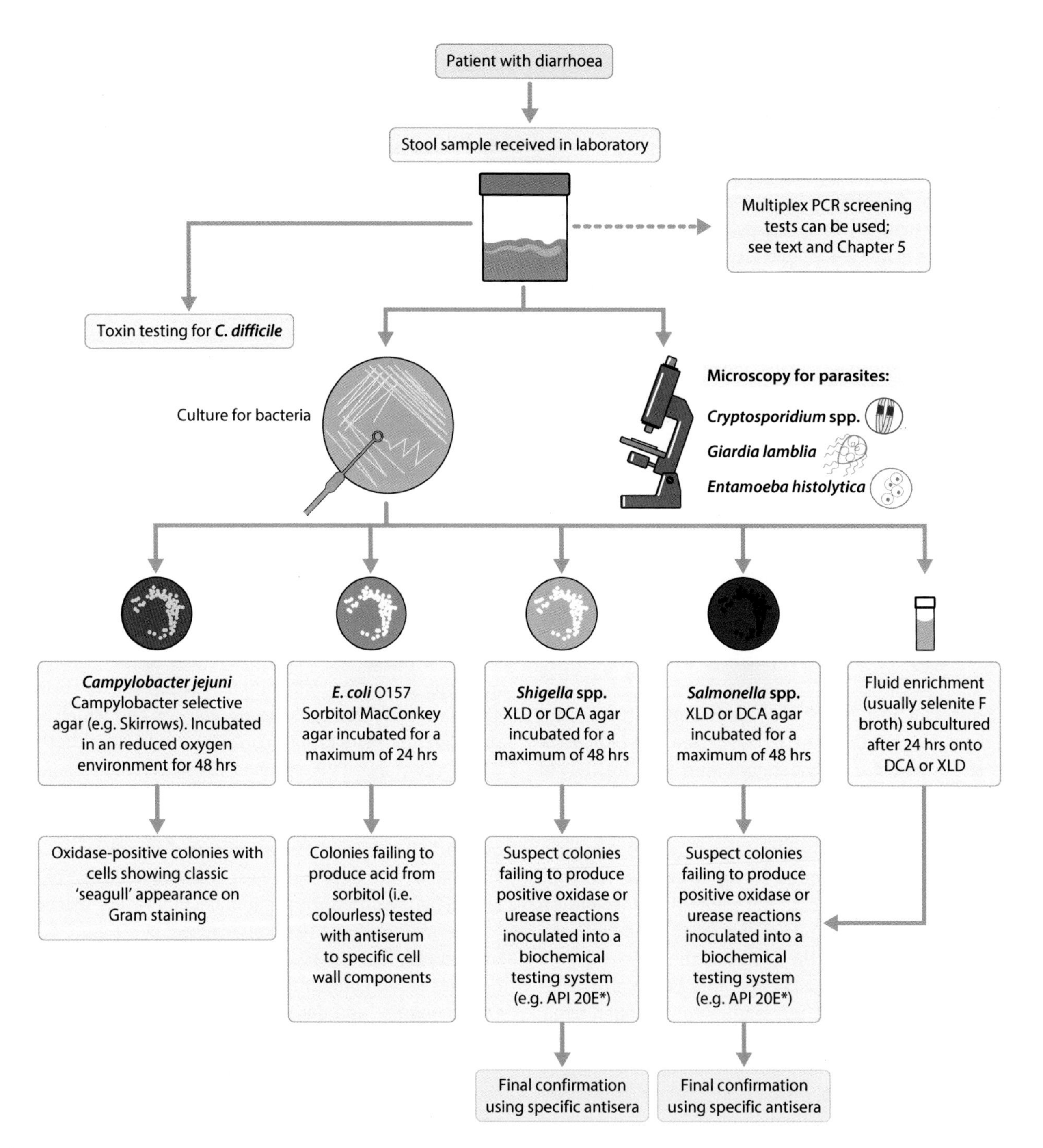

Figure 11.16 When a stool specimen is submitted to the laboratory, various procedures are based on culture (for bacteria), microscopy and enzyme immunoassay.
* MALDI-TOF technology is used for organism identification too (see Chapter 5).

and anaerobes (see Chapter 4). The antibiotic regime will be modified once organisms have been identified from aspirates, pus or blood culture.

The decision to give an antifungal agent as empirical treatment requires reflection as to the likely status of the 'normal bowel flora'. In the intensive care unit patient not responding to appropriate antibacterial cover, an antifungal should be given if *Candida* has been isolated from sites such as endotracheal tube secretions or a catheter specimen of urine.

CLOSTRIDIUM DIFFICILE

An outline of the diagnostic methods and treatment options for *Clostridium difficile* is shown in **Figure 11.17**. An EIA test is used to screen the stool specimen for the presence of the glutamate dehydrogenase (GDH) enzyme specific for this organism. This test is sensitive and specific and has a negative predictive value of over 99%. GDH-positive specimens are tested by an EIA for toxins A and B; if positive, the patient has a toxin-secreting organism in that stool sample. (Certain laboratories test GDH-positive/EIA toxin-negative samples by PCR to determine if the organism has the genes for the two toxins. If this test is positive, it shows that this organism has the potential to switch on toxin production.)

Based on clinical assessment, interpretation of the positive result is done in conjunction with the Bristol stool chart (**Figure 11.18**). Stools with a score of 4 or less are considered to reflect normal bowel function. Included in the review is the following:

The patient should already be nursed in a single room, and if not, the reason for this identified.

Current antibiotic prescriptions must be reviewed, and antibiotics stopped where possible.

The appropriate treatment for *Clostridium difficile* is started.

Laxative prescription must be reviewed and stopped.

White cell count, C-reactive protein, creatinine clearance and albumin are monitored.

The Bristol stool chart is reviewed. The form of all motions is recorded on this chart by nursing staff, and the information is used by medical team in their assessment of the patient, including the response to treatment.

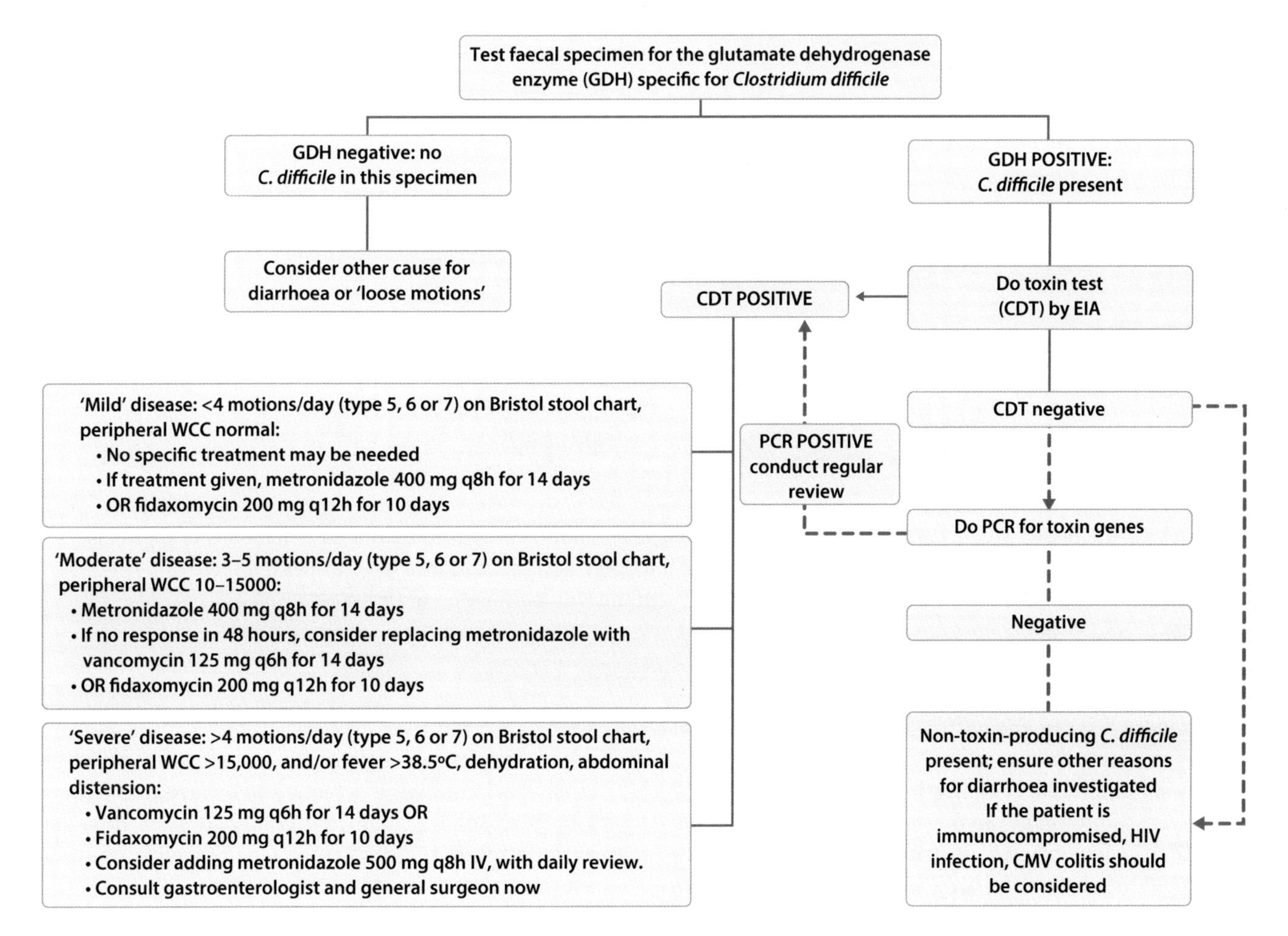

Figure 11.17 A flow diagram of the testing protocol for *Clostridium difficile* infection, the interpretation of results and the various treatment options. These are based on the Bristol stool chart score, and inflammatory markers.

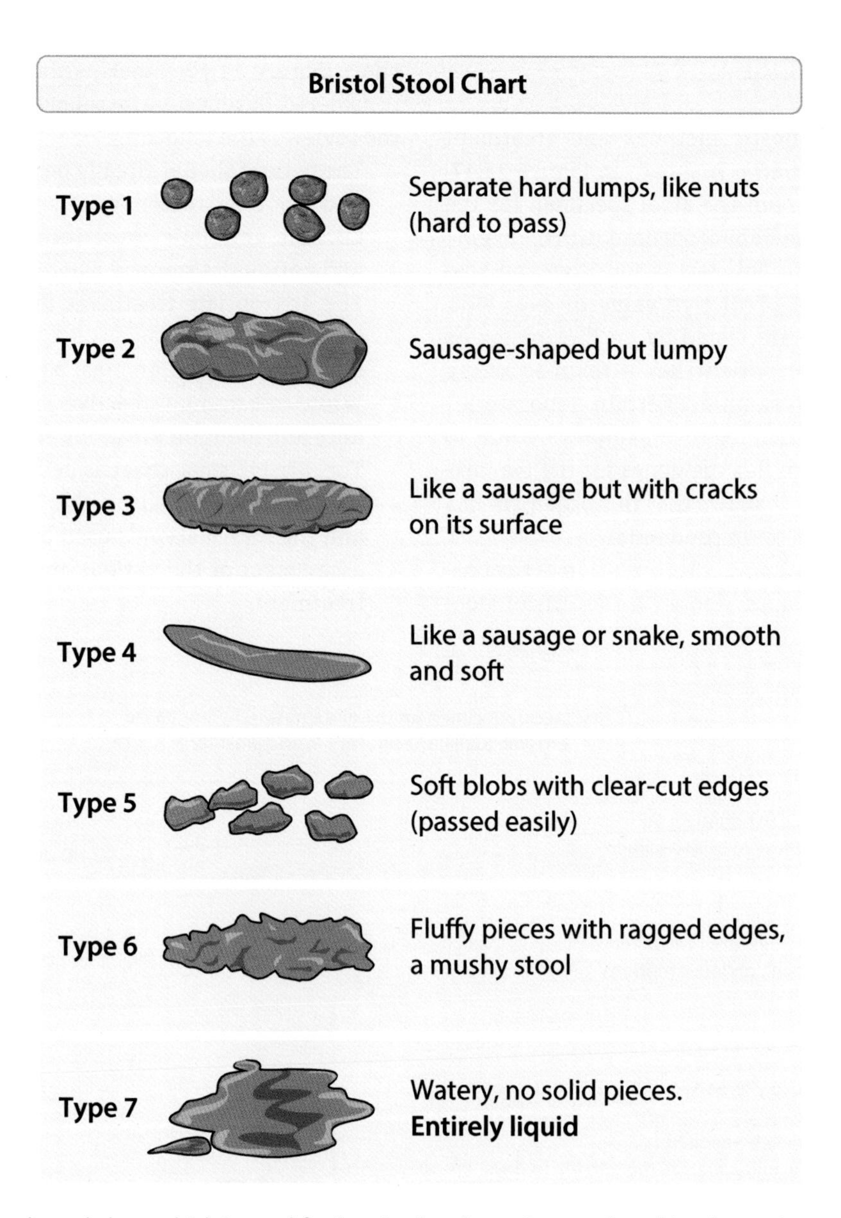

Figure 11.18 The Bristol stool chart which is used for 'scoring' each motion produced by the patient, and is of particular relevance in managing the patient with *Clostridium difficile* infection, and the response to treatment. A score of 5–7 is regarded as being consistent with diarrhoea. (Reproduced with the kind permission of Dr KW Heaton.)

In severe disease, the gastroenterologist and general surgeon must be contacted. Pseudomembranous colitis is an additional complication.

The use of proton pump inhibitors must be reviewed, as their amelioration of gastric acid production is considered to increase the likelihood of re-infection from a contaminated environment.

Certain laboratories in the UK supply extracts of faecal specimens from healthy donors as treatment. These are used to transplant a normal flora, restoring colonization resistance. In addition to each stool specimen being screened to ensure it is pathogen-free, donors have regular blood tests done to exclude other relevant infections. Extracts of stool are given to patients with relapsing infection via nasogastric tube. The response rate is considered to be over 90%.

Chapter 12

Infections of the Liver, Biliary Tract and Pancreas

INTRODUCTION

The anatomical relationships of the liver, biliary system and pancreas are shown in **Figure 12.1**. In the liver, arterial blood enters the sinusoids via the hepatic arterioles, and this blood mixes with that entering via the portal venules. The deoxygenated blood of the portal system is rich with all the nutrients that have been absorbed from the bowel. In addition, it will contain low levels of toxins, including lipopolysaccharide (LPS) of bacteria. From the sinusoid, blood drains into the central vein and the hepatic vein. Bile is secreted by the hepatocytes into the intercellular canaliculi, and drains via the ductules into the tributaries of the bile duct.

The microanatomy of the liver is shown in **Figure 12.2**. The Kupffer cells, which are macrophages, engulf damaged red blood cells and apoptotic cells that need removal from the blood, and ingest and degrade bacterial LPS absorbed from the bowel. They are also important in the phagocytosis of bacteria entering the liver via the portal vein. Stellate cells are subsinusoidal cells which are important in retinoid storage, including vitamin A, and play a central role in fibrosis of the damaged liver.

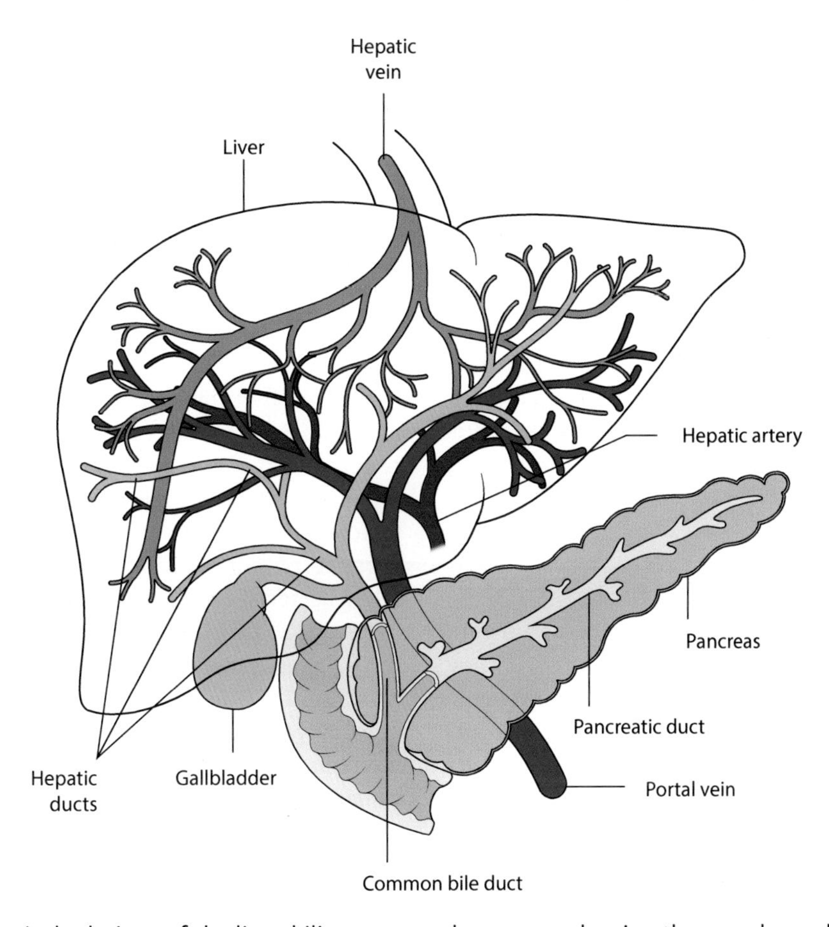

Figure 12.1 The anatomical relations of the liver, biliary tract and pancreas, showing the vascular and duct systems whereby organisms can reach these organs.

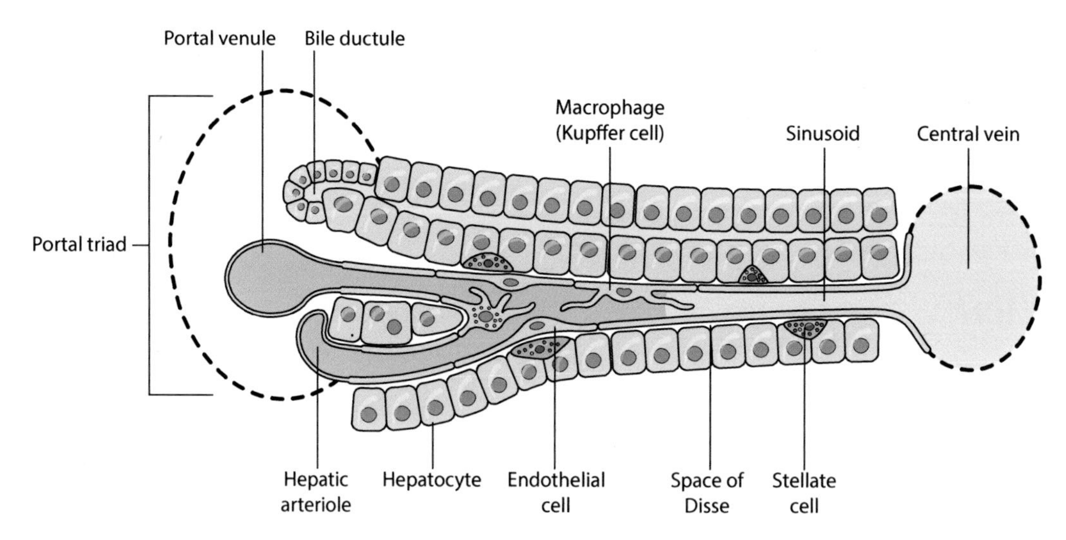

Figure 12.2 The microanatomical structure of the liver sinusoid, and the relations of vascular and ductal systems.

ORGANISMS

The organisms listed in **Figure 12.3** reflect the various routes whereby they reach the liver, pancreas and biliary system.

PATHOGENESIS

There are three main routes whereby organisms reach these organs.

VIA THE HEPATIC ARTERY TO THE LIVER

All blood-borne organisms will reach the liver by this route. Bacteria, including *Staphylococcus aureus*, yeasts, and the malaria parasites are examples. The hepatitis viruses include HBV and HCV; Epstein–Barr virus (EBV) and cytomegalovirus (CMV) can also cause an acute hepatitis.

VIA THE PORTAL VEIN TO THE LIVER

One or more species of bacteria can enter the portal vein from an appendix or diverticular abscess, and if they breach the defences of the liver and enter the systemic circulation, sepsis can arise. These organisms can also settle in the liver to cause an abscess. While there is no strict rule, single abscesses can be polymicrobial, with multiple abscesses caused by a single bacterium. *Klebsiella pneumoniae* is recognized for its ability to cause abscesses on its own.

Salmonella Typhi is an example of exogenous bacteria that will reach the systemic circulation via the portal vein.

HAV and HEV replicate at low levels in the intestinal epithelium, from where a primary viraemia enables them to reach the liver. Parasites such as the larvae of *Ascaris lumbricoides*, and *Entamoeba histolytica* enter the systemic circulation via this route as well, and *Entamoeba* can form liver abscesses. Unlike a pyogenic bacterial abscess, neutrophils are not recruited, as *Entamoeba* induces apoptosis in liver cells and a non-purulent 'anchovy paste' collection forms. Often there is a single abscess in the right lobe of the liver.

VIA THE BILIARY SYSTEM FROM THE DUODENUM

It is likely that low numbers of bacteria enter the common bile duct and ascend into the proximal part of the pancreatic duct and even into the gallbladder, but these organisms are removed by the normal flow of bile and pancreatic secretions and cause no problems. Any compromise of anatomical or physiological integrity will allow bacteria, or yeasts, to establish themselves. Stones or sludge within the system give rise to inflammation in the gallbladder (cholecystitis) and common bile duct (cholangitis). Tumours of the biliary duct can compromise bile flow too. Adult *Ascaris lumbricoides* can enter the common bile duct and cause obstruction to bile flow; this is a consideration in the patient who is from tropical regions of the world where this parasite is endemic.

Acute pancreatitis is often drug induced, but a wide range of organisms are associated with the condition too. These include *Ascaris*, CMV, *Legionella*, *Leptospira*, mumps and *Salmonella* Typhi. Bacteria and yeasts can take advantage of this inflammatory situation if it is prolonged.

Infections of the liver	Infections of the biliary tract	Infections of the pancreas
Hepatitis	*Streptococcus anginosus*	*Streptococcus anginosus*
HAV, HBV, HCV, HEV, CMV, EBV	'Coliforms'	'Coliforms'
Abscess	*Klebsiella pneumoniae*	*Klebsiella pneumoniae*
Staphylococcus aureus	*Pseudomonas aeruginosa**	*Pseudomonas aeruginosa**
Streptococcus anginosus	Mixed anaerobes	Mixed anaerobes
'Coliforms'	*Candida albicans**	*Candida albicans**
Klebsiella pneumoniae		
Mixed anaerobes	*Ascaris lumbricoides*	*Ascaris lumbricoides*
*Candida albicans**		
Entamaeba histolytica		

Figure 12.3 A list of organisms that can cause infections in the liver, biliary tract and pancreas, which can reach an organ by one or more routes. **Candida* and *Pseudomonas aeruginosa* are likely to overgrow in the alimentary canal following prolonged broad-spectrum antibacterial treatment.

Necrosis in the pancreas can result in development of a pseudocyst in the adjacent lesser sac, which contains pancreatic enzymes and necrotic material. Organisms in the duodenum, including *Pseudomonas aeruginosa* and *Candida*, can cross the inflamed barrier into the pseudocyst. Organisms can also access these sites via the pancreatic duct or the blood.

DIAGNOSIS AND MANAGEMENT

Fever, abdominal pain, diarrhoea, cough, right upper quadrant (RUQ) tenderness, jaundice, as well as raised inflammatory markers, liver function tests and serum amylase are markers of a broad range of infections. Key to management is consideration of which organism(s) are likely to be involved, and what microbiological investigations need to be done. Prompt collection of blood cultures, preferably before antibiotics are started, is often the first step. The use of ultrasound and contrast-enhanced computed tomography (CT) imaging is central to the diagnostic process, enabling identification of stones, as well as the size and number of collections. The degree of pancreatic necrosis can also be determined. This is important as a necrosis score of 50% or more directs the appropriate consideration of prophylactic antibiotics. Imaging is also necessary to guide the insertion of drains and obtain fluid for microbiological diagnosis.

Fluid specimens need to be sent immediately to the laboratory for microscopical examination and culture. The microbiologist will discuss positive results with the clinical team, and discuss the appropriate antibiotic regime. Specimens obtained from a site such as a liver abscess that are culture negative should prompt discussion of further testing, including use of the 16S ribosomal ribonucleic acid (RNA) gene identification system.

Difficult infections involving the biliary system, pancreas and adjacent zones of the peritoneum can occur after surgery involving these organs, with the development of fistulas further complicating the situation. It is important that the clinical team involves the microbiologist in these cases. In particular, there needs to be ongoing dialogue to review the relevance of new organisms appearing in specimens, as well as changing antibiotic profiles. In addition to *Candida*, the environmental bacterium *Stenotrophomonas maltophilia* is an example, accessing the bowel in the same way as *Pseudomonas aeruginosa*. This organism is naturally resistant to most antibiotics, with the exception of co-trimoxazole, and long-term use of carbapenems such as meropenem can lead to it being 'selected out'. Although it is considered a relatively less pathogenic organism, its presence in these situations complicates the antibiotic decision process.

HEPATITIS VIRUSES

This section discusses HAV, HBV, HCV and HEV. Features of the genome organization and replication of HAV, HBV and HCV have been discussed in Chapters 1 and 4.

Serological tests for HAV, HBV, HCV and HEV should be done for all cases of hepatitis; in addition, testing for CMV and EBV should also be considered in the adolescent or young adult with acute hepatitis. HEV is considered to be acquired following travel to endemic parts of the world such as South East Asia, the Indian Sub-Continent or

Africa. However, sporadic cases are identified locally, and so lack of a travel history should not exclude testing.

Features of these viruses, and markers used in their diagnosis are given in **Figures 12.4, 12.5**. Serology is used to identify acute infection (HAV, HBV, HEV), chronic infection (HBV), natural immunity (HAV, HBV) and vaccine-derived immunity (HAV, HBV). With HCV, the results of antibody tests are of no use on their own. All patients who have been infected with the virus will be HCV antibody positive, but chronic infection is identified by the presence of HCV RNA in blood by reverse transcriptase-polymerase chain reaction (RT-PCR). Patients who are antibody positive but RNA negative have eradicated the original infecting virus only; immunity is not a feature of this infection.

HAV

HAV is a single-stranded RNA picornavirus, and is transmitted from person to person mainly via the faecal–oral route. There are several genotypes, but only one serotype.

Virus	Incubation period: usual/ [range]	Transmission	Acute	Chronic	Complications	Comments
HAV (1-RNA)	45 [20–60 days]	F/O	Y	N	Acute liver failure (ALF)	Infection confers immunity
HBV (2-DNA)	60–90 [45-180 days]	Blood-borne	Y	Y	ALF, cirrhosis, hepato-cellular carcinoma (HCC)	HBsAb confers immunity
HCV(1-RNA)	60–90 [45–180 days]	Blood-borne	Y	Y	Cirrhosis and HCC	No natural immunity
HEV(1-RNA)	45 [20–60 days]	F/O	Y	(N*)	Acute liver failure	Pregnancy: significant mortality
CMV	Several weeks	Saliva/respiratory	(Y)	(Y)	Acute and reactivated infection is important in pregnancy and the immunocompromised patient	
EBV	Several weeks	Saliva	(Y)	(Y)	Reactivated infection is important in the immunocompromised	

F/O: faecal–oral route of transmission.

*SOT: solid organ transplant patients, in particular amongst the immunocompromised, can have chronic infection

Figure 12.4 The viruses associated with acute and chronic hepatitis.

Virus	Acute	Past infection	Natural immunity	Chronic infection
HAV	HAV IgM	HAV IgG	HAV IgG	Not known
HBV	HBcIgM	HBcIgG	HBsAb	HBsAg+, HBeAg+ (High risk)* HBsAg+, HBeAb+ (Low risk)*
HCV	HCV RNA**	HCV antibody	Not conferred by infection There are 6 genotypes	HCV RNA detected
HEV	HEV IgM	HEV IgG	Unclear status	SOT: ongoing presence of RNA in blood
CMV	CMV IgM Atypical lymphocytes	CMV IgG	N/A	Relevant in the immunosuppressed
EBV	Monospot positive Atypical lymphocytes	EBV IgG to NA (EBNA) (NA: nuclear antigen)	N/A	Relevant in the immunosuppressed

* Patients who are HBsAg positive, but who are negative for both HBeAg and HBeAb should be considered as 'high-risk' until their viral DNA status has been determined.

**HCV RNA: Note that acute HCV disease is not often recognized, as acute infection is usually asymptomatic.

Figure 12.5 The serological and nucleic acid markers used in the diagnosis of viral hepatitis.

Travel to lower-income parts of the world carries a risk. Transmission occurs via contaminated water or food, including fresh fruit and vegetables and undercooked molluscs. Outbreaks occur in other groups of patients that include injecting drug users (IDUs) and men-who-have-sex-with-men (MSM).

Most infections are asymptomatic, especially in children. The incubation period is about 45 days (range 20–60) to acute infection, which is diagnosed serologically by the detection of HAV immunoglobulin (Ig) M antibodies; these may still be present for 6–10 weeks after the acute infection. Individuals are infectious 2 weeks before symptoms and signs appear, but then rapidly become non-infectious. IgG antibodies appear, and provide natural immunity for life, as does vaccination (**Figure 12.6**).

HBV

Most HBV infections are asymptomatic, especially in children. The incubation period is about 60–90 days (range 45–180). Acute infection is diagnosed by the presence of IgM antibodies to the core antigen (HBcIgM), as well as the presence of HBsAg and HBeAg (**Figure 12.7a**). HBcIgG appear, followed by HBeAb and HBsAb (**Figure 12.7b**). HBcIgG is a life-long marker of past infection, but these antibodies are not protective.

There is a window period of several months between loss of HBsAg and the appearance of HBsAb, and that is why follow-up 6 months after acute infection is done, to determine whether seroconversion to HBsAb, and development of natural immunity, has occurred. If seroconversion is shown, the virus infection is considered as being cleared from the liver.

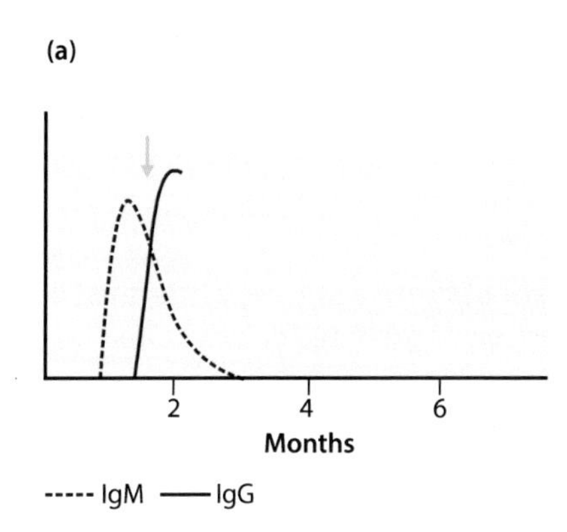

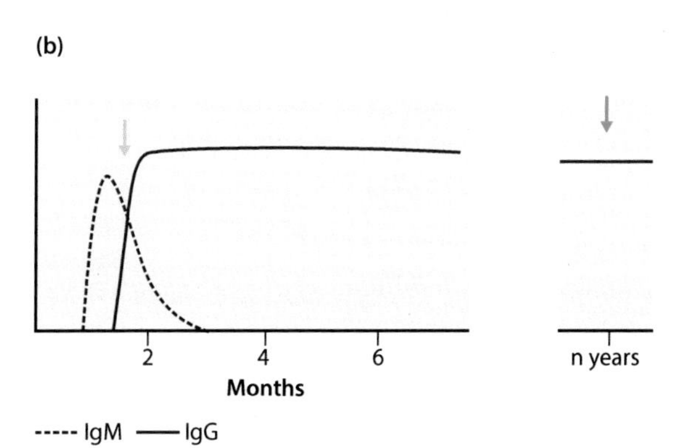

Figure 12.6 (**a**) The patient with acute HAV infection, and HAV IgM antibodies are detected (yellow arrow); (**b**) HAV IgG antibodies are detected 'n' years later, and identify natural or vaccine-induced immunity (green arrow).

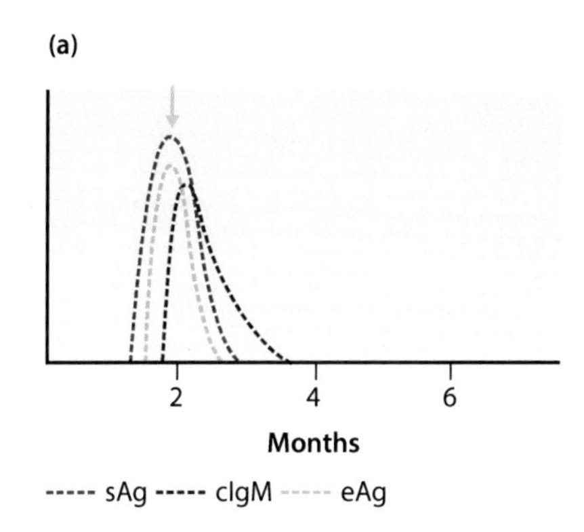

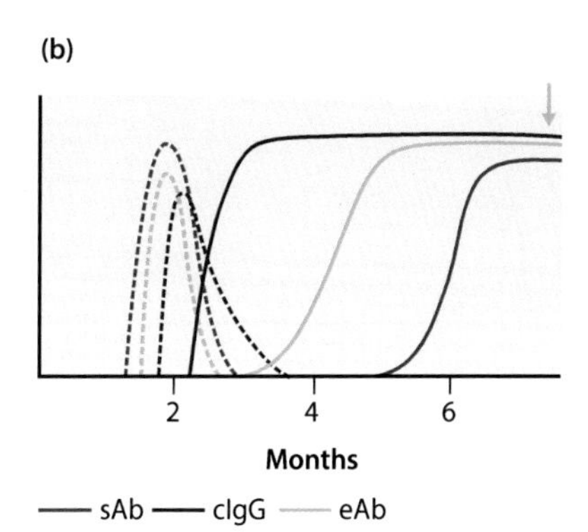

Figure 12.7 (**a**) Acute HBV infection is confirmed by the presence of HBsAg, HBeAg and HBcIgM (yellow arrow); (**b**) 6 months after acute infection the absence of HBsAg and presence of HBsAb confirm seroconversion and natural immunity (green arrow). HBcIgG antibodies are the marker of past infection. HBeAb are also present.

Certain individuals are unable to clear the virus and become chronic carriers, remaining HBsAg positive. The prevalence of chronic carriage varies significantly, from 0.1–2.0% in the USA and Europe to over 8% in South East Asia and China. It is the age at which the acute infection occurs that determines the likelihood of chronic carriage. If this occurs perinatally, the carriage rate is over 90%, in the age range of 1–5 years it is 10–20%, dropping to 5% in immunocompetent adults. This reflects the problems the immature immune system has in dealing with this virus. The presence of HBeAg shows ongoing high levels of virus replication, and 'high-risk' carrier status (**Figure 12.8a**). However, most individuals eventually seroconvert and produce HBeAb, giving them 'low-risk' carrier status (**Figure 12.8b**).

This is an important categorization. For example, babies born to HbsAg+/HBeAg+ mothers are protected by being given hepatitis B virus immunoglobulin (HBIG) and the first dose of the HBV (surface antigen) vaccine at birth, whereas babies whose mothers are HBsAg+/HBeAb+ are given the first dose of the vaccine at birth (Chapter 16). It is important to emphasize that following acute infection, the HBsAg-positive individual is considered infectious until HBsAb appear.

HCV

HCV is a member of the Flaviviridae, and has a 1-RNA genome. It is an enveloped virus with two glycosylated proteins, E1 and E2, and the core protein forms the capsid. Antibodies produced following acute infection do not confer immunity. There are two reasons for this. First, there are six major genotypes, and at least 25 subtypes of the virus, and second, a hypervariable region of the genome, coding for about 30 amino acids in the E2 protein, gives rise to continually changing amino acid sequences, and thus changing antigenic determinants. This enables the virus to 'escape' the immune system, as the heterogeneous antibodies produced against the changing E2 protein are not protective. The virus is blood-borne and IDU is a particular risk. Transmission is uncommon via sexual intercourse. Estimates of perinatal transmission vary between 0 and 4%.

Acute infection is usually asymptomatic and is seldom confirmed, but an incubation period of 60–90 days (range 45–180) is recognized. Only a minority of individuals clear the infection, and are HCV antibody positive and HCV RNA negative on testing (**Figure 12.9a**). Viraemia persists in 50–85% of infected individuals, which is confirmed by the presence of HCV RNA (**Figure 12.9b**).

HEV

HEV is a 1-RNA virus and is the only member of the Hepeviridae, and the hepevirus genus; it is transmitted by the faecal–oral route. The infection is associated with travel to the Middle East and South East Asia, but it is recognized that HEV infection in higher-income countries is more common than previously thought. The source of these infections is not clear, but may be from pork or wild animal meat. Patients with acute hepatitis must be tested for HEV, irrespective of their age or travel history.

Most infections are asymptomatic, especially in children. The incubation period is about 45 days (range 20–60 days), and acute infection is diagnosed by detection of HEV IgM antibodies, which may be present for 6–10 weeks after the acute infection. IgG antibodies are a marker of past infection (**Figure 12.10**). There are

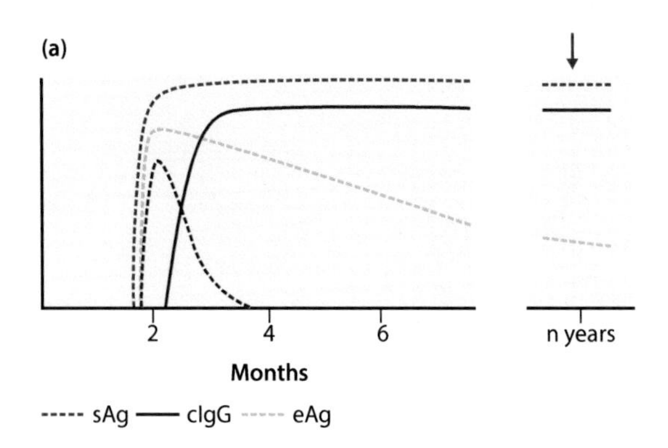

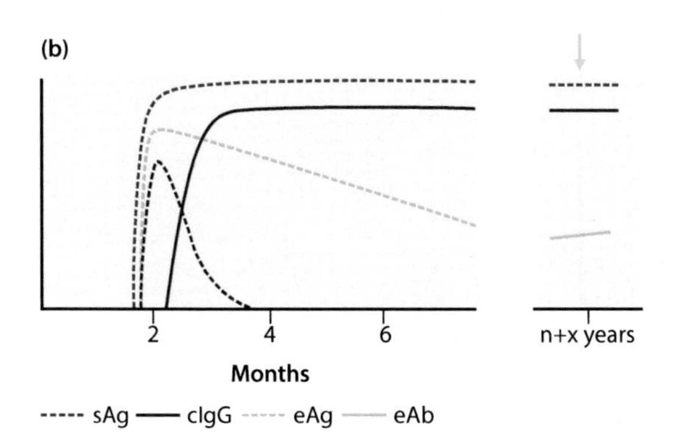

Figure 12.8 (**a**) The individual who does not seroconvert to natural immunity is HBcIgG, HBsAg and HBeAg positive, and this is the finding when chronic carriage is identified after 'n' years (red arrow); (**b**) most chronically infected individuals eventually lose the HBeAg and produce HBeAb at 'n+x' years (amber arrow).

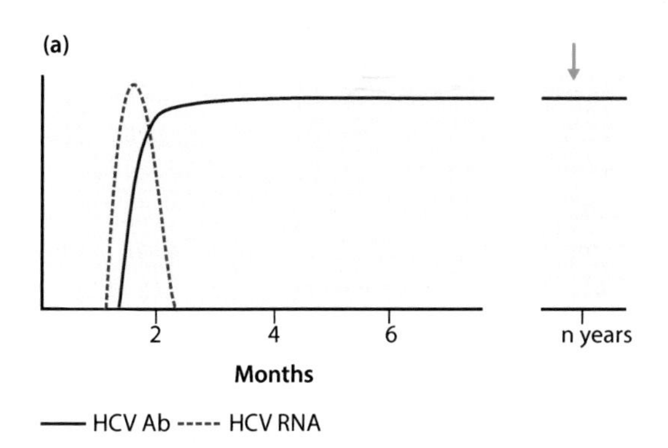

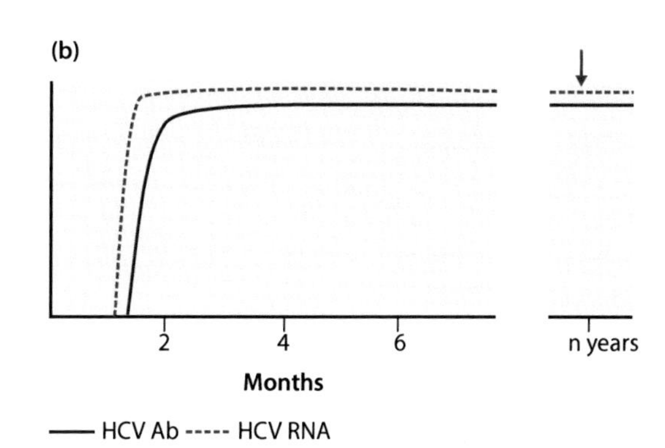

Figure 12.9 (**a**) When an individual is screened at 'n' years the presence of hepatitis C virus (HCV) antibodies and absence of HCV ribonucleic acid (RNA) shows the infection was cleared (green arrow); (**b**) the presence of HCV RNA shows chronic infection (red arrow).

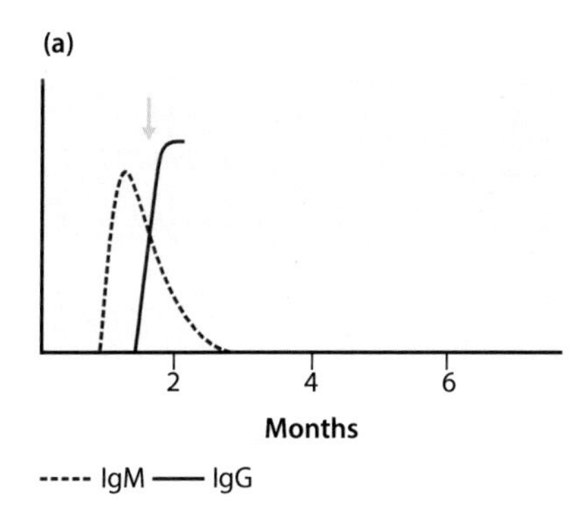

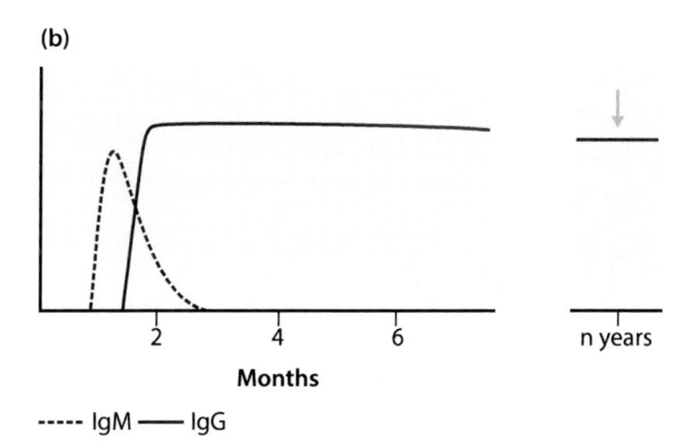

Figure 12.10 (**a**) The patient with acute hepatitis E virus (HEV) infection; HEV IgM antibodies are detected (yellow arrow); (**b**) HEV IgG antibodies are detected 'n' years later, and identify past infection (amber arrow).

five genotypes of HEV, and one major antigenic determinant in immunity; the degree of long-term protection following infection is unclear. Most clinical infections are mild, but severe infections occur in pregnancy, with a mortality rate of up to 25%. Chronic HEV infection can occur in the immunocompromised, especially the solid-organ transplant (SOT) patient. This is confirmed by the ongoing presence of HEV RNA in blood.

AN OUTLINE OF THE TREATMENT OF CHRONIC HBV AND HCV INFECTION

This is a rapidly changing science. Patients are assessed in terms of their virological markers, as well as stage of liver disease based on biochemical and histopathological markers, comorbidities, and coinfection with HBV, HCV or HIV.

There are complex treatment protocols, with drug interactions, side-effects and resistance creating challenging management situations, and recommendations are therefore regularly updated. Molecular testing is central to the management of these patients, with determination of viral load in plasma by PCR (HBV) or RT-PCR (HCV). In addition, the HCV genotype(s) that the patient is infected with influences the effectiveness of treatment and the drug combinations used. Examples of the various drugs used in treating chronic HBV and HCV are listed in Chapter 4, **Figure 4.1**.

HBV

In addition to the parameters discussed above, the e antigen status of the patient is important, with the key marker of treatment success being conversion from HBeAg+ to HBeAb+.

Treatment is usually for a minimum of 24–48 weeks, with peginterferon being used as the first-line agent. Nucleoside and nucleotide analogues are second-line agents that can be continued long term. The aim of treatment is to suppress the viral load in blood below detectable levels, and achieve biochemical and cellular remission in the liver.

HCV

Positive treatment outcomes are influenced by genotype, detection of disease at the earliest opportunity, younger age and female sex. Treatments range from 12 to 48 weeks, and the outcome depends on the time it takes for the viral RNA to become undetected in blood. Once treatment is stopped, a sustained virological response (SVR) is when the viral RNA remains undetected after 12 or 24 weeks, depending on the regime used. Treatment was centred on peginterferon combined with ribavirin. Due to cytokine-associated side-effects with interferon, combinations of drugs that target the specific enzymatic functions of the virus are now being increasingly employed.

Chapter 13

Infections of the Skin, Soft Tissues, Joints and Bone

INTRODUCTION

Infections of the skin and soft tissues include a wide range of clinical situations and organisms (**Figure 13.1**). Many infections arise following a breach of the skin, emphasizing the importance of this natural barrier. Once the skin is compromised, organisms can enter the deeper soft tissue. A surgical incision breaches the barrier, and surgical site infection (SSI) is an important entity, involving bacteria such as *Staphylococcus aureus*, including methicillin-resistant organisms (MRSA). The bite of a dog or cat introduces members of their oral flora, including *Pasteurella multocida* and *Capnocytophaga canimorsus* into the tissues, and can cause serious local and systemic infection. *Eikinella corrodens*, a member of the mouth flora of humans, is important in injuries of soft tissue and joints of the aggressor sustaining a 'clenched fist' injury. Not infrequently, these individuals present to

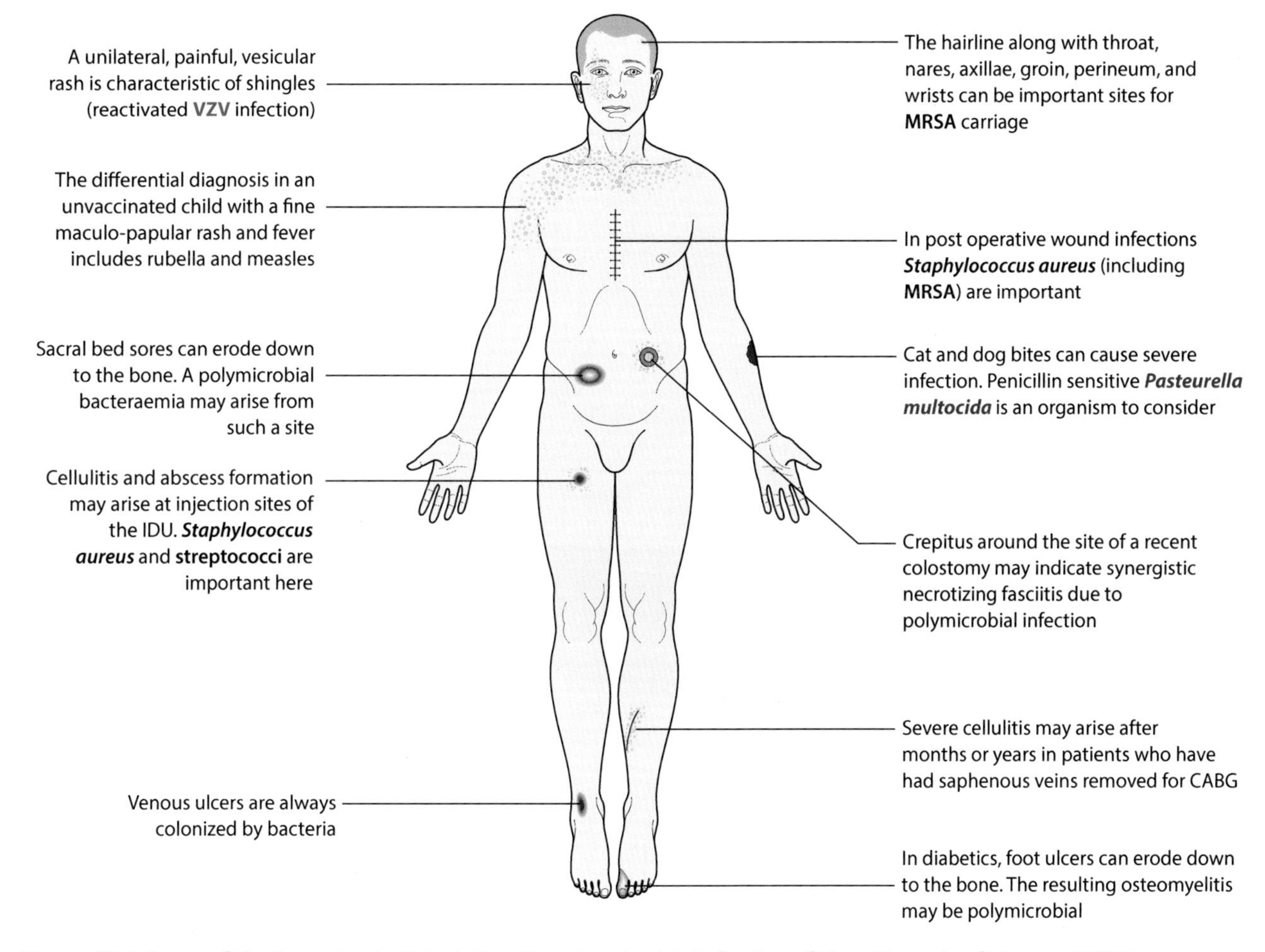

Figure 13.1 Some of the important clinical situations involved in infection of the skin and soft tissue. (CABG: coronary artery by-pass graft.)

hospital some days after the incident. (HBV, HCV and HIV need to be considered here too.)

Staphylococci and streptococci are commoner causes of infections of the joints and bone. *Haemophilus influenzae* b was a common cause of arthritis and osteomyelitis in children under 5 years. However, such invasive disease is now rare in countries where Hib vaccination is practised.

In the individual with a chronic joint infection, it is always important to consider a wider range of organisms, and this includes mycobacteria. Some examples of organisms that cause bone and joint infections are shown in **Figure 13.2**. Joint replacement is common in orthopaedic surgery, and when infection occurs, this can be particularly challenging to treat.

In addition to local infections, the skin is an important site for the manifestation of systemic disease. Some of the terms used to classify skin lesions are shown in **Figure 13.3**. Small, non-blanching lesions progressing to larger petechial and then ecchymotic lesions prompt the consideration of meningococcal sepsis. In the immunosuppressed patient, new skin lesions should be seen by the dermatologist and biopsied for histological and microbiological investigation. Organisms can be deposited in the skin from a focus elsewhere in the body. *Cryptococcus neoformans* is one example, as is *Fusarium*, which is also an environmental fungus.

ORGANISMS

A list of bacteria commonly associated with the various conditions discussed in this chapter is shown in **Figure 13.4**. Note that the majority of these are the colonizing or exogenous colonizers of the body (Figure 2.1).

Spirochaetes such as *Treponema pallidum*, the causative agent of syphilis, and *Borrelia burgdorferi*, the causative agent of Lyme disease, produce skin rashes. Lyme disease is transmitted from animals such as deer by ticks (a zoonosis), and the rash migrates out from the site of the tick bite and is termed erythema migrans. Arthritis is another manifestation of Lyme disease.

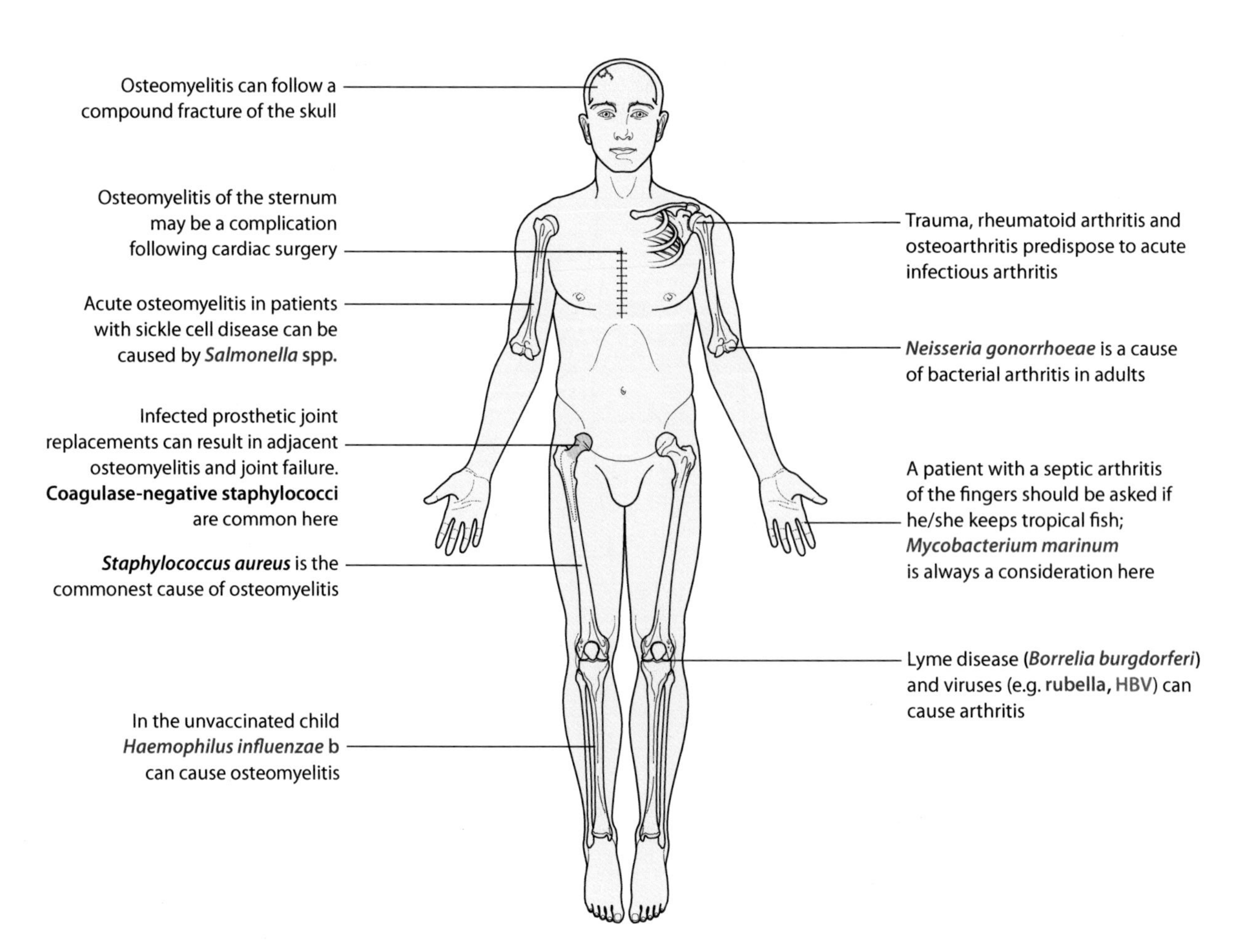

Figure 13.2 Examples of infections of the joints and bone.

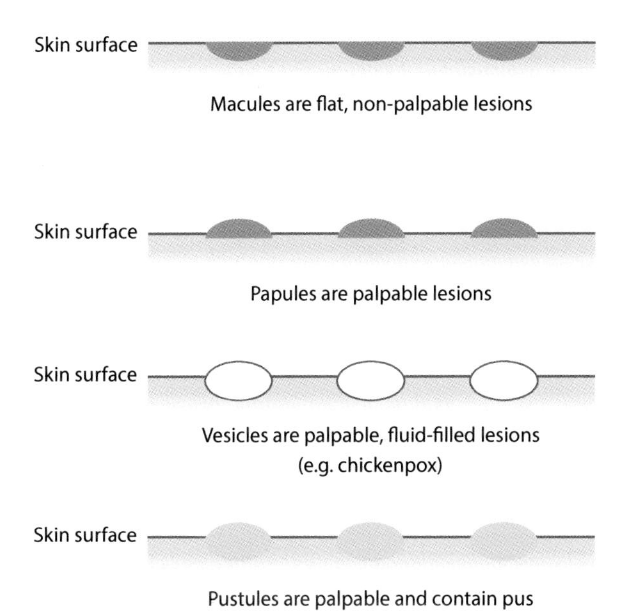

Figure 13.3 The differentiation of macules, papules, vesicles and pustules.

Skin and Soft Tissues	
Impetigo	***Staphylococcus aureus*, *Streptococcus pyogenes***
Folliculitis	***Staphylococcus aureus*,** *Pseudomonas aeruginosa* **(hot-tub folliculitis)**
Furuncles/carbuncles	***Staphylococcus aureus***
Paronychia	***Staphylococcus aureus*, *Streptococcus pyogenes*,** *Pseudomonas aeruginosa*, *Candida albicans*
Ecthyema/erysipelas	***Streptococcus pyogenes***
Cellulitis	***Staphylococcus aureus*, *Streptococcus pyogenes***
Necrotizing fasciitis	***Streptococcus pyogenes***
Synergistic necrotizing fasciitis	**Mixed** anaerobes, ***Staphylococcus aureus*,** 'coliforms' (e.g. *Proteus mirabilis*)
Native joint arthritis	
Gram positive	***Staphylococcus aureus*, *Streptococcus pneumoniae*, *Streptococcus pyogenes***
Gram negative	*Haemophilus influenzae*, *Neisseria gonorrhoeae*, 'coliforms'
Cat and dog bite	*Capnocytophaga canimorsus*, *Pasteurella multocida*, anaerobes
Human bite	**Oral streptococci,** *Eikinella corrodens*, anaerobes
Fish tank arthritis	*Mycobacterium marinum*
Prosthetic joint infections	
Gram positive	***Staphylococcus aureus*, *Staphylococcus epidermidis* (CNS), *Corynebacterium*, *Enterococcus faecium***
Gram negative	*Escherichia coli* ('coliforms'), *Pseudomonas aeruginosa*
Yeasts	*Candida albicans*, *Candida tropicalis*
Osteomyelitis	
Post operation (e.g. CABG)	***Staphylococcus aureus*, *Staphylococcus epidermidis* (CNS), *Enterococcus faecium***
Haematogenous	***Staphylococcus aureus*,** *Mycobacterium tuberculosis*, *Brucella abortus* **(vertebral)**
Diabetic foot osteomyelitis	**Usually polymicrobial (Mixed** anaerobes, 'coliforms', *Pseudomonas aeruginosa*, **streptococci, enterococci)**

Figure 13.4 The range of organisms to consider in skin, soft tissue, joint and bone infections.

A number of viruses have the skin as their target organ. Herpes simplex (HSV1 and HSV2) and varicella zoster virus (VZV) produce characteristic vesicular lesions. In the case of HSV these lesions are usually restricted to the genital or oral regions, but in chickenpox, lesions are widespread over the upper body, arms and head. These viruses can persist in the dorsal root ganglia and have the potential to cause reactivated disease. With VZV this presents as shingles or herpes zoster, which has a characteristic dermatomal distribution, depending on which nerve root the latent virus is reactivated from. With measles, rubella and parvovirus B19, a skin rash is useful to prompt their consideration in the diagnosis, although the rash of the latter viruses is not readily distinguishable.

PATHOGENESIS

INFECTIONS OF THE SKIN AND SOFT TISSUE

An intact skin surface, its relative dryness, desquamation of cells, a surface pH between 5.0 and 6.0 and an endogenous flora of coagulase-negative staphylococci and diphtheroids are all barriers to infection. In addition, sebum produced by the sebaceous glands is converted to free fatty acids by the normal flora of the skin, which inhibit the growth of pathogens such as *Streptococcus pyogenes*.

An outline of the structure of the skin and a range of skin infections including impetigo, staphylococcal scalded skin syndrome, infection of the hair follicles, ecthyma, erysipelas and cellulitis is shown in **Figure 13.5a–d**. Hot-tub folliculitis is caused by *Pseudomonas aeruginosa*. Outbreaks have occurred amongst groups of young children using an inadequately cleaned plastic paddling pool in warm summer weather.

Bacteria enter the skin via minor abrasions, surgical incisions or via the hair follicles. It is likely that cellulitis can also occur as a result of an occult bacteraemia. *Streptococcus pyogenes* can enter the blood from the pharynx or from lesions around the toes such as 'athlete's foot', and if it reaches skin or soft tissue where anatomy and physiology are compromised, a focus of infection can be established. Not infrequently a young man may, for example, have a fall and bruise an elbow. Several days later he presents to hospital with a serious soft-tissue infection and sepsis. It is likely that a haematoma in a seemingly innocuous bruise becomes the focus of this infection. The normal anatomy of soft tissue in the area where leg veins have been removed for cardiac bypass surgery is compromised, and is a potential site where *Streptococcus pyogenes* can establish itself. Chronic venous ulcers are an important condition, especially in the elderly patient, and this is highlighted in Chapters 3 and 17.

FASCIITIS AND MYOSITIS

Necrotizing fasciitis of the limbs is often caused by *Streptococcus pyogenes*. This condition can also arise after abdominal surgery, when a number of organisms work synergistically, including the streptococci, 'coliforms' and anaerobes of the bowel; *Staphylococcus aureus* may also be involved. Polymicrobial infections can occur in sites such as the groin of the intravenous drug user (IDU), with abscesses, fasciitis and gangrene being complications.

Once an organism with the pathogenic properties of *Streptococcus pyogenes* enters a fascial plane, there is little resistance to dissection by the expanding bacterial population in this zone separating skin from muscle. The resulting inflammatory response compromises the neurovascular bundles lying within the fascial plane. The skin over the affected area progresses from a red and painful cellulitis to a dusky red colour, and then becomes grey and painless with fluid-filled bullae. Invasion of the deeper fascia and progression to myositis can arise (**Figure 13.6**). The speed with which this process can take place is remarkable, and underlines the need for a full and ongoing assessment of the patient with severe cellulitis.

Gas gangrene caused by *Clostridium perfringens* is uncommon, but can occur in a soft tissue injury of the lower limb where perfusion is poor. This environment can enable the anaerobe to multiply and, by producing a range of potent histotoxins, the infection progresses to gangrene. The presence of crepitus in any area of soft tissue is an alert to consider such gas-forming anaerobes originating from the bowel.

It is probable in synergistic soft-tissue infections, where for example an abdominal surgical wound is contaminated with bowel flora, that 'coliforms' use up available oxygen in the tissue, enabling mixed anaerobes to exploit the situation. Factors such as poor nutrition, obesity and diabetes contribute to the development of this synergistic gangrene. Fournier's gangrene is a serious soft tissue infection involving the scrotum, which can spread to the abdominal wall. Predisposing factors include local trauma and diabetes. This is a polymicrobial infection, usually involving organisms of the bowel flora.

SEPTIC ARTHRITIS

Septic arthritis is more common in a joint affected by rheumatoid disease or where there is a history of trauma; diabetes, old age and malnutrition are also contributory factors.

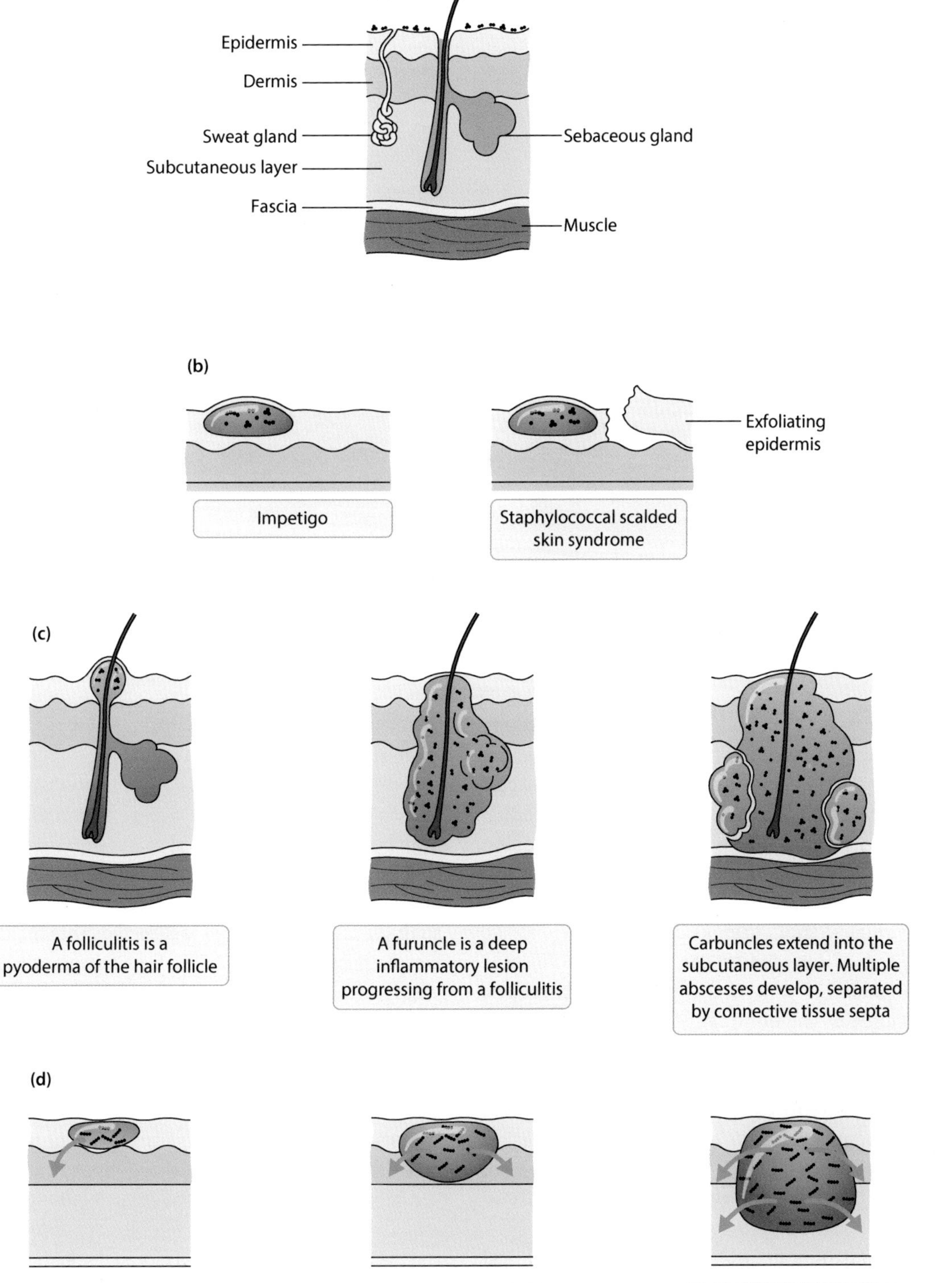

Figure 13.5 (**a**) The structure of the skin. (**b**) Impetigo and staphylococcal scalded skin syndrome. (**c**) Infections in the hair follicles. (**d**) Ecthyma, erysipelas and cellulitis.

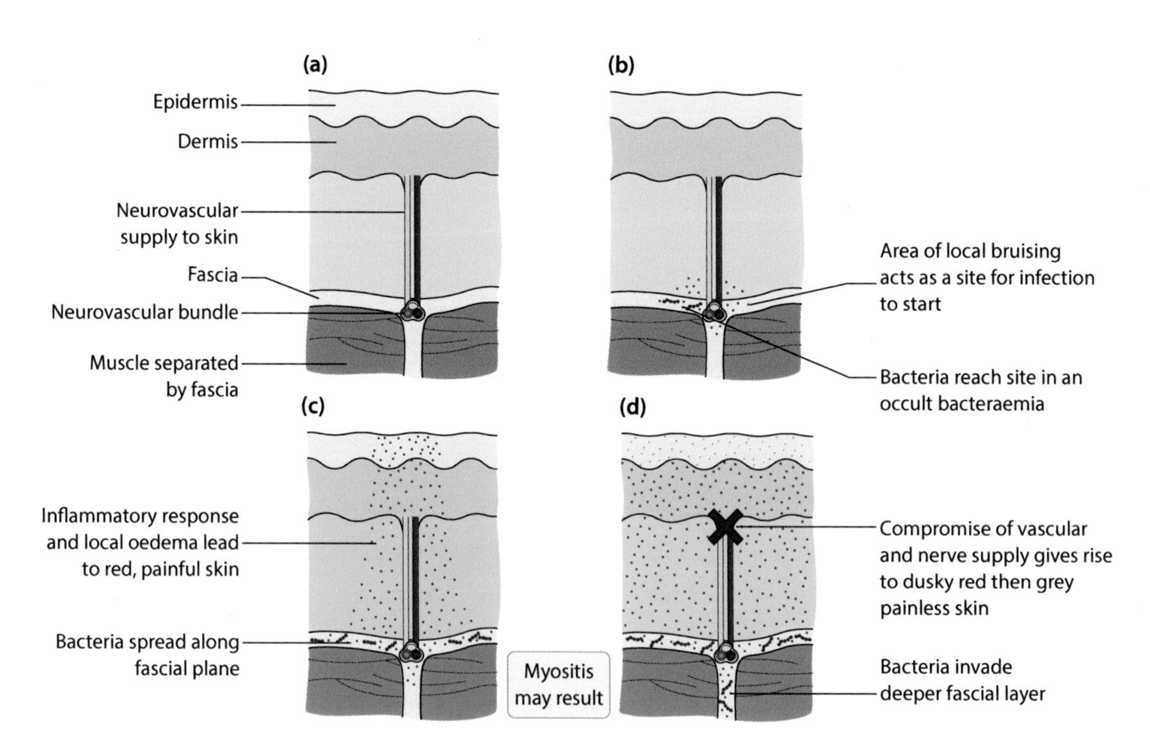

Figure 13.6 The mechanism whereby fasciitis arises. (**a**) Skin and muscle are separated by the fascial plane. (**b**) Bacteria can reach the fascial plane via the blood. (**c**) They spread along the fascial plane. (**d**) The neurovascular supply to the skin is damaged and invasion of the deeper fascial planes can occur.

An outline of the structure of a joint and routes whereby bacteria can enter is shown in **Figure 13.7**. The synovium lacks a basement membrane and bacteria in an occult bacteraemia can access the joint space. Their reproduction results in an inflammatory response with invasion of neutrophils in large numbers. Damage to the joint can occur; *Staphylococcus aureus*, for example, produces a chondrocyte protease that destroys cartilage.

A wide range of organisms can cause native joint arthritis. One alert is to ask the patient with septic arthritis of a joint of the hand if they keep tropical fish or work in a pet shop that stocks them. *Mycobacterium marinum* can cause septic arthritis, and in the immunocompromised patient, can seed to other bones and joints via the blood. Unless this 'fish tank' association is made, repeat fluid aspirates processed for standard microscopy and culture are reported as 'no growth'. The specimen must be processed for mycobacteria; *Mycobacterium marinum* grows optimally in the laboratory at 30°C.

Infections of prosthetic joint devices can arise as a result of organisms such as coagulase-negative staphylococci being introduced at the time of operation or by the haematogenous route (**Figure 13.8**). Prosthetic joint infections can be classified into acute and chronic, and details of this are outlined in **Figure 13.9**.

Once bacteria are established within the prosthetic joint, they will reproduce and organize into a biofilm over a period of days. When a biofilm is established, antibiotic cure is increasingly unlikely to be effective. Cure can often only be achieved by removal of the prosthetic components, and extensive debridement. Comprehensive sampling at operation maximizes the culture of the offending organism, so that antibiotic therapy is directed. This includes down to bleeding (bone) tissue, and later re-implanting a new prosthesis with 'cement' impregnated with antibiotics that are known to have activity against the offending organism.

OSTEOMYELITIS

Bone may become infected following direct introduction of bacteria into this tissue by trauma or surgery. Osteomyelitis can also arise from the haematogenous route. An outline of this process in children is shown in **Figure 13.10**. Capillary loops in the metaphysis of a long bone may develop small haematomas as a result of some external force such as trauma. Organisms from an occult bacteraemia can settle in this haematoma and initiate infection. The resulting medullary abscess can then extend, via the Haversian canal system, through the cortex of the bone. A sequestrum of necrotic tissue remains within the bone.

In children, the fibrous periosteum of the bone is not breached and a subperiosteal abscess forms. In adults, the

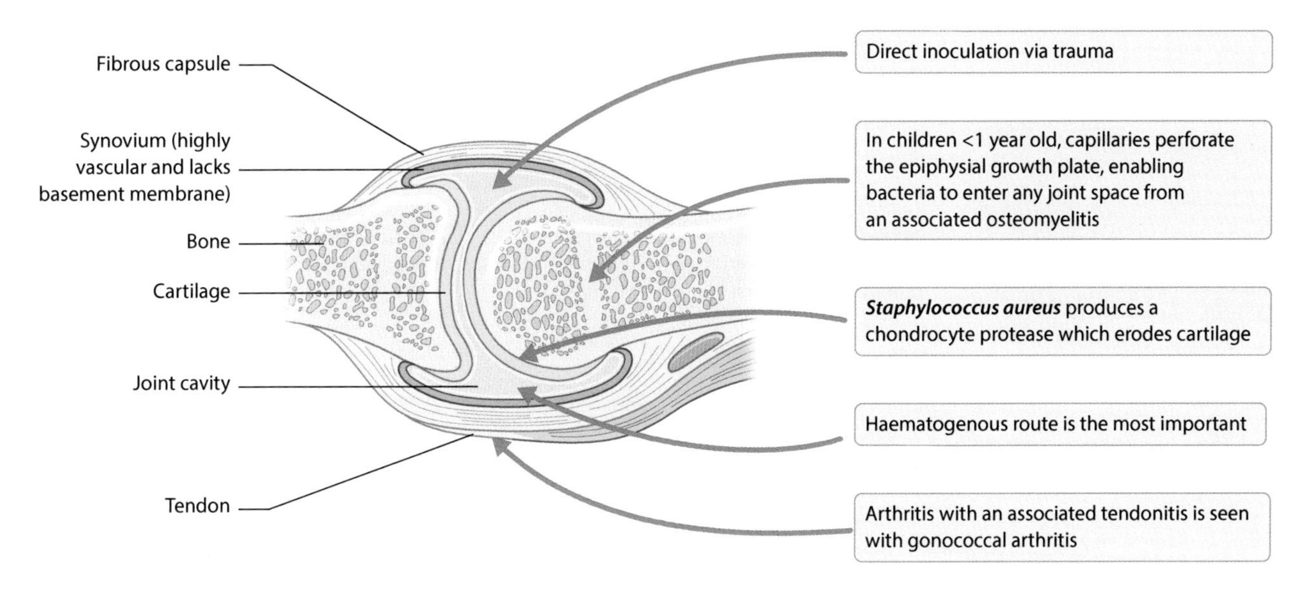

Figure 13.7 The structure of a joint and the routes whereby bacteria reach a joint.

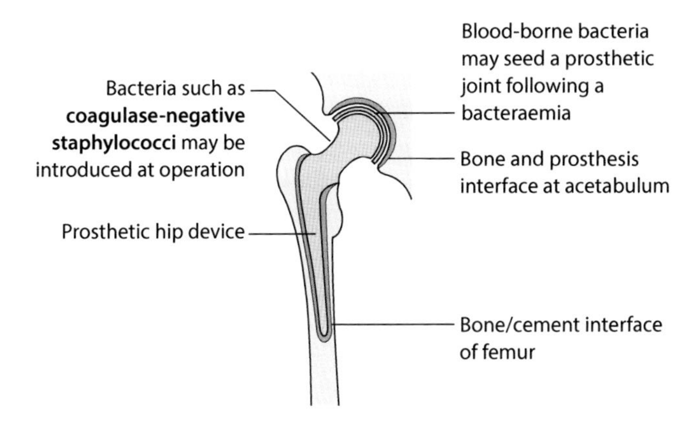

Figure 13.8 Bacteria can be introduced into a prosthetic joint at the time of operation or via the blood.

	Acute	Chronic
Time after operation	**<4 weeks**	**≥4 weeks**
Length of symptoms	**<3 weeks**	**≥3 weeks**
Clinical features	**Acute pain, fever, red and swollen joint**	**Chronic pain, loosening, skin sinus to prosthesis**
WCC in joint aspirate	**>10,000 × 10^6/L**	**>3,000 × 10^6/L**
Biofilm state on prosthesis	**Absent to forming**	**Present**
Surgical treatment options	**Debridement and change mobile parts**	**Complete removal, debridement and re-implantation** **(Can be done as single-stage or two-stage procedure)**
More common organisms	***Staphylococcus aureus* (MSSA/MRSA)** *Escherichia coli, Klebsiella pneumoniae* *Enterobacter aerogenes*	**Coagulase-negative staphylococci** ***Corynebacterium*** ***Propionibacterium*** *Candida*

Figure 13.9 Features of infections of prosthetic joint, and important causative organisms.

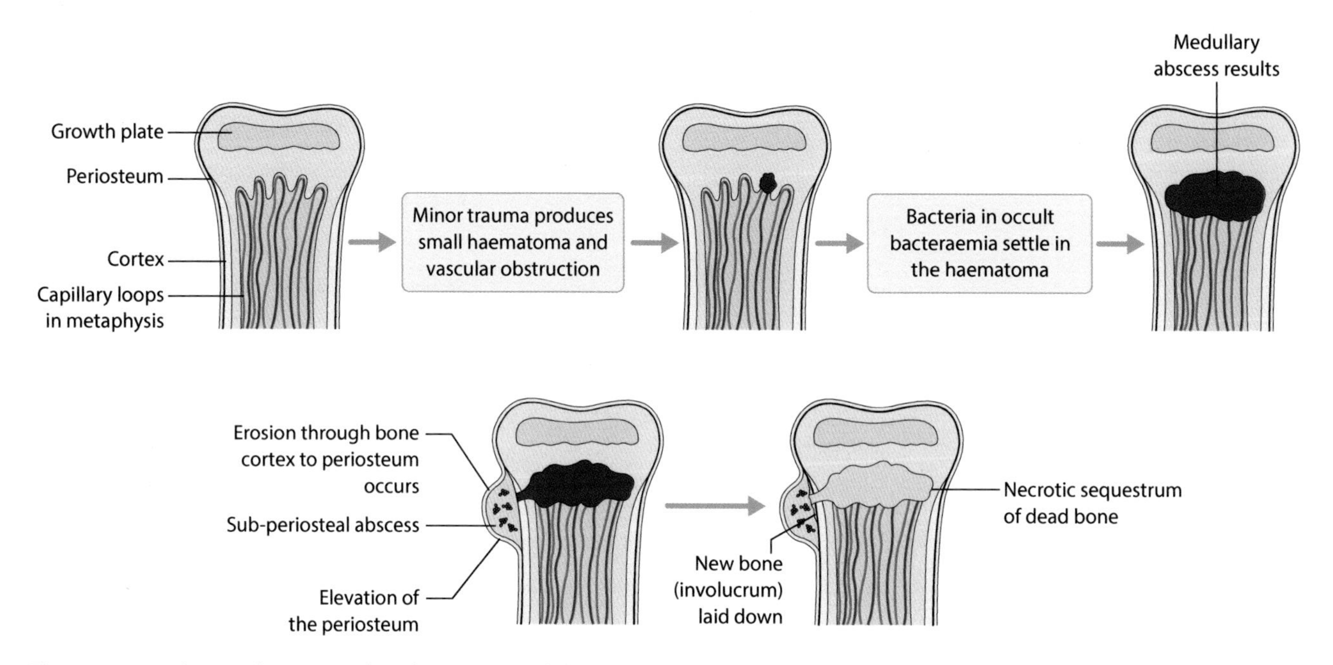

Figure 13.10 The mechanisms whereby osteomyelitis develops in a child.

periosteum is breached and soft tissue abscesses occur. New bone, termed the involucrum, is laid down around the breach on the surface of the bone. In children less than 1 year of age, the infection in the metaphysis of any bone can erode through the growth plate and cause an associated septic arthritis. After this age, this type of septic arthritis only occurs where the joint space encloses the metaphysis, such as the hip joint.

Infection of bone is classed as either acute or chronic. Chronic infection may have continued for weeks and months before the diagnosis is made. Whether acute or chronic, osteomyelitis is a major problem as it is difficult to eradicate bacteria from the sequestrum, and even after successful treatment of an acute infection, bacteria may survive in a 'dormant' state to cause disease many years later.

Spondylodiscitis is the infection of the vertebral bone and the intervertebral disc. These sites can become infected from a blood source, and endocarditis is one important entity that is always considered; *Staphylococcus aureus* is a common culprit. The range of organisms is considerable and includes *Mycobacterium tuberculosis*, and, uncommonly *Brucella*, which would be associated with a relevant travel history.

DIABETIC FOOT INFECTIONS

As a result of peripheral neuropathy and vascular insufficiency, minor trauma leads to a breach in the integrity of the skin on the toes and the sole of the foot to form small ulcers. These become colonized with bacteria such as *Staphylococcus aureus*, *Streptococcus agalactiae* and *Streptococcus dysgalactiae*. Cellulitis develops around small ulcers, with extension into the soft tissue. Unless this is prevented, colonization with 'coliforms' and anaerobes occurs, and *Pseudomonas aeruginosa* can be introduced from domestic water sources. Reproduction of this polymicrobial flora then extends down to bone, initiating osteomyelitis that progresses to a chronic infection.

DIAGNOSIS AND MANAGEMENT

Common skin conditions such as furunculitis are usually straightforward to diagnose and can be treated with topical fucidic acid, which is active against *Staphylococcus aureus*. Paronychia may require incision and drainage, and a specimen is collected for culture and identification of the culprit. A kitchen worker with recurrent paronychia of fingers nails caused by an enterotoxin-producing strain of *Staphylococcus aureus* can potentially introduce the organism into food during its preparation. If this is stored incorrectly, for example at room temperature, the organism will reproduce, with toxin causing acute vomiting within 1–4 hours in those who have consumed this contaminated food.

The patient with a skin rash needs prompt assessment in relation to infection control risk too. Measles and rubella can be a consideration, but more frequently it is the vesicular rash of VZV that warrants this attention. In addition to chickenpox, the patient with exposed herpes zoster is a risk to susceptible individuals, noting that that those with a covered rash pose a risk to immunocompromised individuals, which includes staff in this group too.

The key to diagnosing bacterial infections of skin, soft-tissue, joint and bone centres on collection of blood for culture. In the setting of cellulitis of the limbs, this may be the only way to reliably identify the causative agent. Fluid aspirated from an infected native joint should be processed promptly in the laboratory. The white cell count will show a neutrophilia, and while this count can be low early in an infection with several thousand cells/μL, the usual count is in excess of 50,000 cells/μL. The Gram stain result can reveal the organism. Relevant details need to be entered on the request form; the patient with septic arthritis of a joint of the hands who keeps tropical fish could have an infection caused by *Mycobacterium marinum*.

For the diagnosis of herpetic lesions, a vesicle should be carefully opened, and a swab collected for HSV and VZV PCR. The swab is placed in a white-topped standard container. Some laboratories request that a few millilitres of sterile saline is added to the container.

The treatment of cellulitis of the limbs caused by *Streptococcus pyogenes* can be challenging; the scenario of the young man referred to on page 198 is an example. Based on renal function, the optimum therapy in the patient who is not allergic to β-lactams is high-dose benzylpenicillin and clindamycin. The latter is given as it stops protein synthesis and toxin production. However, a situation can be reached when these antibiotics have decreasing effectiveness in the setting of developing necrotizing fasciitis. The involvement of the surgical team from the earliest opportunity is critical and ongoing. For the penicillin-allergic patient, daptomycin (or a glycopeptide, usually vancomycin) is used with clindamycin. Dosing must be maximized from the outset, ensuring that serum levels are therapeutic.

Management of a prosthetic joint infection is directed by the senior orthopaedic surgeon on duty. Under no circumstances should an aspirate of the joint be undertaken by anyone other than a member of the orthopaedic team experienced in doing this. The initial intervention should be collection of a set of blood cultures, although it is not common for bacteria to enter the blood from a prosthetic joint. In most circumstances the patient is clinically stable, and the orthopaedic team can do an aspirate. Where possible the patient is taken to the theatre as soon as appropriate, so that full operation can be done, with deep biopsies within the joint being taken for microbiological and histopathological examination. Operations range from retention of the prosthesis, to one-stage or two-stage exchange, after extensive debridement. The new prosthesis is then implanted.

In the non-septic patient, the orthopaedic team will usually initiate a broad-spectrum combination of vancomycin and meropenem after operation, unless the immediate or previous microbiology results direct otherwise. Once an organism is identified, the necessary narrow-spectrum antibiotics are used. As most infections are caused by gram-positive bacteria, antibiotics for relevant organisms identified include benzylpenicillin, flucloxacillin, vancomycin, teicoplanin and daptomycin. Rifampicin is often added as a second agent, especially as it is considered to have better penetration into biofilm. The quinolones such as ciprofloxacin are used to treat infections caused by susceptible gram-negative bacteria, as their oral bioavailability and tissue penetration is good. In certain patients where the prosthesis cannot be removed, long-term suppressive antibiotic treatment is given.

These antibiotic recommendations generally apply to native bone and joint infections too. In addition to the use of a quinolone such as ciprofloxacin in the treatment of infections caused by gram-negative bacteria (and methicillin sensitive *Staphylococcus aureus* [MSSA]), clindamycin is often used as the oral agent to treat susceptible gram-positive bacteria. Its oral bioavailability and tissue penetration are good. In all situations the correct dose of antibiotics must be given, and the body mass index of the patient must be taken into account. Treatment ranges from 4 weeks for an uncomplicated native joint infection, to 6 weeks for acute osteomyelitis; chronic infection can be treated for 3 months or longer.

Diabetic foot infections are a diagnostic challenge. Superficial swabs of an infected foot ulcer will grow a range of bacteria and are of dubious value, and the diabetic team and podiatrist should involve the microbiologist. If the patient is clinically stable, antibiotics should be stopped for 2 weeks. After this, deep tissue samples, including bone, are taken. The microbiologist needs to ensure that all relevant organisms are fully identified, and that the antibiotic susceptibility tests done include the necessary range of antibiotics, so that treatment can be targeted and effective. Where possible, oral antibiotics should be used; the minimum duration of treatment is usually 6 weeks.

Chapter 14

Infections of the Central Nervous System

INTRODUCTION

The brain and spinal cord are protected by the skull and spinal column. Three connective tissue layers, the pia mater, arachnoid mater and dura mater, separate the nervous tissue from bone (**Figure 14.1**). Between the first two layers in the subarachnoid space is the cerebrospinal fluid (CSF), which acts as a shock absorber. It is produced by the choroid plexus of the ventricles, exiting by the foramina of Luschka and Magendie, and then circulates around the brain and spinal cord. CSF is reabsorbed by the arachnoid granulations, which extend into the superior sagittal sinus, one of the great vessels draining the brain. The blood–CSF barrier consists of capillary endothelial cells resting on a basement membrane. The tight junction between these cells is such that constituents of the plasma, such as albumin, are unable to cross into the CSF under normal circumstances. The blood–brain barrier is the boundary between the vasculature and the brain tissue.

Any pathology in the brain that increases intracranial pressure can have disastrous consequences, as the bony skull and fibrous supports of the brain do not allow room for any significant expansion. Brain swelling thus leads to compression, herniation of the brain and brain cell death. The pathological process that occurs, for example in meningitis, is outlined in **Figure 14.2**. Here a simplified 'brain' is shown enclosed in the skull. The inflammatory response in the subarachnoid space and release of cytokines results in loosening of the tight junctions between vascular endothelial cells. This allows albumin into the CSF, producing vasogenic oedema. Toxic substances from bacteria and neutrophils, and compromise in the supply of oxygen and nutrients to brain cells, result in cytotoxic oedema. Vasogenic and cytotoxic oedema both contribute to brain swelling, which obstructs the outflow of CSF from the ventricles. The pressure behind this obstruction forces fluid from the CSF into the interstitial compartment, compounding the swelling. If medical intervention is not prompt, brain damage or brain death are likely outcomes.

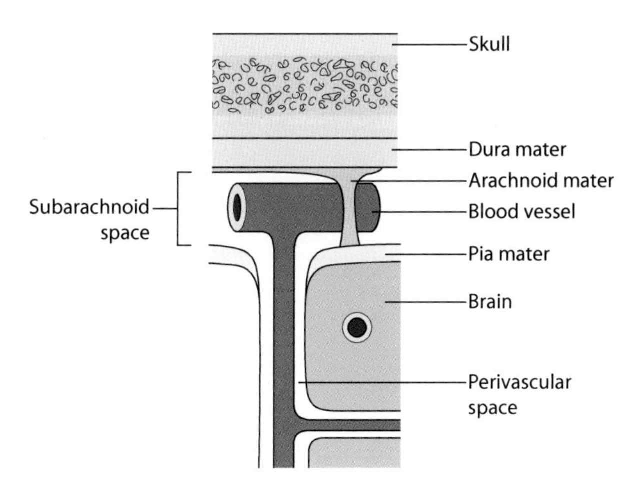

Figure 14.1 The various connective tissue layers that surround the brain and spinal cord. The subarachnoid space contains the cerebrospinal fluid.

Organisms can reach the brain and spinal cord by a number of routes. The brain has an extensive arterial and venous system (**Figure 14.3**). The arterial system can transport organisms in septic emboli from distant sites of infection such as a lung abscess or an infected heart valve. Bacteria such as meningococcus enter the blood from the nasopharynx and reach the brain via the arterial system; if they manage to enter the CSF, meningitis is the likely result. Septic thrombophlebitis can occur in the main venous structures associated with the brain, such as the cavernous sinus, venous sinuses and the tributaries of the jugular veins. Within the skull lie the nasal sinuses and mastoid air spaces, separated in places from the brain by thin bone. Infection here may erode through the bone into the brain.

The array of infective conditions of the central nervous system (CNS) is impressive, as is their clinical presentation. Meningitis, encephalitis, brain abscess, subdural and epidural abscesses and cavernous sinus thrombosis are examples. Clinical symptoms may be non-specific, such as fever, headache and vomiting; the elderly patient can present with confusion or increased confusion. There can be a specific neurological deficit caused by a lesion at a particular site in the brain. Encephalitis due to herpes simplex virus (HSV1) may manifest as unusual behaviour,

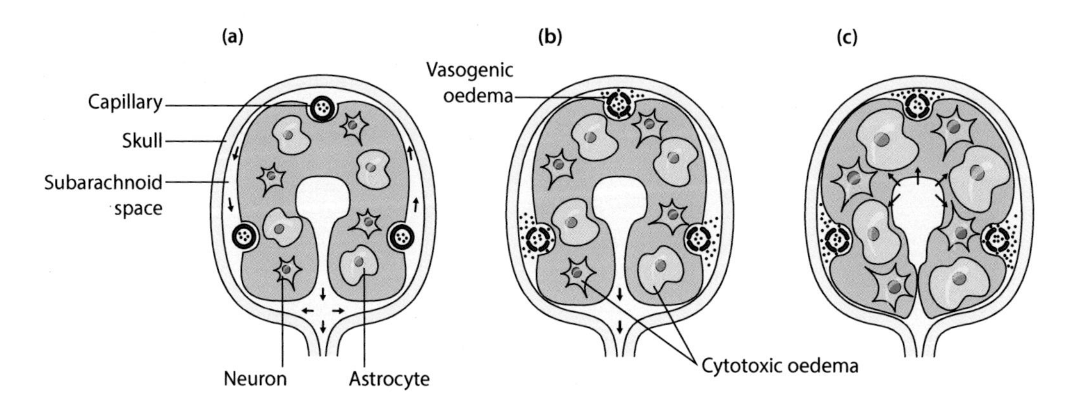

Figure 14.2 (**a**) The 'normal brain'. (**b**) Inflammatory conditions such as meningitis result in vasogenic and cytotoxic oedema. (**c**) Obstruction of cerebrospinal fluid outflow forces fluid into the interstitium, resulting in interstitial oedema.

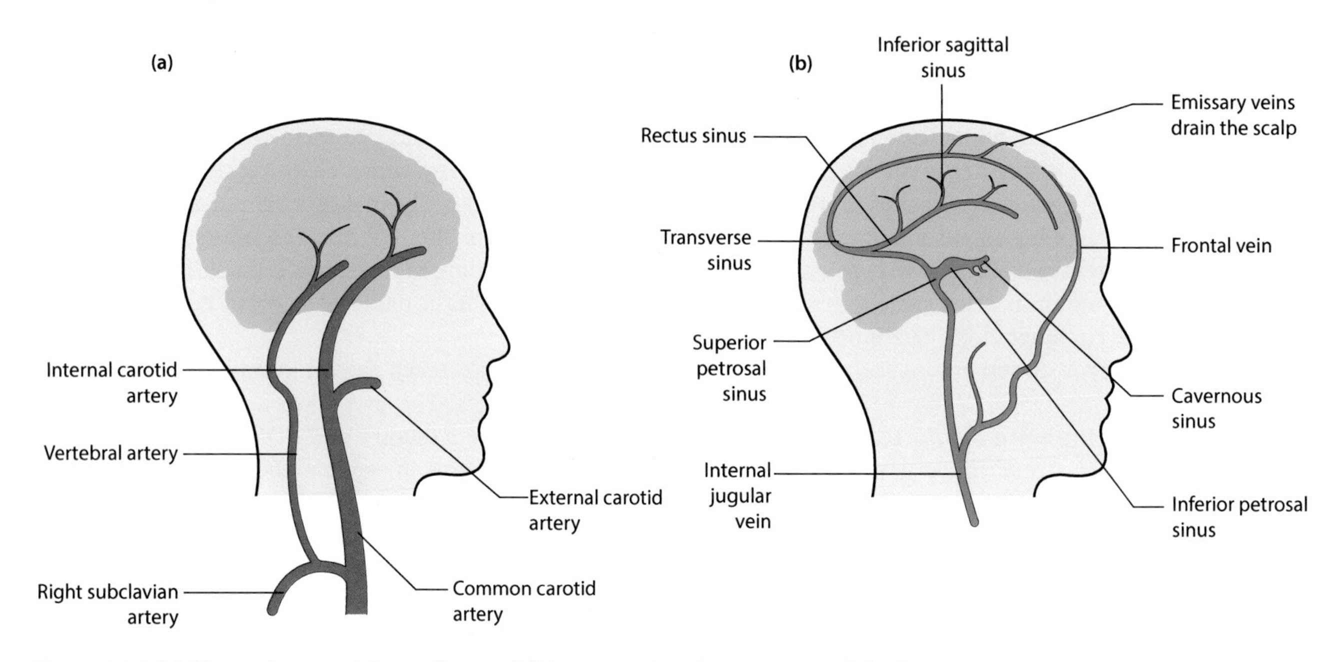

Figure 14.3 (**a**) The major arterial supplies, and (**b**) venous drainage systems of the brain.

characteristic of the replication of the virus within cells of the temporal lobe. Toxins are responsible for botulism and tetanus.

The range of organisms that can cause disease in the CNS is considerable and includes bacteria, fungi, viruses, protozoa and trematode parasites. Examples include the environment yeast *Cryptococcus neoformans*, an important cause of meningitis in acquired immunodeficiency syndrome (AIDS) patients, and viruses such as HSV1, varicella zoster virus (VZV) and the enteroviruses. When cysts of the pork tapeworm *Taenia solium* settle in the brain, neurocysticercosis arises; the resulting inflammation can cause seizures, which may be the first manifestation of this infection.

ORGANISMS

A list of bacteria to consider in various clinical settings is shown in **Figure 14.4**. *Listeria monocytogenes* is acquired from contaminated food; dairy products, including soft cheese, as well as patés, not subjected to food safety checks can be sources. It should be considered in the immunosuppressed patient. It can cause maternal sepsis and chorioamnionitis, usually in the last trimester

Meningitis	Encephalitis	Brain abscess (as source)
		Otitis media/mastoiditis
Streptococcus pneumoniae	Enterovirus	*Streptococcus* spp.
Haemophilus influenzae	HIV	*Streptococcus pneumoniae*
Neisseria meningitidis	HSV1	*Moraxella catarrhalis*
	VZV	*Bacteroides*
*Listeria monocytogenes**	Rabies virus	
*Streptococcus agalactiae***		**Sinusitis**
Enterovirus	**The arboviruses**	*Streptococcus pneumoniae*
HSV2	Eastern equine encephalitis virus (EEV)	*Haemophilus influenzae*
	Japanese encephalitis virus (JEE)	*Bacteroides*
	Western equine encephalitis virus (WEE)	
	Zika virus	**Dental/lung abscess/bronchiectasis**
Cryptococcus neoformans		
		Streptococcus anginosus
Naegleria fowleri		*Bacteroides*
		Fusobacterium necrophilum
		Infective endocarditis
Neurosurgery infections (EVD/shunts)	**Toxin-mediated disease**	*Staphylococcus aureus*
		Streptococcus anginosus
Coagulase-negative staphylococci	*Clostridium botulinum*	
Corynebacterium spp.	*Clostridium tetani*	**The immunocompromised patient**
Staphylococcus aureus		
Klebsiella pneumoniae		Non-tuberculous mycobacteria
Pseudomonas aeruginosa		*Aspergillus*
Stenotrophomonas maltophilia		*Candida*
Candida		*Nocardia*

Figure 14.4 Examples of the wide range of organisms that can cause infection in the central nervous system, including the anatomical site they occur or originate from. (*Consider especially in the immunocompromised and neonate, **usually associated with neonatal meningitis.)

of pregnancy, whereby it gains access to the fetus. While meningitis is uncommon in the mother, both blood and CSF from the neonate can grow *Listeria*. Neonates born prematurely or subjected to prolonged rupture of membranes are at risk of invasive disease, including meningitis, with *Streptococcus agalactiae* (GBS). This is acquired by aspiration of endogenous vaginal flora of the mother during delivery (Chapter 16). Pathogens such as *Neisseria meningitidis*, *Haemophilus influenzae* b, and pneumococcus can be members of the exogenous 'colonizing flora' of the nasopharynx.

The patient with chronic suppurative lung disease arising from aspiration of oral bacteria may develop a brain abscess via an embolic episode from the lung. Ventriculoperitoneal (VP) shunt infections are usually caused by endogenous skin organisms such as the coagulase-negative staphylococci, which can contaminate the shunt at the time of insertion. These are also important organisms associated with external ventricular drain (EVD) infections, where a range of bacteria and yeasts are relevant. These patients often have extended stays on intensive care/high-dependency units, where ventilator-associated pneumonia is treated with repeat courses of broad-spectrum antibiotics such as meropenem. The environmental bacterium *Stenotrophomonas maltophilia* is not infrequently 'selected out', and goes on the cause an EVD infection. This organism can only be reliably treated with co-trimoxazole. Isolates resistant to this agent are not uncommon, and create a difficult management situation when the drain cannot be removed for neurosurgical reasons.

Naegleria fowleri is a free-living amoeba found in fresh warm water. It is a very uncommon cause of meningitis usually occurring in otherwise immunocompetent young adults several days after swimming in a lake; it is, with exception, fatal.

PATHOGENESIS

MENINGITIS

Meningococcus and *Haemophilus influenzae* b are exogenous colonizers of the nasopharynx and, under certain conditions, enter the blood. This ability to enter the blood depends on various host and organism factors, as shown in **Figure 14.5**. It is probable that meningococcus loses its capsule in order to be effectively transported across the epithelium. *Haemophilus* appears to cross the epithelium by traversing the junction between cells. Once in the blood these bacteria must be 're-capsulated' in order to reduce any chance of being phagocytosed.

In the capillary system of the brain, bacteria can attach to the endothelium by adhesins. A few organisms enter the subarachnoid space and the process of meningitis is initiated. The example discussed here is *Neisseria meningitidis* (**Figure 14.6**). These few bacteria initially have an advantage as CSF has low levels of complement and immunoglobulin (Ig) G.

Cytokines produced by the few activated macrophages and T cells in the CSF activate endothelial cells to express a surface selectin protein and secrete interleukin (IL) 8, which stimulates neutrophils within the venule to express integrin proteins. Neutrophils are arrested by the selectin/integrin reaction and crawl between the endothelial cells to reach the CSF, to be primed to consume and destroy the meningococcus. The relative increase in the number of white blood cells (WBC) can be massive, from 1–2/µL to over 1000/µL. In addition, loosening of the intercellular junction that enables entry of neutrophils also allows albumin and other plasma proteins to flow into the CSF. The end result is the typical picture of bacterial meningitis, with large numbers of bacteria, neutrophils and a raised protein level. Metabolism of glucose by WBC and bacteria results in significant lowering of CSF glucose. It is the intense inflammation adjacent to the arachnoid

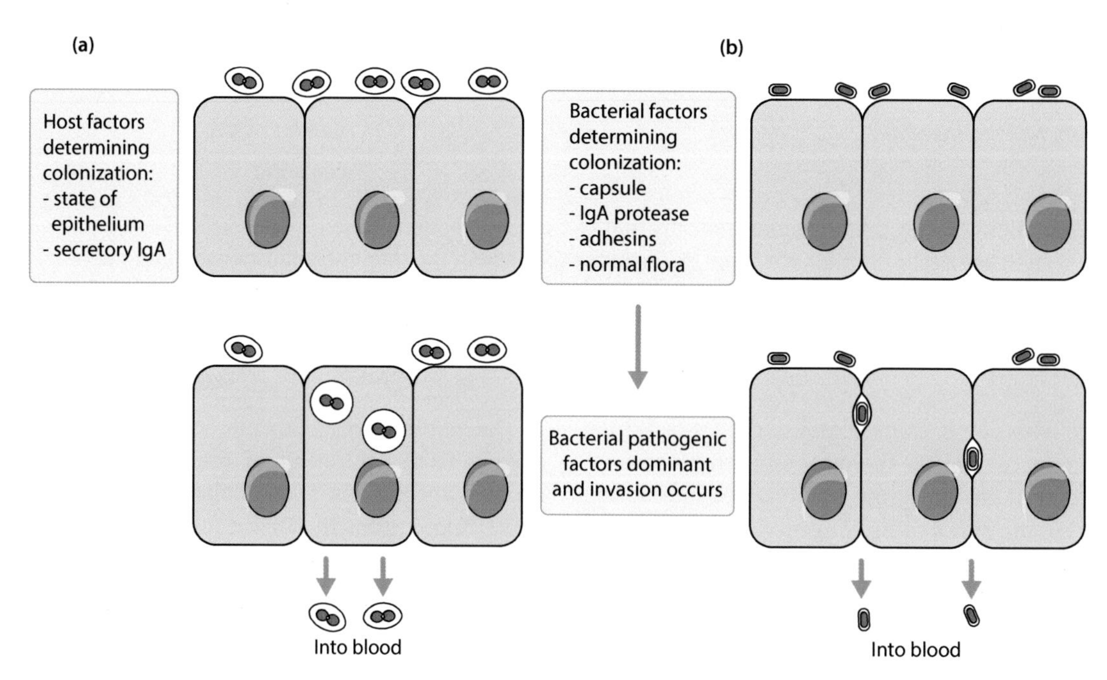

Figure 14.5 Bacteria such as (**a**) meningococcus and (**b**) *Haemophilus influenzae* b enter the blood from the nasopharynx. Various host and bacterial factors influence the process.

mater and pia mater that gives rise to symptoms of headache, photophobia, neck stiffness and seizures.

Via a viraemia, viruses replicate in the endothelial cells first, or cross directly to reach the glial cells surrounding the neurons or the neurons themselves. In the CSF a macrophage–lymphocyte response occurs. As neutrophil recruitment does not occur, there is limited disruption of the blood–CSF barrier, so influx of albumin from the blood is limited and CSF glucose levels are normal or marginally reduced.

Cerebral malaria is one of a number of serious complications of *Plasmodium falciparum* infection. This centres on the sequestration of infected erythrocytes in the microvasculature of organs such as the brain. The *Plasmodium falciparum* erythrocyte membrane protein 1 is anchored in the surface membrane of the red blood cell (RBC). In the capillary system this increases adherence of infected RBC to endothelial surface proteins such as CD36. Local cytokine activation upgrades expression of adhesion proteins, and increases RBC sequestration, neutrophil accumulation and platelet activation. There is compromise of delivery of oxygen and glucose to cells, as well as removal of waste metabolites, which results in the symptoms of cerebral malaria.

BRAIN ABSCESS

Once bacteria enter brain tissue, for example from an infected embolus lodged in a small vessel, a cerebritis will develop over a number of days as the inflammatory response attempts to control bacterial growth. Progression to a central area of necrosis results in an abscess, which becomes enclosed in a collagen capsule laid down by fibroblasts. Around this expanding abscess is a zone of tissue oedema and the whole structure exerts a mass effect, compressing the brain against its fibrous supports and

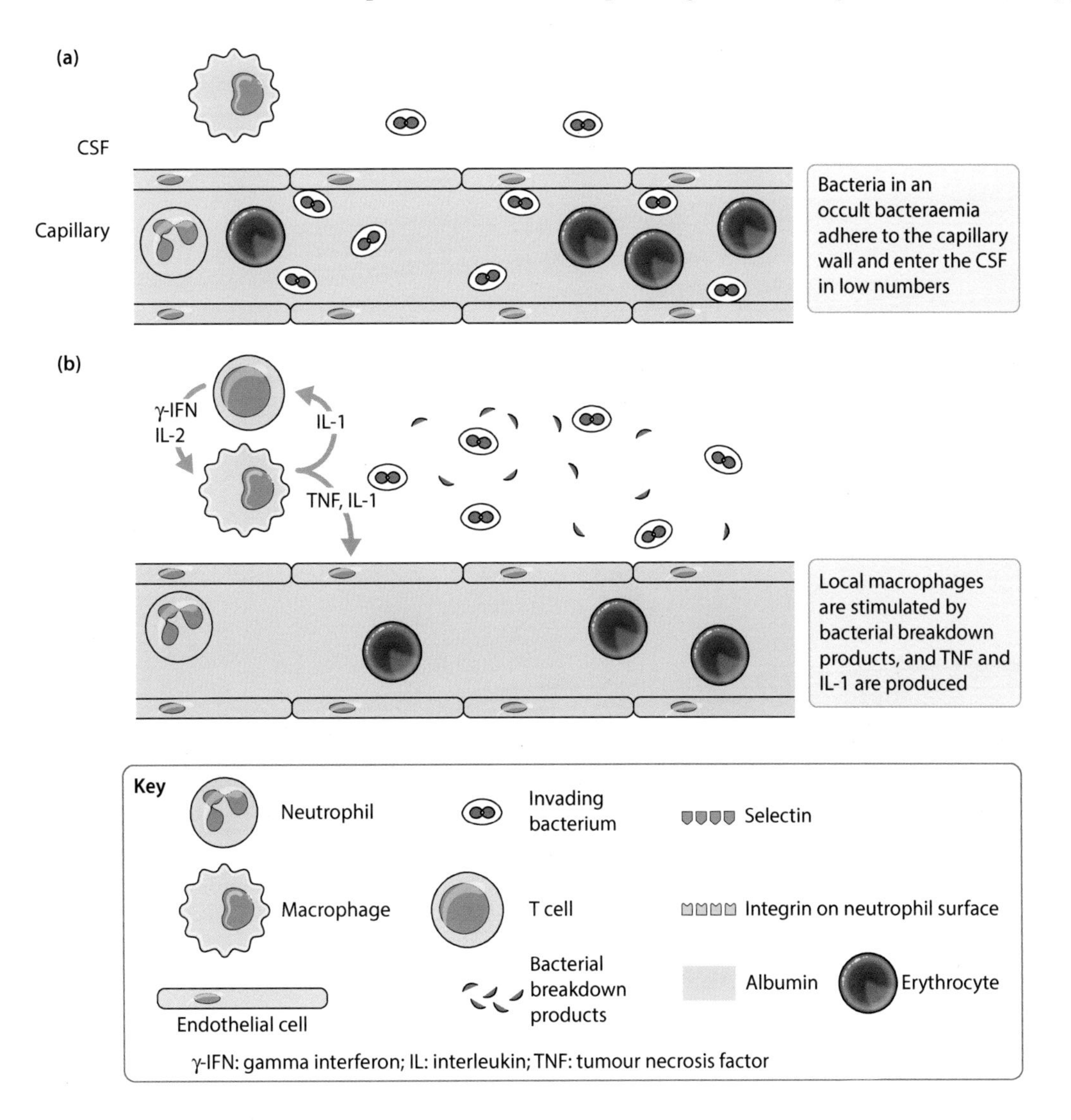

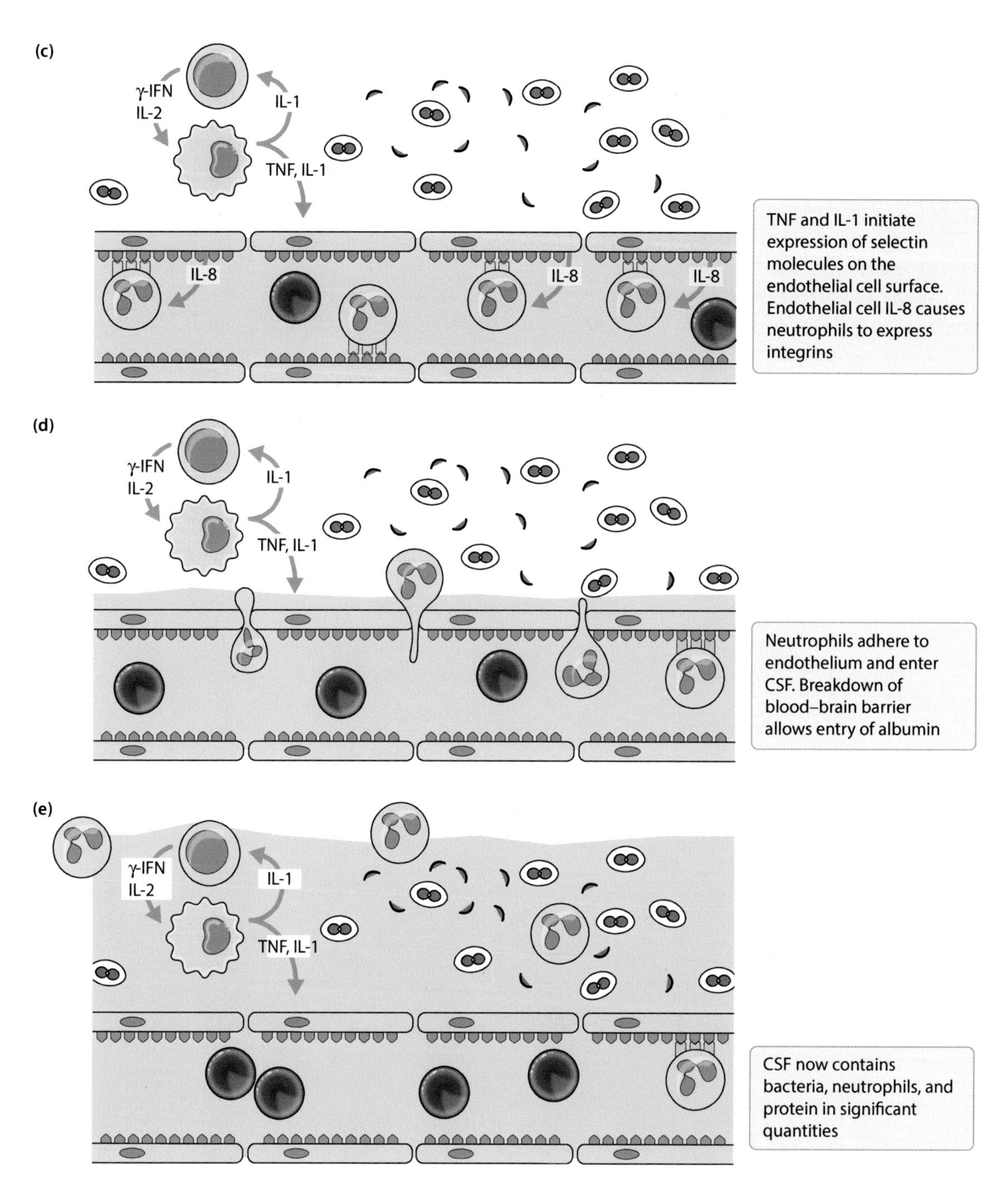

Figure 14.6 A simplified outline of the stages that occur in meningococcal meningitis, which take place over a period of 12–24 hours. (**a**) Entry into the cerebrospinal fluid (CSF); (**b**) stimulation of the immune response; (**c**) margination of neutrophils; (**d**) entry of neutrophils and albumin into the CSF; (**e**) the last stage following entry of neutrophils and albumin into the CSF.

the skull (**Figure 14.7**). A lumbar puncture (LP) should not be done in these circumstances. Removing fluid from the subarachnoid space in the spine will increase the pressure difference between the brain and the cord, exacerbating any pressure effect. The resulting herniation of the brain can have catastrophic consequences.

More centrally situated abscesses can rupture into the ventricles. The blood supply to the central white matter is less than elsewhere in the brain and the action of the fibroblasts in laying down the collagen capsule is incomplete. The weaker medial side of the capsule can then rupture, releasing pus into the ventricles.

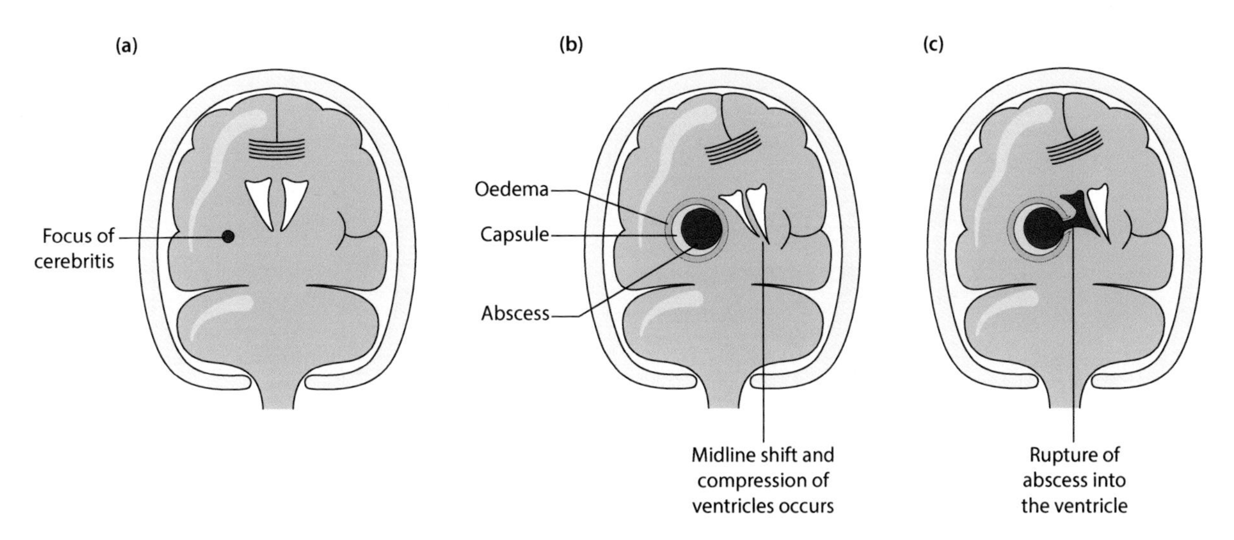

Figure 14.7 (**a**) A focus of cerebritis. (**b**) This progresses to a brain abscess, which is surrounded by a collagen capsule and tissue oedema. (**c**) An abscess can rupture into the ventricle.

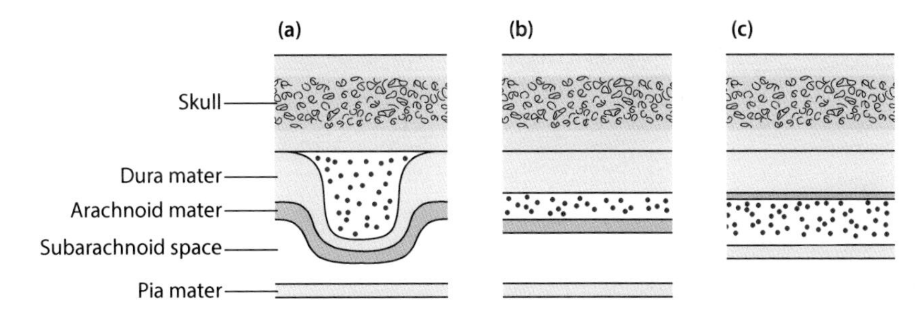

Figure 14.8 Infection in the connective tissue layers surrounding the brain and spinal cord can give rise to: (**a**) extradural; (**b**) subdural; and (**c**) subarachnoid (meningeal) infections.

SUBDURAL AND EXTRADURAL INFECTIONS

Bacteria can enter both the subdural and extradural spaces (**Figure 14.8**). In the subdural space, organisms can spread easily in the potential space between the arachnoid and dura mater. Extradural infections tend to be more localized because of fibrous attachments of the dura to the skull. Infections can also occur in relation to the spinal cord (**Figure 14.9**). In both extradural and subdural infections, the mass effect at the site of the infection can give rise to compression of the spinal cord, creating a neurosurgical emergency.

VENTRICULOPERITONEAL SHUNT AND EXTERNAL VENTRICULAR DRAIN INFECTIONS

VP shunts are inserted to relieve hydrocephalus. Non-communicating hydrocephalus may result from a congenital abnormality or it can be acquired. Fibrosis and subsequent blockage of the CSF outflow channels can develop as a result of an intraventricular haemorrhage in a premature neonate, or following bacterial ventriculitis. In order to relieve the hydrocephalus, a shunt is inserted into a ventricle, which exits the skull, and via a subcutaneous route, drains CSF into the peritoneum. Organisms may be introduced at the time of operation and initiate shunt infection some time later. Usually skin bacteria such as coagulase-negative staphylococci are involved. Patients may present with a headache and low-grade fever, and may have symptoms and signs of peritonitis, indicating colonization of the shunt throughout its course (**Figure 14.10**).

External ventricular drains (EVDs) are inserted during neurosurgical operations following removal of space-occupying lesions or head trauma. These EVD are used to monitor intracranial pressure and relieve hydrocephalus by draining CSF. Regular sampling of EVD CSF samples is performed to monitor the white cell count and differential, and determine the appearance of bacteria or yeasts to guide appropriate treatment.

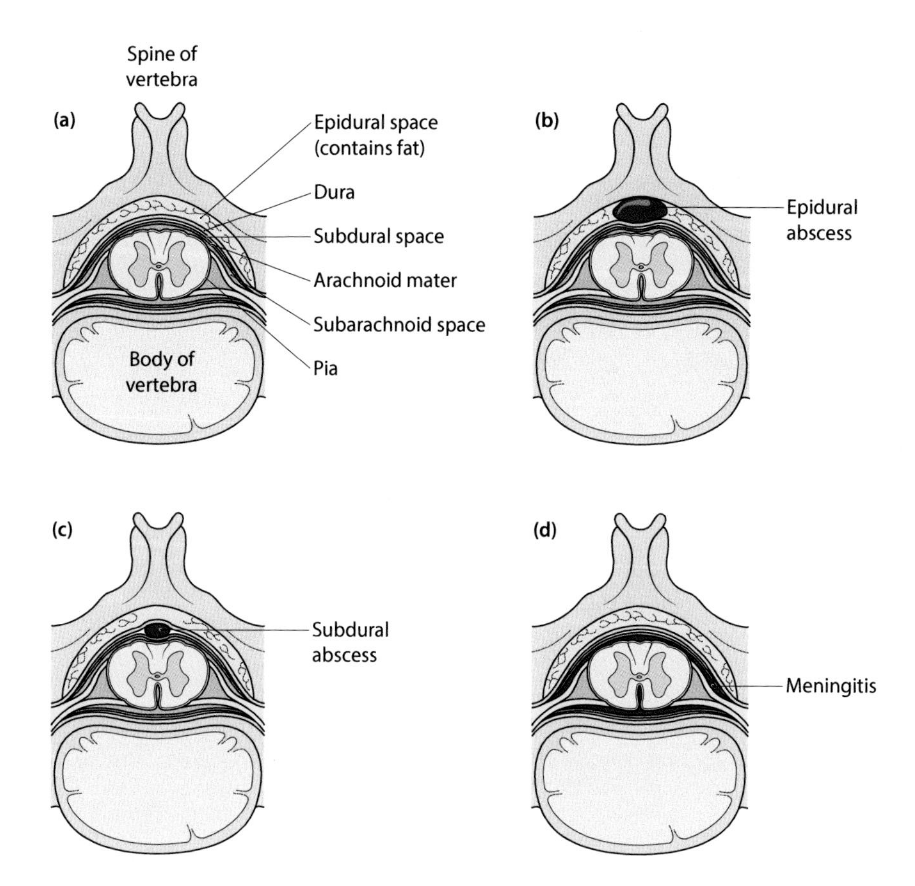

Figure 14.9 (**a**) Outline of the structure of the spinal column; (**b**) epidural; (**c**) subdural; and (**d**) subarachnoid (meningeal) infections.

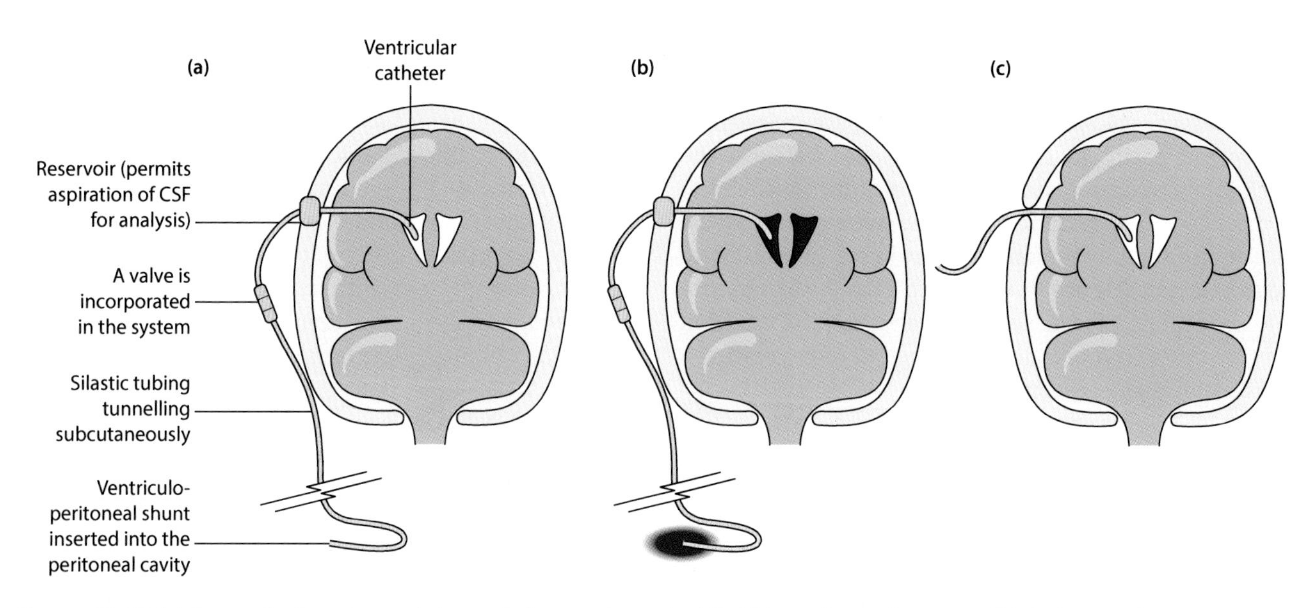

Figure 14.10 (**a**) A ventriculoperitoneal shunt drains cerebrospinal fluid into the peritoneum. (**b**) A shunt infection can extend from the ventricles to the peritoneum. (**c**) Usually the ventricular portion is externalized.

TOXIN-MEDIATED DISEASES

Although tetanus and botulism are uncommon infections, they are considered here because of their unique toxin-mediated pathogenesis. Botulism is usually food borne, while tetanus is associated with soft-tissue infection, where conditions are suitably anaerobic to enable inoculated spores from soil to germinate. This infection should be considered in the elderly patient who was vaccinated decades earlier, and, with waning immunity, presents with symptoms such as trismus ('lockjaw'), risus sardonicus of the face muscles and abdominal rigidity.

TETANUS

Clostridium tetani is a spore forming, motile gram-positive obligate anaerobe. The spore is very stable and can survive in soil and animal manure for years. When anaerobic conditions exist in a soft-tissue injury, the spores germinate and the vegetative cells produce the toxin. An outline of the action of the toxin is shown in **Figure 14.11**. The active component enters the terminals of lower motor neurons, where it inhibits neurotransmitter release. It also travels in a retrograde direction to reach the cell bodies and terminals of inhibitory cells in the CNS.

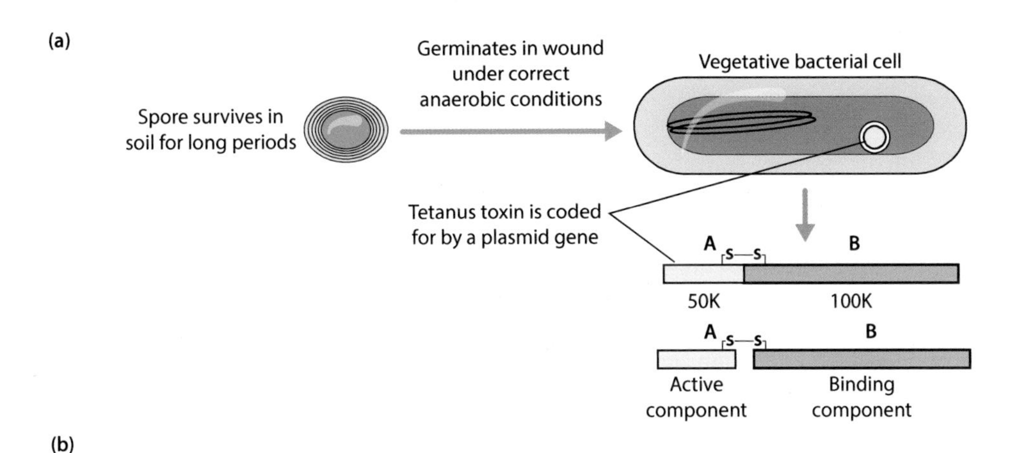

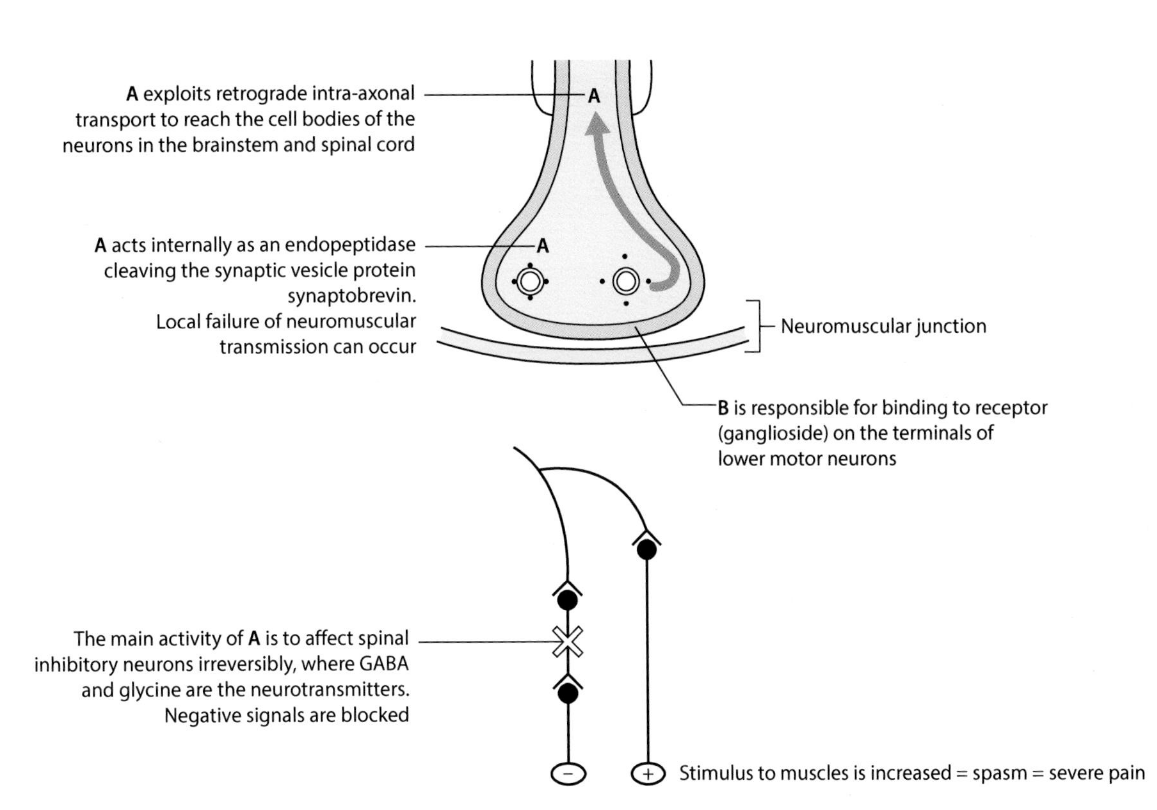

Figure 14.11 Some features of (**a**) the tetanus toxin and (**b**) its mode of action. (GABA: γ-amino-n-butyric acid.)

By irreversibly inhibiting the action of glycine and γ-amino-n-butyric acid (GABA)-mediated inhibitory neurons, the motor neurons are left relatively unaffected and spastic paralysis occurs.

BOTULISM

Botulism is an intoxication caused by *Clostridium botulinum*. The bacterial spores survive cooking, then germinate in food that is stored incorrectly, producing toxin. When the food is consumed the toxin is absorbed into the blood. The active component of the toxin inhibits the release of acetylcholine at the neuromuscular junction and flaccid paralysis arises (**Figure 14.12**).

DIAGNOSIS AND MANAGEMENT

The clinical diagnosis of infection in the CNS may be straightforward. The patient with fever, headache, neck stiffness, vomiting, photophobia and the typical rash of meningococcal infection is likely to have sepsis and meningitis caused by *Neisseria meningitidis*. However, on many occasions the diagnosis is not straightforward. Patients present with new confusion or increased confusion, or behavioural changes considered to be compatible with a viral meningitis. On occasion the patient with a CNS infection caused by *Listeria monocytogenes* can present with symptoms similar to herpes encephalitis. Seizures can be the presentation of an uncommon infection, neurocysticercosis caused by the larval form of *Taenia solium* in the wrong host.

In the setting of bacterial meningitis, the collection of blood cultures before antibiotics are given is of critical importance, as the causative organism is frequently isolated from blood. If the LP is delayed, antibiotics will affect the likelihood of isolating the organism from CSF.

Radiological investigations such as computed tomography (CT) and magnetic resonance imaging (MRI) scans are an essential part of diagnosis and can determine the degree of brain swelling in meningitis, as well as the size and site of a brain abscess. These procedures are also used to rule out mass lesions before a LP is conducted.

LABORATORY DIAGNOSIS

The need to perform a LP can be different for various age groups. When meningitis is considered in a previously healthy young child, a LP may not be

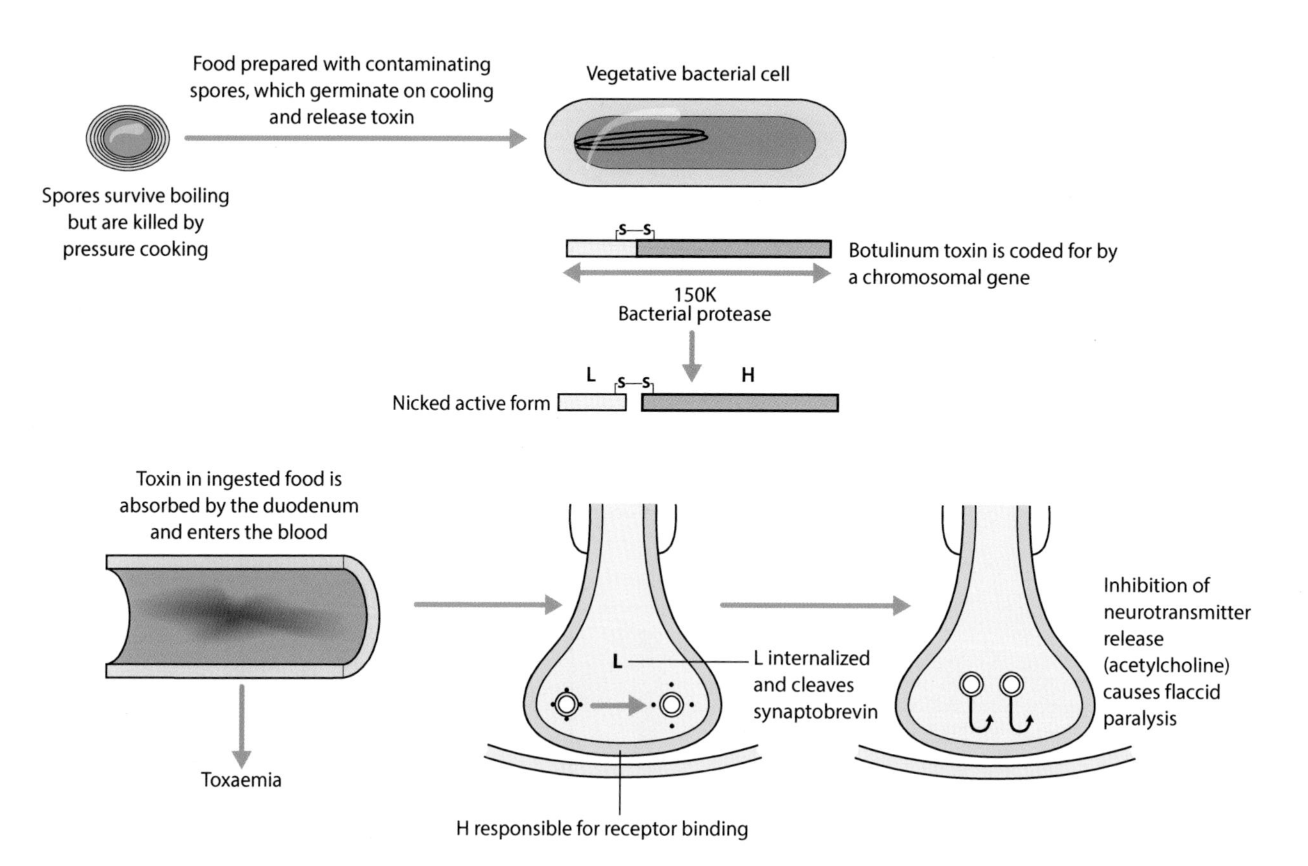

Figure 14.12 Some features of (**a**) the botulism toxin and (**b**) its mode of action.

done, and following collection of a blood culture, and ethylenediamine tetra-acetic acid (EDTA) blood for polymerase chain reaction (PCR), an antibiotic active against meningococcus, pneumococcus and *Haemophilus*, usually cefotaxime or ceftriaxone, is administered. In adults an LP is always collected, unless clinical assessment and imaging direct otherwise, so that the appropriate range of diagnostic tests can be done.

It is essential once CSF or pus from an abscess has been collected that it is sent to the laboratory immediately for microscopy and culture. It is usual to collect three samples of CSF; the first two are sent for protein and glucose determination and the third to the microbiology department. Microscopical and biochemical parameters used in the interpretation of CSF specimens are shown in **Figure 14.13**. The differentiation between normal CSF values and those representative of bacterial, viral and tuberculous meningitis is useful. In bacterial and tuberculous meningitis, the protein is raised and the glucose reduced. Neutrophils are the predominant cell type in bacterial meningitis and monocytes in tuberculous meningitis. In viral meningitis, a normal glucose and

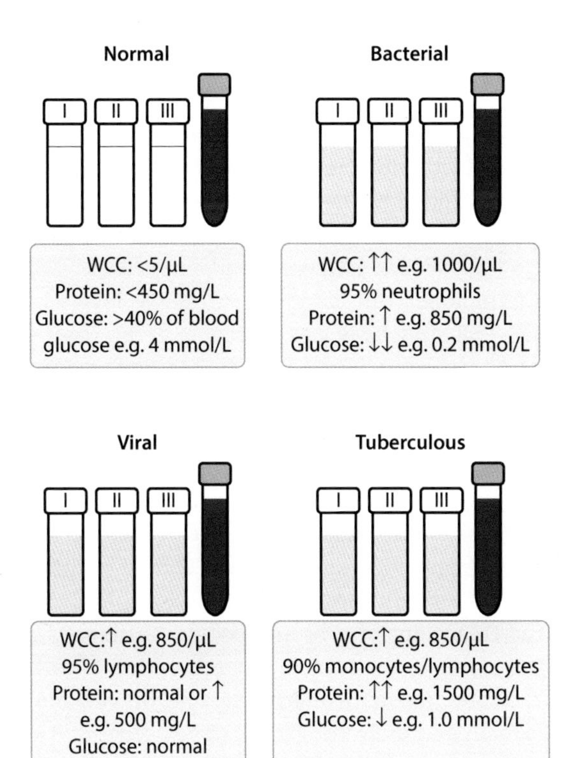

Figure 14.13 When collecting cerebrospinal fluid (CSF), three specimens should be taken, as well as blood for glucose determination. The features of a normal, bacterial, tuberculous and viral CSF are shown.

moderately raised protein in the setting of a low to moderate lymphocytosis is the usual finding.

The Gram stain of the CSF can confirm the diagnosis within a matter of minutes, and direct the necessary treatment (Chapter 1, **Figure 1.5**). If present, this stain will also show yeasts, and if so, the India ink stain for *Cryptococcus* is done. The large capsule of these yeasts shows up clearly against the black background of the ink (**Figure 14.14**). This result will provide the earliest identification of the organism, and is an alert for human immunodeficiency virus (HIV) infection.

If the Gram stain does not reveal organisms, the sample is processed by PCR, which includes meningococcus, pneumococcus, HSV1, HSV2, VZV and enterovirus. In the setting of tuberculous meningitis, stains for mycobacteria are seldom, if ever, positive. Taking into account the length of time that *Mycobacterium tuberculosis* takes to grow in the laboratory, the use of the mycobacterial PCR must be discussed with the microbiologist.

It is important that the clinical team liaise with the microbiologist to ensure that all the necessary tests are considered for what is a precious specimen. The volume of CSF may not be adequate for all the tests requested. 'Left over' samples in the separate Blood Sciences department often need to be retrieved.

CT and MRI scans are essential in the diagnosis of brain and other CNS abscesses. Scanning enables the neurosurgeon to visualize and enter the abscess to drain fluid, relieving intracranial pressure. The specimen obtained is vital for microbiological and histopathological examination. CT scans are also used to determine if an abscess is responding to treatment (**Figure 14.15**).

The diagnosis of tetanus and botulism is outlined in **Figure 14.16** and **Figure 14.17**, respectively. Both are essentially clinical diagnoses, but in the case of botulism the toxin may be identified in blood and in suspected food.

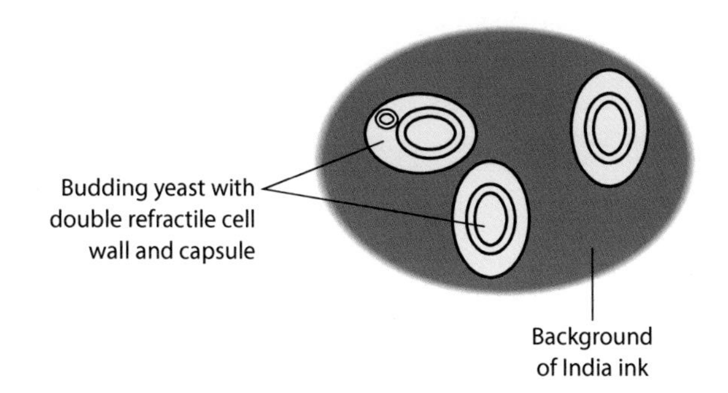

Figure 14.14 The capsule of *Cryptococcus neoformans* shows up clearly when cerebrospinal fluid is stained with India ink.

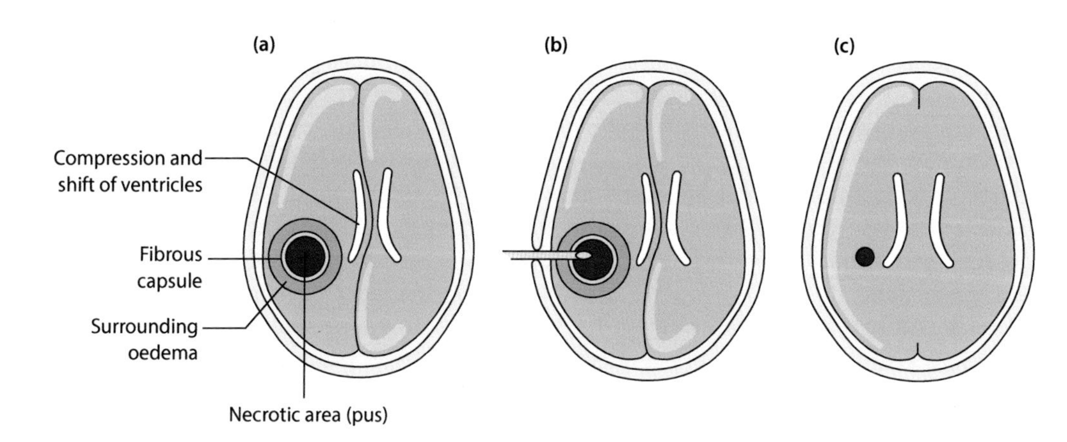

Figure 14.15 Computed tomography scans are important in: (**a**) the diagnosis; (**b**) surgical intervention; (**c**) follow-up management of a brain abscess.

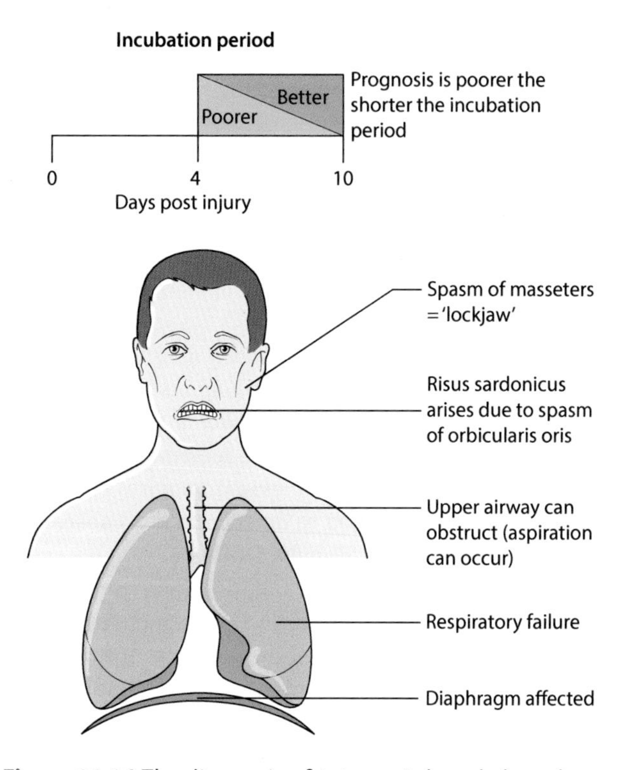

Figure 14.16 The diagnosis of tetanus is largely based on clinical evidence.

TREATMENT

The agents used to treat meningitis need to be able to cross the blood–CSF barrier and achieve therapeutic levels there. The antibiotics named in the 'mock' guidelines (Chapter 4) are examples of such agents. For 'community-acquired' meningitis, pneumococcus, meningococcus and, in the unvaccinated child less than 5 years of age, *Haemophilus influenzae* b would be considered. The third-generation cephalosporins, cefotaxime or ceftriaxone, achieve reliable levels in the CSF. Ceftriaxone is appealing as the regime for the adult patient is 2 g q12h. Depending on local susceptibility patterns, benzylpenicillin is also useful for meningococcus and pneumococcus.

Taking into account relevant allergies, benzylpenicillin or amoxycillin should always be included when *Listeria monocytogenes* is a consideration, for example in the neonate or immunocompromised adult; this organism is inherently resistant to cephalosporins. It should be noted with *Listeri*a and *Streptococcus agalactiae* that gentamicin is given with the β-lactam. This aminoglycoside is usually included as the second agent for synergy. This is primarily to treat the infection in the blood, as penetration into the CSF is average at best.

The quinolones can achieve reasonable levels in the inflamed CSF, and have been used to treat meningitis caused by gram-negative bacteria including *Pseudomonas aeruginosa*. These should be considered when there is no appropriate β-lactam, for example. The length of therapy used in the treatment of bacterial meningitis can range from 7 days for meningococcus to 3 weeks for *Listeria monocytogenes*. The causative agent can include viruses such as HSV and VZV, and aciclovir should be given at least until PCR results are available.

In the treatment of a brain abscess, antibiotics must cross the blood–brain barrier. A combination of ceftriaxone or cefotaxime with metronidazole or vancomycin, ciprofloxacin and metronidazole can be used empirically. These regimes are tailored once the results of bacteriological investigations are available, and specific bacteria can be targeted. Treatment is given for a minimum of 6 weeks. Vancomycin predose levels must be maintained between 10 and 20 mg/L (preferably 15–20 mg/L).

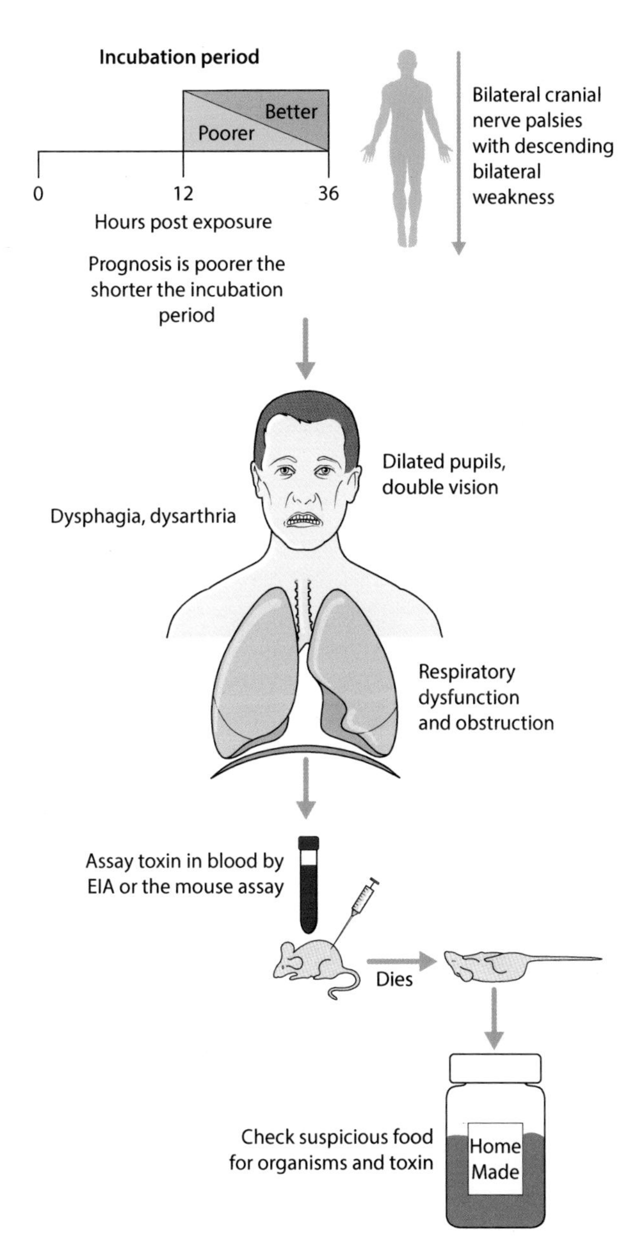

Figure 14.17 In addition to clinical evidence, the presence of botulism toxin can be confirmed in the blood of patients and in the likely food source, from where the organism can be isolated.

In VP shunt and EVD infections, vancomycin and meropenem are the combination frequently used initially, and the regime tailored according to the organisms isolated. In VP shunt infections this requires externalization of the shunt. Often most of the shunt has to be removed, with the part draining the ventricles being externalized.

Where there is access to the ventricles via an EVD, vancomycin and gentamicin are directly administered by the intrathecal route. The daily dosing regime is based on the volume of CSF being drained and the size of the ventricles. These regimes involve ongoing communication between the neurosurgery team and microbiologist.

With toxin-mediated disease, ventilator support is the mainstay. With tetanus, surgical debridement of infected tissues is done, and benzylpenicillin or metronidazole administered.

Chapter 15

Infections of the Eye

INTRODUCTION

Conjunctivitis is a common disease, and can be caused by infection, allergy or chemical irritation. Infectious conjunctivitis affects all age groups and resolution usually occurs following application of topical antibiotics. Infections occur in other parts of the eye, including the cornea, as well as the anterior and posterior chambers. Structures such as the orbit and cavernous sinuses can be involved, and because of their proximity to the brain, life-threatening illness can occur. Any infection of the eye or its associated structures in the setting of pain and vision loss is an emergency and the ophthalmologist needs to be consulted. The gross anatomy of the eye and its adjacent structures is shown in **Figure 15.1**.

Trachoma, which is usually found in lower-income parts of the world, is caused by the A, B, C serovars of *Chlamydia trachomatis*. It is estimated that at least 50 million people are affected, and 1 million are blinded. The organism is spread to the eye by the hands and flies, and the chronic conjunctivitis that arises leads to scarring, corneal ulceration and subsequent vision loss. Serovars B, Ba, D–K are associated with sexually-transmitted infection and neonatal conjunctivitis. The newborn acquires the organism from the cervix of the infected mother during vaginal delivery.

In West, Central and East Africa and parts of Central America, river blindness is a disease that is estimated to occur in at least 20 million individuals. It is caused by the microfilaria of the parasite *Onchocerca volvulus*. This parasite is spread by the blood-sucking simulid blackfly. The microfilarial stage of the parasite invades the anterior chamber of the eye, with corneal ulceration and fibrosis, leading to blindness. The antiparasite drug ivermectin has had a significant effect in treating the condition in recent years.

ORGANISMS

The organisms to consider in eye infections are shown in **Figure 15.2**. This shows bacteria that are either commensal, colonizing or exogenous. *Pseudomonas aeruginosa*

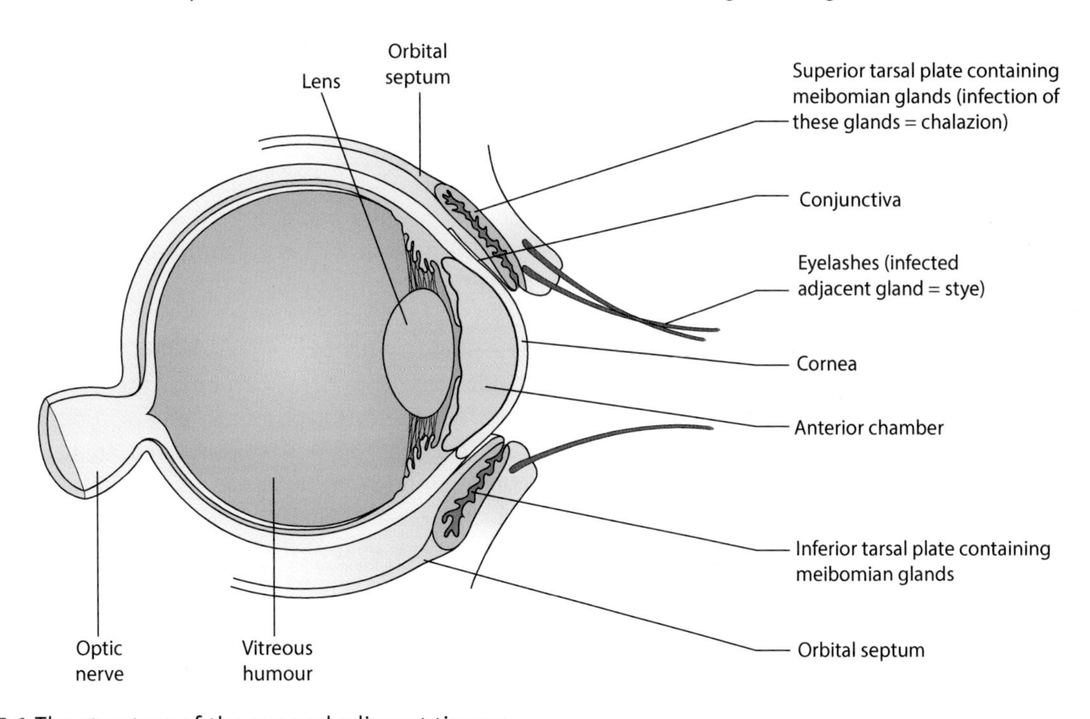

Figure 15.1 The structure of the eye and adjacent tissues.

Conjunctivitis	**Keratitis (cornea)**	**Endophthalmitis**
		Post cataract surgery
Staphylococcus aureus	***Staphylococcus aureus***	**Coagulase-negative staphylococci**
Streptococcus pneumoniae	***Streptococcus pneumoniae***	***Enterococcus faecalis***
Streptococcus pyogenes	***Streptococcus pyogenes***	***Enterococcus faecium***
Haemophilus influenzae	*Pseudomonas aeruginosa* **(CLW)**	*Pseudomonas aeruginosa*
Neisseria gonorrhoeae	*Neisseria meningitidis*	**Post traumatic**
Chlamydia trachomatis **(A, B, Ba, C: trachoma)**	Adenovirus	*Bacillus cereus*
Chlamydia trachomatis **(B, Ba, D–K: STI)**	**Enterovirus 70**	**Coagulase-negative staphylococci**
Adenovirus	HSV	**Environmental gram-negatives**
Enterovirus 70	VZV	
HSV		**Endogenous source (IE, abscess, CVC)**
VZV	*Candida*	***Staphylococcus aureus***
	Aspergillus	***Streptococcus anginosus***
		Streptococcus pneumoniae
		Klebsiella pneumoniae **(e.g. liver abscess)**
	Acanthamoeba **(CLW)**	*Candida*
		Aspergillus
	Onchocerca volvulus	

Figure 15.2 The range of organisms associated with infections of the eye. (CLW: contact lens wearer; IE: infective endocarditis.)

and the protozoal parasite *Acanthamoeba* are acquired from domestic water supplies, and can cause keratitis in the contact lens wearer (CLW). When *Pseudomonas aeruginosa* is identified in specimens from the hospitalized patient, a likely water source needs to be considered.

PATHOGENESIS

Bacterial conjunctivitis is characterized by an inflamed red eye and associated discharge. Severe infection with significant eyelid oedema, extreme hyperaemia and a profuse purulent discharge needs special mention. In this setting, bacteria such as *Neisseria gonorrhoeae* and *Neisseria meningitidis* may be involved, and the massive release of lytic enzymes from dead and dying neutrophils damages the cornea. Ulceration and perforation of the cornea here, or following trauma, allow organisms to reach the anterior chamber and from there the vitreous humour, producing endophthalmitis (**Figure 15.3**). In the hospital setting, *Pseudomonas aeruginosa* can also initiate an aggressive keratitis, and can be of particular importance in the intensive care unit (ICU) patient who has had superficial trauma to the cornea during their management. With keratitis there is usually some degree of pain and vision loss; the ICU patient will not usually be able to relay these symptoms.

Endophthalmitis is infection of the vitreous humour, associated with pain, headache, photophobia and vision loss. Organisms can enter the posterior eye following penetrating trauma, and soil-derived organisms such as *Bacillus cereus* need to be considered. When penetrating injury is caused by plant material, unusual environmental bacteria can be isolated. In postsurgical endophthalmitis, the coagulase-negative staphylococci are particularly relevant. Organisms can also reach the posterior chamber via the blood and an intravascular source such as endocarditis is likely. A candidaemia arising from an infected central venous catheter is another example, as is a liver abscess caused

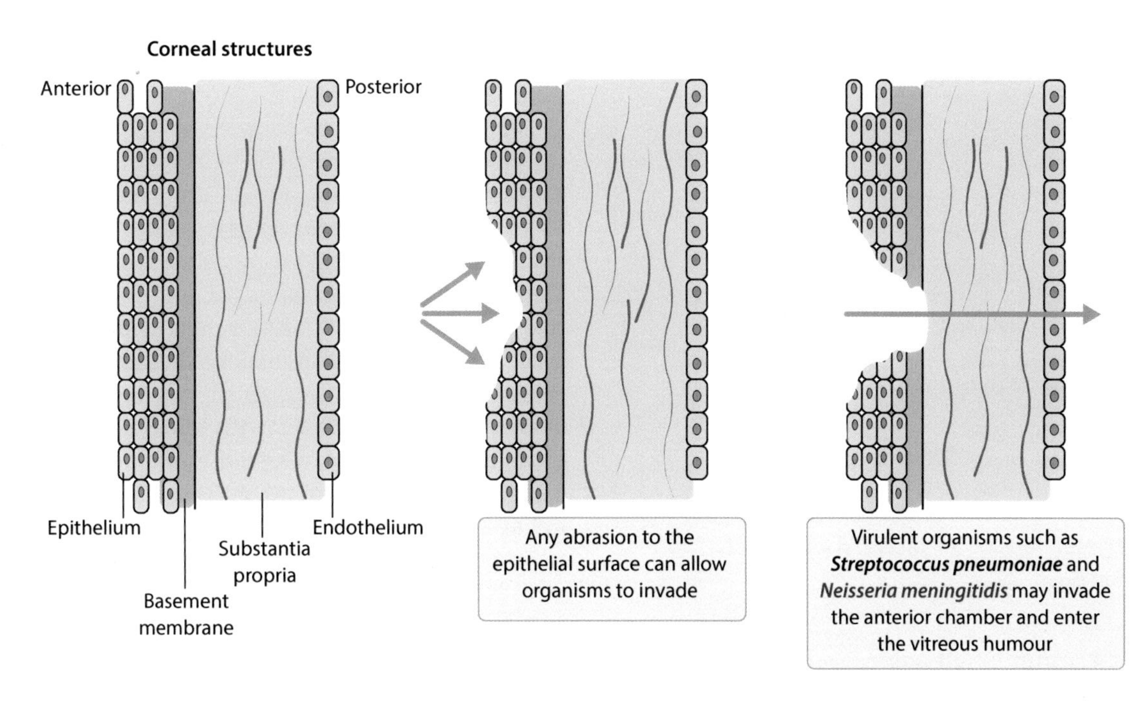

Figure 15.3 Abrasion of the epithelium allows organisms to invade the cornea and beyond.

by *Klebsiella pneumoniae*. The ophthalmologist must be contacted promptly.

Orbital cellulitis and cavernous sinus thrombosis may be difficult to distinguish from each other. Cellulitis usually arises as an extension of infection in a nasal sinus into the orbit. The tissue separating the orbit from the ethmoid air cells is thin, and can be easily damaged by trauma. Orbital cellulitis is associated with dark red skin over the eye, headache and fever. Pain associated with movement of the eye is also present, reflecting the inflammation in soft tissue that surrounds the eye within the orbital cavity. Because of the space restrictions within the cavity, forward displacement or proptosis of the eye occurs. Cavernous sinus thrombosis arises as a result of extension of thrombophlebitis in the facial veins. The patient is severely ill with headache, fever and chills. While the condition may be difficult to distinguish from orbital cellulitis, palsies of the IIIrd, IVth, and VIth cranial nerves, and bilateral signs may be helpful in the diagnosis. The main structures of the cavernous sinus are shown in **Figure 15.4**.

Eyelid abscesses such as a stye or chalazion are suppurative infections of the glands of the lid; infections of the entire lid are usually unilateral, and occur anterior to the orbital septum (**Figure 15.1**). While pain, swelling and erythema occur, movement of the eye is usually normal and pain free, distinguishing this entity from orbital cellulitis.

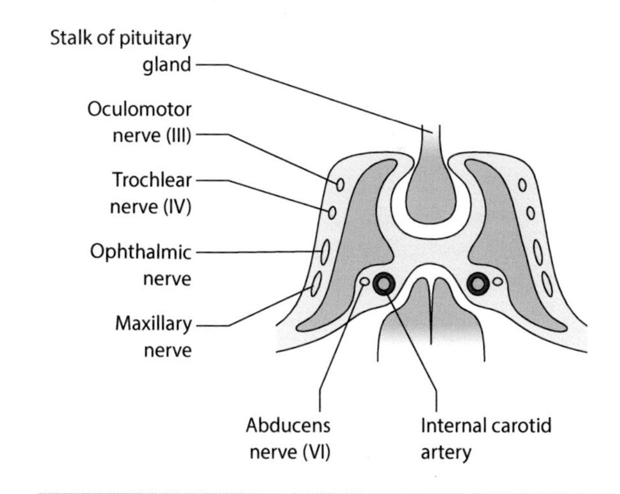

Figure 15.4 The structure of the cavernous sinus and some of the important structures that lie in it.

Dacrocystitis is infection of the lachrymal gland, and canaliculitis is infection of the nasolachrymal duct (**Figure 15.5**). While common pathogens such as pneumococcus and *Streptococcus pyogenes* are usually responsible, chronic infections can be caused by *Actinomyces israelii*. Inflammation in the duct produces

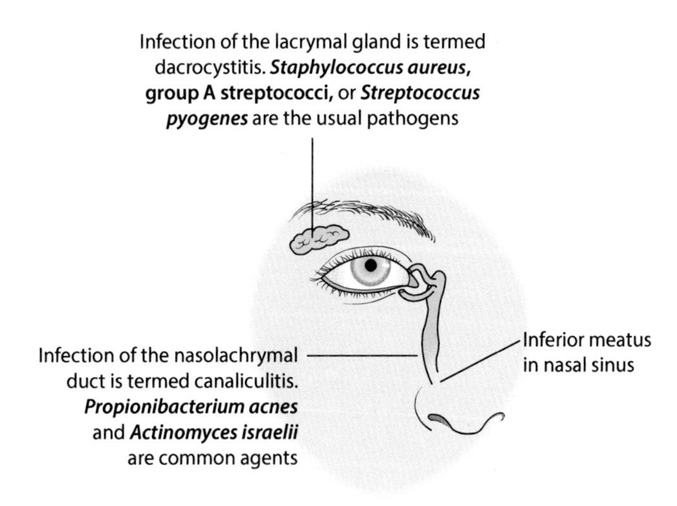

Figure 15.5 Infections of the lacrymal gland and nasolacrymal duct.

a nodule, consisting of an outer layer of fibrous tissue enclosing a purulent collection of neutrophils, which themselves surround 'sulphur granules'. These granules contain conglomerations of the actinomycete bacteria, and their presence in an excised nodule is diagnostic.

DIAGNOSIS AND MANAGEMENT

The various causes of conjunctivitis may be differentiated clinically (**Figure 15.6**). The method of specimen collection is shown in **Figure 15.7**. For chlamydia, the lower eyelid is fully everted, and any pus removed, using a standard 'transport' swab, can be submitted for bacterial culture. The chlamydia polymerase chain reaction (PCR) swab is then rubbed carefully on the conjunctival surface to collect cells containing the intracellular bacteria.

For the diagnosis of keratitis it is essential to refer the patient to the ophthalmologist; with the appropriate anaesthesia, corneal scrapings are collected. These are used to prepare a smear on glass slides for Gram staining, and are also used to inoculate a range of culture media, including chocolate agar, and Sabouraud's selective agar for the isolation of fungi. In the case of CLWs, corneal scrapes are examined for the characteristic cysts of *Acanthamoeba*. This organism can be cultured in the laboratory, by providing it with killed 'coliform' bacteria on an agar plate to scavenge; the presence of the motile amoebae and characteristic cysts is observed. PCR testing at independent or NHS/Public Health England laboratories is more practical for both *Acanthamoeba* and the viruses cited in **Figure 15.2**.

It is possible to differentiate preseptal from orbital cellulitis on clinical grounds, for in the latter situation there is proptosis of the eye and movement is painful (**Figure 15.8**). In the setting of endophthalmitis, aspirated vitreous fluid is sent promptly to the laboratory for microbiological investigation. In all cases of suspected endophthalmitis, orbital cellulitis and cavernous sinus thrombosis, blood cultures are collected.

TREATMENT

Because management of eye infections is the domain of the ophthalmologist, examples of the range of treatments used are given here for information only. Apart from uncomplicated conjunctivitis, all other cases should be referred to the ophthalmologist.

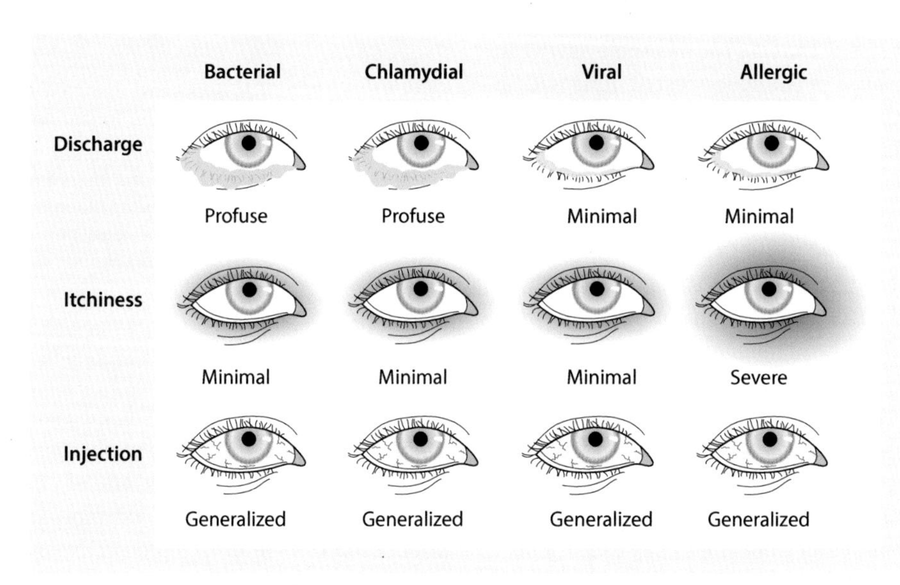

Figure 15.6 Some of the clinical features of bacterial, chlamydial, viral and allergic conjunctivitis.

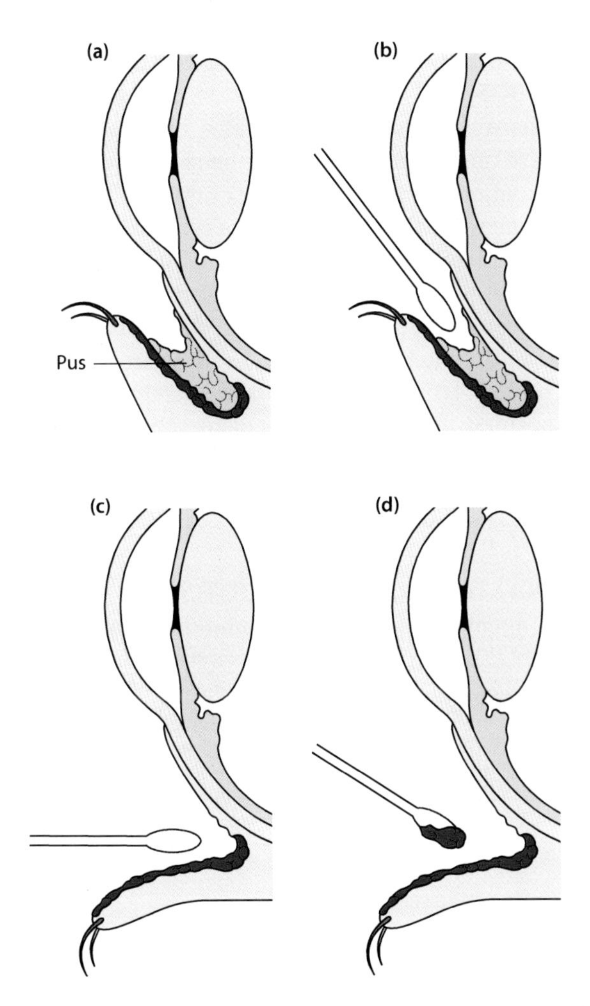

Figure 15.7 (**a**) When there is pus present, (**b**) a standard swab is used to collect the specimen for microscopy culture and sensitivity; (**c**) after removal of any residual pus with sterile gauze the chlamydial/gonococcal polymerase chain reaction swab is used; (**d**) this is rubbed on the palpebral conjunctival surface to obtain cells that will contain the intracellular *Chlamydia*.

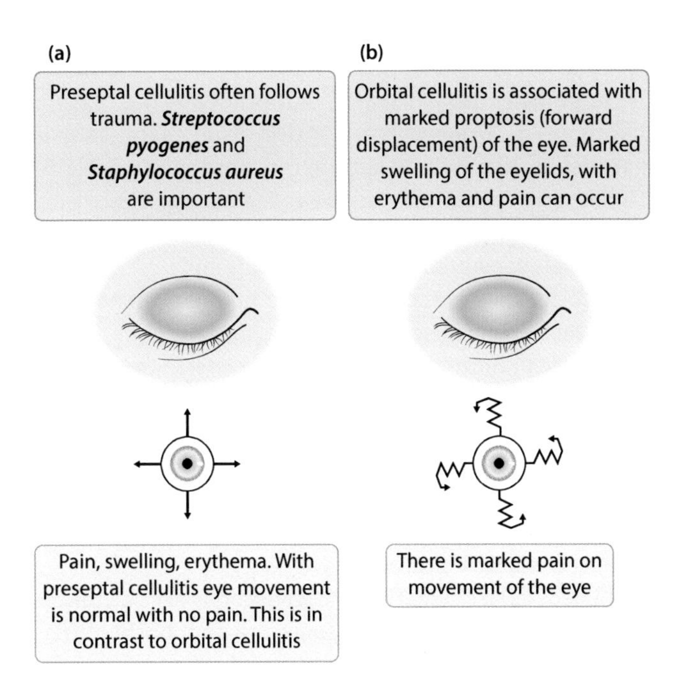

Figure 15.8 It is possible to differentiate (**a**) preseptal from (**b**) orbital cellulitis.

CONJUNCTIVITIS

In most instances topical eye drops or ointment can be used to treat acute bacterial conjunctivitis. In the UK chloramphenicol is widely used; it has broad-spectrum activity and is useful for most of the common bacteria considered, with the exception of *Pseudomonas*. The opinion that chloramphenicol should be avoided because of the risk of aplastic anaemia is not considered to be well founded. Aminoglycosides are also useful and have activity against *Pseudomonas*, as do quinolones, for example ciprofloxacin and ofloxacin. The aminoglycosides and these quinolones are considered to have equivocal activity against streptococci, but with topical application the concentration of these agents is considered high enough to be effective. In Ophthalmology departments, topical cefuroxime and gentamicin can be used as these cover most of the common bacterial pathogens. In the adult, gonococcal conjunctivitis can be treated with systemic antibiotics and a single 500 mg dose of intramuscular ceftriaxone plus 1 g oral azithromycin is used. Regular saline washes of the eye are used to remove purulent exudate.

Chlamydia infections also require systemic antibiotic therapy as the infection is not limited to the eye but can involve the lungs (neonates) and the genital tract. For adults, azithromycin can be given as two 1 g doses a week apart. Alternative options, given for 7 days, are doxycycline 100 mg q12h or levofloxacin 500 mg q24h.

KERATITIS

The antibacterial agents discussed above are used here as well. Herpetic keratitis can be treated with systemic aciclovir, while fungal keratitis can be treated with topical amphotericin B, fluconazole or itraconazole. *Acanthamoeba* keratitis is treated with topical chlorhexidine, which has activity against the trophozoites and cysts, and oral itraconazole or voriconazole are used too.

ENDOPHTHALMITIS

For postoperative infections, intravitreal injection of a glycopeptide (e.g. vancomycin) and an aminoglycoside (e.g. amikacin) is appropriate for most of the bacteria under consideration. In the setting of trauma, the organism that is isolated will influence the regime that is used. When contamination from plant or soil material is likely, it is considered prudent to include clindamycin initially to ensure cover for *Bacillus cereus*. Penetration of antibiotics into the vitreous humour from the blood is variable. Metronidazole, rifampicin, chloramphenicol, co-trimoxazole and the quinolones can achieve useful levels, but their use in this setting depends on the preference of individual ophthalmologists. In this severe infection both systemic and topical agents are used.

Where endophthalmitis has arisen as an embolic event from a condition such as infective endocarditis (IE), the antibacterial or antifungal regime must reflect the requirement of the agents to penetrate the relevant tissue to effect eradication there too.

The coagulase-negative staphylococci are common organisms associated with postsurgery endophthalmitis. The oxazolidinone linezolid is a useful agent to include when treating this infection.

ORBITAL CELLULITIS AND CAVERNOUS SINUS THROMBOSIS

Often the antibiotics used to treat a brain abscess are used here, and a combination of benzylpenicillin, cefotaxime and metronidazole is one example. Cefotaxime or ceftriaxone with clindamycin is another option. The good oral bioavailability of clindamycin and a 'once daily' or 'twice daily' regime of intravenous ceftriaxone make this a useful combination to consider in the outpatient parenteral antibiotic therapy (OPAT) setting in the patient with orbital cellulitis who can be safely discharged from hospital. At least 2 weeks of treatment should be considered. If an organism such as *Pseudomonas aeruginosa* is a consideration, the carbapenem meropenem, or the quinolone ciprofloxacin, is used.

DACROCYSTITIS AND CANALICULITIS

Topical agents that cover the common organisms are reasonable. In the setting of *Actinomyces*, the sulphur granule may have to be excised surgically.

Chapter 16

Infections of the Genital Systems

INTRODUCTION

A wide range of organisms are sexually transmitted infections (STI) and include gonococcus, *Chlamydia*, *Treponema pallidum* (syphilis), the protozoon *Trichomonas vaginalis*, hepatitis B (HBV) and human immunodeficiency virus (HIV). With the male patient presenting to the sexually-transmitted diseases (STD) clinic with a first episode of gonococcal urethritis, the diagnosis and treatment may be straightforward. However, this person must have been infected during sexual intercourse, and could have infected others. Sexual partners need to be traced in order to identify and eliminate infections in them as well.

Infections are often asymptomatic, especially in females. In the case of *Chlamydia*, it is likely that less than 10% of infected individuals seek medical advice. The most common STI of the anogenital tract is caused by human papilloma viruses (HPV), with HPV16 and 18 being causes of cervical carcinoma. Here asymptomatic disease is predominant.

The more sexual partners a person has, the greater the chance that they will acquire at least one STI. Vaginal, anal and oral sex need to be considered in the assessment of the patient. Gonococcus can cause urethritis, cervicitis and proctitis, as well as pharyngitis which is usually asymptomatic. This organism can cause disseminated infection, including septic arthritis. Acute pelvic inflammatory disease (PID) is a complication of both gonococcal and chlamydial infection.

Men-who-have-sex-with-men are also at risk of organisms that cause infections in the alimentary canal. *Shigella flexneri* can be spread between individuals by sexual practice, and can result in severe bloody diarrhoea. While these situations are uncommon, they highlight the importance of infectious disease notification in identifying an outbreak situation with such an organism.

Patients with an STI should be referred to the STD clinic promptly. In addition to the necessary expertise there, all records and tests use anonymized patient codes, which gives the patient full reassurance about confidentiality.

Vaginal discharges include bacterial vaginosis, often associated with the gram-variable bacterium *Gardnerella vaginalis*, candidiasis caused by *Candida* and trichomoniasis caused by *Trichomonas vaginalis*. Bacterial vaginosis and candidiasis are associated with an imbalance in the ecology of the endogenous flora of the vagina.

The newborn child is at risk of infection arising from acquisition of organisms from the normal vaginal flora, importantly *Streptococcus agalactiae* and *Escherichia coli*. These bacteria can colonize the upper airways of the newborn, with the potential to invade and cause neonatal sepsis and meningitis. *Listeria monocytogenes* can cross from the bowel of the expectant mother, usually in the third trimester, and can cause maternal sepsis, chorioamnionitis and miscarriage. Reaching the fetus from the maternal blood or during birth, it can also cause neonatal sepsis and meningitis.

Postpartum infections, endometritis and septic abortions are other important conditions to consider, as are infections arising from gynaecological procedures such as vaginal and abdominal hysterectomy. Not infrequently, *Streptococcus pyogenes* is associated with a postpartum infection, and must be considered from the outset, so that the correct specimens are collected and appropriate antibiotics prescribed.

An outline of the female genital tract and the organisms that make up the normal vaginal flora are shown in **Figure 16.1**. Organisms can reach the uterine adnexae via the uterus and Fallopian tubes, as occurs in PID. Because of the proximity of the introitus to the anus, it is not surprising that the vaginal flora includes organisms of faecal origin. It is important to appreciate the changes in the physiology and bacterial flora of the tract in the life of the female, as shown diagrammatically in **Figure 16.2**. It is relevant to note that the vaginal epithelium in the pre-pubertal female is not keratinized and will support the growth of gonococcus. Any vaginal discharge in a prepubertal girl must be examined for gonococcus, as well as chlamydia, to rule out sexual abuse.

The normal vaginal flora also varies during the menstrual cycle, with a mixed bacterial flora in the follicular phase and lactobacilli dominating the luteal phase (**Figure 16.3**). In the luteal phase, lactobacilli use the glycogen in sloughed vaginal epithelial cells, with lactic acid end-products being important in maintaining the acid milieu of the vagina.

ORGANISMS

A list of the organisms and clinical conditions is shown in **Figure 16.4**. A key organism to consider is HIV. Older age is not a reason for excluding this virus in the clinical assessment, as individuals in their 80s are being identified as HIV positive. A negative test done at the appropriate time is so reassuring, and provides the opportunity to discuss future risks that the patient should consider.

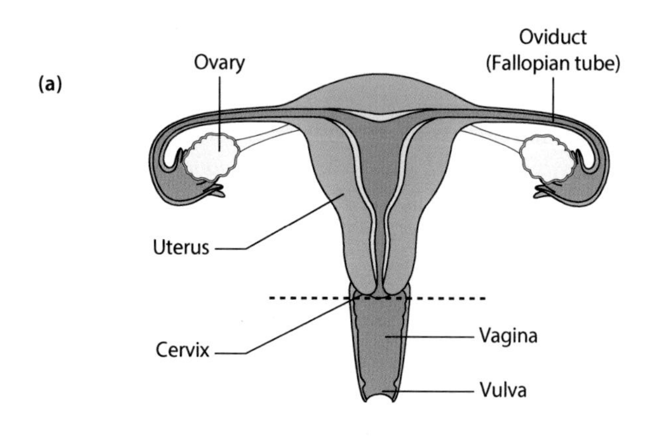

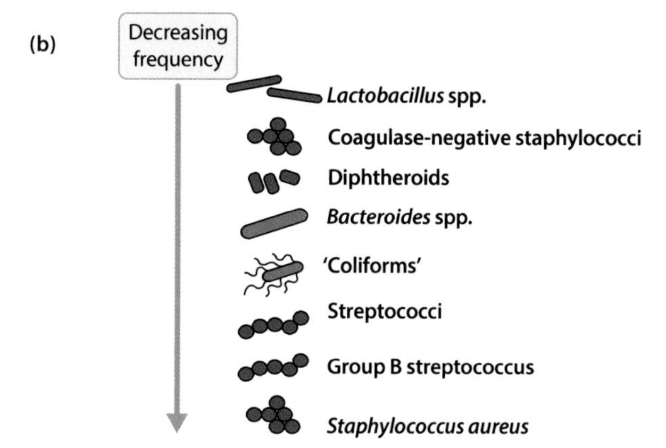

Figure 16.1 (**a**) An outline of the female genital tract. (**b**) Bacteria that make up the normal vaginal flora.

Key infections are associated with pregnancy, and of particular importance is screening at the first Antenatal clinic (ANC) visit, at about 12 weeks gestation. This is to identify those expectant mothers chronically infected with HBV, HIV or syphilis, and determine immunity to rubella virus infection.

PATHOGENESIS

GONORRHOEA

The causative agent of gonorrhoea is *Neisseria gonorrhoeae*, a gram-negative diplococcus, which can infect the columnar and cuboidal epithelium of the cervix, urethra, anal canal and pharynx (**Figure 16.5**). In the endocervix and Fallopian tubes, the organism attaches to columnar epithelial cells by adhesins and the cell wall outer membrane Opa proteins. Following attachment, invasion of the epithelium and subepithelium occurs. The strong inflammatory response is associated with neutrophil invasion, microabscess formation and pus.

Infection in the female can be asymptomatic, but in males an acute urethritis is the usual presenting symptom. Gonococcus can reach the epididymis of the male to initiate epididymitis, and the Fallopian tubes and adnexa in the female, where it is involved in PID. In the newborn, acquisition of the organism at birth can give rise to ophthalmia neonatorum.

CHLAMYDIA TRACHOMATIS

Chlamydia are obligate intracellular bacteria. There are several serological groups or serovars, with the A, B and C serovars associated with ocular trachoma and the D–K serovars associated with inclusion conjunctivitis (including ophthalmia neonatorum) and urogenital disease. *Chlamydia* causes urethritis, proctitis and epididymitis in the male, urethritis and cervicitis in the female and ophthalmia neonatorum in the newborn.

Chlamydia have a complex life cycle within infected epithelial cells. Following entry into cells, the elementary

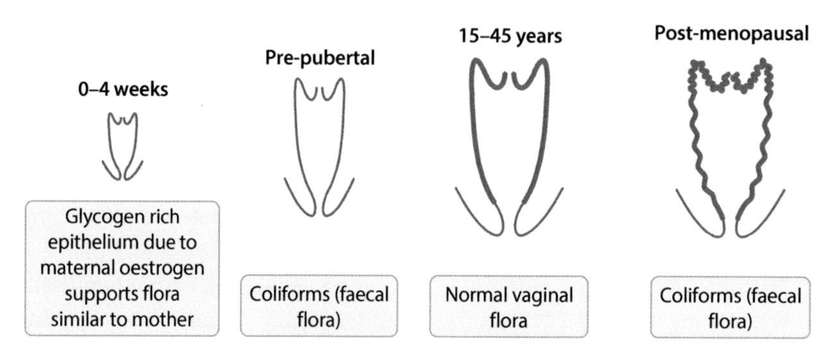

Figure 16.2 Changes that occur in the bacterial flora of the vagina at the various ages.

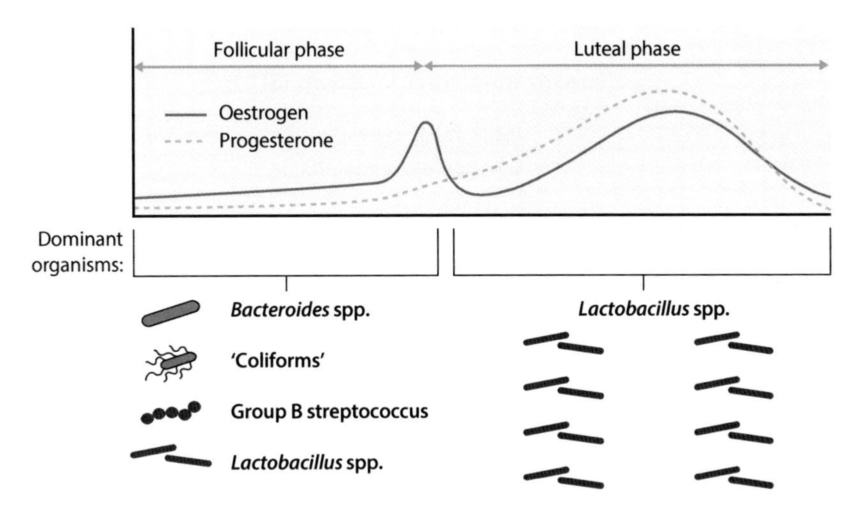

Figure 16.3 The menstrual cycle, with the changes that occur in the bacterial flora during the cycle.

Usually considered to be part of the normal vaginal flora

Lactobacilli
Staphylococcus aureus* (MSSA and MRSA)
Streptococcus agalactiae* (GBS)
Anaerobes (both gram-positive and gram-negative)*
Gardnerella vaginalis* (in low numbers as part of the normal flora)
Candida (in low numbers as part of the normal flora)

'Coliforms'*

(*Streptococcus pyogenes*, never a member of the normal flora, but should always be borne in mind)
* Can be important in post-surgical infections, malignancies of the genital tract

Vaginal discharge

Bacterial vaginosis	***Gardnerella vaginalis*** ***Mobiluncus* spp. (anaerobes)**
Candidiasis	*Candida*
Trichomoniasis	*Trichomonas hominis*

Vulvovaginitis (including in pre-pubertal girls)

Haemophilus influenzae
(*Neisseria gonorrhoeae* [in pre-pubertal girls])
Streptococcus pneumoniae
***Streptococcus pyogenes* (GAS)**

Enterobius vermicularis (pin worm)

Pelvic inflammatory disease

Acute:	***Chlamydia trachomatis*** ***Neisseria gonorrhoeae***
Chronic:	***Chlamydia trachomatis, Gardnerella vaginalis*, mixed "coliforms", mixed anaerobes, (*Actinomyces*)**
Actinomycosis associated with IUCD use	***Actinomyces israelii***

(*Continued*)

(*Continued*)

Sexually transmitted infections (STI)	
Acute pelvic inflammatory disease	*Neisseria gonorrhoeae* *Chlamydia trachomatis*
Cervicitis	*Neisseria gonorrhoeae* *Chlamydia trachomatis* HSV1, HSV2
Adult conjunctivitis	*Neisseria gonorrhoeae* *Chlamydia trachomatis*
Ophthalmia neonatorum	*Neisseria gonorrhoeae* *Chlamydia trachomatis*
Gastroenteritis (MSM)	*Shigella flexneri*
Genital ulcers	HSV1, HSV2 *Treponema pallidum* *Haemophilus ducreyi*
Pharyngitis	*Neisseria gonorrhoeae* HIV (usually asymptomatic)
Proctitis	*Neisseria gonorrhoeae* *Chlamydia trachomatis*
Syphilis	*Treponema pallidum*
Trichomoniasis	*Trichomonas hominis*
Urethritis: gonococcal non-gonococcal	*Neisseria gonorrhoeae* *Chlamydia trachomatis*

Viruses

Hepatitis A virus	**HAV** (anogenital/oral sex)
Hepatitis B virus	
Human immunodeficiency virus	
Cervical carcinoma	HPV 16, 18
Genital warts (condylomata acuminata)	HPV 6, 11

Associated with pregnancy, maternal infection and/or transmission to the fetus, newborn and neonate

Antenatal screening

HBV
HIV
Rubella virus
Treponema pallidum

Other important infections

Asymptomatic bacteriuria in the mother during pregnancy
***Streptococcus agalactiae* (GBS)**
Listeria monocytogenes
VZV

Figure 16.4 The range of organisms making up the normal flora of the vagina, and those associated with infections in the female and male.

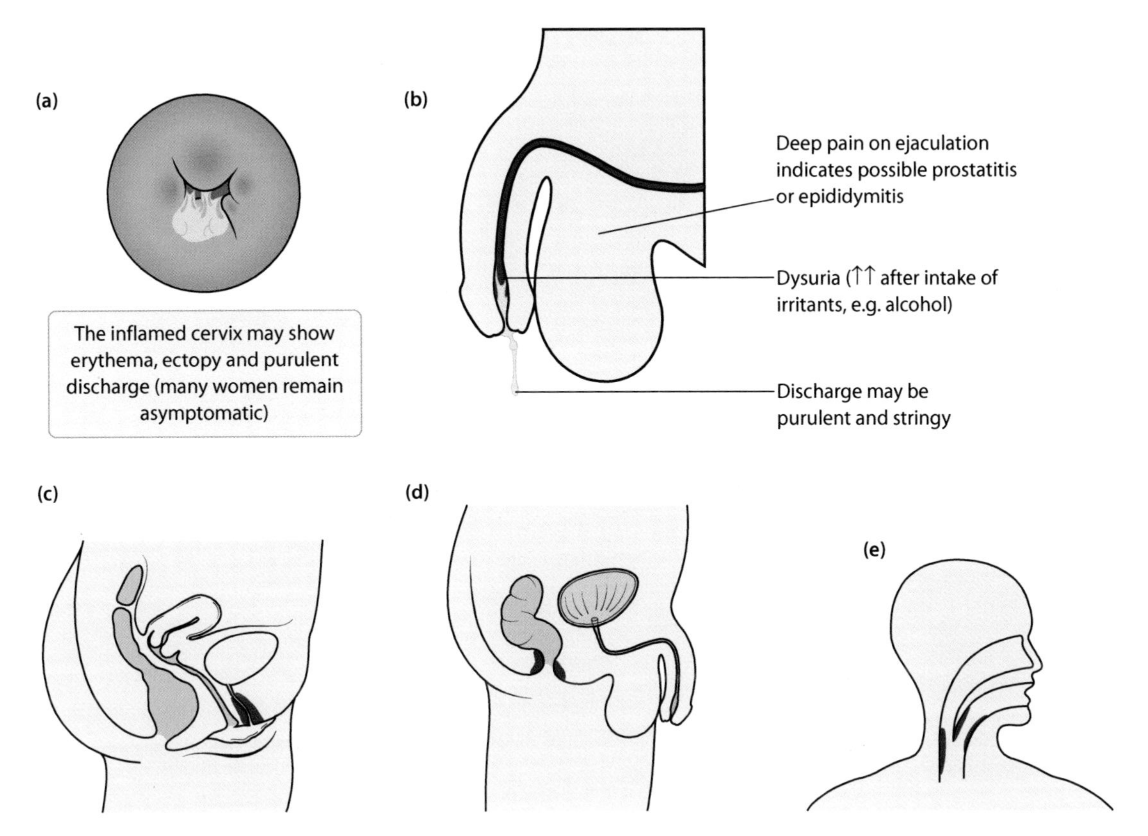

Figure 16.5 In gonococcal disease the organism can cause infection in the cervix (**a**), urethra and epididymis (**b**), female urethra (**c**), anus (**d**) and pharynx (**e**).

body converts to a reticulate body, the reproductive form of the organism. Reproduction occurs in an endosome, a membrane-limited structure within the infected cell. Reticulate bodies produce the elementary bodies that are then released as the extracellular infective form of the organism (**Figure 16.6**).

PELVIC INFLAMMATORY DISEASE

Acute PID arises following ascent of gonococcus and *Chlamydia* through the uterine cavity to the Fallopian tubes and adnexae (**Figure 16.7**). Both subclinical and chronic disease can then occur, where bacteria of the vaginal flora are also involved. Inflammation and fibrosis damage the Fallopian tubes, obstructing the normal movement of ova to the uterus, and ectopic pregnancy can result. When both tubes are affected, infertility can arise. A pelvic abscess is also a complication of PID, and when on the right side, this and an ectopic pregnancy need to be differentiated from disease of the appendix.

VAGINITIS

Vaginal discharges are associated with *Gardnerella vaginalis*, *Candida* and *Trichomonas vaginalis*.

Gardnerella vaginalis is a gram-variable facultative organism associated with bacterial vaginosis. Other organisms include *Mycoplasma hominis* and the anaerobe *Mobiluncus*. They have the ability to overwhelm the lactobacillus population of the vagina, with the resulting inflammatory response producing the discharge. *Gardnerella* can be detected on gram-stained preparations adherent in large numbers to vaginal epithelial cells, termed 'clue' cells.

Candida infections are associated with factors that change the normal flora and physiology of the vagina. Broad-spectrum antibiotics alter the vaginal flora, allowing yeasts to overgrow. The oral contraceptive is another predisposing factor, and by increasing glycogen stores in vaginal epithelial cells it enables *Candida* to outgrow the lactobacilli. In addition, oestrogen induces *Candida* receptors on the surface of vaginal epithelial cells.

Growth of *Trichomonas* in the vagina is optimal in a less acidic environment, and explains the tendency for symptoms to exacerbate in progesterone-dominant states such as pregnancy and menstruation. A purulent foul-smelling frothy discharge may be present in up to half of infected individuals.

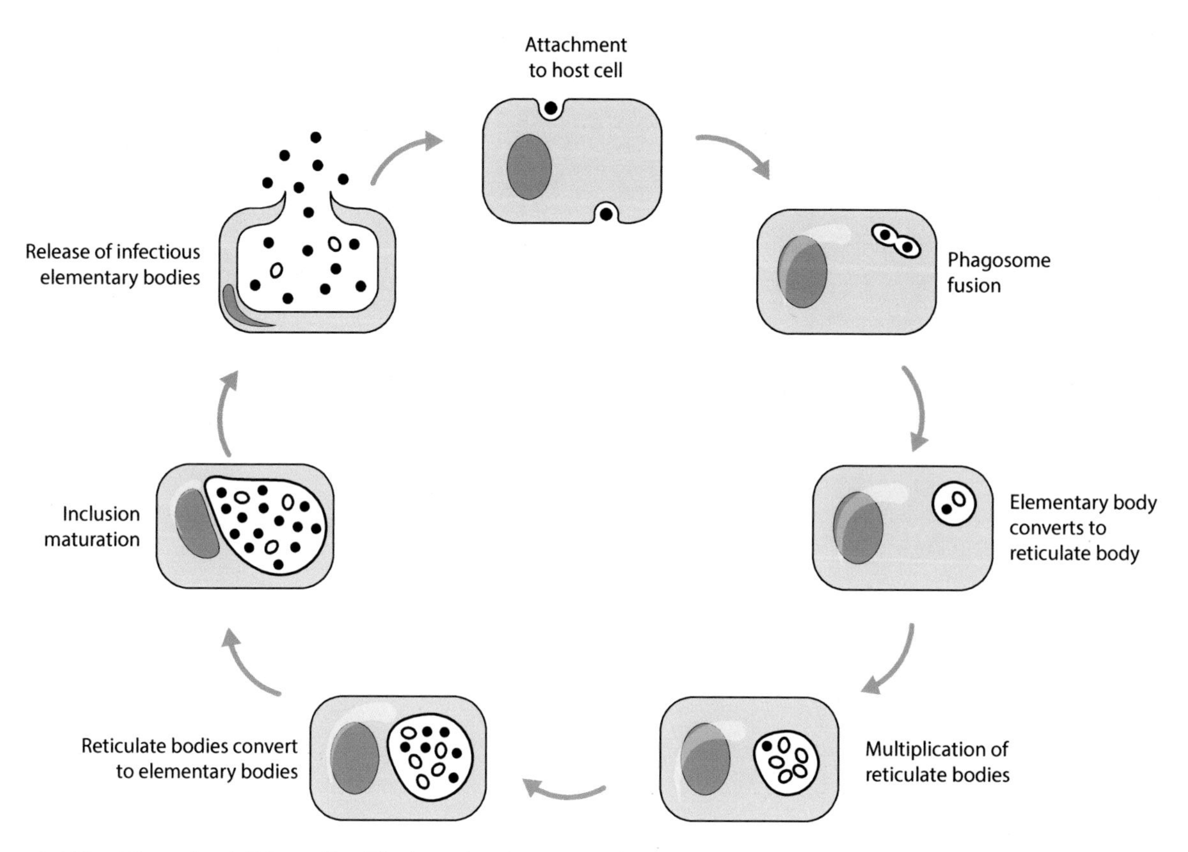

Figure 16.6 The life cycle of *Chlamydia*. (Black circle: elementary body; white circle: reticulate body.)

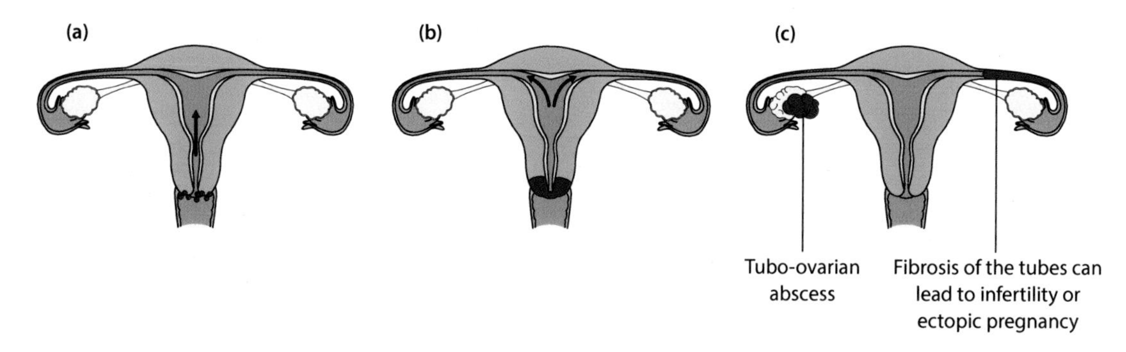

Figure 16.7 Pelvic inflammatory disease. (**a**) Bacteria ascend the uterus; (**b**) from here they enter the Fallopian tubes; (**c**) infection in the tubes results in fibrosis and a tubo-ovarian abscess can also develop.

POSTPARTUM INFECTIONS, SEPTIC ABORTIONS AND GYNAECOLOGICAL INFECTIONS

In postpartum infections and septic abortions, organisms of the vaginal or bowel flora, including anaerobes, are important. *Streptococcus pyogenes* should always be considered. Damage to the myometrium, retained products of conception and, in the case of a septic abortion, uterine perforation, enable organisms to establish a focus of infection (**Figure 16.8a**). Caesarean section wounds may also be a focus of infection, and with *Streptococcus pyogenes*, necrotizing fasciitis and toxic shock syndrome can occur. In hysterectomy operations, vaginal organisms may establish infections such as cuff cellulitis, cuff abscess or pelvic abscess (**Figure 16.8b**), despite the use of prophylactic antibiotics. Obesity, diabetes and difficult operations predispose to infection (see Chapter 4).

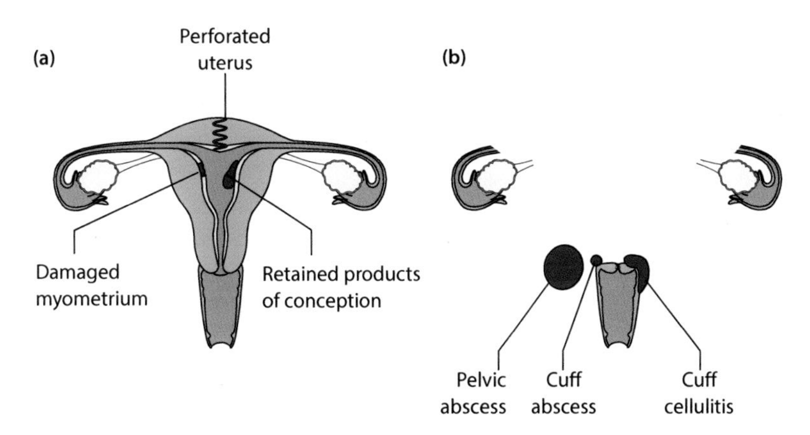

Figure 16.8 (**a**) Some reasons why infection may arise postpartum or as a result of an abortion. (**b**) Some types of infection that may arise following a hysterectomy.

DIAGNOSIS AND MANAGEMENT

This section centres on the management of the patient with a vaginal or urethral discharge. When a patient is seen in the A&E department it is prudent to contact staff in the STD clinic who will give the necessary advice, and visit the patient as necessary. As outlined below, specific procedures for these infections are used in STD clinics to maximize the microbiological diagnosis. There is an on-site biomedical scientist from the microbiology laboratory who performs the necessary tests in a designated laboratory suite at the STD clinic.

THE PATIENT WITH A VAGINAL DISCHARGE WHERE A STI IS UNLIKELY

Where the patient is in a stable relationship and/or a STI is unlikely, it is reasonable to consider bacterial vaginosis, candidiasis and trichomoniasis, recognizing that the latter is transmitted by intercourse; an outline of the investigation pathway is shown in **Figure 16.9**. These conditions can be diagnosed clinically, and determining the pH of the discharge is useful and is a straightforward bedside test. Swabs should always be collected appropriately, as outlined in **Figure 16.9**.

When empirical treatment is given, this is usually metronidazole for bacterial vaginosis and trichomoniasis. *Candida* is treated with either a topical clotrimazole or oral azoles such as fluconazole.

SEXUALLY TRANSMITTED DISEASES: GONORRHOEA AND CHLAMYDIA

Gonococcus and *Chlamydia* are the two infections routinely screened for at STD clinics. In addition to obtaining a Gram stain result promptly on a specimen, chocolate agar plates are inoculated and incubated at 37°C in a 5% CO_2 atmosphere. These optimum growth conditions are essential for culturing gonococcus, so that its antibiotic susceptibility profile can be determined. Resistance to penicillins, quinolones such as ciprofloxacin and the tetracyclines is a serious problem. (Currently, a single intramuscular dose of ceftriaxone [500 mg] can be given empirically. Options for *Chlamydia* include doxycycline for 7 days; azithromycin as a 1 g dose 1 hour before or 2 hours after food is another option.)

Combined molecular tests for *Chlamydia* and gonococcus are done with specific swabs that access the appropriate site of the relevant anatomy.

THE MALE PATIENT WITH A URETHRAL DISCHARGE

An outline of the testing procedures, and example of a swab set, for male patients attending the STD clinic is shown in **Figure 16.10**. The narrow section of the swab is inserted 2–3 cm into the urethra, rotated several times, removed and snapped off into the transport container (**Figure 16.10 a,b**). A plastic loop can then be inserted to the same depth, rotated and the material collected used to directly inoculate an agar plate for gonococcus culture (**Figure 16.10 c–e**). A smear is dried onto a glass slide, and Gram stained. The presence of neutrophils containing gram-negative diplococci can usually be considered as diagnostic of *Neisseria gonorrhoeae* infection in the male patient (**Figure 16.10 f,g**).

In the A&E setting, 5–10 mL of a 'first-catch' urine collected into a sterile white-topped container at least 1 hour after the last micturition is a useful non-invasive specimen for *Chlamydia* and gonococcus testing by polymerase chain reaction (PCR) (**Figure 16.11**).

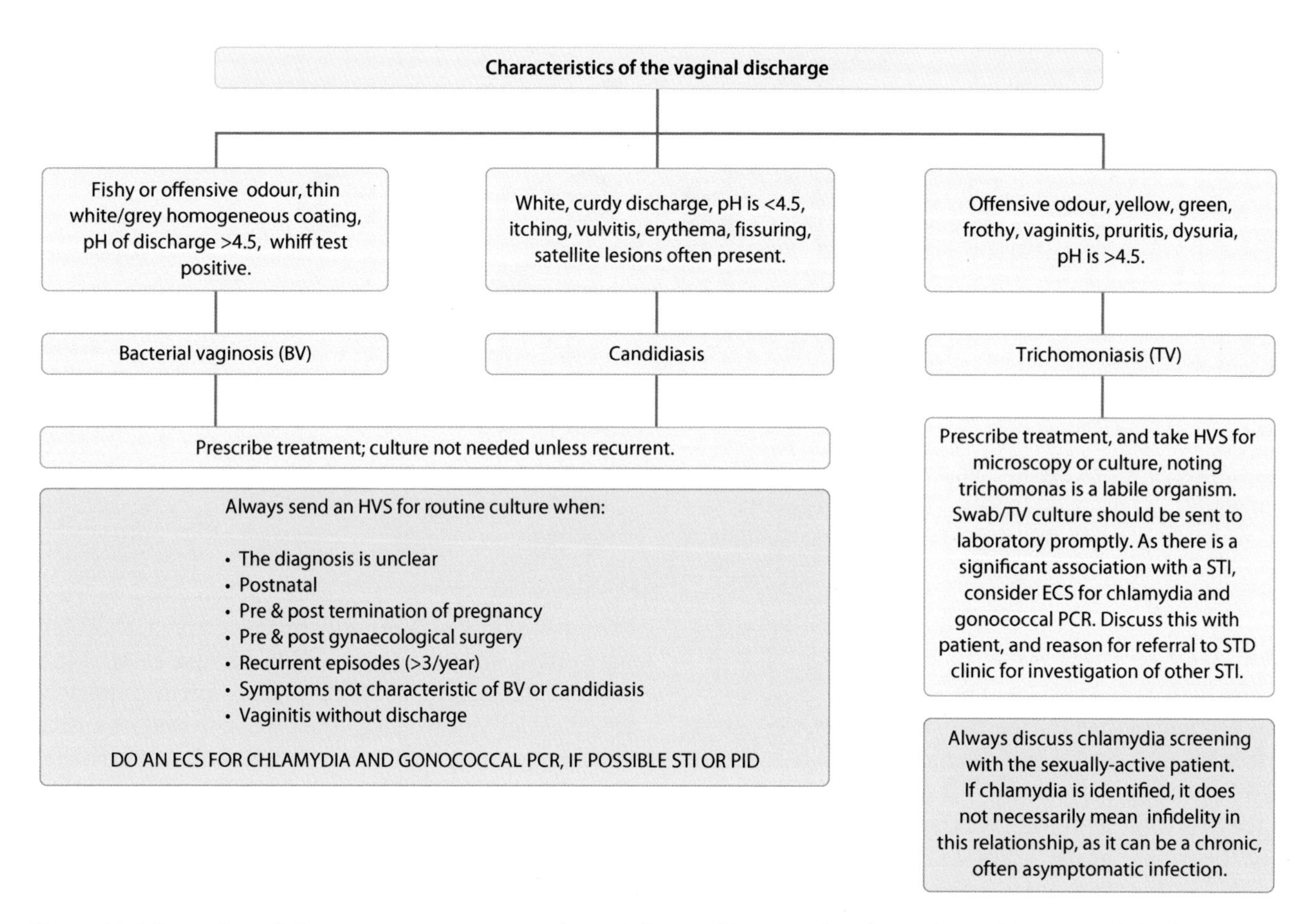

Figure 16.9 An outline of the management process that can be used in assessing the patient with a vaginal discharge.

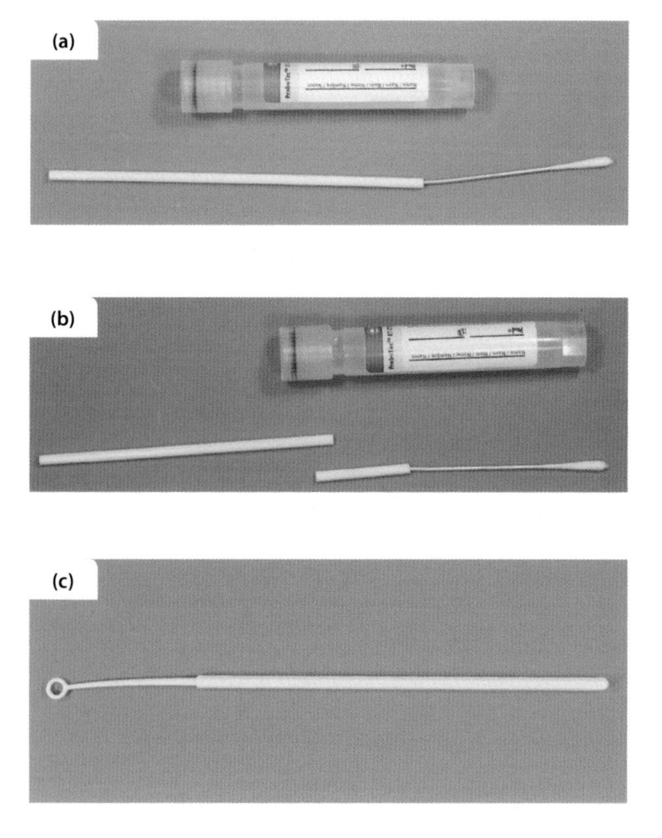

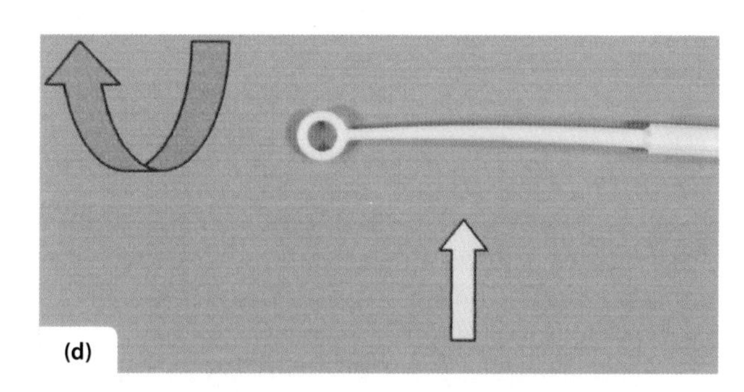

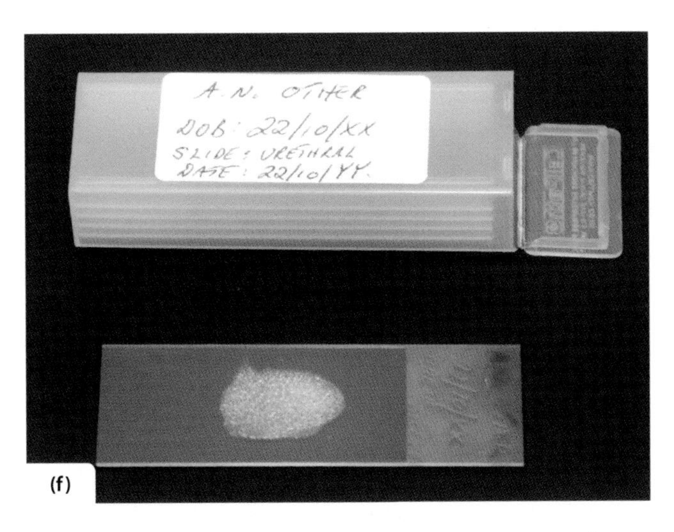

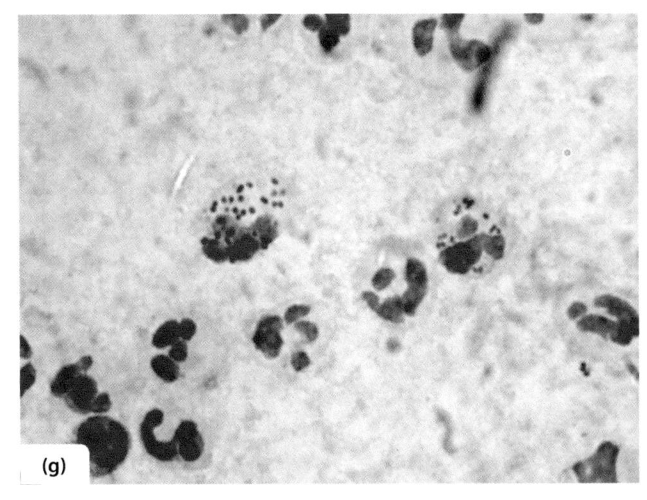

Figure 16.10 The procedure for obtaining specimens from the male patient in the sexually-transmitted diseases clinic, and initial processing, as described in the text.

Figure 16.11 A 'first-catch' urine can be collected into a sterile container for chlamydial and gonococcal polymerase chain reaction testing.

THE FEMALE PATIENT WITH A VAGINAL DISCHARGE

An outline of the process for collecting specimens in the female patient is shown in **Figure 16.12**. An example of a swab set for chlamydial and gonococcal PCR is shown in **Figure 16.13**. Under speculum examination, the large swab is used to remove discharge to visualize the cervical os, and the smaller diameter swab is inserted into the endocervix, rotated several times, removed and the sample end 'snapped off' into the transport container. An endocervical swab (ECS) is collected for direct inoculation onto culture medium for gonococcus. A Gram stain of a dried smear is performed.

Direct examination of vaginal discharge can be done. In addition to microscopy, a sample of the material is placed on a glass slide. When a few drops of 10% potassium hydroxide are added, and a 'fishy odour' is apparent, this is supportive of the diagnosis of bacterial vaginosis (**Figure 16.14**).

STREPTOCOCCUS PYOGENES IN FEMALE AND MALE GENITAL INFECTION

It is prudent to always consider this organism in the female patient with an infection of the genital tract. The infection can range from vulvovaginitis/pruritis in the prepubertal girl, a seemingly innocuous vaginal discharge or a postoperation wound infection in gynaecology or obstetrics.

In all cases the isolation of this organism from a specimen must be communicated by the microbiologist to the clinical team. There needs to be immediate review of the patient, including those who have been discharged home. The patient needs to be assessed in person, and

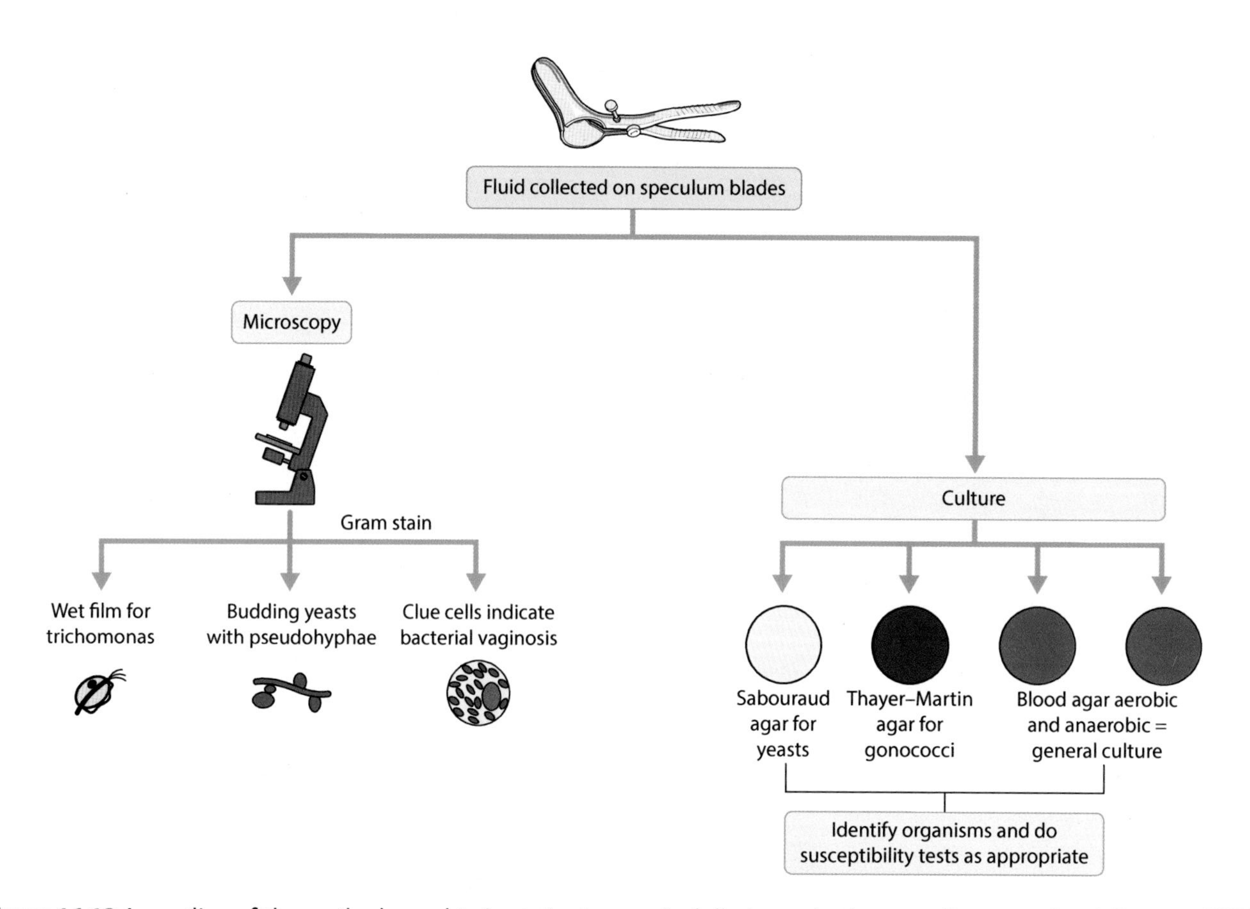

Figure 16.12 An outline of the methods used to investigate a vaginal discharge in the sexually-transmitted diseases (STD) clinic.

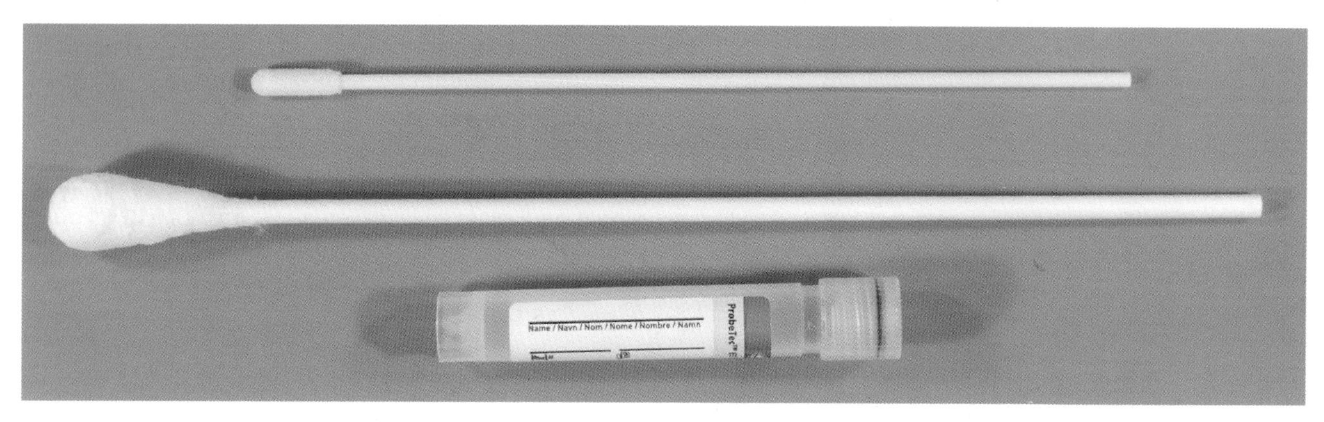

Figure 16.13 For the polymerase chain reaction test in the female patient, the large swab is used to clear discharge and enable visualization of the cervical os. The narrow diameter swab is then inserted into the endocervical canal, as described in the text.

after collection of further specimens as indicated, the appropriate antibiotic is either confirmed or prescribed.

Streptococcus pyogenes can cause balanitis in the male patient. When the foreskin is involved too, this is termed balanoposthitis. When this organism is identified, an appropriate antibiotic is prescribed. It is worthwhile to identify possible sources of the organism in order to give appropriate advice.

PELVIC INFLAMMATORY DISEASE

In most circumstances the diagnosis of PID is based on the clinical evidence and use of imaging. The patient

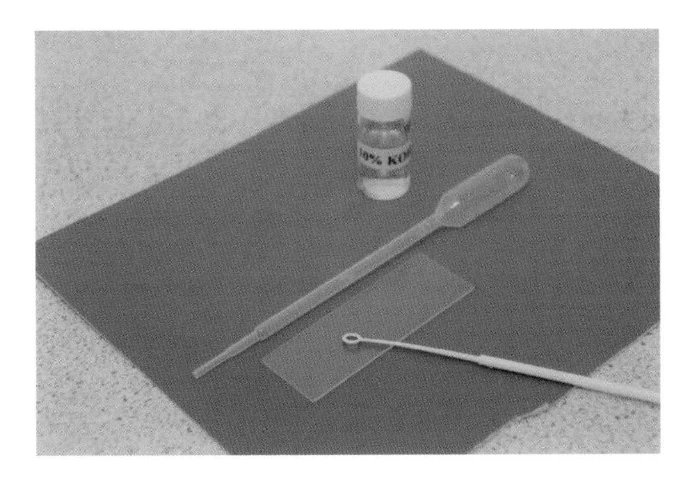

Figure 16.14 Using a sterile plastic loop, a sample of vaginal discharge is placed on a glass slide. Addition of 1-2 drops of 10% KOH, with production of a "fishy odor", supports the diagnosis of bacterial vaginosis.

may have a cervical discharge, and internal examination shows cervical motion tenderness and a palpable pelvic mass; fever, raised white cell count (WCC), C-reactive protein, and erythrocyte sedimentation rate are also relevant. It is appropriate to collect endocervical swabs for gonococcus and *Chlamydia*, and in the hospital setting, blood for culture.

INFECTIONS IN PREGNANCY, THE FETUS, NEWBORN AND NEONATE

Certain infections in pregnancy can be transmitted to the fetus, to the newborn in the perinatal period, the neonatal period (1st month of life) and beyond. In addition to HBV, HIV, syphilis and rubella virus, asymptomatic bacteriuria, and infections caused by *Streptococcus agalactiae* (GBS), *Listeria monocytogenes* and varicella zoster virus (VZV) are important too.

THE ROLE OF ANTENATAL SCREENING

An example of the usual result obtained in the expectant mother is shown in **Figure 16.15**. This result excludes infection with HBV, HIV and syphilis and confirms immunity to rubella virus. If the person is considered to be at ongoing risk of infection (e.g. HIV) during the remainder of the pregnancy, further testing should be done.

Any abnormal result requires specific action for that organism. In the case of the mother being susceptible to rubella, she is immunized after the pregnancy, as measles, mumps, rubella (MMR) is a live virus vaccine.

A summary of the high-risk periods of transmission for these four infections, and potential consequences without intervention during pregnancy and the perinatal and neonatal periods are shown in **Figure 16.16**. Here, the mother is well, and will be referred to the relevant specialist when the abnormal result is identified.

HBV

The main period of transmission of this blood-borne virus (BBV) is in the perinatal period during passage of the child through the birth canal (**Figure 16.16a**). Mothers are identified as being 'high-risk' or 'low-risk' carriers, and the newborn is appropriately managed with HBV immunoglobulin (HBIG) and the first dose of the vaccine, as discussed in Chapter 12.

After completing the vaccination schedule, with presence of protective HBsAb confirmed, babies are also screened at 12 months to show that HBsAg is absent.

HIV

There is increasing risk of transmission of HIV to the fetus from the middle of gestation onwards, and in the perinatal period. Without intervention, about 25/100 babies born to infected mothers will acquire the virus; in addition, breast feeding is a clear ongoing risk in the neonatal period and beyond (**Figure 16.16b**). The prescription of antiviral therapy during pregnancy, and to the newborn reduces transmission to ≤2/100 babies born.

After birth, and the appropriate antiviral management, it is essential to confirm that the child has not acquired HIV from the mother, and to do this both molecular (PCR) and serology tests are done. After delivery, and then at 6, 12 and 24 weeks, the key test is examining samples of deoxyribonucleic acid (DNA) extracted from peripheral white blood cells of the child for HIV proviral DNA; a negative result shows the virus is absent. The essential positive control in the test is detection of HIV proviral DNA in a sample of the mother's blood examined in parallel, which uses the same primers in the PCR amplification process.

The final test to confirm that transmission has not taken place is done at 18 months of age. A blood specimen must be negative for HIV ribonucleic acid (RNA) by PCR, and for antibodies to HIV (by this time maternal antibodies acquired during gestation will have disappeared).

TREPONEMA PALLIDUM (SYPHILIS)

The risk of transmission of the organism is throughout gestation, and the perinatal period (**Figure 16.16c**). The otherwise healthy expectant mother who is treponemal antibody positive is likely to be in the latent stage of disease, recognizing that the transmission risk is highest during active disease in primary and secondary syphilis.

Antibody-positive mothers are promptly treated; benzylpenicillin is one option. The newborn must be carefully examined after delivery to exclude evidence of congenital infection. Serology tests are done to exclude active infection, and the child is screened up to the age of 12 months to show loss of passively acquired treponemal IgG antibodies from the mother.

RUBELLA VIRUS

The vast majority of expectant mothers will be protected by routine MMR vaccination, so a rubella IgG antibody negative result is uncommon. If acute rubella infection does occur during pregnancy, the effects on the fetus are most pronounced in the first trimester and can be devastating (**Figure 16.16d**). Miscarriage can also occur.

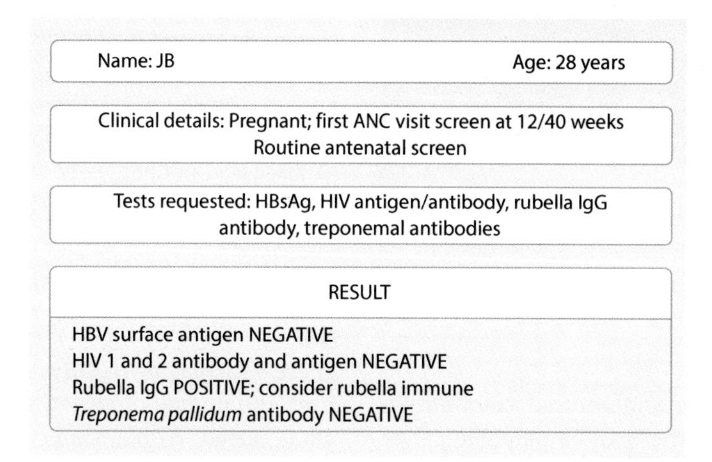

Name: JB　　Age: 28 years

Clinical details: Pregnant; first ANC visit screen at 12/40 weeks
Routine antenatal screen

Tests requested: HBsAg, HIV antigen/antibody, rubella IgG antibody, treponemal antibodies

RESULT

HBV surface antigen NEGATIVE
HIV 1 and 2 antibody and antigen NEGATIVE
Rubella IgG POSITIVE; consider rubella immune
Treponema pallidum antibody NEGATIVE

Figure 16.15 The usual serology result obtained when the expectant mother is screened at the first (12-week) antenatal visit. This shows the patient is not chronically infected with hepatitis B virus (HBV), human immunodeficiency virus (HIV) or syphilis, and is immune to rubella virus infection. Routine screening for rubella in England was discontinued in 2016.

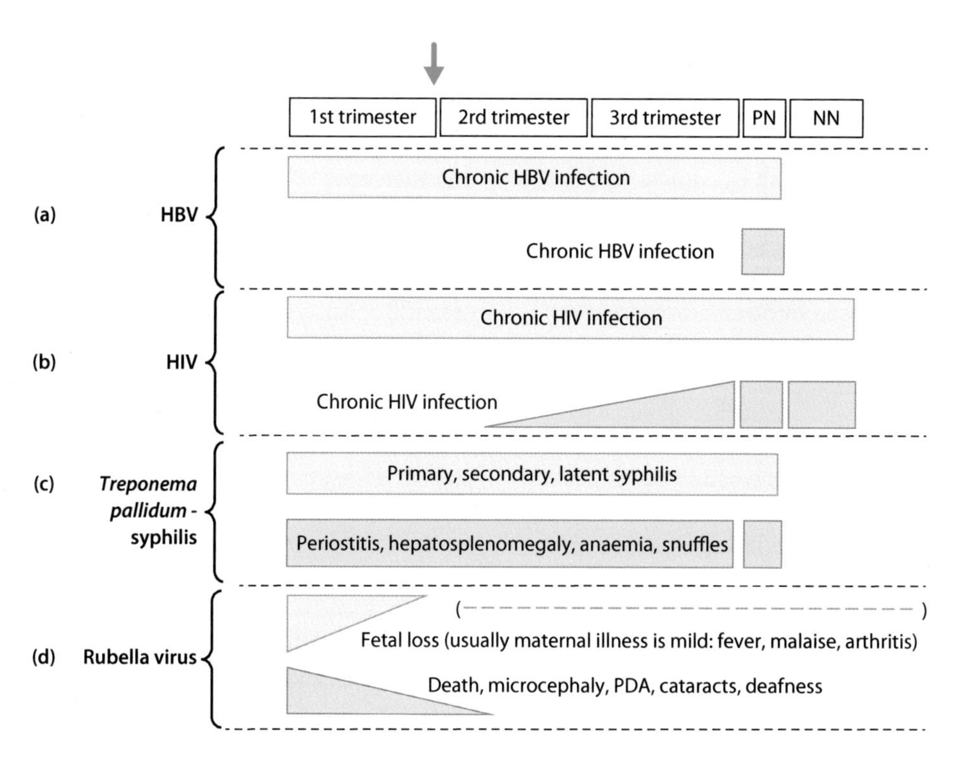

Figure 16.16 The period of infectivity/transmission of a maternal infection to the fetus, or during the perinatal period (PN), and neonatal period onwards (NN) of: (**a**) hepatitis B virus (HBV); (**b**) human immunodeficiency virus (HIV); (**c**) *Treponema pallidum*; (**d**) rubella virus. The consequences of each infection are shown. (Mother: upper blue panel; fetus, newborn, neonate: lower red panels. Green arrow: first antenatal clinic visit. PDA: patent ductus arteriosus.)

Susceptible expectant women who develop a rash-like illness or who are exposed to a person with such a rash-like illness are closely monitored for the appearance of rubella (and parvovirus B19) IgM antibodies. If these antibodies are detected, the mother is counselled and closely followed up to determine and manage the likely outcome. If acute rubella occurs in the mother after 20 weeks gestation, there is no increased risk of congenital abnormalities in the child. (The same applies to acute infection with parvovirus B19.)

In England in 2016, the decision was taken to stop testing for rubella as part of antenatal testing. There are several reasons for this:

- Rubella infection levels in the UK are so low they are defined as eliminated by the World Health Organization.
- Thus, rubella infection in pregnancy is very rare in the UK.
- Being fully immunized with the measles, mumps and rubella (MMR) vaccine before becoming pregnant is more effective in protecting women against rubella in pregnancy.
- The screening test used can potentially give indeterminate results and cause unnecessary anxiety.

OTHER INFECTIONS THAT POSE A RISK TO THE MOTHER AND FETUS

Asymptomatic bacteriuria, GBS, *Listeria monocytogenes* and VZV are of significance in pregnancy and in the perinatal and neonatal periods.

ASYMPTOMATIC BACTERIURIA

Although the role of bacteriuria in pregnancy and prematurity, as well as intrauterine growth retardation (IUGR), is not well defined, the relationship between asymptomatic bacteriuria and acute pyelonephritis in the mother is undisputed. The risk periods in the mother and fetus are shown in **Figure 16.17a**.

The prevalence of asymptomatic bacteriuria in early pregnancy is about 5%, and if untreated, about 20–40% of patients will develop acute pyelonephritis later in pregnancy. This compares with about 1% of patients who do not have bacteriuria early in pregnancy who later develop pyelonephritis. This infection is usually, but not exclusively, caused by *Escherichia coli* or other 'coliforms'.

For this reason, screening at the first ANC visit is essential, and should also be done in the third trimester. A clean-catch midstream urine (MSU) sample should be obtained, and irrespective of the WCC, the presence of a pure growth of an organism of >10^7 cfu/L is significant. As a single specimen is not diagnostic of asymptomatic bacteriuria, a second MSU is collected to confirm the presence of the same organism, based on its identity and antibiotic susceptibility profile. It is so important that advice about correct specimen collection is given in order to obtain a MSU (see Chapter 10). Antibiotics should be prescribed for 7 days. One to 2 weeks after completion of treatment, a further MSU should be collected to confirm eradication of the infection. At each subsequent ANC visit, a MSU should be collected to check that the bacteriuria has not recurred.

Treatment is based on the susceptibility profile of the organism. In addition to amoxicillin, trimethoprim and nitrofurantoin are considered acceptable to use in pregnancy.

STREPTOCOCCUS AGALACTIAE (GBS)

GBS is a member of the bowel flora, from where it can become part of the colonizing flora of the perineum and vagina. Colonization rates in pregnant and non-pregnant women range from 10 to 40%, with the principle reservoir the rectum. GBS can cause asymptomatic bacteriuria, which can be an indication of heavy vaginal colonization. The usual risk periods for the child are shown in **Figure 16.17b**. The rate of vertical transmission at the time of birth is about 50%, and factors that increase the risk of invasive disease in the child include delivery at less than 37 weeks, amniotic membrane rupture ≥18 hours before delivery and maternal fever ≥38°C during delivery. When (usually neonatal) infection in the child is suspected, blood for culture is collected, and a lumbar puncture (LP) will often be done too. Neonatologists have specific regimes for prophylaxis and empirical treatment; benzylpenicillin with gentamicin is an example.

Indications for prophylactic antibiotics for the mother during labour are as follows:

- Previous infant with invasive GBS disease.
- GBS bacteriuria in the current pregnancy.
- Positive GBS screening culture in the current pregnancy.
- GBS status unknown, and any of the following:
  - Delivery <37 weeks.
  - Amniotic membrane rupture ≥18 hours.
  - Intrapartum temperature ≥38°C.

Carriage of group B streptococcus in non-pregnant and pregnant women is common. Antibiotics do not eradicate GBS vaginal carriage (and are therefore not indicated), as re-colonization will occur with organisms from the rectum; the role of antibiotics is at the time of labour. GBS is not a cause of a simple vaginal discharge in women.

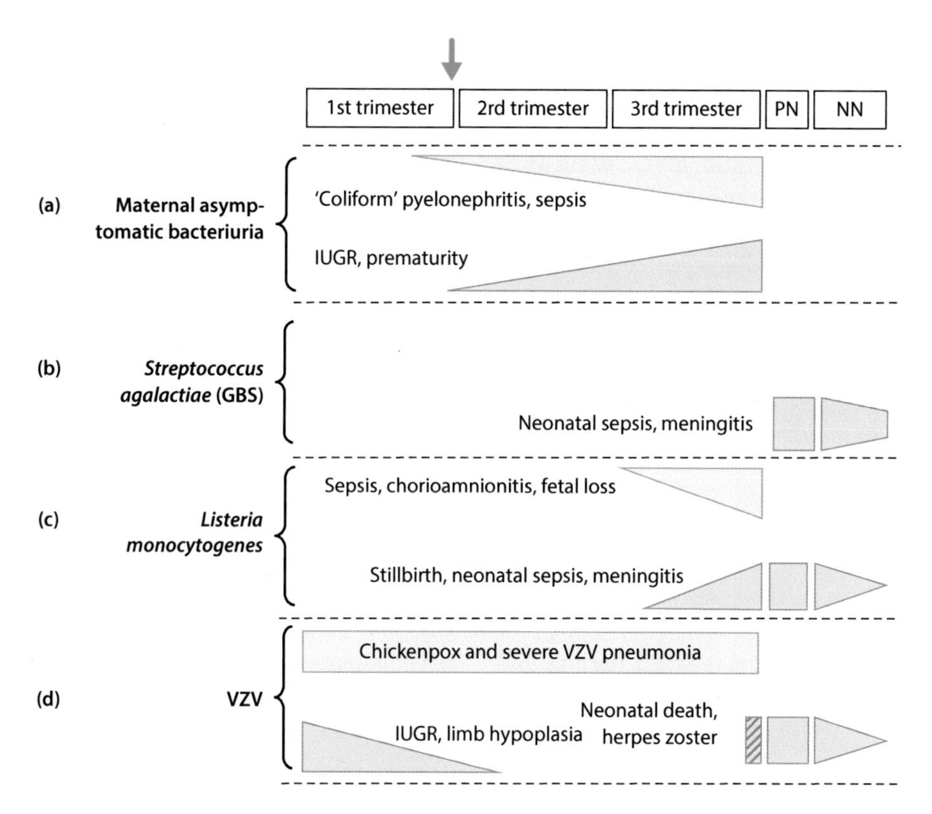

Figure 16.17 Asymptomatic bacteriuria, and infection with three important organisms in pregnancy, or during the perinatal period (PN), and neonatal period onwards (NN). (**a**) Maternal asymptomatic bacteriuria; (**b**) *Streptococcus agalactiae* (GBS); (**c**) *Listeria monocytogenes*; (**d**) varicella zoster virus (VZV). The consequences of each infection are shown. (Mother: upper blue panel; fetus, newborn: lower red panels. Hatched box [VZV] relates to period where passively acquired IgG antibodies would not be present to protect the newborn if the mother develops chickenpox in the week before delivery. Green arrow: first Antenatal clinic visit.)

LISTERIA MONOCYTOGENES

This small gram-positive motile organism is associated with consumption of unpasteurized dairy products and other foods such as paté. Asymptomatic stool carriage probably occurs in at least 1% of the population. Bacteraemia is more common in pregnancy, especially in the third trimester, and relates to the natural immunosuppression in this period, that the organism takes advantage of. *Listeria* invades the blood from the bowel to initiate sepsis in the mother (meningitis is uncommon). From the blood, the organism crosses the placenta to initiate infection in the fetus and amniotic fluid. The risk periods for the mother and child are shown in **Figure 16.17c**.

Blood culture is the diagnostic test in the mother, as are amniotic fluid and placental swabs collected at the time of delivery. When perinatal or neonatal sepsis is suspected blood culture and LP are done. The regime for GBS is appropriate in the child; *Listeria* is resistant to cephalosporins.

VZV

If a mother develops chickenpox in pregnancy, there is a risk to the fetus of limb hypoplasia, IUGR and skin scarring. The incidence of congenital varicella syndrome ranges from 1 to 2% in the first 20 weeks, after which the risk is substantially lower. There is a significant risk of neonatal mortality if the mother develops chickenpox 7 days before to 7 days after delivery (or if the neonate whose mother is seronegative is exposed to VZV from any source within the first 7 days of life). This is because protective maternal IgG antibodies that would usually cross the placenta would not have been produced in that time period. The main risk periods for the mother, fetus and newborn are shown in **Figure 16.17d**.

The seronegative mother who is exposed to a case of chickenpox or exposed herpes zoster, must be appropriately managed within 10 days, the minimum incubation period of the virus. In these circumstances she is given VZV immunoglobulin (VZIG) as passive protection to prevent or ameliorate the infection. Because serious infection can occur in the mother, especially pneumonia, the mother who develops VZV infection must be given aciclovir within 24 hours of the typical rash appearing.

The at-risk newborn is also given VZIG as protection, and is closely monitored for infection; if identified aciclovir is given.

Chapter 17

Infections in a Modern Society

INTRODUCTION

This chapter highlights management of specific patient groups, being those with immunodeficiency, patients requiring renal support, those on the intensive care unit (ICU), the elderly and the injecting drug user (IDU), as well as those without a spleen.

THE IMMUNOSUPPRESSED PATIENT

There are similarities in the way human immunodeficiency virus (HIV) infection and immunosuppressive drugs dismantle the immune system. It is therefore not surprising that infections found in the patient with untreated HIV can be found in the patient on immunosuppressive drugs; key agents are outlined below:

Drug treatments include:

- Antimetabolites, such as cytarabine and the anthracycline daunorubicin, are used in induction chemotherapy for acute myeloid leukaemia (AML). These interfere with deoxyribonucleic acid (DNA) synthesis, and target rapidly dividing cells of the bone marrow, in order to ablate the malignant cells.
- Monoclonal antibodies against cytokines such as tumour necrosis factor (TNF)-α (infliximab).
- Mycophenolate inhibits synthesis of guanosine nucleotides; its effect is most pronounced on the growth of T and B cells.
- Ciclosporin and tacrolimus inhibit the production of interleukin 2, essential for the maturation and proliferation of T cells.
- Steroids such as prednisolone inhibit cell proliferation, and methotrexate acts as an anti-folate agent. These are more general immunosuppressive agents, but their use is no less important to consider.

Increasing age, especially with comorbidities of diabetes, renal failure and malnutrition also compromise the function of the immune system. The relevant parts of the immune system to consider are shown in Chapter 2, **Figure 2.7**.

THE PATIENT INFECTED WITH HIV

Through mucosal surfaces, HIV is taken up by dendritic cells, which interact with macrophages and T cells. Replication in these cells leads to a population of activated, infected CD4+ T cells that proliferate and disseminate to the lymphoid tissues of the body. Here there are a large number of target cells for virus replication, and the inflammatory response gives rise to the symptoms of acute HIV infection, including generalized lymphadenopathy. High titres of virus are detected in the blood, and there is depletion of CD4+ T cells. This initial infection is then curtailed by partial immune control, involving uninfected HIV specific CD4+ T cells, HIV-specific CD8+ cytotoxic T cells and neutralizing antibodies, resulting in control of replication. This is reflected in a drop in viral titre and recovery of CD4 cell count.

However, as the virus has integrated itself into the genome of target cells in a latent form, low levels of replication extend the population of infected CD4+ T cells. The stage is reached when cell-mediated control of virus replication is lost, the CD4 count starts to fall and the rate of virus replication increases. This gradual widening of the infection amongst the CD4 cells leads to loss of a coordinated cytokine response, affecting the activation of memory T cells, and the ability of the body to activate CD8-T cytotoxic cells. Loss of activated T cells that can recognize and kill cells latently infected with organisms such as *Mycobacterium tuberculosis* and *Toxoplasma gondii* allows reactivation of these organisms. Progressive deterioration of the immune response disables the ability to control ubiquitous and usually innocuous organisms, including fungi (*Cryptococcus neoformans* and *Pneumocystis jirovecii*), or the environmental mycobacterium *Mycobacterium avium*.

An effective neutralizing antibody response to extracellular virus is not made. The reason for this is that the reverse transcriptase is highly error prone, leading to widespread amino acid substitutions in the viral envelope proteins. This is turn generates continuing changes in epitopes of the envelope proteins gp120/gp41. These epitopes remain heavily glycosylated with carbohydrate residues added during the post-translation glycosylation process taking place in the endoplasmic reticulum. Glycosylation significantly decreases the immunogenicity of these proteins, while their unique ability to bind to the CD4 and CCR5 receptors is not diminished. As HIV infection gradually overcomes the T4 cell population, the stimulation of B cells and memory B cells is lost, and the number of immature B cells increases; there is eventual exhaustion of an effective antibody response.

DIAGNOSIS

The time scale of events in the untreated HIV infected individual, including the World Health Organization (WHO) staging criteria, is shown in **Figure 17.1**, with symptoms, signs and infections based on this shown in **Figure 17.2**. Any undiagnosed patient who fits these criteria should be counselled for an HIV test.

Neurological manifestations are an example of the presenting symptoms and entities to consider include:

- Aseptic meningitis is part of the acute HIV syndrome, with headache, fever, neck stiffness, nausea and vomiting. There may be neuropathies of cranial nerves V, VII, VIII. The CD4 count is >200 cells/μL, and CSF has the parameters of a viral meningitis.
- *Cryptococcus* meningitis and *Toxoplasma* encephalitis usually manifest at a CD4 count of <200 cells/μL. Fever, headache and confusion are present and in the case of toxoplasmosis, seizures can occur. With cryptococcal meningitis, the CSF protein is raised, glucose decreased, and the heavily capsulated yeast can often be seen on India ink staining. Progressive multifocal encephalopathy caused by the JC papovavirus and central nervous system (CNS) lymphoma manifest at this stage.
- Cytomegalovirus (CMV) encephalitis with the subtle presentations of confusion, apathy and withdrawal usually occurs with a CD4 count below 50 cells/μL.
- It should be noted that neurosyphilis can arise at any stage of infection.

MANAGEMENT

Once the patient has been identified as infected by the HIV antigen/antibody test, regular monitoring of the CD4 count and HIV viral load is done.

Management of the patient with HIV infection is done by the consultants in sexually-transmitted diseases (STD) and infectious diseases, and consultant virologists have a central role. The decision to initiate treatment centres on the CD4 count, but it is now usual to initiate antiretroviral therapy (ART) earlier in the course of the disease at higher CD4 counts. A combination of at least three agents is used (see Chapter 4, **Figure 4.1**); it is worthwhile to refer to Chapter 4, **Figures 4.13, 4.14**, in relation to the structure and replication of HIV. In addition, active and

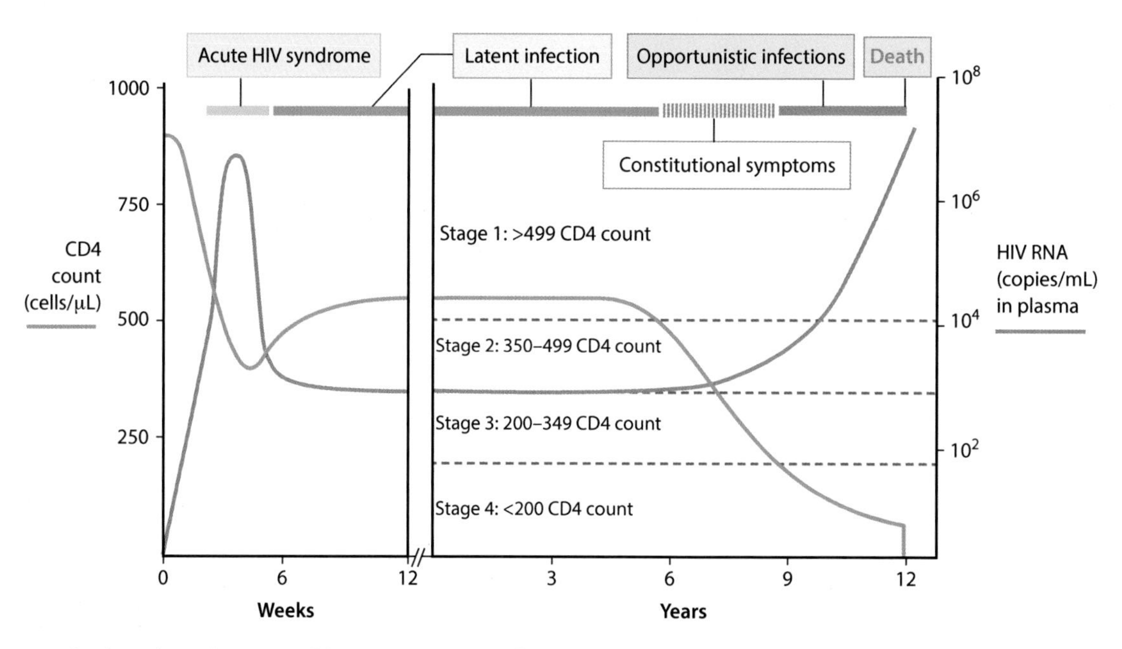

Figure 17.1 The key clinical stages of human immunodeficiency virus (HIV) infection in the untreated individual, showing the trends in the CD4 count and virus replication, based on viral load in blood. The World Health Organization (WHO) stages based on the CD4 count are marked.

Stage 1 (CD4 count >499)	Stage 2 (CD4 count 350–499)	Stage 3 (CD4 count 200–349)	Stage 4 (CD4 count <200)
	Constitutional symptoms		**Specific conditions**
Asymptomatic	Moderate weight loss	Severe weight loss	HIV wasting syndrome
Persistent generalized lymphadenopathy	Recurrent oral ulceration	Oral hairy leucoplakia	Invasive cervical carcinoma
	Stubborn seborrheic dermatitis*	Diarrhoea >1 month	Lymphoma (cerebral, B cell, NHL)
		Unexplained persistent fever	HIV cardiomyopathy
			HIV nephropathy
	Infections		
	Recurrent pharyngitis, sinusitis, otitis media	Severe bacterial infections	*Pneumocystis* pneumonia
	Herpes zoster (VZV)	Pulmonary tuberculosis	Chronic HSV infections
	Fungal nail infections	Persistent oral candidiasis	Cryptococcal meningitis

Figure 17.2 Symptoms, signs and infections within the WHO stages of HIV infection. These are important alerts for considering HIV infection at the earliest opportunity when they are present. (*Seborrhoeic dermatitis is a dry itchy, flaky [dandruff-like] rash on the scalp, face and chest.)

latent infections need to be appropriately identified and treated.

Examples of asymptomatic/latent infections include chronic hepatitis (HBV and HCV), *Treponema pallidum* and *Toxoplasma gondii*.

Treatment is complicated, centered on long-term ART. The challenges can be considerable; the HIV patient chronically infected with HBV and HCV poses drug interaction and side-effect issues, and these all need to be taken into account. The overall aim is reducing replication so that the viral load is below the limit of detection, and the CD4 count returns to a level that enables reasonable immune function to be restored. This, along with judicious use of prophylactic antibiotics, significantly reduces the risk of infection by organisms that characterized the acquired immunodeficiency syndrome (AIDS) epidemic.

An ongoing problem is the development of resistance to these antiviral agents. This adds another management arm to the process, which is the identification of resistance markers by sequencing technologies.

THE HAEMATOLOGY AND ONCOLOGY PATIENT

In this group of patients, the neutropenic leukaemia patient exemplifies the challenges in the diagnosis and treatment of infection. Within 4 or 5 days of initiation of induction chemotherapy, the neutrophil count in these patients drops below 1.0×10^9/L. In the ensuing neutropenic period the patient is at risk of overwhelming infection. Chemotherapy is not only acting on bone marrow, but all high turnover cells of the body will be affected, and mucositis of the upper alimentary canal is an example. Some possible sites of infection in the neutropenic patient are identified in **Figure 17.3**.

Haematology units will have specific protocols for the empirical treatment of fever in the neutropenic patient (**Figure 17.4**). These will be influenced by microbiology results, of which blood culture is the most important. Usually a broad-spectrum agent with *Pseudomonas* activity such as piperacillin/tazobactam is used first. If the temperature does not settle within 48–72 hours, this is replaced with meropenem. Based primarily on blood cultures, vancomycin is usually given for colonization of the central venous catheter (CVC) with staphylococci. There is then a low index of suspicion of fungal infection, exemplified by invasive pulmonary aspergillosis. High-resolution computed tomography (HR-CT) scans are done, and serology tests for fungal galactomannan antigen are used too. Antifungal agents include voriconazole, caspofungin and

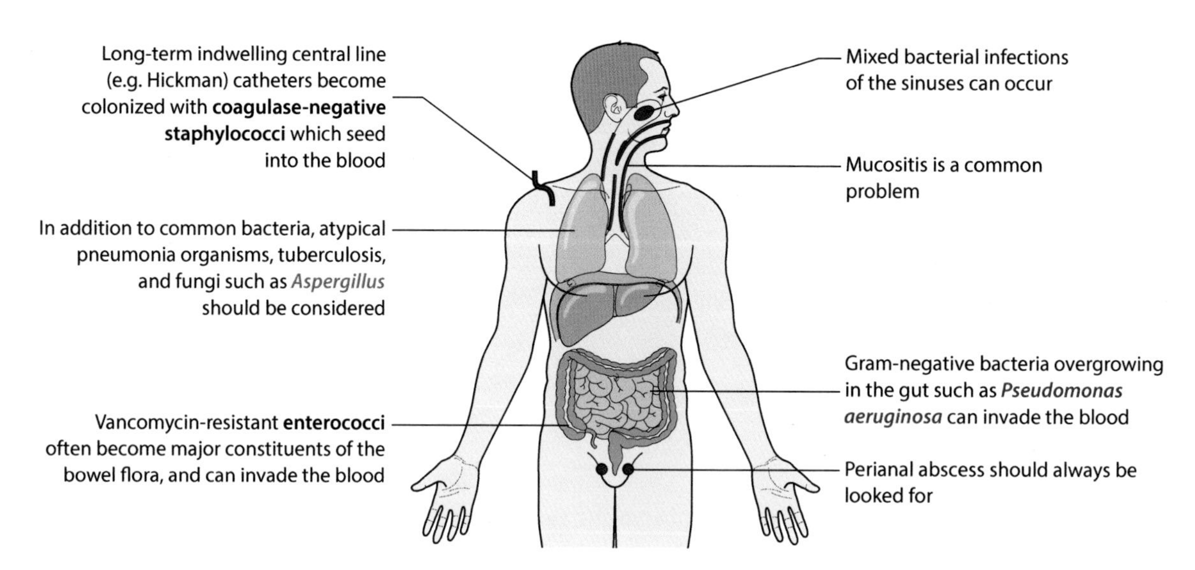

Figure 17.3 Some sites where infection can arise in the neutropenic leukaemic patient.

liposomal amphotericin B. Blood cultures that identify a colonized CVC prompt the decision for its removal and replacement.

THE SOLID ORGAN TRANSPLANT PATIENT

Successful transplantation of a solid organ depends on the degree of cross-matching between the donor organ and the recipient, and the extent of the immunosuppression that has to be given in order for the transplanted organ to survive and function. Immunosuppression will put the patient at risk of a wide range of organisms that include bacteria, fungi, viruses and parasites. Donor and recipient are screened serologically for a number of organisms, including HBV, HCV, HIV and CMV. The identification of any of the blood-borne viruses (BBVs) in a potential donor is a contraindication to transplantation, while CMV is not. Close monitoring of CMV DNA by polymerase chain reaction (PCR) would be done, along with ganciclovir treatment.

Fever can be caused by organ rejection, infection or drugs. Rejection may be accompanied by myalgia and arthritis, thus mimicking infection. It is important that all possible sites of infection are considered and the appropriate specimens taken. In solid organ transplant patients infections can in general be divided into those that occur in the first month after the operation and those that occur in the next 5 months, the immunosuppressed period (**Figure 17.5**). In the first month, infections that arise are a result of the transplantation itself and the intensive care period that follows. Surgical wound infections, abscess formation, ventilator-associated pneumonia, bacteraemia arising from long-term central lines and urinary tract infections are examples. For each type of transplant, local complications may arise. In renal transplantation, infection in the urinary tract is a recognized complication. In liver transplantation, liver and peritoneal abscesses can occur. These relate to the various anastomoses that have to be made including re-routing of the biliary tract drainage. In heart and lung transplantation, mediastinitis can be a problem. The anastomoses of the lung and the ablation of the cough reflex are all important contributors to infection here.

THE RENAL PATIENT

HAEMODIALYSIS

Patients on haemodialysis are at risk of infection at the site where a dialysis catheter is inserted. These long-term lines can become colonized with staphylococci, enterococci, 'coliforms', *Pseudomonas* and yeasts, resulting in infection around the insertion site, bacteraemia or fungaemia. Any fluid around the site, as well as peripheral and central line blood culture sets, should be collected.

In haemodialysis patients important organisms to consider are the BBVs, HBV, HCV and HIV. All patients on haemodialysis need to be regularly screened for these viruses even though modern dialysis machines are such that cross-contamination is unlikely. However, patients who are known to be infected with one or more of these viruses are dialysed on a separate machine. It is routine practice to give patients on haemodialysis the HBV vaccine. The overall response to the vaccine may not be particularly good, reflecting the immunosuppressed nature

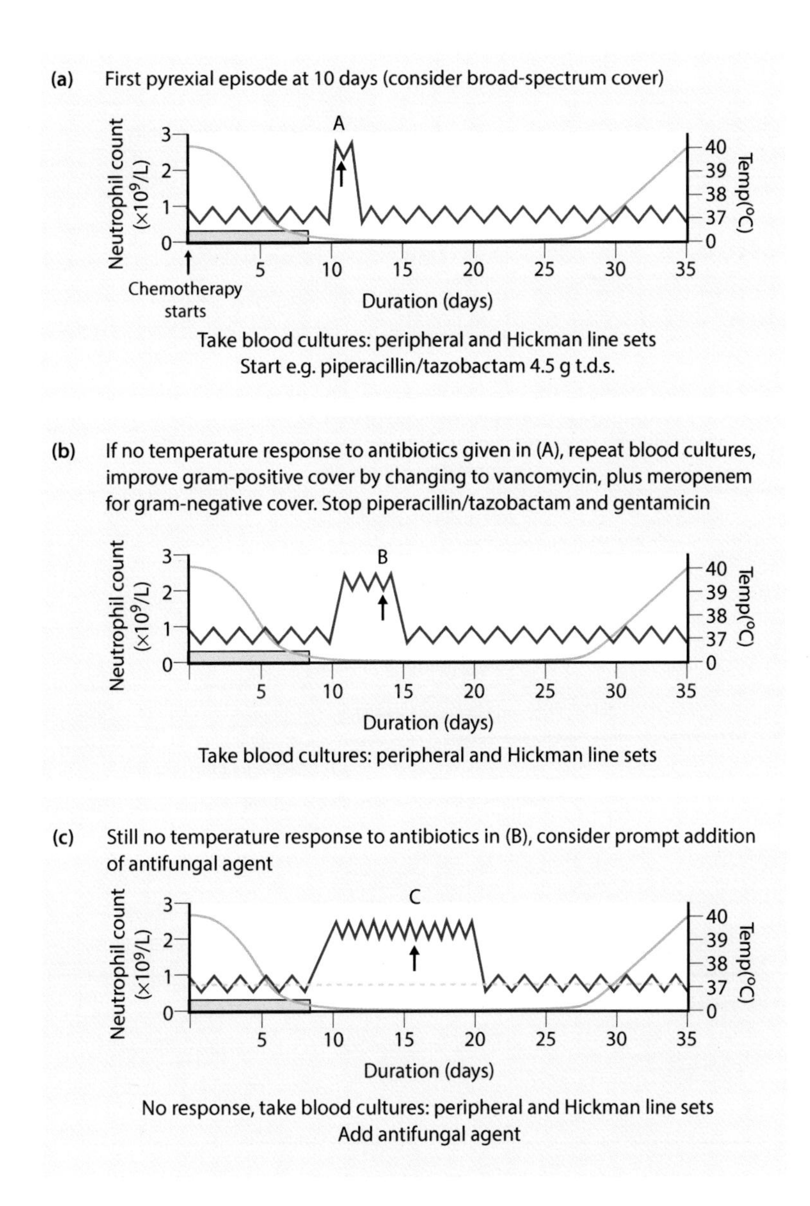

Figure 17.4 The time course of events in the management of the neutropenic patient with a fever. (**a**) Fever at day 10 (A) is treated with a broad-spectrum agent; (**b**) if there is no response, antibiotics are changed (B), with vancomycin being given for possible staphylococcal line infection; (**c**) if there is still no response an antifungal agent is added (C).

of these patients in chronic renal failure. Patients who have been on holiday and had dialysis in other units, and those outside the UK in particular, require dialysis on a designated machine until additional screens have excluded a BBV infection. Screening for carbapenemase producing Enterobacteriaceae (CPE) must be done.

CHRONIC AMBULATORY PERITONEAL DIALYSIS PERITONITIS

Chronic ambulatory peritoneal dialysis (CAPD) has been one of the major advances in the management of patients with end-stage renal failure. CAPD peritonitis is a recognized complication of the process, and up to half of patients on CAPD can have an episode of peritonitis in the first year. Recurrent infection may result in termination of CAPD in certain patients, necessitating their return to haemodialysis.

CAPD peritonitis arises as a result of several factors. These patients are considered to be immunocompromised as a consequence of their renal failure. However, the most important factor is the disruption of the integrity of the abdominal wall by the long-term catheter passing into the peritoneal cavity. Organisms can enter the peritoneum by subcutaneous passage or they may contaminate the dialysis fluid itself. Dialysis fluid has a

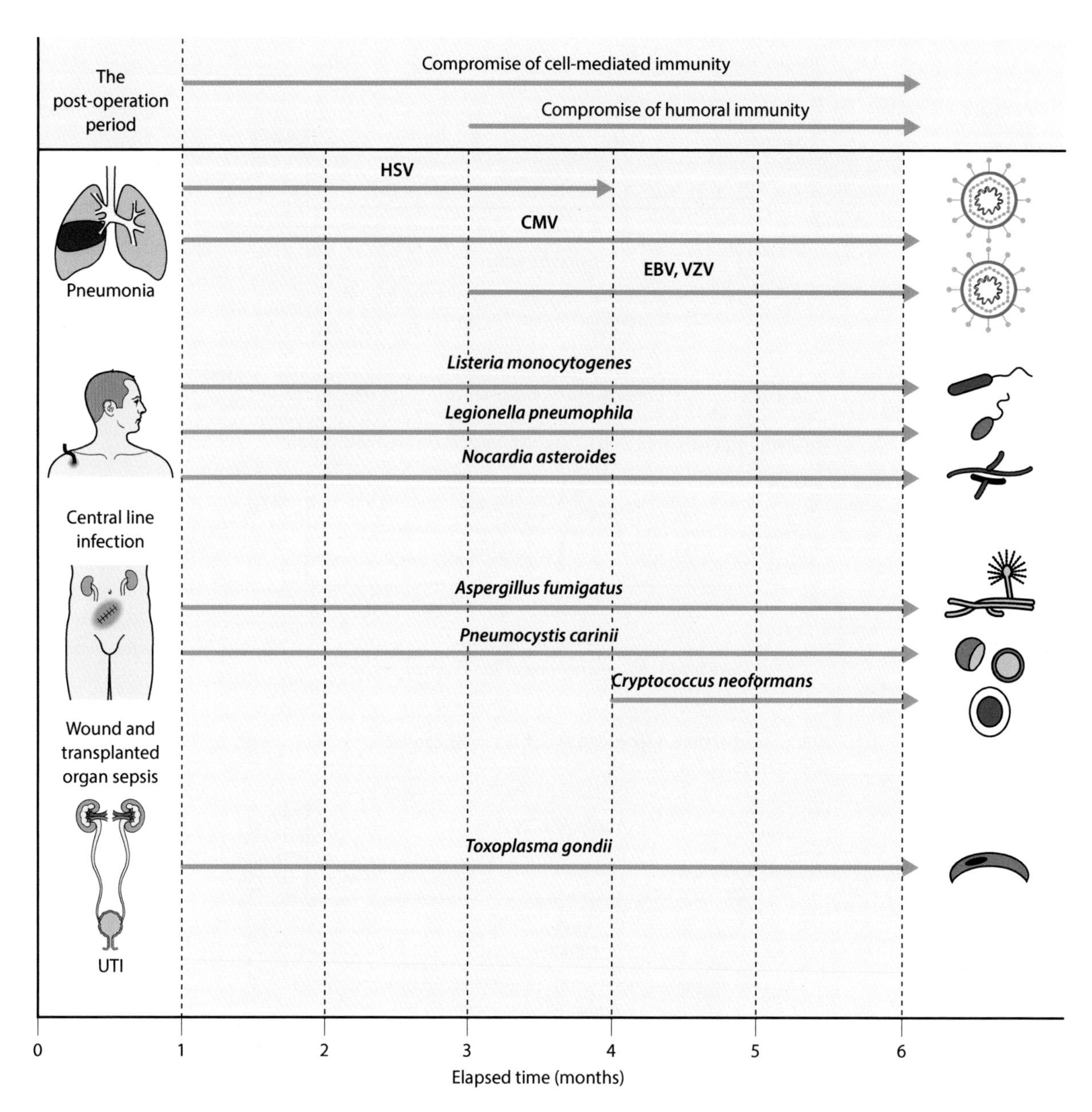

Figure 17.5 In the solid organ transplant patient the main infection risk is within the first 6 months. This can be divided into the postoperative period and the immunocompromised period. Relevant clinical situations and organisms are shown. (Modified, with permission from Rubin RH *et al.* [1981] Infection in the renal transplant recipient. *American Journal of Medicine* **70**:406.)

low pH and high osmolality, which is likely to impair the functioning of macrophages and neutrophils. In addition, the fluid is likely to affect the normal physiology of the bowel, and bacteria may cross the bowel wall into the peritoneum. The repeated changes of the dialysis fluid wash out opsonins such as complement factor C3.

Patients with CAPD peritonitis may present with abdominal pain and tenderness, nausea and vomiting, and the dialysis fluid is cloudy. At this stage it is usual to prescribe empirical antibiotics by the intraperitoneal route. Either vancomycin or teicoplanin and an aminoglycoside such as gentamicin are given. These antibiotics cover the common bacteria found in CAPD peritonitis. Diagnosis relies on the collection of fluid for microscopy and culture (**Figure 17.6**); the white cell count is raised above the normal value of 100/μL and the differential count usually shows a predominance of neutrophils. The Gram stain may identify bacteria or yeasts. On occasion there may repeatedly be no growth from the CAPD fluid, and other organisms such as mycobacteria need to be considered.

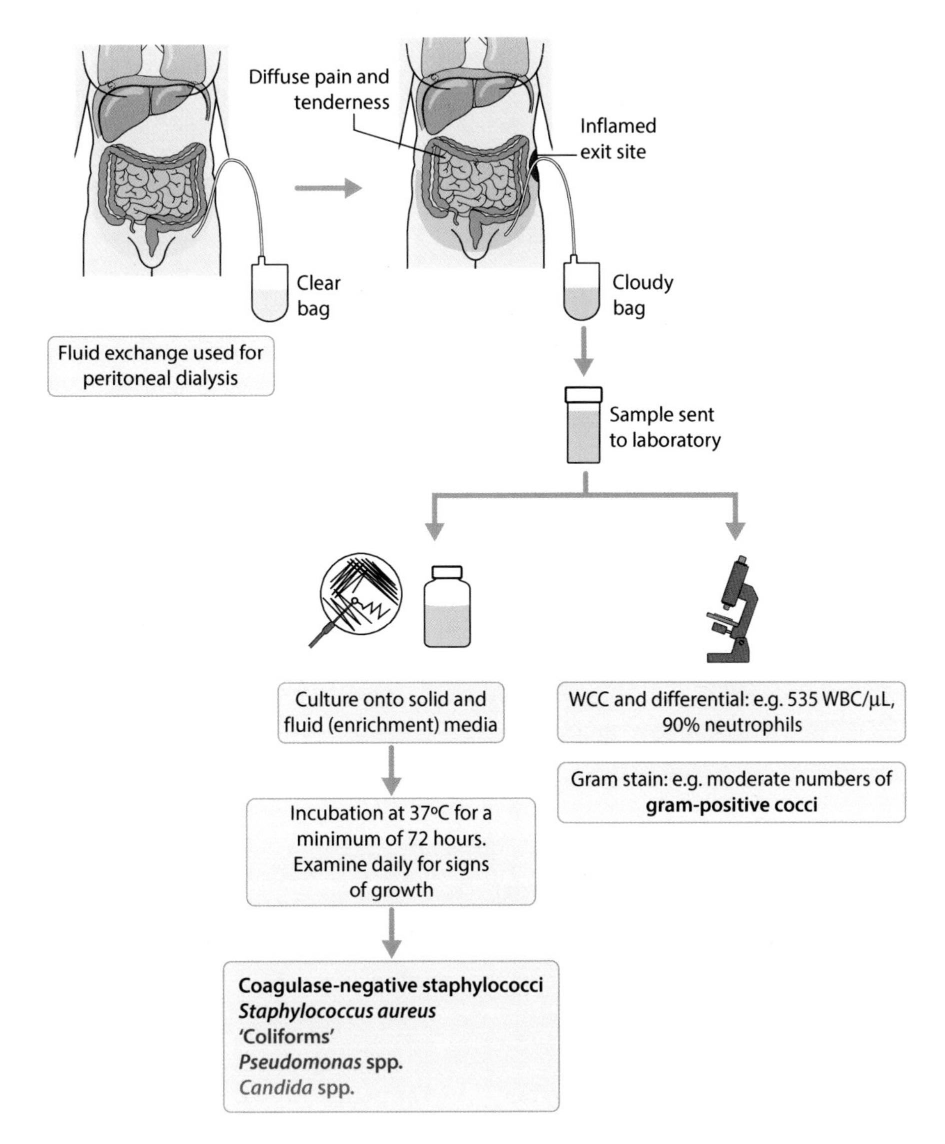

Figure 17.6 Peritonitis in the chronic ambulatory peritoneal dialysis (CAPD) patient. In the setting of symptoms and cloudy bags, CAPD fluid should be sent for microscopy and culture. Some common organisms are shown.

Occasionally algae have been known to contaminate CAPD fluid.

THE INJECTING DRUG USER

For the IDU, social exclusion, poverty, malnutrition and drug use all contribute to weakening the individual's immune status. Repeated injections into the groin to access the femoral vein, for example, compromise the integrity of the skin and soft tissues. Repeated introduction of bacteria from the flora of the groin gives rise to cellulitis and abscess formation; septic thrombophlebitis of the femoral vein can also occur. *Staphylococcus aureus*, streptococci of the *Streptococcus anginosus* group and anaerobes should always be considered in these infections.

It is likely that injected drugs contain particulate impurities which, travelling at speed, impinge on and damage the endothelium of the tricuspid valve. The resulting deposition of fibrin and platelets at this site is ideal for bacteria in the blood to settle in and initiate endocarditis. Microbial contamination of drugs can also be a major issue. *Clostridium novyi* is one example and it can cause fulminating systemic infection.

Some of the common sites of infection and organisms to consider in the IDU are shown in **Figure 17.7**. Blood cultures should be collected, and any tissue or pus that is obtained should be examined promptly in the laboratory.

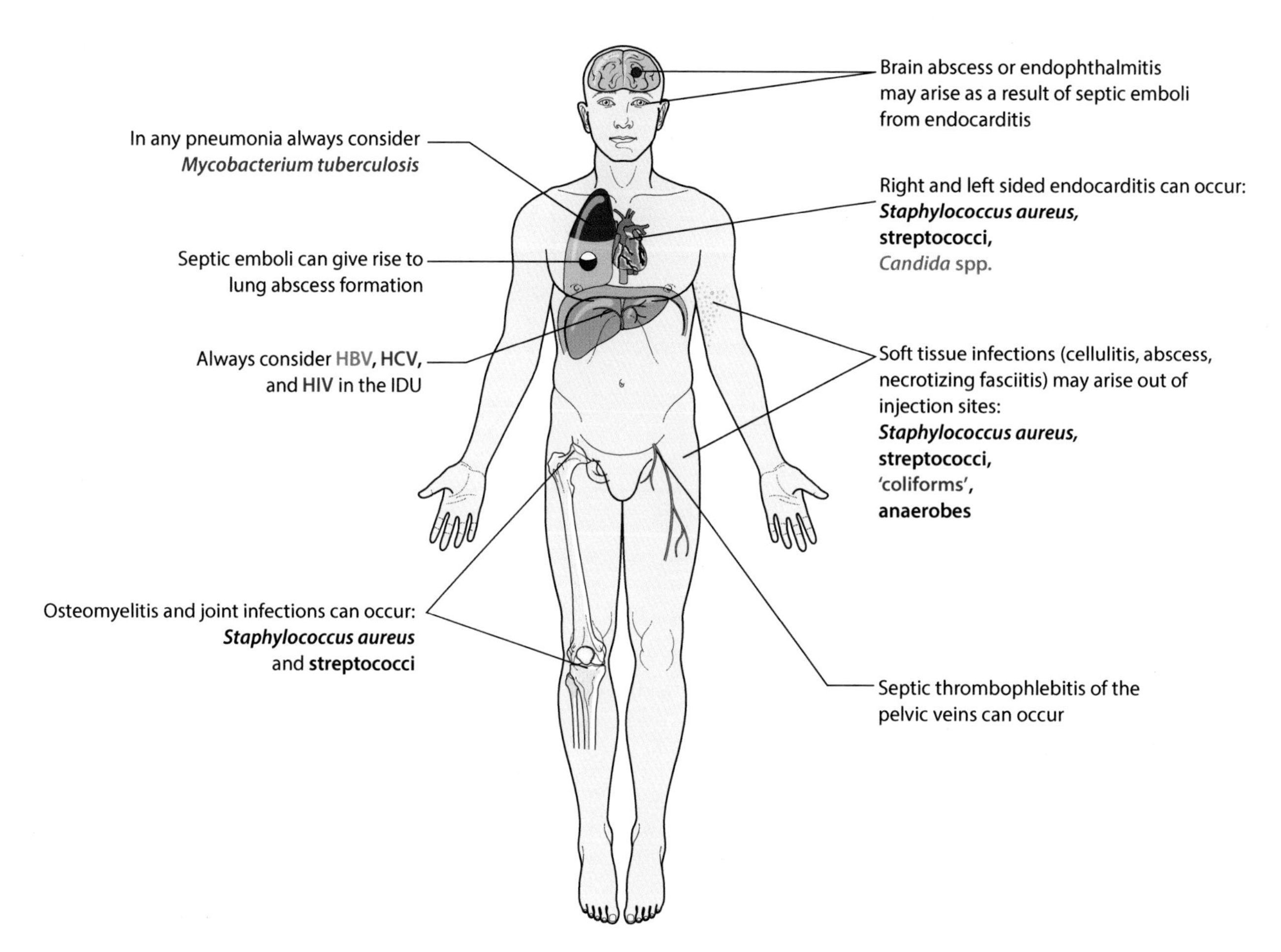

Figure 17.7 Some of the sites and organisms to consider in the injecting drug user.

Acute endocarditis is a medical emergency, and after the collection of three sets of blood cultures over a 20–30 minute period, high-dose antibiotics appropriate for *Staphylococcus aureus* are given, unless microbiological evidence dictates otherwise. Urgent referral to the cardiologist and cardiothoracic surgeons should be made in the setting of deteriorating cardiac function.

The prescription of antibiotics in this group of patients can be a problem. Peripheral intravenous access is usually difficult as veins are often fibrosed as a result of repeated drug injection and a central line may be needed. This may create a further problem, as the IDU now has a direct route to inject illicit drugs brought in to hospital by friends or relatives. This can be of particular concern when the patient has just had a valve replaced for infective endocarditis. The physician and medical microbiologist need to consult regularly on the best regime for each individual patient and can even consider the use of antibiotics that have good oral bioavailability.

The patient who is HBV, HCV or HIV positive is referred to the STD physicians for appropriate management. Follow-up of contacts who may be at risk of acquiring these viruses is needed.

THE INTENSIVE CARE UNIT PATIENT

There are many reasons for patients being admitted to an ICU. Severe community-acquired pneumonia and meningococcal sepsis are examples. Patients admitted to the cardiothoracic ICU after routine heart surgery usually require less than 24 hours of intensive care to stabilize them, which is essentially ventilatory support.

The patient will have key anatomical defences of the body compromised, which allow endogenous and colonizing bacteria to gain access to sites that they would usually be excluded from.. The endotracheal tube (ETT) used for ventilation provides a direct route for bacteria to enter the lungs. The ongoing requirement for arterial and venous access provides a route for bacteria to initiate line infections and line-associated bacteraemia. A nasogastric tube can obstruct the opening of a nasal sinus, with sinusitis resulting. Permanent catheterization of the bladder provides a route for bacteria and yeasts to enter the body. Bacteria isolated from blood, venous and arterial access sites, ETT or urine should be used to guide treatment. The microbiologist should provide this information by a ward-based service and take part in the decisions about the antibiotics to use in each patient.

THE ELDERLY PATIENT

The elderly patient group comprises the majority of patients admitted to hospital, so it is worthwhile examining the key points that relate to this patient group.

The elderly patient can be malnourished, and with poor food intake in hospital, this will be exacerbated with increasing stay, which is confirmed by a continuing deterioration in serum albumin levels. Wound healing is compromised, as is the ability of the individual to combat infections such as *Clostridium difficile*.

In elderly patients with dementia it can be difficult to assess their mental status in response to a CNS infection. Elderly patients can also have a degree of age-related nuchal rigidity and the examination for meningism is not straightforward. It can be difficult for these patients to localize irritation and pain. Ongoing athlete's foot can be missed, unless the toes of both feet are fully examined; this dermatophyte infection can cause distressing discomfort. It can also be a route whereby *Streptococcus pyogenes* gains access to the skin, soft tissues and blood. The elderly, confused patient can be difficult for nursing staff to manage, especially at night when staffing levels on the ward are at a minimum. The confused, unsupervised patient with diarrhoea may not only inadvertently be spreading *Clostridium difficile*, but outbreaks of *Salmonella* have been linked to such situations. This emphasizes the need for the ward (patient) kitchen and staff facilities to be locked at all times.

A particular challenge arises in the elderly hospitalized patient with a fever of unknown origin. In addition to infection, malignancies, connective tissue disease and drug-induced fever have to be considered. Repeat and prolonged courses of antibiotics with a broadening spectrum are given, with the expectation that the microbiologist will always be able to make an antibiotic recommendation. This creates a sense that 'something has been done' and delays a full review.

The senior member of the clinical team needs to discuss the patient with the infectious diseases physician or microbiologist. If infection is the likely cause, all the past microbiology results have to be fully collated and reviewed. Infective endocarditis, tuberculosis and abdominal collections need to be considered, and the appropriate diagnostic tests initiated. An informed decision to stop or withhold antibiotics should be taken at this stage. If not done already, the HIV test must be appropriately considered.

Full examination of the patient is imperative. Chronic venous ulcers may be covered with compression bandages, and if the legs have been 'recently bandaged', especially when the patient has been admitted from the community, the (incorrect) decision that the ulcers do not need to be examined is made.

There are difficulties in obtaining specimens from the elderly patient. Confusion and incontinence can make collection of a midstream urine or stool specimen a problem. There needs to be close cooperation with nursing staff to ensure that a timely and best quality specimen is obtained. Collection of blood for culture should be performed with assistance when doing this procedure in the confused or demented patient.

Excretion of antibiotics diminishes with age, and it is important to take into account renal clearance as well as body mass when determining the dose and frequency of antibiotics. Broad-spectrum antibiotics such as the fluorinated quinolones and second- and third-generation cephalosporins are recognized as risk factors for *Clostridium difficile* infection. It is important to ensure that their use is appropriate and that clear indications are documented in the clinical notes with a stop date on the prescription. Nitrofurantoin is a first-line agent used to treat cystitis; however, it is not excreted in the renal tract when the glomerular filtration rate is <45 mL/min.

Patients in their 80s have been newly diagnosed with HIV, showing that there is no upper age limit to consider HIV infection. With the appropriate counselling, which may involve difficult discussions with family members, it is reasonable to have this test done. It should be emphasized that a negative test, provided exposure was not in recent weeks, is very useful in excluding an important range of organisms.

The older patient on immunosuppressive treatment is at risk of a wider range of infections. With ongoing diarrhoea and/or colitis, CMV should be considered, and the patient with unresolving pneumonia may be infected with *Pneumocystis jirovecii*.

A summary of key issues in identifying infection in the elderly patient is shown in **Figure 17.8**.

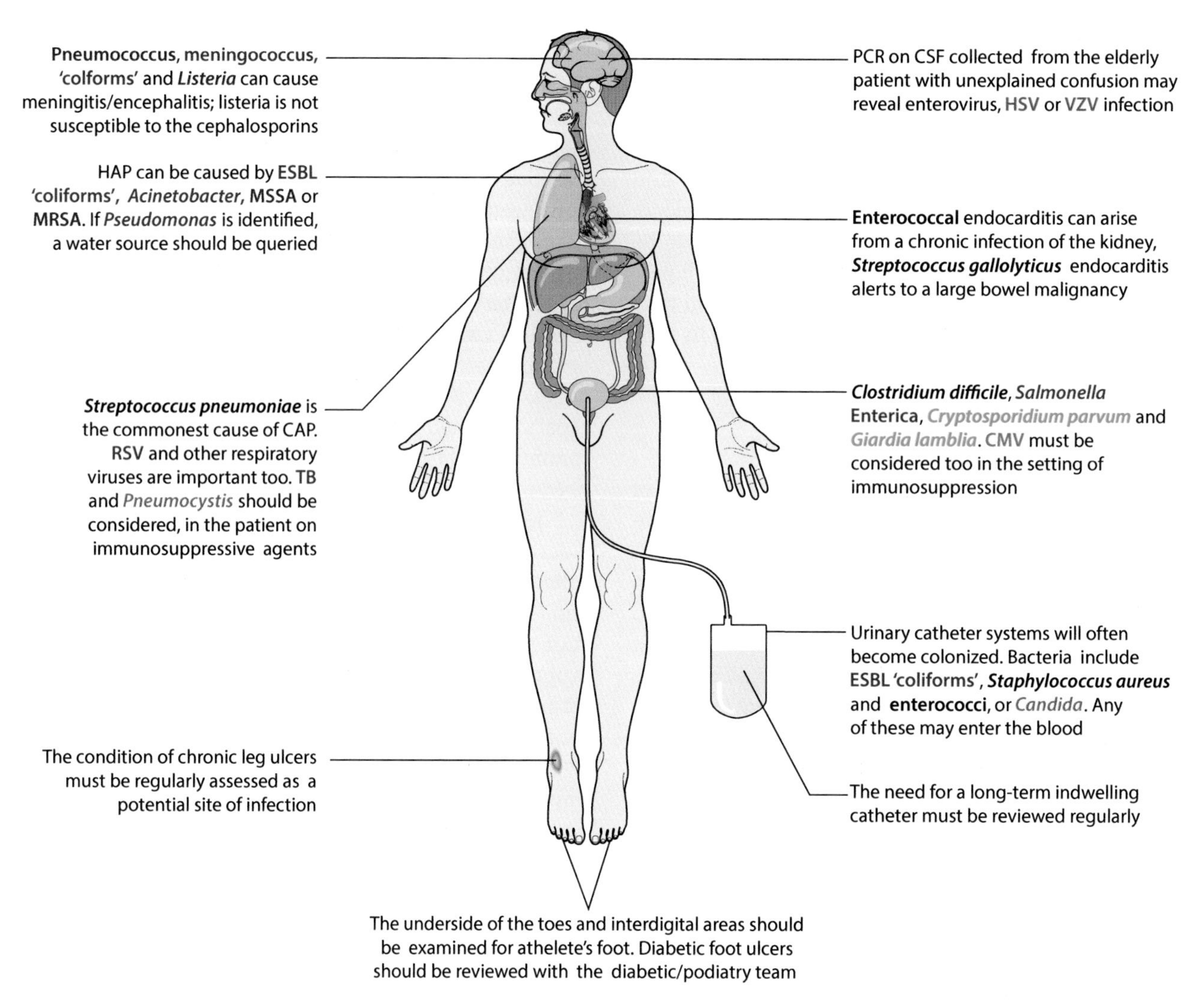

Figure 17.8 Examples of infections to consider in the elderly patient.

THE PATIENT WITH AN ABSENT OR DYSFUNCTIONAL SPLEEN

The absence of a spleen predisposes the individual to infection by the capsulated bacteria *Haemophilus influenzae*, *Neisseria meningitidis* and *Streptococcus pneumoniae*, and immunization against these bacteria is necessary. If a splenectomy is done as an elective operation, immunization should be completed 4 weeks before the procedure. The spleen may need to be removed as a result of extensive bleeding following a surgical mishap, or a road traffic accident, where immunization should be initiated at least 2 weeks after the procedure, or after the patient has been discharged from the ICU for this time. These timings optimize the antibody response.

For adults, the recommendation is:

Time 0: 23-valent pneumococcal polysaccharide vaccine, meningococcal C–*Haemophilus influenzae* b conjugate vaccine and meningococcal B vaccine.

Time 2 months: meningococcal A, C, W135, Y conjugate vaccine and booster dose of meningococcal B vaccine.

The 23-valent pneumococcal vaccine is repeated every 5 years. Influenza vaccine should be given annually, before the 'flu season. Prophylactic antibiotics are recommended, which for adults should be life long. The following individuals without a spleen are considered at higher risk of pneumococcal disease:

- Adults over 50 years of age.
- Patients with previous invasive pneumococcal disease.
- Patients who have undergone a splenectomy for a haematological condition, where there is ongoing immunosuppression.

For adults, penicillin V, 500 mg q12h or clarithromycin 250 mg q12h can be used. When the patient is nil-by-mouth after an operation, an appropriate intravenous antibiotic is given. Patients with an absent or dysfunctional spleen are at increased risk of severe falciparum malaria and guidance should be given on appropriate malaria prophylaxis and the need for close adherence to it, as well as preventative steps to limit contact with the mosquito vector, before travel to an endemic area is undertaken. Babesiosis, caused by the protozoon *Babesia* is transmitted by ticks. It is endemic in parts of rural Scotland and the east coast of the USA, so a relevant travel exposure has to be considered here.

All animal bites need to be assessed and treated promptly, to reduce the chance of infection by the gram-negative *Capnocytophaga canimorsus*, which can lead to fulminant sepsis in these patients. Other bacteria of the oral flora of animals include *Pasteurella multocida* and anaerobes. Antibiotics are always prescribed in these situations; in the patient with no penicillin allergy, co-amoxiclav is considered the antibiotic of choice.

The patient should be advised to obtain and wear a medical alert bracelet or chain that identifies their absent or dysfunctional spleen status. These can be obtained on-line for a small fee.

Index

Pages numbers followed by c refer to the classification figures.